```
QK
926     Free, John Brand
F7
          Insect pollina-
          tion of crops
```

DATE DUE

43839

Insect Pollination of Crops

Insect Pollination of Crops

by

John B. Free

*Rothamsted Experimental Station,
Harpenden, Herts, England*

1970

ACADEMIC PRESS · London and New York

ACADEMIC PRESS INC. (LONDON) LTD
Berkeley Square House
Berkeley Square,
London, W1X 6BA

U.S. Edition published by
ACADEMIC PRESS INC.
111 Fifth Avenue,
New York, New York 10003

Copyright © 1970 By ACADEMIC PRESS INC. (LONDON) LTD

All Rights Reserved
No part of this book may be reproduced in any form by photostat, microfilm, or any other means, without written permission from the publishers

SBN: 12–266650–X
Library of Congress Catalog Card Number: 72-117141

PRINTED IN GREAT BRITAIN BY ABERDEEN UNIVERSITY PRESS LTD

To: Nancy, Nicola,
    Anthony and Mark

There's a whisper down the field where the year has shot her yield
And the ricks stand gray to the sun,
Singing: "Over then come over, for the bee has quit the clover,
And your English summer's done."
Rudyard Kipling, "The Long Trail"

Preface

In ancient times, measures were taken which ensured that the fig tree was pollinated and bore fruit, so the use of insects to pollinate economically-important crops is not a new subject. Nevertheless, interest in insect pollination has increased considerably during the past few decades and arrangements for insect pollination are now part of standard management practices when growing many crops. It has recently been estimated that, in the U.S.A. alone, a million honeybee colonies are rented annually for pollination services. With increase in hybrid seed production it is likely that demands for pollination will become greater still in the near future.

Basic information on using honeybees and other insects as pollinators, and the benefit that is derived from doing so, is scattered through the literature and this is the first attempt to condense the information into one volume. I hope this book will be useful to growers, plant breeders, beekeepers, advisory and research workers, and help to provide a foundation for continued progress.

The first part is concerned with the pollinating insects that can be used by man to supplement wild pollinators, and ways in which this can be done most effectively. In the second part, each crop is discussed in turn and, when applicable, attempts are made to give available information on: flower structure as it relates to pollination; the crop's pollination requirements; the insect species that normally pollinate it and the ability of honeybees to do so; the increased crops that have been obtained using honeybees for pollination; the behaviour of honeybees and other insects on the crop; production of nectar and pollen by the crop and its food value to honeybees; and recommendations on the number of honeybee colonies needed per hectare.

Although there are many facets of the pollination of different crops that are similar it is very striking how many crops have their own unique pollination problems. I have to give the available facts about all plants that produce edible or other useful products other than ornamental flowers or timber. Attention has been drawn to the lack of information about some important crops. Plants that are known to be entirely, or almost entirely, wind-pollinated are not included, even though honeybees may collect pollen from some of them.

Most work has been done on crops in temperate climates and information on tropical ones is usually sparse. It is evident that often only a small proportion of flowers of many tropical crops produce fruit or seed, and although poor nutrition may be partly responsible, lack of sufficient pollination also seems important. When there is a paucity of information about a crop, I have sometimes felt it desirable to include preliminary studies which would benefit from more adequate data. I hope to revise this book eventually and I will always be glad to learn of new findings. Because many crops are known by different "common" names in different parts of the world, the Latin name is given in the text; the common names can be found by referring to the plant index.

I have many debts of gratitude to express and especially to Miss Cita D. Cooper who has borne the brunt of the typing, to Miss Elizabeth Robinson for library and card index work, to Mr. P. Reed for checking the manuscript and proofs, to Mrs. Mary Mullings for translating Russian and German papers, and to Mrs. Doris E. Jolly for additional secretarial help. The assistance afforded by the abstracting services and library facilities of the Bee Research Association has been invaluable. Without them much important information would have been lost to me. Several of the photographs have kindly been supplied by Dr. W. P. Nye (Agricultural Research Service, Wild Bee Pollination Investigations, Utah State University, Logan, Utah, U.S.A.), one by Dr. M. V. Smith (University of Guelph, Ontario, Canada), and one by Dr. J. Simpson (Rothamsted Experimental Station). Permission to reproduce the many illustrations borrowed from books and journals is gratefully acknowledged. My colleagues Dr. C. G. Butler and Dr. J. Simpson have encouraged me throughout the preparation of this book and have kindly read and criticised the manuscript. Finally, it is a pleasure to acknowledge that several of the ideas incorporated into this book have been obtained from numerous discussions with bee research and advisory workers in Britain and North America during the past several years.

August 1970   JOHN B. FREE

Contents

Preface		vii
Conversion Factors		xi
1	Introduction	1

PART I: INSECT POLLINATORS

2	Foraging Behaviour of Bees	15
3	Organization of Honeybee Colonies	50
4	Management of Honeybee Colonies for Pollination	65
5	Using Bumblebees as Pollinators	90
6	Using Solitary Bees as Pollinators	102
7	Pollination in Enclosures	119

PART II: CROPS NEEDING INSECT POLLINATION

8	Cruciferae	135
9	Malvaceae	151
10	Sterculiaceae	169
11	Linaceae	179
12	Rutaceae	183
13	Vitaceae	189
14	Anacardiaceae	193
15	Papilionaceae: *Medicago*	198
16	Papilionaceae: *Trifolium*	215
17	Papilionaceae: *Vicia*	242
18	Papilionaceae: *Phaseolus*	257
19	Other Papilionaceae	265

20	Grossulariaceae	278
21	Myrtaceae	288
22	Passifloraceae	292
23	Cucurbitaceae	297
24	Umbelliferae	314
25	Rubiaceae	318
26	Compositae	322
27	Vacciniaceae	336
28	Solanaceae	349
29	Chenopodiaceae	359
30	Polygonaceae	363
31	Lauraceae	367
32	Euphorbiaceae	370
33	Moraceae	374
34	Rosaceae: *Pyrus* and *Prunus*	380
35	Rosaceae: *Fragaria*	417
36	Rosaceae: *Rubus*	422
37	Liliaceae	426
38	Other Families	430
	References	444
	Author Index	507
	Animal Index	523
	Plant Index	527
	General Index	539

Conversion Factors

1 centimetre (cm)	=	0·39 inch
1 inch	=	2·54 centimetres (cm)
1 foot	=	30·48 centimetres (cm)
1 metre (m)	=	1·09 yards
1 yard	=	0·91 metres (m)
1 kilometre (km)	=	0·62 miles
1 mile	=	1·61 kilometres (km)
1 square metre (m^2)	=	1·196 square yards
1 square yard	=	0·836 square metres (m^2)
1 hectare (ha)	=	2·47 acres
1 acre	=	0·405 hectares (ha)
1 kilogram (kg)	=	2·205 pounds
1 pound	=	0·454 kilograms (kg)
1 ounce	=	28·35 grams (g)

Chapter 1

Introduction

Pollination Requirements of Crops

Plant species of economic importance are either self-fertile and set fruit or seed with their own pollen (self-pollination), or self-infertile and need to receive pollen from other plants of the same species (cross-pollination). Some self-fertile species are automatically pollinated with pollen from their own flowers, but often the flowers are so constructed that either wind or insects are needed to transfer pollen from their anthers to their stigmas. Moreover, self-fertile plants may produce more fruit, or seeds of better quality, when cross-pollinated than when self-pollinated, and various devices often favour cross-pollination to self-pollination even when the latter can occur. Wind is the principal pollinating agent of agricultural grasses and a few other species, whereas most agricultural and horticultural crops that have conspicuous, coloured and scented flowers are adapted for insect pollination.

The task required of an insect pollinator will depend upon whether the plant species is self-fertile and partially self-pollinating, or self-fertile and not self-pollinating, or self-infertile, and the efficiency per insect visit will vary accordingly. Perhaps surprisingly, the categories into which some economically important crops fall are not known for certain. Thus, the extent to which flowers of many self-fertile varieties of fruit trees benefit from insect visits and the extent to which they are self-pollinating still need investigating.

Insect pollination may give advantages other than increasing the yield of a crop. An abundance of pollinators sets a greater proportion of early flowers of some crops (e.g. *Vicia faba*) resulting in an earlier and more uniform crop. Insect pollination of other crops increases not only the quantity but also the quality of the fruit (e.g. *Cucumis melo*, *Fragaria* × *ananassa*).

Attempts to select autofertile varieties of economically important species have not in general been successful because inbreeding and loss of hybrid vigour has had a detrimental effect on yield. In fact, if anything, the emphasis is now toward the growing of hybrid seed. Whereas cross-pollination between plants or trees is mostly beneficial, and, of course, essential for self-infertile plants and producing hybrid seed, when attempts are being made to produce pure seed, cross-pollination from another strain or variety is detrimental; for many species, much more knowledge is needed about the amount of contamination likely to occur in different circumstances, although it seems that the only way to guarantee pure seed is to produce it in a glasshouse. Rarely, as with European varieties of *Cucumis sativus*, the flowers have to be isolated from bees to prevent pollination, seed set and the development of misshapen fruit. All the different requirements present their own problems which will be discussed in the relevant sections.

Types of Pollinating Insects

The most important pollinating insects are solitary bees, bumblebees and honeybees. Insects other than bees are often recorded visiting flowers of commercial crops and are essential pollinators of a few (e.g. *Theobroma cacao*, pages 169–178), but, in general, they lack sufficient body hairs and the necessary behaviour patterns, and probably few transfer pollen from the anthers to the stigmas of the flowers they visit. Furthermore, unlike the bees which forage consistently to obtain sufficient food for their young, most other insects forage to satisfy their own immediate needs only, and feed on a variety of foods other than from flowers. Hence it is assumed that they perform only a supplementary role in pollination, although detailed studies of their behaviour are generally lacking. Probably, many of the most important, supplementary pollinators are various Diptera, including those belonging to the genera *Eristalis, Syrphus, Platycheirus, Rhingia, Calliphora, Lucilia, Sarcophaga, Bibio, Dilophus* and *Bombylius*.

Under natural conditions there is usually no very great concentration of one flower species in any one place, and the native insect population that is visiting them is probably sufficient to pollinate them. However, when many acres are occupied by a single flowering crop, and moreover certain localities are favoured for growing particular species, there may be too few wild insect pollinators, and however favourable the other factors involved in the production of a normal seed or fruit crop may be, the yield may be limited by lack of pollination. The tendency of fields to become enlarged, with accompanying dilution of the wild pollinator population, aggravates the situation. Moreover, the size of the populations of wild bees and other insects varies from year to year and place to place so they cannot be relied upon.

It is also evident that clean and intensive cultivation of the land has destroyed many natural food sources and nesting sites of wild pollinating insects; in pioneer conditions in North America wild pollinators were adequate to pollinate the small areas of cultivated crops, but as land clearing progressed they became insufficient. It is supposed that there has been a decrease in the numbers of our wild insect pollinators as a result of applications of insecticides and herbicides. Although this may well be true, there is no sound evidence to support this supposition; any such evidence would be very difficult to obtain. Planting of large areas of a single crop tends to provide ample forage for a limited period of the season only, and there may be little or no forage available to pollinating insects at other times.

Practical methods for utilizing solitary bees to pollinate *Medicago sativa* have been developed in the past few years, and together with bumblebees and blowflies they are sometimes used to pollinate crops in glasshouses and cages. However, for most crops an insufficiency of wild pollinators can only be compensated for by using honeybee colonies. Because the honeybee produces honey and wax it has been kept by man in hives for many years and can be made readily available in considerable numbers as, and when, needed. Unfortunately, in many countries honeybee colonies tend to be concentrated near large towns and their distribution is not ideal for helping to pollinate agricultural crops, but they can be moved to crops needing pollination and in recent years this practice has grown considerably. Honeybees are particularly useful because they will visit and pollinate such a very large proportion of economically important plant species.

The honeybee that is indigenous to Europe and Africa and that was introduced to many other parts of the world including North America, is *Apis mellifera*. Throughout the book this is the species of honeybee concerned unless otherwise indicated. The eastern honeybee (*Apis cerana*) is also kept in hives in India, China and Japan, but the giant honeybee (*Apis dorsata*) and dwarf honeybee (*Apis florea*) have not proved to be amenable to this.

It is extremely difficult to evaluate the pollinating services of the honeybee, but those who have attempted to do so have estimated that its value as a pollinator is several times that as a producer of honey and wax, and the extremely important contribution honeybees make to a prosperous agricultural economy is now generally recognized. The pollinating potential of a single honeybee colony becomes evident when it is realized that its bees make up to 4 million trips per year and that during each trip an average of about 100 flowers are visited.

Honeybee colonies for pollination should be used as efficiently as possible, and especially as too few are available. Studies have been started to find the best way to distribute colonies on crops, to increase their foraging, particularly on the crop concerned, to increase the proportion of flower visits during which pollination occurs, and to cause the bees to distribute pollen more efficiently. When the crops to be pollinated are not especially attractive to bees, it is most important that the colonies should be managed as efficiently as possible to induce the bees to visit them.

Various cultural methods, including use of fertilizers, presence or absence of irrigation, correct spacing of plants, and use of wind breaks, have been tried to improve the attractiveness of crops. Growers can help increase the efficiency of their pollinating insects, including wild pollinators by, whenever practical, arranging that competing crops do not flower at the same time, and by restricting the planting of any one crop to avoid too much dilution of the available pollinator populations. Arranging for the peak flowering period to coincide with the seasonal peak in numbers of the most important wild pollinators, and even arranging to provide other food sources while their populations are increasing, should also be seriously considered. Growers must at all times avoid any cultural practices harmful to wild pollinators or honeybees. It has become very evident during recent years that plant breeders should ensure that new strains are attractive to bees, even if they do not select specifically for this. These are all efforts to improve the honeybee and flower relationship. In fact many crops have special pollination problems solely because they are being grown in parts of the world where their natural pollinators are absent. It is possible that much can be done to remedy this situation by judicious introduction of pollinators to new areas (see Bohart, 1966).

Determining the Need for Insect Pollination

There are a number of ways to find whether a particular crop benefits from insect pollination. It is impossible to keep pollinating insects away from an area of crop without caging it, and the most common procedure is to enclose parts of the crop in insect-proof screen cages (Fig. 1) and put honeybee colonies in some but not others. Other plots, of the same area as that covered by a cage, are left exposed. Hence comparison is made of the three treatments: (a) caged with bees, (b) caged without bees, (c) not caged and visited by pollinating insects including honeybees from nearby colonies. Sometimes insecticide is applied to plots caged without bees to eliminate all pollinating insects. Ideally both the caged plots with and without bees should be treated with insecticide if the effect of the bees alone is being determined, but this often entails confining the bees to their colonies for a day or so, with a consequent loss in pollinating efficiency. Sometimes additional screen cages are used with a mesh size that excludes large insects, such as bees but not small ones from entering; however, it is difficult to know precisely what effect the use of cages with different size mesh does have on the insect population. Attempts are sometimes made to determine the effects of shading and reducing wind inside cages by using cages without tops or without sides, but as these treatments probably also reduce bee visits, it is again difficult to evaluate the results obtained.

Because of the possible effect of the cages themselves on plant growth and fruit or seed yield (page 121), experiments in which attempts are made to determine the effect of pollination by comparing only the yield of plots caged to exclude insects with the yield of uncaged plots are not satisfactory, and are only referred to in subsequent chapters when little other evidence about a particular crop is available. Even the comparison of the three basic treatments above, is not entirely satisfactory. A comparison between plots caged

Fig. 1. Pollination cages. (Photo: J. Simpson.)

with bees and uncaged plots indicates the effect of the cage, but the uncaged plots may be visited by more or fewer honeybees than the caged plots as well as by other insects which might be more efficient pollinators. A comparison between plots caged with and without bees indicates: the effect of bee pollination alone when both plots with and without bees have been treated with insecticide; the effect of bees and any other insects present in pollinating when only the plots caged without bees are treated with insecticide; and the additional effect of bee pollination to that of other possible insects present when neither plots caged with bees nor those without bees are treated with insecticide.

Results from cage experiments may be assessed by finding the number or weight of seeds, pods or fruit per cage, or by finding the percentage of flowers that set fruit or seed.

Unfortunately, the increased yield or other benefits obtained in cages with bees may show little relation to that obtained by moving colonies to a crop of the species concerned. There may be several reasons for a smaller response: pollinating insects including honeybees may already be present; the species concerned may be relatively unattractive and bees may not work it when there is a choice of others; only a small proportion of bees that visit the flowers may act as pollinators and the larger proportion present in cages may give more pollinating bees per flower than in the field crop; or the behaviour of bees may be such that they do not normally pollinate the flowers when visiting them for nectar only, but because the caged flowers provide the only source of pollen for the enclosed colony there are relatively more bees obtaining pollen from the flowers and pollinating them, than occurs in the open field. Thus, it is only by studying the behaviour of bees while they are foraging on the crop concerned that the results of cage experiments can be properly evaluated. Furthermore, it is helpful if the result from cage experiments can be assessed in conjunction with other methods of estimating the need for pollination.

One such method is to put a group of colonies in the centre of a crop and demonstrate: (a) that the number of honeybees on the crop decreases with distance from the colonies and (b) that there is a corresponding decrease in seed or fruit production. Often the area of a crop is not large enough or sufficiently homogeneous to show such a decrease, and if this is so, colonies may be put at one end of it and the number of foragers may then decrease toward the other. However, in order to check that any decrease in yield is due to diminishing numbers of bees and not other factors, this experiment should be repeated the following year with colonies at the opposite end of the field.

Another possible way of determining the value of insect pollination involves varying the pollinator population and determining the set of flowers at different stages of the flowering period; a field a few kilometres away should be used as a control.

A precise way to find the value of a flower visit is to enclose flowers, while still in the bud stage, in muslin or paper bags, remove the bags when the flowers open and watch them continuously until they are visited by insects, then bag them again and later determine the seed or fruit set. However, this method is time consuming and can only be used when bagging flowers of the species concerned is neither detrimental to set, nor likely to increase self-pollination.

It is also possible to determine the value of honeybees and other insects by surveys in which attempts are made to correlate the yield of crops with the number and type of insects foraging per unit area on them, or even with the number of honeybee colonies present within a certain radius. Such a survey needs to be extensive to be valuable and most have not been.

In combination with all the above methods it is advisable to hand pollinate some of the flowers to determine the maximum set possible under the conditions concerned. Indeed, hand pollination of experimental plots in a sample of fields can be used to find whether inadequate pollination is a factor limiting yield (see Butler *et al.*, 1956). Attempts to evaluate honeybees and other insects as pollinators should also, when possible, be supplemented by observations on their foraging behaviour and examination of their bodies for pollen. It should be remembered that plants of varieties that are clones (i.e. propagated vegetatively) are likely to show less variation in their pollination requirements than plants of varieties that are not.

Concentration of Honeybee Colonies Needed

There is often a considerable difference between the number of colonies that can be profitably maintained for honey production and the number necessary for pollination in an area, especially as in many districts they will be needed for pollination for only a few weeks each year.

Estimates of the number of colonies necessary to pollinate a given area of crop are usually based on the experiences and assumptions of growers and beekeepers, rather than on experimental results. The rate of $2\frac{1}{2}$ colonies per hectare has been quoted extensively ever since Hendrickson (1916) and Tufts (1919) suggested it as being suitable for pollinating fruit orchards. However, the honeybee population needed will obviously depend on the concentration of flowers, their attractiveness, and competing insects and crops, and will of course vary from species to species and place to place. Furthermore, the percentage of flowers of a crop that are open may differ greatly during the different stages of its flowering period, and the number of colonies needed will vary accordingly. When it is known that a crop is difficult to pollinate, usually because it is a species relatively unattractive to honeybees, more than $2\frac{1}{2}$ colonies per hectare are sometimes recommended. In contrast, when a crop is attractive, but its flowers are comparatively sparse, or can be self-pollinated, fewer than $2\frac{1}{2}$ colonies per hectare are suggested (e.g. *Cucumis melo*, *Trifolium repens*).

Ways of measuring the honeybee population needed to pollinate a crop are only just beginning to be formulated; they are usually based on the working rate and pollinating efficiency of bees on the crop concerned and hence the number of foragers needed per unit area of crop to pollinate it (page 67). However, such estimates are far from easy to make for some crops, particularly for tree fruit crops where very few of the flower visits made result in pollination (page 397) and when weather during flowering is likely to be variable. In order to facilitate additional calculations of this sort, I have given as much relevant information as I can find for each crop.

The provision of colonies for pollination should always be done as a commercial undertaking with proper safeguards for both beekeeper and grower. The beekeeper is either paid a fixed rent for the hire of his colonies or is given an agreed share of the crop produced, sometimes allocated on the number of colonies he has provided per hectare. Whichever method is used, several factors should be taken into consideration before the final agreement is reached. These include: the strength of the colonies, the number required, the distance they have to be transported, the way the colonies are distributed in the crop, the time they are there, the amount of management they will require, the need for feeding supplementary protein and carbohydrate, the loss or increase of possible honeycrop and the danger to colonies from insecticides. It is hoped that with more efficient management of colonies for pollination fewer will be needed per unit area.

There are various ways of estimating the size of the pollinator population on a crop. Sometimes the actual numbers are needed, but at other times the relative numbers will suffice, depending upon the objective. Attempts to make estimates by collecting samples in insect nets (e.g. Linsley, 1946) or suction traps are open to the objection that many or most of the insects collected might not be visiting the flowers, and many of those that are might not be pollinating them. Moreover, sweeping with an insect net is likely to disturb foraging. However, both these methods are of value in permitting positive identification of the insects present.

A common method of determining population size is to demarcate strips about 1 m wide and 90 m long, along the edges of crops, and count the insects while walking at a consistent pace along the strip. Some observers find that holding a metre-long stick over the strip facilitates counting. Because the number of foragers tends to decrease from the edge to the centre of a field, counts made at the edge do not give a true picture of the pollinator population in the field as a whole; however, because of possible damage to his crop, the grower is sometimes not willing for counts to be made inside it, so there is no alternative to making counts at its edge. Data so obtained are, of course, suitable for comparative studies. When it is possible to work within a crop, one way used is to count the pollinators present in demarcated areas of the same size, scattered at random throughout it (e.g. Vansell and Todd, 1946). Some sort of orientation mark is necessary for an observer to find these areas and this might lead to a larger concentration of pollinators than usual (page 45). This objection can be overcome by using a portable device through which the observer can rapidly survey randomly chosen areas of a metre square (Smith and Townsend, 1952).

The number of pollinator visits per flower is often of basic importance. One of the simplest and most effective methods of discovering this involves tagging a number of easily visible flowers and recording the number of visits they receive during a given period of time (Levin *et al.*, 1968).

Techniques for Studying Foraging Behaviour

When studying foraging it is often necessary to be able to identify some or all of the bees from a given colony and there are several ways of doing this. Choice of method is determined by the study being made and the material available.

It is possible, although tedious, to mark all or many, of the foragers of a colony with spots of fast-drying paint; a different colour can be used for different colonies. A method by which bees are immobilized against a string net, and which allows about 20 bees to be marked per minute has been described (McDonald and Levin, 1965).

Strains or races of honeybee with different body colours can be employed. Much use has been made of the recessive genetic mutant "cordovan" in studying the foraging behaviour of honeybees (e.g. Taber, 1954; Peer, 1955; Levin, 1959; Lee, 1961). In this strain, parts of the body that are normally black are cordovan brown and the bees can be readily distinguished from other strains in the field. Moreover, different lines of cordovan bees have different numbers of abdominal yellow bands and so can be distinguished from each other. The total honeybee population in an area can be estimated by moving cordovan colonies of known strength to a crop and determining the proportion of bees foraging on it that are cordovan (Taber, 1960).

Another method of marking the bees of a colony is based on findings that, when a honeybee colony is fed even a small quantity of syrup containing radio-active isotope (e.g. phosphorus 32, gold 198 or carbon 14), the majority of the bees soon contain it (e.g. Nixon and Ribbands, 1952). The distribution of foragers from these colonies is determined by collecting foragers in the field and examining them for radioactivity (e.g. Courtois and Lecomte, 1958; Levin, 1960; Lee, 1961).

Bees can also be identified by spraying them with a stain at the hive entrance (Smith *et al.*, 1948) or by forcing them to walk through a device, known as a "marking block", at the hive entrance (Fig. 2) in which they pick up fluorescent powders (Musgrave, 1950; Smith and Townsend, 1951; Free *et al.*, 1960). The latter technique not only allows the foragers to be identified, but, because they leave some of the fluorescent dust on flowers, it is also possible to determine where they have been foraging by using a fluorescent lamp at night on the crop concerned. Fluorescent or other coloured powder and especially methylene blue dye, have also been dusted on flowers and their distribution by bees and other insects determined (page 164). However, this technique is not an entirely satisfactory way of determining the area over which bees might distribute pollen, as the ease with which the powder and pollen are transferred might differ considerably.

It is often necessary for an observer to be able to recognize individual bees and to achieve this, marks of different colours and combinations of colours or symbols, often based on letters of the alphabet and numerical figures, are painted on the bees' thoraces and abdomens. Small numbered discs which can be glued to bees' thoraces are available commercially. Marking often entails immobilizing bees with anaesthetics or by chilling; anaesthetizing bees with nitrous oxide or carbon dioxide alters certain aspects of their physiology and behaviour but using chloroform or chilling them does not appear to do so (Ribbands, 1950; Free, 1963).

Fig. 2. Beehive with marking block at entrance.

Apart from actually observing or collecting bees from a crop it is possible to discover the species they have visited by examining the pollen in their honey-stomachs, on their bodies or in their pollen loads. Use is often made of devices known as "pollen traps" (Figs. 3 and 4) which scrape the pollen loads from a proportion of the pollen-gatherers as they enter their hives (e.g. Smith, 1963). Although this method is suitable for obtaining general or relative information about the foraging of a colony it does not give exact data, because the percentage of pollen loads removed by a trap depends upon their size, which differs, in addition to other factors, with the different flower species from which they originate (see Synge, 1947). Another method entails using a suction device to take periodic samples of pollen-gatherers at the hive entrance (Nye and Mackensen, 1965).

Fig. 3. Beehives with pollen traps beside *Trifolium pratense* field.

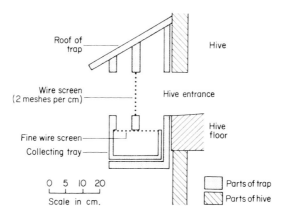

Fig. 4. Diagrammatic cross-section of pollen-trap attached to Langstroth hive.

The above is but a brief survey of the techniques available. Some will be discussed more fully in appropriate places but for more general information Lecomte (1962a) should be consulted.

Part I

Insect Pollinators

Chapter 2

Foraging Behaviour of Bees

Bees visit flowers to collect nectar and pollen. They are attracted to flowers and recognize them by their colour, shape and odour (bumblebees, see Kugler, 1943; honeybees, see Ribbands, 1953; Frisch, 1965).

Bees are able to distinguish only four qualities of colour, yellow, blue-green, blue and ultraviolet (honeybees, Frisch, 1914; bumblebees, Kugler, 1943); and when they are working flowers of one colour only, they become conditioned to it and do not visit flowers of a different colour. However, when a species has flowers of more than one colour, bees readily change between them and so probably ignore colour as a distinguishing feature (e.g. Darwin, 1876; Mather, 1947). Bees are also able to learn the general shape of flowers, and the general form of plants (Darwin, 1876; Manning, 1956) but their visual acuity is small and they readily move between tall and stunted plants of the same species, and between flowers at different stages of opening.

However, bees have a highly developed sense of smell and can readily be trained to associate forage with a particular scent or mixture of scents (Frisch, 1919; Ribbands, 1955). Their threshold for scent perception is usually much lower than that of man, and honeybees and bumblebees can become conditioned to flowers that man cannot smell (e.g. *Vaccinium myrtillus*, Frisch, 1919; *Echium vulgare* and *Linaria vulgaris*, Kugler, 1932). Although the general form of a plant or flower, and especially the colour of a flower, guide bees to it from a distance, when a bee is close to a flower, scent provides the stimulus to alight (e.g. Frisch, 1919; Knoll, 1926). When a strange scent is added to flowers, foragers are often discouraged from visiting them (e.g. Clements and Long, 1923; Butler, 1951; Frisch, 1965). Once bees have become conditioned to a species they will even continue to visit flowers denuded of petals, although in reduced numbers (e.g. Vansell, 1942).

Although the petals are the most conspicuous feature of an entomophilous flower, the adroecium itself may be modified to supplement, or supplant, the petals in attracting insects, either visually or odoriferously (see Percival,

1965). Petal expansion, nectar secretion and scent production reach their maximum to coincide with anther dehiscence so the nectar foragers pick up mature pollen on their bodies. The petals wither and scent and nectar production cease soon after a flower is fertilized. In unpollinated flowers nectar secretion persists for longer than usual.

Nectar and its Collection

The nectaries, or nectariferous tissue, which secrete nectar, may occur in many parts of the flower including the receptacle, petals and sepals and the bases of the filaments and pistil. They are not merely passive valves but are secretory glands with a characteristic, active metabolism (e.g. Frey-Wyssling et al., 1954; Shuel, 1956). Nectar secretion is influenced by the maturation of the stigma and stamens, and also often by the age of the flower and is usually greater on the first day, or first few days, a flower is open than later (Shuel, 1961). Nectar secretion by some species is of very limited duration.

The threshold temperature necessary for nectar secretion and the temperature above which it ceases also differs with different species and helps determine where crops of different species can be grown commercially. Irrespective of temperature, nectar secretion is greater on a sunny than a dull day, reflecting the fact that the nectar sugars are products of photosynthesis, which in turn is influenced by sunlight (Shuel, 1955a). Soil moisture, atmospheric pressure, size of nectary and position of the flower on the plant may also influence the amount of nectar secreted.

The composition and properties of nectar have been reviewed by Beutler (1953), Shuel (1955a) and Percival (1965). Nectar contains mostly sugar, but small amounts of other substances contribute to its aroma and the characteristics of the honey prepared from it. These include organic acids, volatile oils, polysaccharides, proteins, enzymes and alkaloids. The three main sugars of nectar are sucrose, fructose and glucose; minor sugars present in various species include maltose, raffinose, melibiose, trehalose and melezitose. Wykes (1952a) analysed the nectars of sixty species and concluded that for any one species the proportion of the different sugars present tended to remain constant. This was confirmed by Percival (1961) who found that 828 species she examined had nectars of a constant composition. The nectar of flowers with deep corolla tubes and protected nectaries contained mostly sucrose with small amounts of glucose and fructose, whereas the nectar of shallow flowers with unprotected nectaries contained little sucrose and mostly glucose and fructose.

The mouth parts of the bee come together to form a tube (see Snodgrass, 1956) through which nectar or other sweet liquid is sucked (Fig. 5). In the anterior part of the abdomen, the alimentary canal is enlarged into a crop or honeystomach in which the nectar is temporarily stored. The maximum nectar-carrying capacity of the honeystomach of the honeybee is about 70 mg but average nectar loads are about 20–40 mg (see Park, 1922; Frisch, 1934; Ribbands, 1953; Dade, 1962), depending to some extent on the attractiveness of the nectar, temperature, and the previous experience of the bee concerned.

Fig. 5. Honeybee collecting sugar syrup.

Wykes (1952b) found that honeybees preferred solutions of single sugars in the following descending order: sucrose, glucose, maltose and fructose. A mixture of equal parts of glucose, sucrose and fructose was preferred to a solution of any single sugar of the same concentration or to a mixture of these sugars in different proportions. This latter finding is surprising as few nectars have equal proportions of the three main sugars and most are either sucrose-dominated or fructose-glucose dominated. However, Furgala *et al.* (1958) found that the proportions of sucrose, glucose and fructose were more similar in *Melilotus alba* where they formed 36, 27 and 24% of the total solids, than in *Medicago sativa*, *Trifolium hybridum* and *Trifolium pratense* and they stated that honeybees preferred the former species.

When the sugar concentration of nectar is below a certain level (which has been estimated at 20%) the amount of energy needed to evaporate its water content to produce honey may make its collection uneconomical, and it is obviously to the bee's advantage to collect nectar with the greatest amount of sugar as quickly as possible; hence the most important factors influencing the attractiveness of nectar are its abundance and sugar concentration. This has been demonstrated by experiments in which the concentration of sugar syrup presented to bees and the ease with which it could be collected was varied (e.g. Frisch, 1934) and by correlating the abundance of bees visiting crops with the quality and quantity of nectar present (e.g. Kleber, 1935; Vansell, 1934; Butler, 1945a; Shaw, 1953). For example Vansell (1942) once saw honeybees desert *Prunus domestica* for *Arctostaphylos manzanita* at 10.00 h but return to *P. domestica* in mid-afternoon when the *A. manzanita* nectar was somewhat exhausted. On another occasion honeybees were abundant on *Brassica alba* but not on *Citrus sinensis*, when the nectar sugar concentrations were 43 and 20% respectively; by 15.00 h the mustard nectar was partly exhausted, the orange nectar concentration had risen to 28% and the bees had changed from mustard to orange.

There may be great differences in the average sugar concentration of nectar of different species (e.g. *Prunus avium* 12%, *Trifolium pratense* 22%, *Citrus sinensis* 30%, *Trifolium repens* 41%, *Brassica rapa* 51%; see Percival, 1965). Different varieties of the same species may also differ greatly in nectar concentration. But although different species and varieties may have nectar with different average sugar concentrations, even within a single flower, and especially shallow open ones, the sugar concentration is subject to considerable fluctuations as a result of exposure to wind and rain, and changes in temperature and relative humidity. Hence the attractiveness of a species may differ at different times of the day and at different stages of flowering.

In fact, although a flower species may exhibit a characteristic daily secretory rhythm which is closely followed by the abundance of nectar-gathering bees visiting it, this rhythm may be influenced by the aging of the flowers, by the amount of reabsorption that occurs and by concentration variations which depend upon changes in relative humidity (see Beutler, 1930; Kleber, 1935). Beling (1929) discovered that bees become conditioned to the time of daily periodic nectar production of the particular species they are visiting and spend the remainder of their day in the hive. When the time approaches at which nectar is available, they congregate near the hive entrance.

Park (1932) found that the concentration of nectar collected by bees was altered only very slightly while a bee was *en route* to its hive and the alteration was a decrease, not an increase as had previously been assumed. The average decrease was about 1% and was greater for nectar of high than of low concentration. It has recently been demonstrated that this is because the bees dilute the nectar with saliva from their labial glands, the dilution being greater with the more concentrated food (Simpson, 1964; Free and Durrant, 1966a).

Visits by bees and other nectar-gathering insects may also increase nectar secretion. Pedersen and Todd (1949) showed that the sugar concentration of nectar in flowers bees had visited was less than in flowers bees had not visited; Wykes (1953) discovered that periodic removal of nectar from flowers increased the total amount of nectar and sugar secreted, although the sugar concentration was lower. Bogoyavlenskii and Kovarskaya (1956) reported that flowers from which nectar was removed three times per day produced more sugar than those from which it was removed once a day. Pedersen (1961) suggested that sampling the nectar concentration of flowers should indicate the amount they have been visited and hence the likelihood that they had been pollinated.

Pollen and its Collection

The digestible portion of pollen is mainly composed of protein, fat and carbohydrate with various inorganic substances (see Todd and Bretherick, 1942). It is eaten by various adult insects, especially those belonging to the orders Hymenoptera, Diptera and Coleoptera, and pollen, or its derivatives, forms an important part of the larval food of solitary and social bees whose hairy bodies are well adapted for pollen collection. Bumblebees and honeybees possess special modifications for packing pollen (Fig. 6) which is transported back to their colonies as pollen pellets in the pollen baskets, or corbiculae, on their rear legs (see Hodges, 1952; Snodgrass, 1956). The two pollen pellets collected by a bee during its foraging trip are referred to as a pollen load. The size and weight of a load differs greatly with different crops and average weights of from 8 to 29 mg have been reported, although between 14 and 20 mg would seem to be the more usual (e.g. Park, 1922; Parker, 1926; Maurizio, 1953).

Todd and Bishop (1941) calculated that ten average size pollen loads are necessary to provide the protein to rear one honey bee and 2 million pollen loads, or 20 kg pollen, to rear the brood produced by a strong honeybee colony per year. Others have reported similar findings. For example, Schaefer and Farrar (1946), and Louveau (1954) estimated that a colony needs 18–23 kg and 25–30 kg of pollen per year, and Wafa (1956) calculated that in Egypt a colony collects an average of 16 kg pollen per year (range 11–31 kg) with a monthly average ranging from 0·4 kg in October to 2·4 kg in August.

Synge (1947) offered bees a choice of pollen of different species in a Latin Square arrangement inside the feeder of their hive and found that on four of the five occasions the bees were offered *Trifolium repens* and *Trifolium pratense* pollen the *T. repens* was preferred. The exception was obtained with a colony that had shown a foraging preference for *T. pratense*. Levin and Bohart (1955) provided a selection of pollen about 90 m from a thirty-colony apiary before any natural pollen supplies were available, and were able to demonstrate consistent differences in the preferences for pollen of different species; thus they found that *Brassica alba* and *Trifolium pratense* were usually very attractive while *Medicago sativa* had a relatively poor attractiveness. Doull (1966) showed that pollen from one species of *Eucalyptus* was less

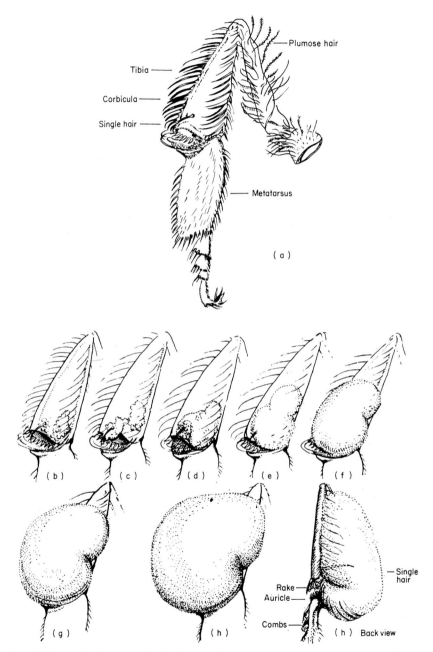

Fig. 6. The corbicula, or pollen basket. a, outside of right 3rd leg; b to h progressive packing of pollen load (after Hodges, 1952).

attractive than pollen from others. These experiments indicate that the pollen itself can determine the bees' tendency to select it, irrespective of the ease with which it can be collected. The selection of pollens did not seem to be influenced by their age, colour, moisture or protein content. Some pollens have a greater nutritional and biological value to honeybees than others, giving greater longevity and greater development of brood food glands, ovaries and fat bodies (Maurizio, 1950) and in France the nitrogen content of the pollen collected undergoes an annual cycle which reaches its maximum in May and June (Louveaux, 1959). However, there is no evidence that bees select pollen for its nutritive value.

Levin and Bohart (1955) mentioned that there seemed to be a correlation between the intensity of a pollen's odour and its selection, but because such evidence was rather subjective they could not be sure of this. If intensity of odour is responsible for a pollen's attractiveness, and this odour could be identified and synthesized it could possibly be most important not only for increasing the attractiveness of crops but also making pollen substitutes, which are fed to colonies during dearth conditions, more acceptable. Louveaux (1959) reported that pollen contains phytosteroles that attract bees. Initial experiments by Taber (1963a) demonstrated that a hexane or ethyl ether extract of pollen contained materials which are very attractive to foraging bees and initiate the behavioural response of packing the corbiculae. When this attractant was removed from pollen, bees would not collect the residue even though it contained over 97% of the total dry weight including most of the nutritive substances. In contrast bees collected non-nutritive cellulose when the attractant was added to it. Robinson and Nation (1968) compared the consumption, inside hives, of a basic artificial diet to which various extracts of pollen had been added, and found that the addition of the acetone soluble lipid fraction of pollen significantly increased the amount of food taken. Lepage and Boch (1968) and Hopkins et al. (1969) isolated and identified from bee-collected pollen a C_{18} straight chain trienoic acid which was very attractive to honeybees; thus a dish of cellulose powder to which this acid had been added received about fifteen times as many visits from foraging bees as control dishes containing cellulose powder only. Further progress in these studies may be expected in the near future.

Parker (1926) discovered that bees collect pollen from different plants at different times of the day. Synge (1947) pointed out that before pollen is made available to bees it is necessary for the anthers to dehisce and the flower to open, and that which ever process occurs last is the limiting one. The anthers of some species dehisce in the bud (e.g. *Trifolium pratense, Trifolium repens, Vicia faba*), while the anthers of others dehisce after the flowers open (e.g. *Cucurbita pepo, Ribes nigrum*) and anther dehiscence and flower opening of others occurs almost simultaneously (e.g. *Brassica alba*). Whereas the pollen of most species is available throughout the greater part of the day, the peak period of pollen presentation differs widely and tends to be characteristic of the species concerned; correlations have been established between the time of day a pollen is most abundant and its collection by honeybees (e.g. Synge, 1947; Percival, 1950, 1955).

When the anthers dehisce in the bud the time the flowers first open will determine when pollen is available. Some flowers open for one day only; others open on a number of successive days, closing each night. The time of opening may vary with the age of the flower. For example, *Vicia faba* flowers begin opening at about 14.00 h on the first day, but 11.00 h on the second day, and 08.00 h on the third day (Fig. 7). In association with this, most *Vicia faba*

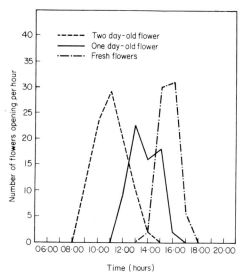

Fig. 7. Time of opening of *Vicia faba* flowers of different ages (after Synge, 1947).

pollen is available in the afternoon when flowers are first opening. Flowers of other species whose anthers dehisce in the bud mostly open in the morning and yet others open fairly continuously throughout the day so that a steady supply of pollen is available. When flower opening is not the limiting factor, the time of anther dehiscence regulates pollen collection. For example, "sleep movements" of *Ranunculus acris* occur very early and very late in the day and anther dehiscence which reaches its peak between 09.00 h and 10.00 h governs pollen presentation. Sometimes simultaneous dehiscence of all the anthers of a flower occurs (e.g. *Cucurbita pepo*, *Brassica oleracea*, *Ribes nigrum*) but anther dehiscence of other species occupies some days (e.g. 1–2 days, *Prunus cerasus*; 1–3 days, *Fragaria × ananassa*; 1–4 days, *Rubus fruticosus;* 1–5 days, *Prunus persica;* 2–7 days, *Pyrus communis;* 2–9 days, *Rubus idaeus* and 6–13 days, *Helianthus annuus*). The amount of pollen produced per flower also differs greatly with different species, but no obvious connection between the amount per flower and the tendency of bees to collect it exists.

Although the regular rhythm of pollen presentation, which is characteristic of a species, is probably less variable than the nectar availability rhythm, it is also subject to fluctuations in association with changing weather conditions. Hence meteorological factors not only influence the flight of bees directly, but also indirectly through the production of pollen and nectar in the flowers. Temperature seems to be particularly important as a factor limiting both flight and pollen availability (e.g. Hambleton, 1925; Wafa and Ibrahim, 1957a, 1958). Foraging flights may occur at 12 to 14°C (54 to 57°F) in early spring but slightly higher temperatures are needed later in the year and especially on cloudy days (e.g. see Lundie, 1925; Parks, 1925).

In general, in Western Europe, with increase in temperature from 10 to 30°C (50 to 86°F), there is an increase in pollen collection as a result of a steady increase in the numbers of stamens ripening and presenting pollen. Light intensity, rain and relative humidity are also important; however, in the field it is very difficult to evaluate the effect of one of these factors independently of the others (see Synge, 1947; Ribbands, 1953; Percival, 1955). Pollen collection is more intense when a favourable day follows unfavourable foraging weather than when conditions have been uniformly good. This is probably partly because the bees respond more readily to improvement of conditions, and partly because the pollen need of their colonies is greater.

Bees collecting pollen sometimes deliberately scrabble over the anthers to do so, but others become dusted with pollen incidentally, when it is available, while they are collecting nectar. Although many bees that collect pollen incidentally comb it from their bodies into their pollen baskets, others, which make no attempt to pack it into their pollen baskets, scrape it from their bodies and discard it. This behaviour was first recorded by Synge (1947) for bees visiting *Helianthus annuus* and has since been observed on this species and also on *Taraxacum officinale, Rubus idaeus,* and *Brassica napus* (Free, 1968a, 1968b; Free and Nuttall, 1968a). These species produce abundant pollen and its rejection is easily observed; perhaps similar, but less obvious, behaviour occurs on species whose pollen is less abundant (see Percival, 1955). Certainly, on the same crop and at the same time, some nectar-gatherers collect pollen loads while others do not. Presumably nectar-gatherers that retain pollen and pack it into their pollen baskets receive a greater stimulus to collect pollen than those that collect nectar only, but not as much as those that deliberately collect pollen. Some nectar-gatherers with large pollen loads may discard pollen from their bodies; presumably their corbiculae are filled to capacity but not their honeystomachs. The size of nectar and pollen loads varies greatly. Averages of 10 to 40 mg have been recorded for nectar and 7 to 20 mg for pollen (e.g. Park, 1922; Parker, 1926). There is an indication that bees collecting both nectar and pollen do not collect as much of either type of forage as bees collecting one type only (Hassanein and Banby, 1956).

Irrespective of whether foraging bees discard or pack pollen into their pollen baskets, their bodies are often covered with much pollen. Usually there is about twice as much on a bee's thorax as on its abdomen and pollen-gatherers tend to have more than nectar-gatherers (Free, 1966). This is one reason why pollen-gatherers are more likely to pollinate the flowers. Lukoschus (1957) and Skrebtsova (1957a) found that the amount of pollen varied with the species and variety the bees were working. Thus, excluding any pollen in their corbiculae the amount ranged from an average of 47 thousand pollen grains on the bodies of bees working *Fagopyrum emarginatum* to 4·2 million grains on bees working *Prunus avium*. Bees visiting one variety of *Ribes grossularia*, *Fragaria × ananassa* or *Rubus idaeus* had about twice as much pollen as those visiting other varieties of the same species. It has been demonstrated that part of this variation is due to a negative correlation between the size of pollen grains and the number adhering to bees' bodies, but undoubtedly most is because of differences in the amount of pollen produced by flowers of different species. Indeed, some species (e.g. *Ribes nigrum*) produce so little pollen that bees rarely collect pollen loads from them.

Flower Preferences of Different Bee Species

Loew (1885) classified species of bees as monotropic, oligotropic and polytropic according to whether they restrict their visits to one plant species, work a few closely related species or visit several species. Usually the relationship between bee and flower species is behavioural and physiological rather than morphological but some monotropic and oligotropic bees exhibit special morphological adaptations for removing nectar and pollen from the flower species they work. The flowers also often exhibit one or more reciprocal adaptations. Such bee/flower relationships are especially found in the bee families Panurgidae, Andrenidae, Melittidae, Anthophoridae and Megachilidae and among the plant families Cucurbitaceae, Convolvulaceae, Malvaceae, Cactaceae and Onagraceae (see Linsley, 1961).

Even when a bee species is polytropic it may prefer some flower species to others. For example, Linsley and McSwain (1942) found that *Anthophora linsleyi* preferred *Salvia carduacea* pollen to any other, and collected it exclusively in two of three locations where studies were made; Chambers

(1946) discovered that *Andrena varians* visited fruit trees relatively more than did three other species of the same genus that were foraging in the same locality at the same time; Free (unpublished) captured 107 female *Andrena armata* on return to their nests and 82 of them had been visiting *Acer pseudoplatanus*. The strong preferences of the solitary bees *Megachile rotundata* and *Nomia melanderi* for *Medicago sativa* pollen has been exploited with extremely beneficial results (Chapter 6).

Many solitary bees are only active during a short season, and occur in areas where there are few flower species and so have little choice but to specialize; in contrast, social bees are not limited to one brood generation and a brief foraging period, but the colony as a whole often forages throughout the flowering season and the foraging life of an individual bee may embrace the flowering periods of several successive species. As a result they are necessarily polytropic. Even so, preferences exist which may partly be determined by the shape of the flowers. Both bumblebees and honeybees prefer segmented flowers with a disrupted outline and there is some evidence that bilaterally symmetrical flowers are visited more by bumblebees than honeybees (Leppik, 1953; Free, 1970a). Different species of bumblebees sometimes prefer different shaped flowers and seek flowers in different types of habitat. Thus, in Scotland, Brian (1957) found *Bombus pratorum* and *B. agrorum* foraged in sheltered habitats, and visited mostly open flowers and flowers with corollas of intermediate length; *B. lucorum* visited mostly open flowers but foraged in exposed habitats, and *B. hortorum* restricted its visits to flowers with long corolla tubes. *B. agrorum* was more readily discouraged by the presence of other foragers, particularly *B. pratorum* which often alighted directly upon bees that were already feeding. It would be interesting to discover whether the same preferences are maintained in other parts of the world where these species exist.

Floral preferences of bumblebees are reflected in the types of pollen collected by colonies of different species. Thus, Brian's *B. lucorum* colonies collected their major supply of pollen from *Erica* spp., *Trifolium repens*, and *Lotus corniculatus*, and her *B. agrorum* colonies collected much pollen from *Trifolium pratense* and *Vicia* spp. in addition. Brian remarked that, although *Tilia* spp. was not locally common, small but significant amounts of it were collected by the *B. lucorum* colonies; and in a site in England where *Tilia* is abundant, Free (1970b) found it provided a major supply of the pollen of *B. lucorum* colonies. Indeed, the number of species of pollen collected by a bumblebee colony may largely depend on the abundance of its favourite species; when *B. agrorum* and *B. sylvarum* colonies were put beside a *Trifolium pratense* field they collected only *T. pratense* pollen, whereas a *B. lucorum* colony beside the field collected 43% of its pollen from other species (Free, 1955). Knee and Moeller (1967) found that, in Wisconsin, *T. pratense* was the favourite pollen of *B. fervidus* and *B. nevadensis auricomus*, *Medicago sativa* of *B. americanorum* and other flower species provided favourite pollens of *B. griseocollis*.

Some bumblebee species tend to collect pollen from fewer plant species than others (e.g. see Stapel, 1933; Brian, 1951; Free, 1958a). This is presumably connected with the availability of flower species and the time of year they predominate. Although it is commonly concluded that bumblebees cannot communicate the location of a source of food directly to other members of their colonies, Brian (1951) supposed that the more homogeneous pollen mixture of her *B. lucorum* than of her *B. agrorum* colonies might indicate an attempted mobilization of the foraging force to work a particular crop. This suggestion was supported by Free's (1970b) findings that two *B. lucorum* colonies exploited the local flora in ways too different to be explained by the random wanderings of individual bees. Perhaps the scent of the predominant pollens in a nest induces the bees to seek them when they begin foraging; this could be tested experimentally.

Communication of Floral Sources by Honeybees

The ability of a successful honeybee forager that has found a good source of forage to communicate its location to other members of its colony undoubtedly contributes towards the efficiency of honeybees in exploiting the surrounding flora. The means by which bees do this has recently been reviewed so comprehensively (Frisch, 1965, 1968) that only a summary of the relevant facts will be presented here. When the food source is near the hive the forager performs a "round dance" (Fig. 8a). In this it describes a series

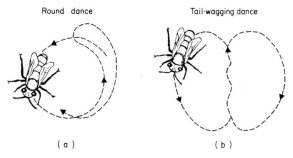

Fig. 8. Honeybee dances. a, round dance. b, tail-wagging dance (after Lindauer, 1961).

of circles on the comb, alternating between clockwise and anti-clockwise directions after every one or two circles (Frisch, 1923). Some of the nearby bees attempt to follow its manoeuvres (Fig. 9) and, every now and then the dancing bee regurgitates a drop of nectar from its honey stomach and offers it to them (Fig. 10). The dance alerts the following bees to the presence of a rich source of food; they learn its odour from the odour of the flowers adhering to the dancing bee's body and from the odour of the food they receive. They leave the hive and seek food of the correct odour in its vicinity.

Fig. 9. Successful honeybee forager performing a communication dance.

Fig. 10. Food transfer between two worker honeybees.

Because of this communication and recruitment, the number of bees visiting a rich source of food can increase rapidly and the more numerous and lively the dances, the greater the number of recruits.

When the food source is distant from the hive, the successful forager performs a "wagging" dance (Fig. 8b) (Frisch, 1946). In this, the bee moves a short distance in a straight line or "run", makes a semi-circle back to the beginning of the straight run, moves to the top of the run, makes another semi-circle in the opposite direction and then keeps repeating the whole figure for several minutes. During the dance, the forager is followed by other bees as in the round dance. The straight part of the dance is characterized by a rapid wagging of the bee's abdomen. The duration of the straight run,

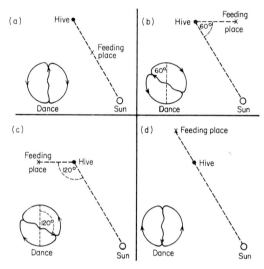

Fig. 11. Relationship between the angle of the straight run part of the dance on the vertical comb and the angle between the feeding place, sun and hive (after von Frisch, 1954).

or other factors associated with it, indicate the distance of the food source from the hive, and the further the distance, the longer the time taken to make the straight run. The straight run also gives information on the direction of the food source. The angle the straight run makes with the vertical is the same as the angle between the feeding place, hive and sun, so that bees must be able in some way to transpose the one for the other (Fig. 11).

In general, round dances are performed when the food source is within 25 m of the hive and tail-wagging dances when it is 100 m or more. Between these distances, dances that are transitional between round and tail-wagging dances are performed, the form of the transitional dance depending on the race of honeybee concerned (Fig. 12).

An illustration of the efficiency of communication by these dances was provided by Lindauer (1952). He observed that forty-nine of fifty-one bees that followed dances and returned with pollen had collected loads of the same colour as the dancing bee they followed. He also recorded that fifty bees that followed dances, later danced themselves; forty-six gave the same dance as the dancing bee they had followed, although two of them had collected forage from the "wrong" flower species and four gave a different dance. Hence although the system of communication is not fool proof it seems very efficient in both giving and receiving information. In natural conditions probably few

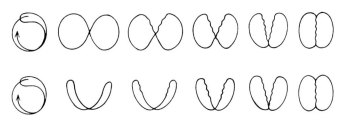

Fig. 12. Diagrammatic transition from round dance to tail-wagging dance. Top: via ∞ form. Bottom: via sickle-shaped dance (after Frisch, 1967).

recruits arrive at the precise spot indicated by a dance, but it is not necessary or desirable that they should. If a food source is worth communicating it is likely to be composed of several plants with a somewhat scattered distribution and if the points of arrival of recruits are also somewhat scattered the crop will be exploited more efficiently.

A honeybee foraging at a rich source of artificial food, such as a dish of sugar syrup, sometimes exposes its Nasanov scent gland (Fig. 13), which is located in the fold between its fifth and sixth abdominal tergites, and so distributes an odour which is able to attract potential foragers searching in the vicinity of the actual site (see Butler, 1969). Bees have been seen to expose their Nasanov glands on flowers in glasshouses (Frisch and Rosch, 1926; Free and Racey, 1966) when exceptional amounts of nectar were probably present, but they have not been reported to do so on flowers in the open. Part of the reason bees expose their Nasanov glands at a dish of syrup may be that it lacks a distinctive odour, and the amount of scenting diminishes when a floral scent is added to such a food source (Free, 1968c; Wenner et al., 1969). Bees also expose their scent glands when collecting water from clean dishes (Free and Williams, 1970a). Communication to water sources probably needs to be more precise than to most natural floral crops and the release of Nasanov gland odour helps to achieve this precision, and may well be its main function during foraging under natural conditions.

It seems probable that there is correlation between the tendency of a bee to dance and to expose its scent gland (Frisch, 1923). Bees may make several visits to a feeding place before they dance on returning to their colony (Lindauer, 1948) and bees do not expose their scent glands until after several trips (Ribbands and Speirs, 1955; Free, 1968c). Such a delay would probably have the biological advantage of not guiding other bees to a very transient food source.

Fig. 13. Honeybees exposing their Nasanov scent glands while foraging on sugar syrup.

Bees recruited to a source of food may be either those that have no previous experience of it, or those that have visited it before but were waiting in the hive until it was again available for collection. Bees of the latter category may also be recruited to a food source by contact with a non-dancing bee that has just returned from it; it has yet to be determined whether this type of recruitment is induced by the "excited" arrival of the forager, or the odour of the food source on its body, or antennal contact, or whether all these factors play their parts.

Floral Preferences of Different Honeybee Colonies

The foragers of a honeybee colony usually visit many flower species during the season, but it is undoubtedly a reflection of their ability to communicate the sources of good forage that at any one time they collect most of their forage from a few plant species, although small amounts from many others. Some species are mostly visited for nectar only, whereas many others are also valuable sources of pollen, and a few are visited for pollen only. Different colonies may utilize the local flora in different ways. Todd and Bishop (1941) and Eckert (1942) noted that the pollen collected by colonies side by side sometimes came from predominantly different sources. This aspect of foraging behaviour was extensively studied by Synge (1947). The number of loads of pollen of representative species collected in the pollen traps of her two colonies that showed the greatest differences are given in Table 1.

TABLE 1

Plant species	Colony E4	Colony K5
Salix	392	3,160
Ulex	2,956	701
Buxus	114	809
Aubrieta	483	1,069
Berberis	424	1,653
Betula	0	2,328
Quercus	2,587	79
Ranunculus	4,967	1,507
Vicia faba	4,979	223
Trifolium repens	67,444	32,415
Rosa	608	2,008
Papaver rhoeas	1,834	8,066
Epilobium angustifolium	1,735	3,576
Trifolium pratense	29,386	45,148
Hedera helix	6	2,610

Similar findings have been reported by Maurizio (1949, 1953), Louveaux (1954), Schwan and Martinovs (1954) and others. It seems that these differences between colonies arise partly through chance differences in the discoveries of scout bees when searching for new crops. Because the majority of bees do not seek forage on their own but are recruited to crops by dancing bees, differences in the findings of scout bees are magnified by the foraging force as a whole. Differential conditioning to certain types of pollen may also help explain differences in the foraging behaviour of colonies. Thus, Louveaux (1954) discovered that colonies which collected most *Brassica napus* pollen also collected most *Brassica sinapis* pollen, although there was an interruption of about 10 days between the flowering of the two species and he suggested that certain colonies have a "taste" for cruciferous pollen. This was confirmed by Free (1963) who found that when a colony was moved to a *Brassica alba* field, bees that had previously collected other cruciferous pollen transferred more readily to *B. alba* than those that had not. These differences between colonies may also be partly genetically determined. Recent work has demonstrated that some colonies not only prefer *Medicago sativa* pollen but this preference can be enhanced by selection (pages 87, 88).

The amounts of pollen of a particular species that is collected by neighbouring colonies tend to remain relatively constant and to rise and fall in unison, presumably in accordance with weather conditions and pollen availability. A typical example is shown in Fig. 14. However, this is not invariably

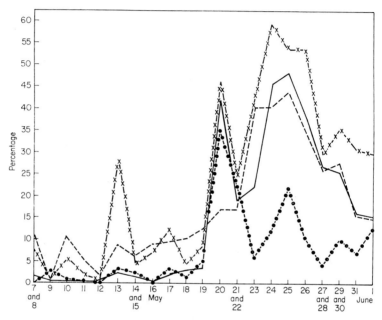

Fig. 14. The amount of *Pyrus malus* pollen collected by different colonies (after Free, 1959).

so; thus a colony that has collected only relatively small amounts of a particular pollen at an early stage of flowering may eventually collect more of it than its neighbouring colonies later on; conversely a colony that has collected a large proportion of its pollen from a particular species may collect a relatively small proportion from it later on.

The number of species a colony visits probably depends very much on the locality. Free (1959) recorded that colonies visit averages of about thirty to forty species for pollen in England, but great variability occurred and some colonies visited twice as many as others. Much of this variation occurs among those species that provide only a small proportion of a colony's total pollen requirements, although a given colony occasionally collects much of its pollen from a species that neighbouring colonies are visiting little, if at all (Maurizio, 1953; Louveaux, 1954; Free, 1959). The principal species from which bees collect pollen tend to be exploited by all colonies. These species are either cultivated plants or very common wild plants.

Comparison of the floral preferences of honeybee colonies have been based almost entirely on pollen collection. Pollen is abundant in some nectars and undoubtedly contributes toward a colony's protein requirements. Even when it is not abundant there is usually sufficient pollen in nectar to determine its source, and it would be interesting to compare the nectar as well as the pollen preferences of neighbouring colonies.

Poorly-adapted Pollinators and Flowers

When a flower and foraging insect are not well adapted to each other insect visits may occur without accompanying pollination. Pollen is generally more readily accessible to visitors than nectar, which is frequently well hidden, and may be available to specialized visitors only. When the adaptability of bee and flower is incomplete it is nearly always the nectar-gatherers that fail to pollinate. Plant species are now grown for food, or other uses, in parts of the world far from where they originated, and sometimes in the absence of their natural pollinators. In such circumstances careful consideration should be given to importing the natural pollinator with the introduced plant species (pages 4, 101, 116). However, mostly the insect pollinators already present, and especially the honeybees and bumblebees, are able to pollinate the crop satisfactorily, although the pollinating efficiency per visit may not be as great.

Honeybees are sometimes too small to "trip" the pollinating mechanism of a flower, or to bridge the gap between anthers and stigma, or to touch the anthers when collecting nectar only. Sometimes their tongues are too short to reach the nectaries of flowers with long corolla tubes, and the relatively few that enter flowers collect pollen only. In fact, with few exceptions, the behaviour of pollen-gathering honeybees is more conducive to pollination than those collecting nectar only.

Although bumblebee workers, even of the same species, vary greatly in size, the relative tongue length of some species is much greater than that of others (e.g. Brian, 1954; Holm, 1966a) and short-tongued bumblebees are sometimes unable to reach the nectar of some flower species with long corolla tubes. Some of these short-tongued species (e.g. *Bombus lucorum, B. terrestris*) often bite holes near the bases of the corolla tubes and obtain nectar through the holes without entering the flowers. Although the mandibles of honeybees are not suitable for biting such holes, they use those previously bitten by bumblebees. Such "robber" bumblebees and honeybees are useless as pollinators.

Many crop plants (e.g. *Vicia faba, Ricinus communis, Helianthus annuus, Cannabis sativus* and *Gossypium* spp) have extra-floral nectaries. Although they undoubtedly served some useful biological function in their place of origin (see Ruppolt, 1961), they seem sometimes to have a detrimental effect on pollination, especially when they secrete nectar before the flowers open because bees that become conditioned to them do not visit the flowers and so do not pollinate.

Constancy during a Single Foraging Trip

Although a bee species may be polytropic, most individuals keep to one flower species only during a single trip. This fidelity of the bee has attracted attention from the earliest times. Darwin (1876) pointed out that constancy to flower species is advantageous to the plant by facilitating cross-pollination. It is also advantageous to the bee, enabling it to forage more quickly than if it had to learn the sites of the nectaries on each flower it visited, and enabling it to keep to a species yielding abundant nectar and pollen.

The flower constancy of bees was originally studied by observing them for as long as possible while they were foraging. Observation alone has the disadvantage that the observer selects and continues to watch only those bees that remain within his field of vision. Furthermore, the results of this method depend to a considerable extent on the abundance of plant species where the bee is working; thus, on a large homogenous crop it cannot help but keep constant as long as it remains. However, early observers (see Clements and Long, 1923) established that even when many flower species were growing together, the honeybee was usually constant but the bumblebee was less so, and hoverflies and butterflies rarely so.

More recently the contents of pollen loads have been used as an index of constancy. Betts (1920, 1935) made the first thorough study of constancy using this method and she found that between 2 and 7% of honeybee pollen loads were mixed. Later and extensive studies (e.g. Percival, 1947; Maurizio, 1953 and Free, 1963) have found that between 0 and 11% of honeybee pollen loads are impure. The proportion of mixed loads does not depend on the abundance of pollen, but pollen from the principal sources of forage commonly occur as minor components of mixed loads. This is also true for bumblebees and solitary bees but these bees are less constant than honeybees. Betts (1920) found that 66% of bumblebee pollen loads were pure. Clements and Long (1923) found that honeybees, *Andrena* spp., *Megachile* spp. and *Bombus* spp. had 87, 64, 54 and 53% pure loads. Brittain and Newton (1933, 1934) reported that honeybees and *Halictus* spp. were very constant but *Bombus* spp. and *Andrena* spp. less so. Some of their data is summarized in Table 2.

All bee species were more constant where there were large areas of an attractive species such as *Pyrus malus* than when there were many species and comparatively few plants of each. This aspect of the constancy of bumblebees was especially studied by Spencer-Booth (1965) who discovered that only 17% of bumblebees collected from gardens and allotments had pure loads compared to 49% of those collected on homogeneous crops. Manning (1956) pointed out that the greater tendency of bumblebees than honeybees to work small patches of flowers could help to explain why they were less constant. Differences in the floral preferences of different bumblebee species and in the type of area in which they forage could help explain why some bumblebee species are more constant than others (Spencer-Booth, 1965; Free, 1970b).

TABLE 2

Genus	Host plant	No. plant species per load				
		1	2	3	4	5
Apis	All species	163 (62%)	73	22	4	1
Halictus	All species	207 (84%)	34	6	0	0
Bombus	All species	50 (59%)	25	8	2	0
Andrena	All species	82 (45%)	70	26	2	0
Apis	Pyrus malus	118 (80%)	25	3	1	0
Halictus	Pyrus malus	41 (72%)	14	2	0	0
Bombus	Pyrus malus	41 (65%)	17	4	1	0
Andrena	Pyrus malus	71 (57%)	47	5	1	0

Determinations of the percentage of pure loads is not an entirely satisfactory way of expressing constancy. Thus, whereas most mixed loads contain only two species of pollen, as many as eight may be present (e.g. Brian, 1954; Spencer-Booth, 1965; Free, 1970b). In contrast, constancy is often greater than the results imply as there is usually a preponderance of one pollen in a mixed load. Furthermore, a pollen load may contain a few grains only of another pollen that may have been transferred when the bee concerned either visited flowers on which wind or other insects had previously deposited pollen of another species, or entered a cloud of wind-blown pollen, or rubbed against parts of its hive and other bees (see Lukuschus, 1957; Skrebtsova, 1957a). Because of these possibilities some authors do not include as mixed loads those that contain only a few foreign pollen grains.

The manner in which different pollens occur within a mixed load may also reflect the behaviour of the bee. When the two or more pollens are segregated into distinct layers, the bee has obviously been more constant in its behaviour than when the pollens are intermingled (see Betts, 1920; Percival, 1947). Finally, because most honeybees and bumblebees collect pollen on some trips but nectar only on others, their overall constancy must be less than that indicated by examining pollen loads alone.

However, despite these disadvantages, examination of pollen loads is certainly the most reliable and objective way of assessing constancy. The constancy of nectar-gatherers can only be determined by observing them. Attempts to determine the constancy by examining the pollen grains in the nectar loads of bees would be unreliable because their honeystomachs are not empty when they leave home to forage; however, this technique might be usefully employed to determine the constancy of other flower-visiting insects.

When visiting many crops, but especially *Trifolium pratense* and *Medicago sativa*, honeybees and bumblebees accumulate pollen in the proboscidial fossa which is located in the posterior ventral part of the head (e.g. Woodrow, 1952; Minderhoud, 1954; Spencer-Booth, 1965). Bees have great difficulty in removing this pollen (Levin, 1955) and it is only dislodged as new pollen accumulates; bees working one crop may still retain the pollen from a previous one (Furgala *et al.*, 1960). Therefore, although examination of the fossal pollen indicates the species a bee has visited in the past it does not give reliable information about its current behaviour.

Inconstant bees may perhaps help in producing hybrids and new species. Betts (1920, 1935) found there was no special phylogenetic relationship, or similarity in floral structure and colour, between the plant species whose pollen comprised the same mixed loads, but the species concerned were merely growing intermingled. Percival (1947) also concluded that plant species from which mixed loads were derived were often very different in structure and in the size and colour of their flowers, and had little in common except they were growing together and had flowers the same height above ground. Hence, for interspecific crossing to occur during a single trip it appears that closely related species must grow near each other.

Mechanism and Effect of Flower Constancy

Although it has long been accepted that flower constancy is important in facilitating cross-pollination between plants of the same species, until comparatively recently its function as an isolating mechanism was not fully appreciated. It was previously assumed that plant species growing in the same area without hybridizing were incompatible but Mather (1947) showed this was not always so. He grew together *Antirrhinum majus* and *A. glutinosum*, which differ in form of growth and size and colour of flowers, and obtained only 3% crossing, although artificial pollination showed they were fully cross-fertile. He observed that this was because honeybees obviously distinguished between the two species and adhered to one or other when working a mixed stand, and so were not responsible for inter-specific pollination.

These experiments were continued by Grant (1949) who grew together three sub-species of *Gilia capitata*. One of these had creosote scented nectar and oval corolla lobes, and the other two had sweet scented nectar and linear corolla lobes. Individual bees kept either to the first or the other two sub-species only, but as the supply of flowers diminished so also did the constancy of the bees. Bateman (1951) grew two swede varieties and two cabbage varieties in the same plot and observed that individual honeybees were very constant to one or other species but less to the different varieties. These findings demonstrated that allied varieties or sub-species may be too alike for bees to distinguish between.

Mather (1947) found that bees did not always discriminate between the F_1 hybrids of his two *Antirrhinum* species, which had coloured flowers, from *A. glutinosum* which has white flowers, and as a result he suggested some differences other than flower colour seemed to be responsible for bees discriminating between *A. majus* and *A. glutinosum*. A consideration of the senses used by honeybees when foraging on flowers (page 15) would indicate that the odours of different species are their most important distinguishing features to bees, and that bees are more likely to respond to a genetical change associated with flower scent rather than to form or colour. It is relevant that closely related plants often have very different odours (Kerner, 1895). However, differences in colour and form undoubtedly help to reinforce those of odour and, as Grant (1949) pointed out, taxonomists use the floral characters of insect and bird pollinated species as distinguishing features more than they do those of species pollinated by other means. Any association between flower colour and nectar production (e.g. *Trifolium pratense* Goetze, 1930; *Phaseolus*, *Melilotus* Smaragdova, 1957) would of course facilitate conditioning. For comprehensive discussions on pollinator flower relationships and on the various mechanisms of isolating plant species the reviews of Grant (1949, 1950), Baker (1961, 1963), Faegri and Pijl (1966) and Baker and Hurd (1968) should be consulted.

Constancy During Consecutive Trips

Compared to the numerous studies of bee constancy during a single trip little work has been done on constancy during a longer period. Many observers (see below) have marked bees visiting a crop and found they returned to it for many successive trips or days; however, these observations gave little indication of the constancy of bees in general because, when the marked bees were not foraging on the crop, they could have been visiting the same species elsewhere, visiting other species, inside their hives or dead.

Although it has often been observed that the relative bee population on different species changes greatly at different times of the day (e.g. Brittain and

Newton, 1933; Vansell, 1934) it was not known whether the same bees were concerned, and because bees conditioned to crops tend to stay at home at times of day when their particular crops are not yielding nectar or pollen, it seemed probable that the foraging populations on different crops were discrete. However, Percival (1947) observed that bees working *Crataegus monogyna* in the afternoon still bore on the dorsal part of their thoraces, orange-coloured pollen from *Sarothamnus scoparius* which they worked in the morning. Ribbands (1949) watched five marked bees, for two or three days each, foraging on a small plot containing five flower species. Three bees visited one species only, one visited two and one visited four species. The bee that worked two species visited *Papaver rhoeas* in the morning and *Limnanthes douglasii* in the afternoon. These results might have been exceptional because of the nature of the plot and the high honeybee population in the immediate area.

To find the general foraging constancy of honeybees on successive trips and days Free (1963) marked pollen-gatherers and removed and identified one pollen pellet from each whenever it returned to its hive. The percentage of pollen-gatherers collecting the pollen they had when marked decreased on each successive day, although the rate of decrease differed greatly in different experiments, and probably reflected changes in foraging conditions. Thus after one day, about 70 to 90% of pollen-gatherers collected their original pollen, but after a week only 40 to 60% did so. Bees that collected the most common pollen tended to be the most constant, presumably reflecting its greater attractiveness and availability.

On days when a particular pollen was temporarily unavailable, bees conditioned to collecting it usually either did not forage or collected nectar only; very few collected pollen from another species. In general, it appeared that bees only began collecting from another species when pollen of the species they had been visiting was unobtainable for longer periods. Perhaps during the first day or so, most bees do not go to those parts of their hives where they might be influenced by the recruiting dances of bees that have found favourable crops. This temporary fixation, together with adaptability over a longer period, seems to explain the results obtained.

In contrast to Ribbands' (1949) bee, no bees consistently collected different pollens at different times of the day. However, many bees returned from some trips with pollen and others with nectar only. Presumably, the times of day at which the bees collected pollen from a crop was governed by the times at which it was presented, and the relative proportions of nectar-gatherers and pollen-gatherers on a crop vary greatly at different times of the day. On some flowers (e.g. *Pyrus malus; Helianthus annuus*) the transition from pollen to nectar collection is easy, and individual bees readily change from one form of behaviour to the other; many of the bees with pollen loads may have been primarily nectar-gatherers that were collecting pollen incidentally. On other flowers (e.g. *Vicia faba; Trifolium pratense*) from which many individuals gather either nectar only or pollen only, pollen-gatherers desert the crop rather than change to collecting nectar only; it is not known whether they forage elsewhere for nectar. If so, their overall flower constancy will be less than when they are collecting pollen only.

Mixed loads probably provide an indication of the behaviour of bees that are dissatisfied with one crop and sampling another. Free (1963) found that few honeybees with mixed loads collected the same mixture again; most collected pure loads of one of the species on subsequent trips, indicating that they either transferred to another crop, or returned to their original one. Percival (1947) found segregated loads in which one pollen was enclosed between layers of another; presumably on these trips bees had sampled a new pollen but returned to the original one.

Although bumblebees are, on average, less constant during a single trip than honeybees, Free (1970b) found that the day to day constancy of some bumblebee species did not differ greatly from that of honeybees; for example, about 70% of *B. lucorum* pollen-gatherers remained constant over a 10-day period. In contrast to honeybees, some bumblebee foragers collected the same combination of pollen, even in similar proportions, for several consecutive trips, sometimes extending over days; hence even such bumblebees showed some evidence of constancy. However, as with honeybees, the presence of mixed loads sometimes indicated that the bumblebee concerned was transferring from one species to another; sometimes such a transition occurred abruptly, but more often there were one or more trips in which both species were collected.

Nectar-gatherers are more likely to become attached to individual plants or flowers than pollen-gatherers, because the supply of nectar is more likely to be replenished. But it is less easy to obtain relevant information on the foraging constancy of nectar-gatherers, and much of this has been inferred from experiments in which bees have been trained to artificial sources of food. Thus early experiments of Frisch (1923, 1934) showed that the recruitment of honeybees to dishes of sugar syrup increases or decreases in accordance with the abundance and concentration of the syrup available, and when Butler *et al.* (1943) gave honeybees the choice of two sugar concentrations more collected the higher one. Other secondary factors of importance in stimulating lively and lengthy dances, and hence greater recruitment, include a floral odour, a flower-like food container and good foraging weather. When nectar forage is relatively very good the bees exploit it more intensively by collecting larger loads, spending less time in the hive between trips and working longer hours (see Frisch, 1965). The thresholds for recruitment vary with the amount and concentration of nectar available from natural sources and with the demands of the colony; improvement in the quality of a food supply is especially effective in inducing dancing and scent gland exposure (Lindauer, 1948; Ribbands & Speirs, 1955; Boch, 1956). So, preferential recruiting will occur for whatever source of food is most profitable. Ribbands (1949) concluded that the foraging constancy of a bee is determined by the present yield of the crop it is visiting and its memory of past yields, and when a crop is deserted in favour of another the bee may continue to inspect it at intervals. Probably, mostly as a result of different past experiences, different bees respond differently to the same situation, and some dance and expose their Nasanov glands much more readily than others. Probably, for the same reason, there is great variation in the constancy shown by individual bees.

It is apparent from the above results that, although a honeybee forager tends to remain faithful to one species, its behaviour is adaptable enough for it to forsake crops that have become unrewarding and exploit new and beneficial ones. This adaptability is emphasized when colonies are moved to new sites. Free (1963) found that when the flora at the old and new sites was similar, most bees visited the same species at both and the changes that did occur were usually associated with the relative abundance of the different species at the old and new sites. In fact, constancy was similar to that of bees whose colonies were not moved. However, when colonies were taken to areas where one crop predominated most bees that had not foraged on the species concerned tended to do so.

Scout Bees

It is not clear to what extent dissatisfied foragers search for their own crops and to what extent they are directed to them by dancing bees. Certainly not all the bees of a colony are directed to food by others and some search for it themselves (Oettingen-Spielberg, 1949). Such bees are sometimes called "scout bees" and three distinct categories can be recognized: (1) a scout may be a new forager that has left its hive to search for food for the first time when it is attracted to colours, shapes and floral odours (e.g. Oettingen-Spielberg, 1949; Butler, 1951; Lindauer, 1952); however, the majority of new foragers are recruited to crops by successful foragers; (2) a scout bee may be one whose source of forage has failed and who is searching for a new one; (3) a scout may be a forager that is inspecting a depleted source of food to which it is already conditioned, although if the source remains depleted it may become a scout of the second type. Hence a scout bee can be of any age, and scouting is a temporary occupation.

Although at any one time there may be scout bees and crop-attached bees, it seems likely that an individual can readily change from one category to another, and, when a food supply loses its attractiveness, the bees sooner or later, depending on the previous experience of the individuals concerned, seeks other forage to which, if satisfactory they may become temporarily attached. However, each bee continues to inspect, although at increasingly infrequent intervals, its original source of forage. Should its source of forage retain its attractiveness or even increase in attractiveness, it seems unlikely that the majority of bees will ever inspect other crops, although possibly a few do so.

Foraging Distances

There is ample evidence to show that whenever possible bees prefer to forage near their hives, and this preference is enhanced by the tendency of successful foragers to recruit more bees when working near rather than distant food sources Françon, 1938; Boch, 1956). For example, Vansell (1942) found that the number of bees on *Pyrus communis* trees decreased greatly at 60–90 m from the edge of the orchard nearest his colonies, and few bees were found 120–150 m away. Butler *et al.* (1943) observed that dishes of sugar syrup 146 m away from colonies always had more visitors than those 365 m away, although the extent of this difference varied daily. Braun *et al.* (1953) showed that whereas bumblebees and honeybees were evenly dispersed over small fields of *Trifolium pratense* their numbers decreased from the periphery to the centre of large fields. More recently, Haragsim *et al.* (1965) marked colonies in an *Medicago sativa* field with radioactive gold and found the proportion of bees captured that were radioactive decreased with increasing distance from their colonies as follows: 1–200 m, 48%; 2–300 m, 42%; 3–400 m, 38%; 4–500 m, 28%.

Particularly during unfavourable weather, bees tend to forage near their hives. This has been frequently noticed in fruit orchards, where weather during flowering is likely to be changeable. Thus, Sax (1922) observed no bees, and Hootman and Cale (1930) noticed few bees, further than 60 m from their hives in fruit orchards in cool, windy weather, and Nevkryta (1957) found that at 12 to 15°C (54 to 59°F) bees did not visit *Prunus avium* trees more than 125 m from their hives. Butler *et al.* (1943) gained the impression that bees visiting distant dishes of sugar syrup were more easily deterred by unfavourable weather than bees working close to their hives, and that such bees did not work nearer home in these circumstances. This was confirmed by Boch (1956) who found that bees visiting nearer dishes foraged earlier in the morning and later in the evening.

Foraging Areas

Usually, only a proportion of flowers on most plants open at one time so cross-pollination is favoured. Even so bees do not usually keep to an herbaceous plant until they have collected all the food from it, but normally visit only a small proportion of the available flowers on each (e.g. *Viola* spp., Veerman and van Zon, 1965; *Fragaria* × *ananassa*, Free, 1968b), and only a small proportion of flowers per flower head (e.g. *Helianthus annuus*, Free, 1964a; *Trifolium pratense*, Free, 1965a). This behaviour must further facilitate cross-pollination and be especially important for the pollination of self-sterile species. The proportion of flowers visited per plant probably increases with the nectar content, but it has not been definitely established whether bees can determine the presence of nectar or pollen in a flower without alighting. Apparently, a bee is not deterred from visiting a flower on which another has recently foraged (Darwin, 1876; Ribbands, 1949).

Probably, most foraging animals that have a "home" do not range indiscriminately in its locality but, for limited periods at least, forage in a comparatively small area only. This is particularly true of bees and, although they move fairly frequently from one plant to another, they tend to have quite small foraging areas. Early workers observed that many honeybees returned on successive trips to the vicinity of the plants on which they were marked (e.g. Müller, 1882; *Salvia* spp.; Giltay, 1904, Papaveraceae spp.; Bonnier, 1906, *Fagopyrum esculentum*). Bonnier (1909) sprinkled sugar syrup over dead branches for honeybees to collect and found little mixing of the foraging populations occurred between two feeding places 2 m apart and none when they were 20 m apart. MacDaniels (1930a) reported than in fruit orchards bees worked very locally and either kept to a single tree during a foraging trip or moved between adjacent ones, and the practical importance of the size of foraging areas on the fruit set of self-sterile trees was soon realized (page 399).

More extensive observations were made by Minderhoud (1931). He marked honey bees on plots of *Brassica juncea*, *Trifolium pratense*, *Taraxacum officinale*, and *Reseda lutea*, and by plotting their movements found they foraged within an area 10 m². Buzzard (1936a, 1936b) reported that marked bees tended to keep to the same patch of *Melilotus*, *Doryenium* and *Helianthemum*, or the same bunch of *Cotoneaster horizontalis* until the flowers faded, but the area worked varied with the number of flowers available. Honeybees only strayed from one *C. horizontalis* bush to another when the branches of the two were interlaced, and when they did so, quickly returned to their own. Butler *et al.* (1943) marked honeybees visiting a small patch (5 × 7 m) of *Epilobium angustifolium;* 24 h later about 90% of the marked bees seen still foraged within this area and the remaining 10% were foraging 5 m from its centre. Bees marked visiting isolated patches of *Echinops sphaerocephalus* seldom visited neighbouring plots about 17 m away. The number of marked bees diminished each day but some still returned 16 days after marking.

Butler *et al.* (1943) also studied the foraging areas of honeybees on an artificial crop composed of 112 petri dishes containing sugar syrup. They found that provided the supply of syrup did not fail an individual bee tended to confine its foraging to a single dish; only few bees visited dishes other than those on which they were marked and most of these went to neighbouring ones (Fig. 15). However, when the dishes to which bees were conditioned became empty, they searched in the immediate vicinity until they found and foraged on neighbouring ones. Although on subsequent visits many inspected the empty dishes, their tendency to do so diminished with time. Hence, fidelity to area, like fidelity to flower species, can be adapted to meet current

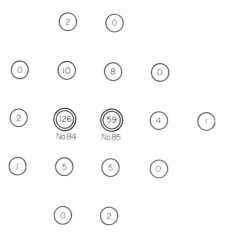

Fig. 15. No. times bees marked on dishes No. 84 and 85 were seen foraging on these dishes and neighbouring ones (after Butler *et al.*, 1943).

circumstances. Experiments with artificial food sources help considerably in understanding the behaviour of bees on flowers, but as Butler *et al.* (1943) stressed, it would be incautious to draw conclusions from experiments with artificial food sources alone, especially as the relative proportions of the time spent actually foraging and in the hive are very different when bees are visiting dishes than when collecting natural forage. As a result of continuing these studies on natural crops Butler (1945a) concluded that whereas in general, nectar concentration determines whether a flower species is visited, nectar abundance determines the proportion of foragers that visit it, but pointed out that it is difficult to separate the effect of these two factors in the field.

Butler (1943, 1945b) suggested that, when competition between foraging honey bees is sufficiently great, a proportion of so-called "wandering" bees, which would be responsible for much of the cross-pollination in orchards, is superimposed on the "fixed" population, and that, furthermore, the proportion of "wandering" bees is greatest near groups of hives. However, such a definite division of the foraging population has yet to be demonstrated.

Singh (1950) observed and plotted the courses of individual bees for as long as possible while they foraged on plots of *Trifolium hybridum, Lotus corniculatus, Aster*, spp., *Taraxacum officinale; Fagopyrum esculentum; Solidago virgaurea* and *Pyrus malus* (Fig. 16). Foraging areas were particularly

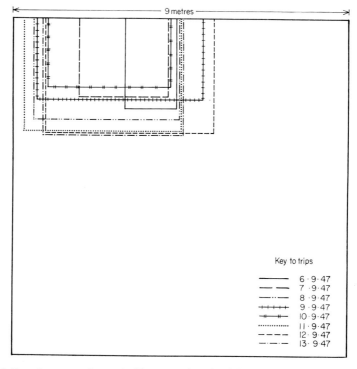

Fig. 16. Foraging areas of a marked bee on a plot of *Solidago virgaurea* (after Singh, 1950).

restricted in size during calm, sunny weather when abundant nectar and pollen were present, but when forage was sparse, either because few flowers were present or the flowers were yielding little nectar or pollen, foraging areas were larger and the bees spent less time per flower and were more easily disturbed. Ribbands (1949) who observed marked bees visiting *Eschscholtzia* spp., *Tropaeolum majus* and *Papaver rhoeas*, also concluded that the size of foraging areas was very variable and was influenced by the number of flowers available, their nectar and pollen content and the amount of competition. Bees working failing sources characteristically became restless and questive and moved from one portion of their foraging area to another and even beyond it. Both Singh and Ribbands found that a bee often tended to alight at or near the same point in its foraging area on successive trips, but its departure points were well separated.

Weaver (1957a) observed that on a crop of *Vicia villosa* an individual bee foraged repeatedly over the same area. At the end of their trips bees made exploratory visits to several flowers before returning to their hives. When nectar was abundant the flowers explored were always close together, but when nectar was sparse they were often scattered far apart. Weaver also confirmed that in improved or settled conditions foraging areas are fairly constant, but as the amount of nectar present diminishes the foraging areas are extended. Free (1960a, 1966a) came to similar conclusions on the effect of the amount of forage available on foraging areas in fruit orchards.

Fluctuations in availability of forage probably help to explain why the longer a bee forages on a crop the greater the area it covers. For example, Singh (1950) found that the size of a bee's foraging area on *Trifolium hybridum* was only about 7 m^2 during a single trip, but increased to between 12 and 46 m^2 during a day's foraging, and to 122 m^2 during 13 days. Free and Spencer-Booth (1964a) found that, in an orchard of dwarf *Pyrus malus* trees, a bee kept to 3 m of a row during a single trip, but covered 340 m^2 during 2 days and 996 m^2 during 8 days.

Weaver's (1957a) observations on the foraging areas of bees working *Vicia villosa* were made near landmarks of various kinds and he pointed out that it is not known how accurately a bee could return to a previous foraging area if no landmarks other than the *V. villosa* plants themselves were present nearby. Levin (1966) found he was able to recover marked bees more easily in *Medicago sativa* fields well provided with landmarks than in fields devoid of them, and tests in which artificial landmarks were put in large fields helped the bees to locate and return to their particular foraging areas. Hence, the size of foraging areas may vary with the abundance of landmarks in the vicinity. Free (1966a) pointed out that in orchards the presence of varieties in discrete rows helps bees to orientate themselves and keep constant to varieties, so that, to encourage mistakes and cross-pollination, this type of planting should be avoided. Perhaps the practice of growing tall plants of another species between two varieties, to avoid contamination between them, is more useful in facilitating the orientation of the bees than in producing a barrier. The presence of wind breaks (page 81) may have a similar effect.

Kikuchi (1963) noticed that when an insect (including a honeybee, bumble-bee, syrphid fly, blowfly, and butterfly) arrives at a flower on which another is working usually either both, or one of them, takes flight. Presumably this sort of disturbance, with its consequent increase in foraging areas, increases with the concentration of insects on the crop and must help to disperse the bees to parts of the crop where there is less competition. However, the presence of this avoiding behaviour depended to some extent on the size of the flower concerned and when the flower was large (e.g. *Helianthus annuus*, *Allium fistolosum* and *A. odorum*) two or more insects often worked the same flower. Weaver (1956a) observed that the tendency of bees to be disturbed while working *Vicia villosa* depended on their foraging behaviour, and bees that obtained nectar from the base of flowers, and so had their eyes and antennae fully exposed at all times, were more easily disturbed than bees that pushed their heads into flowers. A forager often flew threateningly toward another bee which sometimes flew to several metres away, while the aggressor continued to forage nearby.

The little evidence available shows that bumblebees also tend to restrict their foraging to small parts of a large crop (see Free and Butler, 1959). It has been generally supposed that bumblebees have larger foraging areas, and move from plant to plant more frequently, than honeybees, but the few direct comparisons that have been made indicate that, although bumblebees may visit twice as many flowers per trip as honeybees, their foraging areas are similar (Free, 1964a, 1968b).

Contamination of Seed Crops

Probably pollen left by a bee on the edge of its foraging area may be picked up by another bee whose foraging area overlaps, so pollen may be transferred outside the confines of a single foraging area. Even so, because each crop contains a system of overlapping foraging areas, there will be a corresponding series of overlapping pollinating areas, and interbreeding through a self-sterile crop will not be homogeneous. Therefore, when two compatible varieties are grown in adjacent blocks, most intercrossing will occur near to where they adjoin, and will then rapidly diminish with distance. Knowledge of the extent of intercrossing and the distance to which it occurs is most important to growers who wish to produce uncontaminated seed. Brown (1927) and Balls (1929) found that contamination between 2 cotton varieties rapidly diminished from where they adjoined but a little contamination still occurred between plants up to 40 rows apart (page 163). However, with the exception of this work, early information was mostly circumstantial. A thorough investigation was made by Crane and Mather (1943) who grew two self-incompatible varieties of *Raphanus sativus* in adjacent blocks and found that the amount of intercrossing rapidly diminished from 30 and 40% where the two varieties adjoined, to only 1 or 2% where they were separated by 5 m, and this low level of contamination persisted throughout the remainder of the plot. Similar experiments were made by Bateman (1947a) with varieties of *Raphanus sativus* and *Brassica rapa* with similar results. For example, in one experiment contamination was 60% at 6 m, 13% at 24 m but still 6% at 43 m and 1% at 156 m.

Bateman pointed out that the effect of distance on contamination will depend on the foraging behaviour of the pollinating insects and the rate at which contaminated pollen is replaced by uncontaminated. He observed that individual honeybees, solitary bees and hoverflies visiting *B. rapa* apparently moved from flower to flower at random and often re-visited the same plant at irregular intervals; individual honeybees foraging on *R. sativus* tended to work in one direction for a short while but over longer periods they also moved at random. Hence, pollen carried by these foraging insects is likely to be rapidly replaced over a short distance and so could explain the rapid decreases in set with increase in distance separating varieties. It is difficult to associate the small but persistent amount of cross-pollination that occurs over considerable distances with foraging behaviour. Possibly this is partly due to longer flights taken between flower visits by dissatisfied bees.

Williams and Evans (1935) found that *Trifolium pratense* plants grown at 256, 311 and 457 m from a block of another variety showed 45, 34 and 23% crossing. Crane and Mather (1943) grew isolated plants of one *R. sativus* variety at distances from 23 cm to 124 m from a crop of another. In one year there was no crossing beyond 28 m and in another only 0·1% beyond 73 m and nil at 110 m. They concluded that when there is a profusion of flowers a bee keeps to a small area only of about 4–5 m diameter, and when seed crops such as *R. sativus* are grown in large single variety blocks, 90 m is sufficient to prevent contamination. However, if few plants are grown, more intercrossing is likely to occur and a greater isolation distance is desirable.

Minderhoud (1950) marked honeybees on small fields of *Reseda* spp and *Brassica alba* 100 m apart and found that a small percentage of bees visited both, but on large turnip and mustard fields 150 to 350 m apart the marked bees kept constant to one or other field of a pair. Hence, it seems that both the distance between fields and their size may influence a bee's constancy. Minderhoud pointed out that when flowering begins forage is sparse even in large fields and the bees then move readily to another field. He also reported that bumblebees kept to small areas of *Trifolium pratense* and did not spread pollen from small plots of *T. pratense* of thirty to forty plants when these were 50 m or more apart.

The amount of cross-pollination between two blocks separated by 30 m, and between two parts of the same field 30 m apart, may of course be very different. Whereas insects like bumblebees and honeybees, which work methodically from one flower to another in a small area, would be unlikely to transfer pollen this distance in a homogeneous crop, they might more readily fly across a gap of 30 m separating one crop from another. Alternatively, when they have worked their way to the edge of a crop they might merely change their direction, and so be less likely to cross a 30 m gap than 30 m of the field they are currently working. It is not known which alternative is generally more true although Afzal and Khan's (1950) discovery (page 165), that cotton itself is a more efficient isolating barrier than an equivalent area without cotton, favours the former. Perhaps less methodical insects than bees show more tendency to move greater distances, both within and between crops. Obviously more studies on these lines are needed. However, work on many seed crops has confirmed the basic principle that contamination is large where two varieties adjoin, but then rapidly decreases although a small amount of contamination persists for considerable distances. Experiments with different species will be found in the appropriate chapters. *Brassica* seed crops need to be isolated from other varieties of the same species and also from other *Brassica* species with which they may cross (pages 135, 136). In particular, numerous studies have been made to try to diminish contamination of cotton seed crops (page 162–166). The amount of contamination, and the isolation distances necessary for different species, will obviously vary according to several circumstances, including the relative amounts of the two types or varieties, the extent to which self-pollination and self-fertilization occurs, the number and type of pollinating insects and their foraging areas which in turn will depend partly upon the number of flowers present per unit area.

The degree of contamination of a seed crop that can be tolerated depends upon the use to be made of it. "Certified" or "commercial" seed is that sold for producing normal crops; "registered", "stock" or "basic" seed is that maintained by the seedsman for producing commercial seed; "foundation" or "elite" seed is that used for continuing a highly desirable line. Whereas a small amount of contamination is permissible for certified seed, and very little for registered seed, foundation seed must be pure. The distances necessary for isolation vary accordingly. Thus in Canada, the minimum isolation distances are for: certified seed, 46 m; registered seed, 91 m; foundation seed, 183 m. In Britain, the minimum isolation distance for producing certified seed of *Beta vulgaris* and *Brassica* spp. crops is 914 m; and for *Trifolium* spp. and other legumes 193 m for fields under 2 ha, 91 m for fields from 2 to 6 ha, and 46 m for fields larger than 6 ha. Pedersen *et al.* (1969) devised an equation for predicting contamination of *Medicago sativa* crops whose isolation distance, field size, field shape and border size are known. Similar equations for other crops would be most desirable.

While small foraging areas may be advantageous to the seed grower who wants to avoid contamination between varieties and species, they are disadvantageous to fruit growers with orchards of self-sterile varieties which need pollen transferred to them from other trees before they will set fruit (page 403). If possible, allowance should be made for maximum foraging areas in seed crops and minimum in orchards.

A bee has difficulty in grooming certain parts of its body, including the back of its head, the central dorsal part of its first thoracic segment and the first and second abdominal segments (Lukoschus, 1957) and, when a forager leaves its hive is still has an average of several thousand pollen grains on its body, some of which are viable (page 402). Although the pollen on the bodies of foragers leaving their hive is predominantly of the species whose flowers they are working, each bee carries an average of three to four species, and pollen other than the dominant one forms a considerable percentage of the total (Free, unpublished). Pollen-gatherers have less "foreign" pollen on their body hairs than nectar-gatherers. Most of this foreign pollen is probably transferred to bees as they brush up against others in their hive. Even bees of only 4 days old have some pollen on their bodies. There seems to be no reason why at least some of this pollen should not be viable and its presence could help explain the small, but fairly persistent, amount of cross-pollination that occurs on fruit and seed crops some distance from the pollen source.

Although knowledge of the foraging areas of individual bees is important in determining the isolation distance for growing selected strains of seed crops, because of the transfer of pollen from bee to bee in the hive there is always the possibility of some contamination and to obtain pure seed the plants should be isolated in cages or glasshouses, where insects other than honeybees may be used to pollinate them (Chapter 7). Pollen transferred from bee to bee in the hive may cross-pollinate related varieties of many wild species, even when they are not growing close together, and so can help explain the rare instances of hybridization over considerable distances (e.g. Darwin, 1876, Bateman, 1947b) when the bees themselves remain constant, and so could be important in the formation of new species.

Chapter 3

Organization of Honeybee Colonies

Honeybees of Europe, Africa and the New World belong to the species *Apis mellifera*. In southern Asia there are three other species of honeybee: *Apis dorsata*, *A. florea* and *A. cerana*. However, only the last of these three can be kept in hives and, except in India and other tropical Asian countries, there is a tendency for beekeepers to replace it with *A. mellifera*.

Honeybees of the species *A. mellifera* are the world's most important pollinating insects, can readily be made available in considerable numbers and will visit the flowers of most important commercial species. The successful distribution of *A. mellifera* in many sub-arctic, sub-tropical and tropical regions, as well as throughout the temperate zones, is largely because a colony can adjust to seasonal changes and exert considerable control over its internal physical environment. An understanding of certain aspects of the social life of a honeybee colony is essential if efficient use is to be made of it as a pollinator.

During summer a honeybee (*A. mellifera*) colony normally consists of several thousand morphologically identical sterile females (the workers), a single fertile female (the queen) and a few hundred males (the drones) together with a series of parallel wax combs of hexagonal cells containing brood and stores of honey and pollen (Figs. 17 and 18). In the wild, honeybee colonies occur in hollow trees, caves and similar shelters but they can readily be induced to live in man-made hives. The usual hive basically consists of a wooden box containing a series of removable wooden frames in which the bees build their combs. The box is enclosed above and below with a removable roof and a floorboard; the hive entrance, through which the bees enter and leave is located in the floorboard.

The primary division of labour in a honeybee colony is according to caste. Drones fertilize the virgin queens but apart from possibly incubating the brood, they serve no other function; they do not visit flowers but are fed by the workers and feed themselves from their colony's food stores. The queen is specialized for egg laying and all the remaining duties, including comb

Fig. 17. A comb from a honeybee colony covered with worker bees.

Fig. 18. Pollen storage cells.

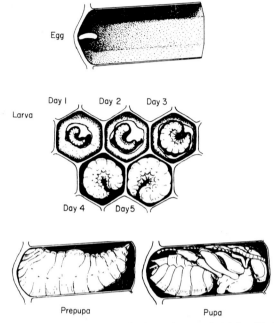

Fig. 19. Stages in the development of a honeybee worker (after Dade, 1962).

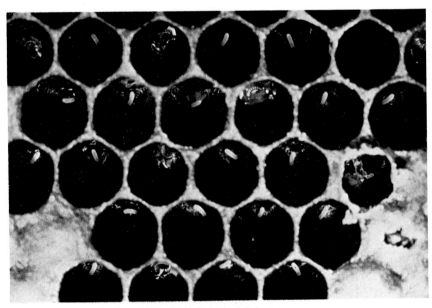

Fig. 20. Cells containing eggs.

Fig. 21. Cells containing larvae.

Fig. 22. Longitudinal section of cells containing pupae.

54 INSECT POLLINATION OF CROPS

Fig. 23. Comb containing brood in all stages of development.

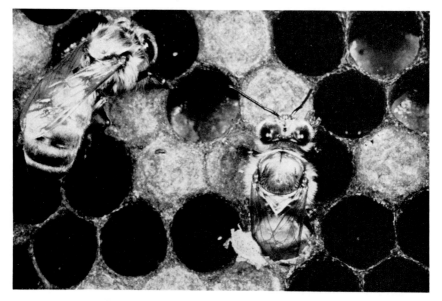

Fig. 24. Honeybee worker emerging from its cell.

ORGANIZATION OF HONEYBEE COLONIES

building, feeding the brood and collecting food, which queens of some of the more primitive insect societies perform at some stage of their lives, are done by workers. Eggs are laid singly, one at the bottom of each cell (Figs. 19 and 20). After about 3 days they hatch into larvae (Fig. 21) which are fed on a special brood food, rich in protein, honey and water. After 5 days of feeding the larva changes into a pupa (Fig. 22) and the worker bees build a canopy of wax over its cell (Fig. 23) so it is completely enclosed. After about another 13 days the adult worker bee emerges (Fig. 24) (see Jay, 1963). The length of life of a worker bee differs greatly under different conditions and different times of the year. Whereas it extends to several months in the winter when little or no brood is present in the colonies, it is reduced to only 4 weeks in June when colonies reach their largest size (Free and Spencer-Booth, 1959).

Duties of Worker Bees

The first detailed studies on the duties of worker bees were made by Rösch (1925, 1931). He gave bees paint marks according to their age, and observed their activities in a glass-walled observation hive. He found that individual bees did not specialize in certain duties but that each did a variety of tasks, the particular task preferred tending to change as the bee grew older. Thus, for the first day or so after its emergence a bee's only task was to clean cells

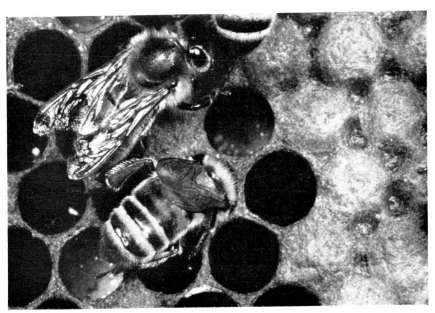

Fig. 25. A worker honeybee feeding a larvae.

so they were ready to receive eggs or stores of food. When between 3 and 13-days-old a bee undertook nurse duty; bees doing nurse duty frequently visit and inspect eggs and larvae in open cells (Fig. 25), and following such an inspection often feed the larvae with a secretion from their hypopharyngeal and mandibular glands which are located in their heads (see Kratky, 1931; Haydak, 1957; Callow et al., 1959); a bee cannot become a nurse until its hypophoryngeal glands have enlarged. In a slightly later age range, when they were between 7 and 24 days old, Rösch's bees developed their wax producing glands, which are located on the underside of the abdomen, and bees of this age built comb. Later still, bees received nectar from successful foragers, packed down the pollen loads that pollen-gatherers had deposited in cells, and removed debris such as mouldy pollen, dead brood and old pieces of comb from the hive. Such bees had finished brood rearing but had not foraged. Just before foraging began some bees became guards. Bees began foraging when 10 to 34 days old and continued to do so for the rest of their lives. In general, the hypopharyngeal glands and wax glands of foragers had shrunk. Thus, although the bees tended to undertake a sequence of tasks, there was a considerable overlap in the ages at which these were done.

Adaptability of Workers to Different Tasks

Rösch's results were confirmed by Lindauer (1952) and Sakagami (1953) who made extensive observations on the behaviour of marked bees of known ages, and also continuously watched the activities of certain individuals for long periods each day. Like Rösch, they found considerable variation in the ages at which individual bees undertook different tasks, and numerous instances of individual bees doing two or more different tasks on the same day (Fig. 26).

Thus, it seemed probable that bees could adjust their behaviour, at least to some extent, to meet the requirements of their colony, and several experiments have been done to investigate this. Artificial colonies have been made by putting together newly emerged workers, but no older workers, and a queen with combs containing brood or empty combs; and it has been discovered that under these circumstances bees of only 2-days-old will feed the brood if necessary, and bees of only 4-days-old will forage, even though they still have large hypopharyngeal glands (e.g. Nelson, 1927; Haydak, 1932). Other experiments by Haydak (1963) showed that nurse bees can be forced to rear brood continuously and that the hypopharyngeal glands of 70% of nurse bees 79–83 days old were still large; however, the weight and longevity of bees reared decreased when the nurse bees were much older than usual. In contrast to the above experiments, when colonies consisting entirely of old bees or foragers, together with a queen and brood were prepared, it was found that the bees nursed the brood only with difficulty at first but soon the hypopharyngeal glands of many of them had enlarged again and the brood was adequately cared for (e.g. Rösch, 1930; Moskovljevic, 1936). Similarly, it has been found that bees can, if necessary, redevelop

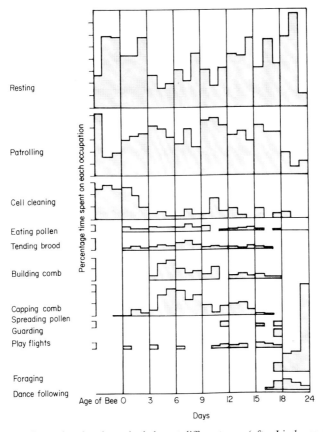

Fig. 26. Duties undertaken by a single bee at different ages (after Lindauer, 1961).

their wax glands and build comb. Hence, the task that needs doing can determine the condition of a worker's glands. However, the converse is probably also true, and when Free (1961) prevented bees from developing their hypopharyngeal glands they omitted nursing duties and foraged earlier than control bees with normally developed glands.

Coordination of Activities of Workers

To maintain the organization and unity of a colony the activities of its members must be co-ordinated in some way, and recent studies on social insect communities have tended to concentrate on discovering how this is achieved. It may partially result from direct stimulation of the individual. Thus, Rösch (1925) noted that during their nursing activities bees often remained idle for hours at a time, or "wandered" about the hive, and Lindauer (1952) observed that such "wandering" bees constantly inspected cells and brood and consequently called this activity "patrolling". He suggested that, when patrolling, each bee gathered directly the information about the tasks needing to be done. It may be supposed that the stimuli encountered release the appropriate behaviour patterns of physiologically suitable bees. Probably, this patrolling also stimulates gland development; thus, Orösi-Pal (1956) showed that the presence in a colony of empty space suitable for comb building stimulated wax gland development and Free (1961) found that the hypopharyngeal glands of bees did not become fully active unless brood was present. Consistent chemical differences have been found in the brood food given to young worker larvae, older worker larvae and queen larvae; if as suggested by Townsend and Shuel (1962) there are differences in the composition of the glandular secretion of different workers, larvae of different ages and castes must tend to stimulate the right category of nurse bee to feed them.

In an undisturbed colony, the brood occurs on the more central combs, and each comb with brood usually has stores of honey and pollen at its edges. On either side of the brood combs, as well as perhaps above and below, there are combs containing stores of honey and pollen only. Younger bees are mostly found on the brood combs but, as they grow older, there is a greater tendency for them to be found on the store combs (Free, 1960b). However, although house-bees tend to remain at the hub of their colonies' activities, and may partially learn the requirements of their colony through direct personal experience, this alone is insufficient to account for the coordination in a large community like the honeybee colony, and it seems that this is largely achieved by the presence in the colony of various pheromones, and also by the transfer of food from one bee to another.

Worker bees are attracted to the vibration, heat and scent produced by a cluster of bees (Lecomte, 1950; Free and Butler, 1955). This attraction, which is a necessary prerequisite of social life, is enhanced when the individuals are able to transfer food with those of the cluster. A bee that begs for food thrusts its tongue between the mouthparts of another bee, and a bee that offers food opens its mandibles, pushes forward the proximal part of its tongue, and regurgitates a drop of food (Fig. 10). During feeding the antennae of both the giver and receiver of food are in constant motion and are continually touching and help the bees to orientate to each other. The food which passes from one bee to the other consists of water, nectar or honey regurgitated from the honeystomach (see Free, 1956).

During summer, there is extensive and rapid transfer of food among the members of a honeybee colony. Nixon and Ribbands (1952) allowed six foragers to collect 20 ml of sugar syrup containing radioactive phosphorus and found that within 5 h, 27%, and within 24 h, 55%, of the bees of their colony were radioactive. Individual bees receive food from bees that are on average older than those to whom they give it, and as bees grow older the average age both of those they feed, and those that they are fed by, also increases (Free, 1957). Hence, food tends to pass through a colony from the older bees, which are foragers, to the younger bees which are in the brood nest. Possibly,

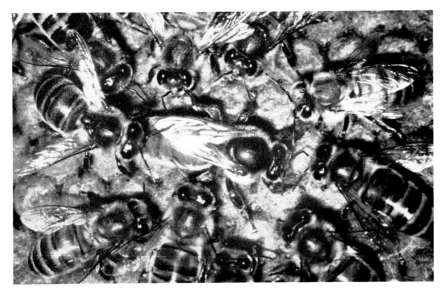

Fig. 27. Queen honeybee surrounded by her "court" of workers.

in this way, all bees can appreciate the quality of food coming into the hive, and perhaps the frequency with which they are offered food enables them to determine the rate at which nectar is being collected. Changes in incoming food supplies affect brood rearing, ripening and storage of honey, wax secretion and comb building; presumably it is partly through food transfer that members of a colony are made aware of the changed conditions and act accordingly.

Transfer of chemical messages, or "pheromones" (see Karlson and Butenandt, 1959), in a honeybee colony may be by direct bodily contact between bees, in their food, or in the hive atmosphere. The pheromones of a honeybee colony that have been given most attention are those emanating from the queen (Fig. 27). It has been shown that the surface of the queen's body carries a substance which worker bees obtain directly by licking, or indirectly

Fig. 28. Queen honeybee being offered food by worker bee.

Fig. 29. Queen honeybee laying eggs in a cell at the edge of the comb.

from other worker bees in regurgitated food, and as a result are inhibited from rearing additional queens, and development of their ovaries is discouraged (e.g. Butler, 1954, 1957; Groot and Voogd, 1954; Pain, 1955). More recent work indicates that odoriferous substances are also involved in complete inhibition (see Butler, 1964, 1967; Allen, 1965; Pain, 1968). The major component of a queen's mandibular gland secretion, that controls queen rearing and worker ovary development, also attracts drones to queens during their mating flights (Gary, 1962); this together with other pheromones also attract worker bees to their queen in a flying swarm (Velthuis and Es, 1964; Butler and Simpson, 1967).

Apart from being a means of communication itself, an important function of food transfer is that it serves as a medium for circulating some of the various pheromones concerned with colony organization. The workers not only feed each other but also feed their queen (Fig. 28) (Allen, 1955) and, one way they might affect the number of eggs the queen lays (Fig. 29) is by regulating the amount of food they give her. They can also influence egg-laying by the number of cells they clean and prepare to receive eggs. However, the relationship is a complex one because pheromones produced by the queen stimulate the workers to rear brood (Free and Racey, 1968a) and to build comb (Chauvin et al., 1961; Free, 1967a).

Why Workers Start to Forage

There is still much to learn about why a bee changes from house-duties to foraging. The temperature in the centre of the brood nest, where the young bees are located, is kept at about 35°C (95°F) and decreases towards the periphery of the colony (see Simpson, 1961). However, as it gets older a bee's metabolic rate increases (Allen, 1959) and the temperature it prefers decreases (Heran, 1952). This may help explain the tendency of bees to prefer to cluster at their colony's periphery as they grow older. Once at the periphery the bees become conditioned to its lower temperature, and their greater ability than younger bees to survive cold and remain active, and fly at low temperatures, is probably further reinforced (Free and Spencer-Booth, 1960). Hence, as a bee gets older it is physiologically better adapted to the life of a forager.

Rösch (1927) supposed that a superabundance of workers performing any particular function would increase the likelihood of the promotion of some of them to the next series of tasks, and Lindauer (1952) suggested that lack of work inside the hive may induce house-bees to become foragers. This seems a likely, although unproven, explanation, but is probably only generally applicable when bees are approaching normal foraging age, and more likely when they have done house duty for some time first (see Free, 1961).

Most bees begin foraging by following dances of successful foragers and very few find crops for themselves (Oettingen-Spielberg, 1949; Lindauer, 1952). It is easy to envisage the possibility that when the amount of incoming nectar is decreased or inadequate, older housebees have increased numbers of bees begging from them, and so are more inclined to follow recruiting dances.

Duties of Foragers

When a bee becomes a forager it collects nectar and pollen, or less often, propolis and water. Propolis is used for cementing up cracks in the walls of the hive and for reducing openings. All the bees that forage for propolis also do cementing work with it inside the hive (Meyer, 1954), so it is easy to see how they could have direct knowledge of their colony's propolis requirements and act accordingly.

Water is not stored in the hive but is collected as needed, either to dilute honey stores, or to evaporate inside the hive to lower the temperature when it becomes too hot. Because the house-bees, not the foragers, use the water, the need for it must be communicated in some way to the foragers and it seems that the transfer of food between bees enables this to be done. Thus, Lindauer (1955) found that, as more water is used in a colony and the need for it increases, those foragers that return to their hive with water or dilute nectar have their loads eagerly and quickly received by other bees, whereas foragers with concentrated nectar have difficulty in finding bees that will accept it. Consequently, the collection of water or dilute nectar is encouraged and the collection of concentrated nectar is discouraged. When water is no longer needed, the foragers returning with it find difficulty in getting their loads accepted by others, but the eagerness with which house-bees accept loads of nectar increases with the nectar's concentration. Kiechle (1961) found that when there was only a small demand for water, few foragers collected it, and these did so only between trips for nectar or pollen; but when water was wanted urgently these bees knew its location and could inform others.

Although some flowers yield pollen only and others yield nectar but little pollen, honeybee foragers can collect both nectar and pollen from most species. There is no strong indication that the tendency of a bee to collect nectar or pollen changes as it grows older (Rösch, 1925; Ribbands, 1952). Indeed, many foragers collect pollen on some trips but nectar on others on the same day, and a forager may change from one type of forage to the other for several days before reverting to its original type (Free, 1963). Probably the type of forage collected depends partly on its availability and, in some flower species, nectar and pollen are most abundant at different times of the same day (Percival, 1965). However, some bees collect nectar only, others pollen only, and yet others both nectar and pollen on the same crop at the same time, so individual foragers probably adjust their behaviour, either wholly or partially, according to their colony's needs.

Factors Causing Pollen Collection

The factors causing nectar collection have been little studied and it is not known whether the amount collected is related to the amount in store. The relationship between the amount of brood present in a colony and its pollen intake and pollen reserves also needs more investigation. However, Nolan (1925) and Todd and Bishop (1941) established that a positive correlation exists between brood rearing and pollen collection at different times of the year. Recently, Cale (1968) reported that he had found positive correlations in three apiaries between the number of eggs present in colonies and the amount of pollen they collected during late spring and summer, and between pollen-gathering and honey production in two of the three apiaries. Probably, therefore, the amount of pollen collected limits brood rearing, especially at certain times of the year. In Scotland the amount of surplus pollen stored in cells suddenly increases from April onwards and reaches a peak of about 1030 cm^2 in June, July and August. There is then a rapid fall and, from October to March inclusive, only about 130 cm^2 of pollen is present (Jeffree and Allen, 1957). It is not clear whether this annual cycle of pollen storage reflects the influence of pollen income on brood rearing, or of brood rearing on pollen income, but because the shape of the seasonal brood rearing curve is similar to that of the pollen storage curve, but has a later rise and fall, the former interpretation is likely to be the correct one. Indeed, Todd and Vansell (1942) found they could induce small colonies to rear brood by feeding them with syrup which had pollen in suspension but could not do so by feeding syrup alone. Todd and Reed (1970) found a positive correlation between the amount of brood in colonies that contained up to 5,000 sq cm brood.

However, it has been shown that the amount of brood in colonies influences pollen collection. Free (1967b) removed brood from colonies and found this caused a rapid decrease in foraging in general and pollen collection in particular, whereas increasing the brood rapidly increased pollen collection. Brood in all stages influenced pollen collection, but the larval stage was especially effective. Although nectar-gatherers transfer their loads to house-bees, often just inside the entrance, pollen-gatherers deposit their loads directly in storage cells. Free (1967b) found that the smell of the brood alone, and contact with bees tending the brood, were each partly responsible for foragers collecting pollen, but actual access to the brood area of the colony was the most important factor. Perhaps, therefore, foragers are normally stimulated to collect pollen by direct contact with the brood. However, the cells in which pollen-gatherers deposit their loads are often near the brood and have been especially prepared to receive it (Free, unpublished). Probably, when a nurse bee has difficulty finding pollen to give to larvae it prepares cells to receive pollen loads, and in this way the number of cells prepared increases with pollen requirements. The amount of pollen collected may, perhaps, then depend on the frequency with which foragers find empty pollen-storage cells and hence on the rapidity with which they can deposit their loads. This hypothesis is in accord with the findings of Veprikov (1936) who removed pollen from colonies and found this increased pollen collection, and of Free (unpublished) who gave additional pollen in shallow dishes just above the brood nests of colonies and found that the nurse bees ate it directly and their colonies collected less pollen.

The presence of a queen alone, irrespective of the brood she has produced, has a direct influence on foraging, as when a queen is removed from her colony pollen collection sometimes rapidly decreases (Free, 1967b). Ribbands (1951) obtained an indication that colonies that were rearing queens and therefore probably contained insufficient queen pheromone, foraged less than colonies not rearing queens. Data of Jaycox (1970) also suggest that the absence of a queen or a queen's pheromones discourages nectar collection but adequately controlled experiments using sufficient colonies are needed. It has recently been shown that a queen's presence encourages small groups of bees to store syrup in combs and to deposit pollen loads (Free and Williams unpublished). It would be interesting to find whether increasing the amount of certain of the pheromones produced by a queen or brood would increase foraging and in particular pollen collection. If so, this could have important practical applications.

No doubt research will help to clarify the ways honeybees communicate with each other and probably many new ways will be discovered. It is to be hoped that it will be possible to exploit at least some of them to increase the pollinating efficiency of the honeybee colony.

Chapter 4

Management of Honeybee Colonies for Pollination

Honeybee colonies for pollination should be managed as efficiently as possible. Recently many studies have been made on how to do this, and there are indications that considerable progress will be made in the near future. However, it is already possible to use colonies more economically and with greater effect than in the past.

Foraging Strength of Different Size Colonies

The object of many beekeeping practices is to obtain full size colonies in time for the flowering of major nectar-producing crops. It is economically even more important to have colonies at full strength for crop pollination. In temperate climates brood rearing is greatly diminished or ceases during winter and increases to a peak in mid-summer (e.g. see Nolan, 1925; Todd and Bishop, 1941; Allen and Jeffree, 1956). Small colonies rear proportionally more brood per bee in spring and continue to rear brood later in the autumn than large colonies (Farrar, 1931a, 1932; Moeller, 1961; Free and Racey, 1968a). A queen's rate of egg-laying may reach a maximum of 1,500 per day in summer, but this varies with different queens (Moeller, 1958) so that fully grown colonies are of different sizes; presumably the amount of honey they store is proportional to their size.

While a colony is growing the number of bees flying from it does not increase disproportionally with its size (Farrar, 1931a; Woodrow, 1934; Gooderham, 1950), but when egg-laying reaches its maximum, the brood/bee ratio diminishes so a greater proportion of bees are available for foraging. Therefore, full size colonies store proportionally more honey in relation to their size than growing ones (Farrar, 1937; Sharma and Sharma, 1950; Moeller, 1961). Although large colonies have a greater foraging population at all times and are to be preferred for pollination work, small colonies are more useful than has sometimes been implied because deterioration in foraging conditions discourages foraging relatively more from large colonies than from small ones (Taranov, 1952; Free and Preece, 1969). Probably the queen's pheromones that encourage foraging are less diluted among the bees of a small than a large colony, and this, together with the greater brood/bee ratio, stimulates a greater proportion of the bees in small colonies to forage. This could help explain why Todd and Reed (1970) found that the amount of pollen gathered per unit of brood decreased with the amount of brood present. Hence, smaller colonies can increase their foraging proportionally less when conditions improve.

When colonies are hired for pollination it is important that they fulfil the minimum size requirements. This will vary with the time of year and in general will be less when fruit trees are flowering than later. The size is usually assessed as the number of combs occupied by bees or brood, or the number of seams between combs occupied by bees or, more rarely, as the number of square inches of brood. Todd and Reed (1970) have recently demonstrated that the brood area of a colony is a valid means of assessing its value for pollination.

Concentration of Colonies Needed

It is of limited value to give general advice on the number of colonies needed per hectare of crop, as this will depend very much on local conditions including the number of honeybees and other pollinators already present, the size of crop, and the presence of competing crops of the same and different species.

One way for a grower to determine the number of colonies required is to increase the concentration of colonies progressively until it makes little or no difference to the concentration of foragers on the crop. Levin and Glowska-Konopacka (1963) increased the number of colonies in a *Medicago sativa* field by stages until there were $7\frac{1}{2}$ colonies per hectare. However, the number of bees per square metre was no greater when $7\frac{1}{2}$ than when 5 colonies per hectare were present, although the population in neighbouring fields increased; hence, 5 colonies per hectare was adequate.

To overcome competition from more attractive crops, attempts are sometimes made to saturate all the bee crops in an area by providing large numbers of colonies. Although good results have sometimes been reported (e.g. see Stapel and Eriksen, 1936; Wilsie, 1949; Thomas, 1951; Raphael and Cunningham, 1960) the increased seed or fruit crop produced must justify the cost of the extra colonies. Moreover, there is little object or necessity in trying to achieve 100% pollination because there is usually a limit to the number of fruits or seeds a plant can support, so that ability to take advantage of additional pollination decreases with increase in the proportion of flowers that are pollinated. When only few flowers are pollinated, the fruits or seeds produced are often larger than when pollination is intense. However, it is best for the grower to ensure that he has sufficient colonies to allow for when conditions are unfavourable to set. Thus, although under ideal circumstances (including fine weather, sufficient pollinizer variety trees and abundant insect pollinators) over-pollination of fruit trees is possible, resulting in small sized fruits and the establishment of biennial bearing, it is wiser to provide sufficient colonies for maximum rather than for minimum pollination; too great a set can be corrected by pruning and thinning, but too small a set cannot. Moreover it has become apparent recently that the receptivity of flowers of many plant species soon diminishes and pollination on the day a flower opens is likely to be the most successful; sufficient pollinators should be present to ensure this happens.

Whereas it is possible for growers of some crops, particularly leguminous ones, to judge the amount of pollination that is occurring by the appearance of the flowers, and if necessary take action to increase the pollinator population, the likely set of most crops cannot be determined until it is too late.

For each plant species it is desirable to determine a formula so that a grower will be able to decide the minimum number of honeybee colonies needed to pollinate his own particular crop.

It has been possible to correlate the number of bees per unit area of cotton with direct evidence of the transfer of pollen (page 161) but probably this is possible with few other crops, and attempts are usually made to relate the number of pollinating insects, often per square metre, or per given number of flowers, with seed set. To be able to do this it is necessary to know the rate at which flowers are visited, the pollinating efficiency per visit, foraging trips per day, the number of hours per day during which foraging occurs, the number of flowers available per day per unit area with different planting systems and different environmental conditions. Information approaching completeness is only available for a few crops, and even then is applicable to limited areas only. Calculations are usually based on the assumption that weather will permit foraging during most of flowering; where this is not usually so, allowance must be made accordingly. Any estimate of the number of colonies needed must also take into account the honeybee population on neighbouring fields of the same crop, and when a crop is grown extensively in the same locality, arrangements should ideally be made on a community basis.

Foraging Efficiency of Colonies and their Distance from Crops

Honeybees will, if necessary, forage a considerable distance from their hives but work close by when suitable forage is available. For example, Eckert (1933) found that, in extreme conditions, colonies stored surplus honey when foraging on a crop 11·3 km away, but in other circumstances, when abundant forage was nearby, foragers were concentrated within 0·8 km of their hives; Peer (1955) observed that some genetically marked bees foraged up to 5·6 km from their hives, but most worked within 4 km, and Lecomte (1960) found that most bees did not forage beyond 600 m from their hives. Theoretically, the nearer colonies are to a source of forage, the less time the bees spend in flying to and fro, and the greater the economy of effort. Attempts have been made to find how much difference this makes in practice.

Eckert (1933) found that during two seasons with good flying conditions (1928 and 1929), colonies located 3·2 km from an irrigated area in which flowering crops occurred produced as much honey as colonies located within the irrigated area, but the gain in weight of colonies more than 3·2 km from the irrigated area decreased with increase in distance (Fig. 30). In a poor

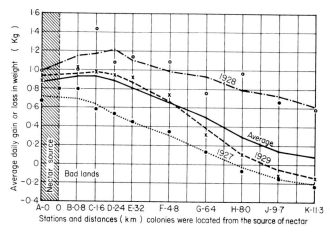

Fig. 30. Changes in weight of colonies situated at different distances from irrigated regions in 1927, 1928 and 1929 (after Eckert, 1933).

season (1927), distance had a greater effect and the crop decreased with distances greater than 0·8 km from the nectar source; however, the gain in weight of colonies situated in the irrigated area and 0·8 km outside during the poor season was less than that of colonies 11·3 km away during the best season. In similar experiments, initiated by Eckert, Sturtevant and Farrar (1935) compared the gain in weight of colonies located within an irrigated area and adjacent to a nectar source with those located 1·9–2·4 km outside the boundary of the irrigated area. In three experiments the average gain in weight of the two sets of colonies was 94 and 69 kg, 33 and 19 kg, and 118 and 94 kg respectively. The latter two experiments were on the same sites in different years; the distant colonies gained only 59% of the weight of the near colonies in the poor honey year but 79% in the good one. Thus it appeared that, in some circumstances, colonies some distance from forage may have diminished crops, but in the parts of the U.S.A. concerned, commercial crops of honey can be obtained economically in favourable seasons even when the colonies cannot be put close to the source of nectar.

Ribbands (1951) extended these experiments in England and demonstrated that even smaller differences in the distances separating colonies from crops may greatly influence honey yields in some circumstances. For example, he found that in favourable conditions (1949) colonies at a *Pyrus malus* orchard and 0·6 km away gained substantially in weight (averages of 6·0 and 4·8 kg) but those 1·2 km away gained an average of only 0·9 kg. In unfavourable conditions (1950) the differences were greater; colonies at the orchard gained slightly in weight (1·4 kg) while those 0·6 km away lost 2·0 kg and those 1·2 km away lost 5·5 kg. Similar, but less pronounced results were obtained in experiments with *Tilia* spp., *Calluna vulgaris*, *Allium cepa*, *Brassica oleracea* by Ribbands (1951) and with *Brassica napus* by Hammer (1961). Svendsen (1964) demonstrated that the pollen yield of colonies also showed an inverse relationship to their distance from crops (Cruciferae spp., *Trifolium repens*, *Trifolium pratense*, and *Centaurea cyanus*). Ribbands (1951) also found indications that the effect of distance was diminished as the quality of the crop improved; he suggested that the greater response shown by his colonies to distance could partly have been because the experiments of Eckert (1933), and Sturtevant and Farrar (1935) were done in dry districts where flying conditions were probably uniformly favourable.

Ribbands discussed the relevance of his results to honey production and concluded that the economical distance for nectar collection sometimes does not exceed 0·4 km. The results are equally relevant to the use of bees as pollinators. Because bees from colonies on the crop itself must perform more pollination, particularly in unfavourable conditions, colonies should be taken to crops needing pollination, and growers should not rely upon bees from colonies situated some distance away. Yakovlev's (1959) results provide a practical demonstration of this. He found that the fruit crop of *Pyrus malus* orchards decreased with increase in distance separating the colonies from the orchard as follows:

Distance (km) between orchard and colonies	0·3	0·9–1·2	1·5–1·9
Fruit crop (kg/ha)	5,100	2,700	2,300

The less time it takes a bee to collect a load, the greater the proportion of its trip it spends travelling between crop and hive. It usually takes much less time to collect a load of pollen than a load of nectar, so shortening the distance separating a colony from a crop probably influences pollen collection even more than nectar collection. However, although the number of trips made by individual pollen-gatherers might be increased by moving their colony closer to a crop, the amount of pollen collected by the colony might not exceed a certain level that was dictated by its requirements. Experiments to investigate this are needed.

Moving Colonies to Crops

In association with the increased demand for pollination services more efficient means of moving colonies have been developed, which include the use of mechanical hive loaders, either of the fork lift or boom type, and various ways of strapping the hive parts together (e.g. see Townsend and Adie, 1952; Hamilton, 1960; Todd and McGregor, 1960). Whenever possible colonies are moved at night but, when daytime moves are necessary and the colonies are likely to be exposed to hot dry conditions, the colonies and hives should be sprayed with water. Bees can cool their colony by evaporating the water on the combs and on their mouthparts (Dunham, 1931; Lindauer, 1954). The higher relative humidity produced by the evaporation and the replacement of body water reserves by drinking makes death by dessication unlikely (Free and Spencer-Booth, 1962).

However, despite precautions, colonies are sometimes damaged during moving. Hutson (1929) reported that moving colonies more than 48 km often killed their brood and such colonies produced less honey than controls. Colonies moved less than 48 km to orchards in a good honey year produced the same amount of honey as control colonies, but in a moderate honey year produced slightly less than the controls. Free (unpublished) analyzed 3 years' honey production records of R. O. B. Manley (Oxfordshire, England), who moved about a third of his colonies to pollinate fruit orchards each spring; the colonies taken tended to be the stronger ones but otherwise they were selected from, and returned to, his apiaries at random. The average numbers of supers of honey (each holding about 23 kg) produced during 3 years in apiaries, from which some colonies were taken to orchards, can be seen in Table 3.

TABLE 3

	Colonies taken to orchards		Colonies not taken to orchards	
	No. colonies	No. supers per colony	No. colonies	No. supers per colony
1954	59	0·13	147	0·23
1955	252	2·75	499	1·97
1956	212	0·18	588	0·30

In 1954 and 1956, both poor honey years, the colonies taken to fruit produced a smaller crop than the rest, but in 1955 a good honey year, the reverse occurred. It appears that during favourable conditions nearness to a good source of forage in spring may have offset any damage done to the colonies while moving them, the larger crop produced by the moved colonies being a reflection of their larger size. However, under poor foraging conditions, moving colonies had a detrimental effect which, because of the size difference between colonies of the two groups, may be greater than the figures imply. Observations are needed on the honey production of numerous colonies of the same size, some of which are moved while others are not.

When a colony is moved to a new site within flight range of the old, and no colonies are left at the old site, most foragers (between 70 and 100%) return to their hive in its new position although many visit the old first. When other colonies are still present near the old site, the bees that return from the new site attempt to join them. However, when a colony is moved to a new site beyond its original flight range, nearly all the bees successfully orientate to their hive in its new position and return to it (Free, 1958b).

Honeybees do not always remain with the colony that reared them and may stray into hives other than their own, particularly during their first flights. This "drifting" behaviour, which is extensive when identical hives are arranged in regular formations, is particularly undesirable because it results in the undue weakening of some colonies and strengthening of others, so lowering the average foraging potential, honey production and pollinating efficiency (see Free, 1958c; Jay, 1965). When hives are arranged equidistantly in rows, bees from the end hives drift less than those near the row centres, no doubt because the end hives are more readily distinguishable, and consequently the end hives gain in bees (Free, 1958c; Jay, 1965). During strong wind, drifting is greater to the leeward hives (Jay, 1965). There is also a tendency for bees to drift to hives nearest to the main line of flight (Otto, 1928; Nekrasov, 1949).

Similar drifting occurs when colonies are moved to new sites, apparently irrespective of whether the foragers are allowed to emerge gradually or in a "body" from their hive at its new site (Otto, 1928; Free, 1958c). Todd and McGregor (1960) reported that when colonies are moved to seed crops the foragers may tend to drift extensively to a few of the colonies. Because bees tend to memorize the position of their own hive in relation to neighbouring ones, when hives are put in a similar formation at old and new sites, they should be kept in the same relative position to each other. Experiments have shown that drifting can be greatly reduced by arranging hives irregularly, facing them in different directions, spacing them well apart, putting them near landmarks and windbreaks, and putting different coloured boards above hive entrances. The first two measures are particularly effective and whenever possible they should be applied when colonies are taken to a crop needing pollination (e.g. Free, 1958c; Fresnaye, 1963; Jay, 1965, 1966, 1969).

Conditioning Colonies to Particular Crops

The foraging constancy of honeybees when their colonies are moved to a new site has already been discussed (Chapter 2). Whenever possible the bees tend to visit the same species as they did before the move, with the result that the proportion of the foragers of a colony that visit any particular species after moving is often related to the proportion that have visited it previously (Free, 1959). Probably the origin of the food stores of a colony plays a part in determining the species which its foragers visit (Free, 1969) but, when one species predominates at the new site, many bees forsake their previous species and change to collecting it. Hence, it would seem important that colonies are not taken to a crop needing pollination until it is flowering sufficiently to be the predominant species in the locality. For many years it was recommended that colonies should not be taken to crops until flowering has begun, or until there are enough flowers for the bees to work, because it was supposed that if the foragers have previously begun visiting other flower species in the locality they will not readily forsake them (e.g. Snyder, 1946; Webster *et al.*, 1949; Coggshall, 1951). These recommendations were based on experience rather than experiments; thus, Karmo (1958) found that bees moved into orchards prior to flowering ranged over the surrounding countryside and did not concentrate in the orchard.

To test these recommendations, Free *et al.* (1960) took one group of colonies to a crop before it flowered and a second group after flowering had begun and determined the relative amount of bees of the two groups that visited it. Experiments were done with *Prunus persica*, *Prunus avium*, *Pyrus malus*, *Lotus corniculatus*, *Medicago sativa* and *Trifolium pratense*, and in each experiment proportionally more foragers (2·3–12·8 times as many) of the second than of the first group visited the crop on the first day of the experiment. These results confirmed those of Free (1959) with *Vicia faba* and *Papaver somniferum*. On subsequent days the relative proportions of bees of the two groups on the crop either remained about the same, gradually became similar, or rapidly became similar. The reason for the latter two circumstances could be because an increasing proportion of the foragers of colonies taken to the crop before flowering visited it, probably because the crop became more attractive than others nearby, and the foragers conditioned to other species were dying off; or it could be because an increasing proportion of foragers of colonies moved after flowering, found more attractive crops elsewhere. *Pyrus communis* is frequently neglected by bees in favour of other crops and, in another similar experiment, Free and Smith (1961) put two colonies in a *P. communis* orchard that was in full flower and found that the mean percentage of *P. communis* pollen out of the total pollen collected decreased from 85 to 49% during the first 6 h as more bees foraged outside the orchard.

It was concluded that, a delay in taking colonies to crop until flowering has begun probably always increases pollination, particularly when the crop has a short flowering period or is less attractive to bees than others in the district. Perhaps in the latter circumstances a few colonies should be released at the crop on successive days of flowering. Insecticidal spraying before flowering begins provides an additional reason for the delay in taking colonies to crops. However, it should be pointed out that, if too great a proportion of the flowers of a crop are open before colonies are taken to it, an important part of the crop may fail to be pollinated, especially as the receptivity to fertilization of flowers of many species soon diminishes.

Shimanuki *et al*. (1967) found that a group of colonies taken to a *Vaccinium macrocarpum* bog a week before peak flowering visited the crop more for pollen than colonies taken at peak flowering. They were unable to account for these results; possibly more attractive crops had come into flower after the first group of colonies had been moved, or possibly the *V. macrocarpum* had become relatively less attractive, partly because of the greater number of insects visiting it. Whatever the reason, their results provide further evidence that the proportion of a colony's foragers visiting a crop can be greatly influenced by management practices.

A flower tends to present most of its pollen at a time of day characteristic of its species (page 23) and *Taraxacum officinale* and *Pyrus malus* present most of their pollen in the morning and afternoon respectively. Free and Nuttall (1968b) found that when colonies were prevented from foraging until the afternoon of the first day they were present in a *P. malus* orchard 64% more pollen-gatherers became conditioned to *P. malus* and 49% less to T. officinale than when they were allowed to forage in the morning. Probably the same principle applies to other crops that present their pollen in the afternoon rather than the morning (e.g. *Pyrus communis, Prunus persica, Trifolium repens,* and *Vicia faba*). However, experiments are needed to prove this. Furthermore, it must be borne in mind that when morning temperatures are high there is a danger that the confined colonies may become overheated. It is probably equally important to ensure that colonies taken to crops whose pollen is presented mainly in the morning (e.g. *Brassica oleracea, Fragaria × ananassa, Ribes nigrum* and *Prunus cerasus*) are released while it is still available. Pollen-gatherers are more valuable as pollinators of all the above crops than bees collecting nectar only. In contrast, nectar-gatherers are the more valuable pollinators of *Helianthus annuus* which presents its pollen early in the morning; hence for this crop it might be advantageous to delay releasing colonies until the peak of pollen presentation is finished for the day.

Karmo and Vickery (1954) stated that colonies moved into a new area may forage in less favourable weather than previously established colonies, but pointed out that if the bees are first released in unfavourable weather they may become conditioned to species that secrete nectar when little or none is secreted by the crop needing pollination. This too is a subject that needs exploring.

According to Karmo (1961a) the threshold for foraging activity is less at the beginning than at the end of foraging for the day, and foraging will cease when conditions are better than those in which it was initiated. Probably this reflects the fact that foraging is governed by light intensity, and temperatures are usually higher at dusk than at dawn. Karmo suggested that the threshold could be artificially lowered by keeping colonies confined to their hives until late afternoon when, upon release, the foragers are eager to work the nearest source they encounter; he supposed that this practice would be especially applicable to crops that were not very attractive to bees. However, any advantage gained in this way must be balanced against such possible disadvantages as loss of foraging time and any damage caused to colonies by confining them; the time of day at which the crop presents pollen must also be considered. Certainly, recommendations to keep colonies destined for pollination work confined for longer periods must be viewed with scepticism.

When a crop is particularly attractive a bee exposes the Nasonov scent-producing gland on its abdomen and the odour produced attracts other bees to the crop (page 29). Frisch (1955) suggested that if we could produce a synthetic scent of these glands, we would have a general attractant for bees which could be of great service to farmers and beekeepers. Recently attempts have been made to identify the components of the Nasonov gland secretion and test their effectiveness (Boch and Shearer, 1962, 1964; Free, 1962a; Weaver, *et al.*, 1964; Shearer and Boch, 1966; Butler and Calam, 1969). Although there is some discrepancy it seems the attractive components of the odour include: geraniol, citral, nerolic acid and geranic acid. However, a synthetic odour is yet to be tested on a field scale; perhaps it might be especially useful in attracting foragers to a crop when colonies are first put in a new location.

Karmo and Vickery (1954) and Karmo (1958, 1961) advised that colonies should be replaced by others as soon as the bees were working in areas or on species other than those intended. They suggested that colonies could advantageously be exchanged between orchards several miles apart every third or fourth day, and could, if necessary, be returned to their original sites 7 to 10 days later. Istomina-Tsvetkova and Skrebtsov (1964) reported that when they replaced colonies in a *Pyrus communis* orchard with the same number of fresh ones, the number of bees foraging increased from 2·2 to 6·4 per 100 flowers. Marucci (1967a) reported favourable results on blueberry crops. However, more experiments on rotating colonies between crops are needed. Karmo (1961) also suggested that colonies could be rotated between different positions in the same large field, but the tendency of foragers to return to their original hive sites and then continue to forage from them would reduce the value of this practice.

In contrast to the above results Levin and Bohart (1957) found that moving colonies from one *Medicago sativa* location to another was of no benefit when approximately equal amounts of *M. sativa* were available at the two locations, but they obtained an indication that such moves might be worthwhile when the colonies were taken from a location where *M. sativa* was predominant and many bees were conditioned to it, to an area where it was relatively less abundant. However, colonies moved to *M. sativa* fields, from locations where there was no alfalfa, collected a greater percentage of *M. sativa* pollen than colonies present at the *M. sativa* fields since the beginning of flowering.

Foraging Range of Colonies

Several workers (e.g. Vansell, 1942, 1952; Karmo and Vickery, 1954; Smith, M. V., 1958) have reported that when colonies are moved into crops in flower, the foraging range of the bees tends to be confined at first to the vicinity of their hives and only gradually expands. Weaver (1957a) reported that during the first few hours colonies were allowed to forage on *Vicia villosa* there was always a much greater concentration of bees near the hives than further away. Karmo (1958) put colonies in a crop of *Vaccinium angustifolium* and found they gradually extended their foraging range to 137 m on the first half-day, to 549 m on the second day and to 686 m or more on the third day. Levchenko (1959) found that when colonies were put in a new location the foraging range of the bees from their colonies extended up to 200 m on the first day, 300 m on the second and third days, and 800 m on the fourth and fifth days; strong colonies expanded their foraging areas more rapidly than weak.

However, such expansion of foraging areas does not invariably occur. Levin (1959) moved colonies of genetically marked bees to the north edge of a *Medicago sativa* field and found the proportional distribution of the bees at 6 sites on the field from 91 to 1609 m distance from the colonies. Most marked bees were found within 320 m of their colonies and very few beyond 640 m. Furthermore, although an increasing proportion of the bees visited a more attractive *M. sativa* field 2 km to the north west, their distribution on the field, to which their colonies were adjacent, did not alter appreciably during the five days of the experiment. Another group of colonies, that had been deprived of their flying bees, and hence contained only inexperienced foragers, were put on the periphery of another *M. Sativa* field; during the next 5 days most foragers kept to within 274 m of their hives, and very few moved to another field 804 m away. In another experiment Levin (1961a) put twenty genetically marked colonies in the centre of a 61 ha field and found the distribution of the foragers at various distances on this and neighbouring fields; another 10 colonies marked with radiophosphorus were put beside them 13 days later and the distribution of both sets of foragers determined. Few foragers were found more than 302 m from their colonies. Foragers from both groups of colonies had similar distribution patterns, and if anything, foragers of the second group tended to disperse more widely. The distribution pattern of the first group was not influenced by cutting 40 ha of *M. sativa* in a neighbouring field and an increase of 70 colonies in the area.

An experiment by Free (unpublished) supported those of Levin; he put a colony in the centre of each side of a rectangular 24 ha field of *Brassica alba* and found that the number of bees foraging diminished toward its centre. After 11 days, the colonies had not expanded their foraging areas and two additional colonies were put beside two of those already present; during the next 3 days the bees of these new colonies were more evenly distributed over the field than bees of the colonies already present. Perhaps the bees from the original colonies remained attached to their individual foraging areas close to their hives, even when foraging was better elsewhere in the field. Other results indicated that bees from colonies located some distance from the field were more inclined to select the best area of the crop than bees from colonies beside the field. Clearly, more work needs to be done to try to discover the factors determining whether colonies' foraging areas do or do not expand. Possibly the concentration of colonies present, and the speed with which nearby flowers become depleted of forage, are important considerations.

Foraging Areas of Colonies

The arrangement of colonies in crops can be important in ensuring uniform distribution of foraging bees but few relevant experiments have been made. The foraging area of a colony depends on many factors, including the amounts of nectar and pollen available per unit area, weather conditions, and the physical features of the area. When colonies are in the middle of the crop, the area of forage available increases with the distance from the colonies; hence, if foragers of colonies in a uniform crop were evenly distributed on it, their numbers would increase with distance from the colonies. However, more bees are recruited to near than to far food sources (Françon, 1939; Boch, 1956) and in poor weather most bees conditioned to foraging on distant flowers remain at home (page 41). Probably because of this more bees are usually found per unit area nearer the colonies (e.g. Mommers, 1948, *Pyrus malus;* Braun *et al.*, 1953, *Trifolium pratense;* Levin, 1959, *Medicago sativa;* Lee, 1961, 1965, *Vaccinium lamarkii*). Šedivý *et al.* (1966) found that bees foraging on a *Medicago sativa* field were especially abundant on parts of the field near their hives in late afternoon and early morning. To overcome the possibility that variation in the supply of forage in different parts of a crop, and various physical features of the area, influenced the distribution of bees, Lee compared only the proportion of genetically and radioactively marked to unmarked bees in locations at different distances from their colonies. Indeed, the foraging population sometimes even decreases as the distance from the colonies increases (e.g. Free and Spencer-Booth, 1963a; Levin and Glowska-Konopacka, 1963).

Although there would be fewer bees per unit area of crop with increased distance from colony groups of all sizes, provided the number of colonies per hectare is constant, the smaller the groups, the closer the groups are together and the greater the overlap of their foraging areas. Ideally, colonies should be distributed singly but it is convenient to both growers and beekeepers to put colonies in groups as large as possible but that still give an equal distribution of the foraging bees. The problem is, therefore, to find the maximum size of groups that can be equally distributed throughout the crop concerned, so that the overlap between adjacent groups is sufficient to prevent the number of foragers decreasing midway between them.

Hutson (1924, 1925) arranged colonies in one *Pyrus malus* orchard in groups of ten to twelve colonies with 183–274 m between groups, and in another orchard in groups of four to five colonies separated by 91–183 m. He counted the bees on trees 23, 46, 69 and 91 m from the groups. The number of bees decreased markedly with distance with the first arrangement whereas, with the second, the distribution of bees at the different locations was approximately equal, presumably because the foraging areas of the adjacent groups overlapped. A similar result (Hutson, 1926) was obtained when groups of twelve colonies were placed 366 m apart in one orchard and groups of five colonies 160 m apart in another. In another experiment, Hutson placed colonies singly every 61 m in one orchard; he found that bees were uniformly distributed on the trees, whereas in a nearby orchard, in which a group of five colonies had been put, the number of bees on trees rapidly decreased with increasing distance from the colonies. Estimates made of the percentage of fruit set at different locations were in general agreement with those anticipated from counts of foragers. Hutson (1926) concluded that it is desirable to place colonies singly, or at least in small rather than large groups. However, despite these results, Philp and Vansell (1932) and Brittain (1933) recommended ten to twenty and five to fifteen colonies respectively in equidistant groups, and Menke (1951) stated that ten to twenty colonies are often grouped together.

Free and Spencer-Booth (1963) ascertained the distribution of bees from eighteen colonies put near the centre of an 4·4 ha *Prunus domestica* orchard. Up to and during peak flowering the number of bees per tree decreased greatly as the distance from the colonies increased, and on the day when fewest bees were foraging the population of foragers was also greatest nearest the hives, although fewer trees were present there (Fig. 31). However, toward the end of flowering when more bees were foraging, the distribution of bees over the trees was about even, and hence the foraging population increased toward the orchard periphery. As well as its nearness,

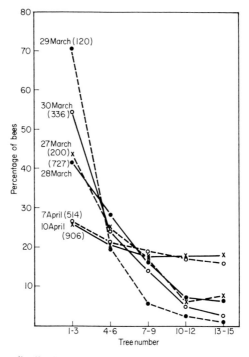

Fig. 31. Percentage distribution of bees on *Prunus domestica* trees at different distances from honeybee colonies; tree number 1 nearest to colonies. Figures in brackets are the average number of bees per count (after Free and Spencer-Booth, 1963).

the relative attractiveness of a particular tree depends on the number of competing insects, partly because they disturb would-be foragers causing them to leave, but mostly because they influence the amount of nectar and pollen available. Probably the number of bees foraging on *P. domestica* trees near the colonies decreased as the trees became less attractive because of the greater number of bees that had been working them. Colonies were also put in a 12 ha orchard of dwarf *P. malus* trees so there were four placed singly, two groups of four, and two groups of nine, in the centres of areas of 0·4, 1·6 and 3·6 ha of orchard respectively (Fig. 32). This was done in each of 2 years. In each year, on parts of the orchard where colonies were put singly, or in groups of four,

MANAGEMENT OF HONEYBEE COLONIES FOR POLLINATION   79

there were as many bees on trees midway between the groups as on trees near to them, but in the 7·3 ha part of the orchard containing groups of nine colonies, bees were fewer midway between the groups (i.e. about 96 m from each) than near the colonies. These results confirmed those of Hutson and showed that at a density of 2.5 colons per ha, nine colonies are too many for one group, whereas the overlap in the foraging range of groups of four suffices to give an even distribution.

Todd and McGregor (1960) found that pollination of *Medicago sativa* was greatest within 91 m of colonies, giving a 55% average pod set, compared to 42% at 137 m. They recommended that colonies should be located at intervals of less than 183 m. They reported that in California it is common

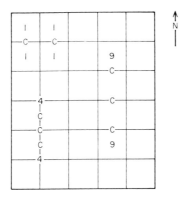

Fig. 32. Plan of *Pyrus malus* orchard. Each square represents 0·4 ha. The figures 1, 4 and 9 indicate number of colonies present (after Free and Spencer-Booth, 1963).

practice to place groups of ten to twelve colonies 161 m apart in rows 161 m apart, the outside row being 80 m from the edge of the field. It seems probable that colonies located near the outside of a field would have a greater tendency to forage on neighbouring crops, but I know of no work that has been done to test this.

Levin (1961b) and Levin and Glowska-Konopacka (1963) obtained indications that the foraging area of a group of colonies was diminished in the direction of a neighbouring apiary, presumably due to increased competition, and that the magnitude of the effect was related to the number of colonies in the neighbouring apiary. They also found that the foraging areas of colonies was influenced by the number of colonies present. Genetically marked colonies were put in the centre of a *Medicago sativa* field and the distribution of the foragers determined. Additional colonies were then dispersed throughout the field so that their numbers were increased successively to $2\frac{1}{2}$, 5 and $7\frac{1}{2}$ colonies per hectare. As the bee population increased, the foraging are as of the original colonies diminished and most of the bees foraged within 91 m of their colony, and few more than 183 m distant (Fig. 33).

The foraging area of a colony and the number of colonies needed depends on many factors, including the amounts of nectar and pollen available per unit area, weather conditions and the physical features of the area, including shelter belts. The optimum size, for both grower and beekeeper, of the group to be used probably depends on the species of crop and many studies of the foraging areas of colonies on different crops are needed. To obtain an even

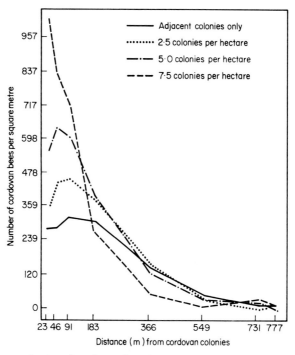

Fig. 33. Average density of cordovan bees following increases in numbers of unmarked colonies (after Levin and Glowska-Konopacka, 1963).

distribution of bees on a crop other features such as the influence of any topographical gradients and wind direction will also have to be investigated. It has been suggested that hives should be sited so that bees can fly out against the prevailing winds and back to the hive in the direction of the wind; presumably this would facilitate scent guidance to the crop.

It is often advocated, particularly for orchard pollination, that colonies should be put in sheltered locations where the sun will shine on them at every opportunity, especially in the early morning, so as to stimulate flight (e.g. see Webster *et al.*, 1949; Cogshall, 1951); Traynor (1966) supposed that placing hives on black tar paper might lead to local increases in temperature and earlier flight. These recommendations are mostly based on supposition rather than experimental evidence. However, Overley and O'Neill (1946) found that 50% more bees flew from a colony protected from wind in a warm sunny location than from a colony of similar size exposed to wind and partial shade. Even if bees do more readily leave sheltered than unsheltered hives during windy conditions, they probably return home again, if they are able, when they meet the wind and those unable to do so may be lost. Hence, it is necessary to compare the return of successful foragers to windy and sheltered hive sites.

Attempts have been made to determine the effect of windbreaks on pollination. Fomina (1961), in the U.S.S.R., found that *Helianthus annuus* flowers produced more nectar and about three times the seed yield 10–20 m from a shelter belt than 400–470 m away, and that *Onobrychis sativa* flowers 10–90 m from a shelter belt secreted more nectar, were visited by more bees, and produced more seed than flowers growing 100–470 m from it. Lewis and Smith (1969) found that the presence of windbreaks in orchards greatly affected the distribution of flying insects and sheltered zones contained three or more times as many individuals of most species as zones not sheltered. However, it was not known how windbreaks influenced the distribution of insects foraging on the flowers. Webster and Wilson (1966) suggested that windbreaks, by reducing the adverse effect of wind on insect visits, could possibly improve the pollination of some tropical crops including *Theobroma cacao*, *Coffea* spp., *Camellia sinensis*, *Citrus* spp., and *Musa* spp. However, whereas windbreaks might be beneficial by providing a warmer, calmer microclimate that encourages nectar secretion, and pollen availability, they might be detrimental by discouraging pollinating insects from visiting the more exposed parts of a plantation. More work is needed to discover this.

"Directing" Bees to Crops

Von Frisch (1923) found that when he trained bees to collect sugar syrup placed on flowers they recruited other bees which then visited flowers of the same species on which no syrup had been provided. Smaragdova (1933) caused bees to visit certain flower species by providing, inside the hive, sugar syrup to which the odour of the flowers concerned had been imparted; she suggested that large scale experiments to "direct" bees to *Trifolium pratense* should be undertaken. Subsequently, many attempts have been made to "direct" bees to certain crops by feeding their colonies sugar syrup containing the scent of the "target" crop. The technique, which is based on the dance language discoveries of Frisch (page 26 to 30), is to immerse flowers of the crop in sugar syrup for some hours, strain off the flowers and feed the syrup to the colonies; often, fresh flowers of the crop are also provided for the bees to stand on while drinking the syrup. Theoretically, bees that take the syrup communicate the odour of the target crop to others which then leave the hive to visit it.

Frisch (1947, 1967) summarized the results of his own experiments. In general, he found that bees could be "directed" to *Trifolium pratense*, *T. hybridum*, *Melilotus alba*, *T. repens*, *Brassica napus*, *Brassica rapa*, *Vicia faba*, and *Fagopyrum esculentum*, although some experiments failed to give a positive result. However, Frisch (1947) and Glushkov (1958) found that only transitory visits of bees resulted from directing them to flowers from which they were unable to obtain food, and that these visits scarcely sufficed for effective pollination. Frisch also stated that it was difficult to direct bees to a particular flower species when they were already obtaining plenty of nectar from other species.

Since Frisch's experiments numerous workers have reported success in directing bees to various crops, including *Pyrus malus*, *Pyrus communis*, *Cucumis sativus*, *Fagopyrum esculentum*, *Trifolium pratense*, *Medicago sativa*, *Gossypium* spp., and *Brassica napus*, and as a result obtaining increases in yield. Recently it has been claimed that reflected ultra-violet radiation is a strong stimulus to foraging bees and it has been suggested that aluminium sheets reflecting ultra-violet light could be used to attract bees to crops (e.g. Braines and Istomina-Tsvetkova, 1956). It has also been reported that bees may be discouraged from visiting a crop by feeding them with a 50% solution of calcium chloride in which flowers of the crop have been soaked.

However, others have reported failure in directing bees (see reviews by Free, 1958c; Glushkov, 1958; Frisch, 1967). Theoretically, such direction should be possible and, furthermore, successful "direction" on the first day only that colonies are present at a crop should help to condition the bees to it. As there is a tendency for experiments of this sort only to be published if they have given positive results, it is likely that there are many experiments which have given negative results that have not been reported. The research workers who have experienced failure have probably taken more precautions, which might contribute to success, than would be likely in a commercial enterprise; probably partly because of this, "directing" bees to crops is not generally practised (e.g. Todd and McGregor, 1960; Frisch, 1967). Indeed, a recent survey (Åkerberg and Stapel, 1966) showed that few European countries practised scent directing, and then only to a limited extent. Perhaps, as Frisch (1967) suggested, the method needs further modification before it is suitable for practical application by growers and beekeepers.

Gubin and Smaragdova (1940) reported that bees could be more successfully induced to visit a crop by training them to a feeder containing scented sugar syrup first, and then gradually moving this to the site of the crop.

Several authors have since claimed that honeybees can be encouraged to visit a crop by putting containers of sugar syrup in the field and covering the syrup with flowers of the crop (e.g. *Trifolium pratense* and *Brassica napus*, Rhein, 1954; *Trifolium pratense*, Pritsch, 1961 and Hansson, 1960; *Cucumis melo*, Nevkryta, 1953). However, it is difficult to do properly controlled experiments on this. Others have reported that spraying a crop with syrup has a similar effect (e.g. fruit trees, Antles, 1953 and Roberts, 1956; *Trifolium pratense*, Blinov, 1960). Cumakov (1955) claimed successful results from spraying both *Trifolium pratense* and honeybee colonies with sugar syrup containing either a little fennel or aniseed oil.

Minderhoud (1948) obtained positive results from spraying *Petunia hybridae* with syrup, but he supposed that bees might work only the sprayed areas of crops. MacVicar *et al.* (1952) found that spraying plots of *T. pratense* with honey solution caused a large but transitory increase in the population of bees, which gathered honey from the leaves and stems as readily as from the flowers; however, this made no difference to seed yields. Stephen (1958) stated that spraying *Pyrus communis* trees and legumes with syrup did not increase the number of foragers visiting them, and that bees in the sprayed areas collected syrup from the leaves, petals and twigs, so that fewer visited the flowers. Kronenberg, *et al.* (1959) put colonies on a crop of *Fragaria × ananassa* and sprayed the plants in one part only with sugar syrup. As a result the plants there were visited by many more bees than plants not sprayed, but this made no difference to the fruit set. Free (1965b) sprayed plots of *Pyrus malus* and *Vicia faba* with sugar syrup and although more bees visited the plots as a result, the effect of spraying was very local and did not increase the number of bees visiting flowers elsewhere in the crop. Furthermore, nearly all the bees that visited the sprayed plots collected syrup only and fewer visited the flowers themselves than in unsprayed plots; as a result less fruit was set in the sprayed *P. malus* plots. Hence it seems that spraying crops with sugar syrup can even be detrimental to pollination.

Increasing the Attractiveness of Crops

Another possible way to increase pollination is to increase the attractiveness of the crop to bees. Long term selective breeding of certain species might produce strains whose nectar or pollen is more accessible to honeybees (page 33). There is little that can be done to improve the amount and quality of pollen produced by a species except by breeding. It might also be possible to select strains that secrete more nectar. Probable genetic variability in nectar secretion has been demonstrated for *T. Pratense* and *M. sativa* (e.g. Anderson and Wood, 1944; Shuel, 1952; Pedersen, 1953; Ryle, 1954). However, because of the pronounced effect of environment on the sugar concentration of nectar, selection based on sugar concentration would be of limited value (e.g. Shuel, 1952; Pedersen, 1956).*

The most effective method of improving a crop may be to select plants directly for attractiveness to honeybees and, therefore, simultaneously for all the factors responsible for this attractiveness.

It is also possible to influence nectar secretion by altering the supply of soil nutrients. Probably all elements which affect plant growth influence nectar secretion. Ryle (1954) and Shuel (1955b, 1957) did factorial experiments with various flower species in which the levels of nitrogen, phosphorus and potassium supplied were varied. They found that the effect of each of these major elements on both nectar secretion and flower production was dependent on the supply of the other two. Shuel (1957) concluded that to obtain maximum nectar production, the level of nitrogen should be low enough to avoid excessive vegetative growth, the level of phosphorus should be sufficient to promote reasonably good flower production but not high enough to reduce nectar secretion, and the level of potassium should not be low enough to drastically limit growth and nectar secretion, nor high enough to inhibit flower production. He pointed out that the fertilizer treatment necessary for optimum nectar secretion varies with different species. Nectar secretion may also depend upon other cultivation practices, especially irrigation and planting distances (e.g. Bogoyavlenskii, 1955; Solov'ev, 1951; Kopel'kievskii, 1964a).

Increasing the Proportion of Pollen-gatherers

Pollen-gatherers are generally more valuable pollinators, and work faster than nectar-gatherers so it would be beneficial to increase their numbers. Because the proportion of foragers that collect pollen increases with the amount of brood in their colonies, it is important that colonies for pollination should contain plenty of brood. Therefore, overwintered colonies which contain brood are preferable for pollinating crops that flower early in the year to colonies established from packages of bees with little or no brood in them (Filmer, 1932; Vansell, 1942).

The dearth of stored pollen in colonies in winter, limits brood-rearing during late winter and spring and hence the size of colonies (Farrar, 1936; Allen and Jeffree, 1956). Feeding colonies with pollen, pollen substitute (4–9 parts soyabean flour to 1 part brewers' yeast to which dry skim milk, commercial casein and dried egg yolk is sometimes added), or pollen and

* Loper and Waller (1970) found that clonal lines of alfalfa that differed in their attractiveness to honeybees also had distinctive differences in their volatile components, so probably the aroma of a flower is important in determining its initial attractiveness, although, of course, continued visitation is determined by the amount and quality of nectar and pollen produced.

pollen substitute, in early spring is a valuable means of increasing a colony's brood-rearing and hence its foraging potential (e.g. Haydak, 1945, 1958; Spencer-Booth, 1960). However, unless the crop which the colonies are to pollinate is a poor source of pollen (e.g. *Gossypium* spp., *Eucalyptus* spp.) providing supplementary pollen during the period for which colonies are present at the crop might be disadvantageous as it might diminish pollen gathering (Free, unpublished). Perhaps giving pollen substitute does not have such an effect; this has yet to be determined.

During the summer Merrill (1924) kept four colonies well supplied with stores and another four colonies with small amounts only he found that the former consistently had more brood, and the latter group did not reach their peak of brood rearing until they were allowed to replenish their stores with natural forage. If this is confirmed it could have important bearing on more efficient management of colonies for pollination, and it would be most interesting to discover how bees become aware of the amount of stores in their colony.

It has been reported that colonies can be induced to increase pollen collection by removing some of their pollen stores and giving them extra combs of brood (e.g. Stapel, 1934; Veprikov, 1936; Rakhmankulov, 1955). However, this method would entail additional manipulation of colonies which by itself might diminish foraging (e.g. Taber, 1963b); furthermore, numerous other colonies would be needed as a reservoir of the additional brood required, and probably this would be less economical and less efficient than taking all the colonies to the crop concerned.

It has also been suggested that removing pollen with a pollen trap may increase its collection. Lindauer (1952) found that a colony with a pollen trap had a greater proportion of pollen-gatherers than two control colonies. But this needs further investigation using more colonies, especially because Free (1963) found that removing pollen from bees' legs at the end of successive trips caused them to change to nectar-gathering, and Lavie (1967) reported that pollen traps reduced the honey yield by 24% and brood rearing by 4%. It has yet to be finally decided whether any reduction in foraging caused by the presence of pollen traps is more than compensated by increased pollen collection. Todd and McGregor (1960) thought that colonies would adjust the amount of brood they reared to the available pollen supply and hence the use of pollen traps would be of questionable value.

Rashad and Parker (1958) found that two colonies with pollen traps produced an average of 15% less brood than four colonies without traps, and this was reflected in a 41% reduction in stored honey. Taking into account the efficiency of the traps (53%), the amount of pollen collected during the year, and the relative amount of brood of control colonies and hence the amount of pollen they needed, it is possible to calculate that the traps increased the total pollen collection by 85%. However, despite the ingenuity of this method the colony variation is too great for these figures to be accepted as typical without more data.

It has been reported that pollen gathering may be increased by placing an empty box of combs beneath a colony's brood chamber (Paddock, 1951); presumably this might be so when there is insufficient room in the brood chamber and immediately above it, but I know of no experimental work to support this contention.

Karmo and Vickery (1957) stated that when pollen was put into pollen dispensers at the entrances to hives (page 413) it appeared to stimulate the colonies to greater foraging activity, and they observed that potential foragers gathered small amounts of pollen, returned to their hives and performed recruitment dances. This suggestion also needs exploring.

In the spring, summer and autumn of 1958, when foraging conditions were generally poor, and in 1959, when foraging conditions were generally good, Free and Spencer-Booth (1961) fed colonies either dilute sugar syrup (40% sugar) or concentrated sugar syrup (62% sugar). Compared to controls, colonies fed concentrated syrup significantly increased the amount of pollen they collected in four of six experiments (average increase of $\times 2 \cdot 0$) and colonies fed dilute syrup did so in 5 of 6 experiments (average increase of $\times 3 \cdot 3$). This increase in pollen collection did not seem to be directly related to any increased brood rearing as the fed colonies increased their brood rearing relative to controls only in 1958; presumably shortage of incoming food limited brood rearing in that year.

The increased pollen collection when colonies are fed sugar syrup mostly results from rapid changes in the behaviour of individual foragers from collecting nectar to collecting pollen. Most of the bees that collect the syrup have not foraged previously and are at that stage of their lives when they normally receive nectar loads from foragers; their absence from the hive entrance probably discourages the foragers from collecting nectar (Free, 1965c).

In other experiments, dilute sugar syrup was fed to colonies put beside crops of *Prunus avium*, *Vicia faba*, and *Trifolium pratense* and it was found that their pollen collection increased by $\times 2 \cdot 2$, $\times 3 \cdot 3$, and $\times 5 \cdot 2$ compared to the controls (Free, 1965d). The proportion of *T. pratense* pollen was also increased by the feeding; probably this reflects a greater tendency of pollen-gatherers than nectar-gatherers to forage near their colonies.

Therefore, feeding sugar syrup to a colony not only increases its pollinating efficiency by greatly increasing the number of its pollen-gatherers but may also increase the proportion of pollen-gatherers that work the crop on which the colony is placed. Some of the increased pollination that has been reported following attempts to "direct" bees to crops by feeding them syrup containing the scent of the flowers of the target crop, could well have come from the syrup feeding alone.

It would sometimes be convenient if combless colonies in cheap containers could be taken to crops needing pollination, and destroyed or left to die when flowering was finished. However, such colonies forage less for pollen than those with comb and brood, and furthermore they tend to abscond (Hudson, 1929, 1930; Filmer, 1932). Although synthetic queen pheromones might be as effective as the presence of a queen in stabilizing a colony and in stimulating comb building and foraging (pages 61, 64), it seems unlikely that a synthetic brood pheromone would be as effective as brood itself. Moreover, pollen-gathering would be deterred until there was sufficient comb prepared to receive pollen loads. However, perhaps these disadvantages will eventually be overcome or outweighed by the advantages.

Although nectar collection seems to extend far beyond immediate colony needs, pollen collection is more directly related to them. This may be partly because man has selected bees that tend to hoard carbohydrate rather than pollen. Successful selection of pollen hoarders, although difficult to do, could result in colonies that collect much pollen irrespective to their needs, and so would be better pollinators than colonies at present available. Perhaps such selection will eventually be associated with the breeding of bees that prefer certain plant species.

Breeding Honeybees for Pollination

There have been frequent claims that some races or strains of bees work at lower temperatures, or under more adverse conditions, than others. While this may well be true, these claims are not supported by statistical evidence. Traynor (1966) argued that although pollen transfer at low temperatures is ineffective because the pollen grains may not germinate, the pollen will be transferred to the stigma and so be ready to grow when temperatures do become favourable; however, anther dehiscence does not usually occur at low temperatures, so only pollen left from a previous favourable occasion will be transferred.

Colonies in the same location at the same time may collect very different proportions of the available pollens (page 31). These differences may be partly due to chance differences in the extent to which bees from different colonies discover and exploit the local flora and, once established, the differences may be reinforced by differences in the odour of the colonies' food stores (Free, 1969). As expected, Nye and Mackensen (1965) found that when colonies were put beside a *Medicago sativa* field some collected a much greater percentage of *M. sativa* pollen than others, but they decided to investigate whether this was caused by a genetic difference between colonies. They found that colonies headed by sister queens that had been inseminated from their brothers were more similar in the proportion of *M. sativa* pollen they collected than were colonies headed by unrelated queens and, in consequence, they decided to select lines showing high and low preference for *M. sativa*.

The percentages of *M sativa* pollen collected in the second, third, fourth, fifth and sixth generations of the high preference line were 40, 50, 66, 85, and 86 and of the low preference line were 26, 15, 8, 18 and 8 (Mackensen and Nye, 1966, 1969; Nye and Mackensen, 1968); furthermore, the much greater tendency to collect *M. sativa* pollen shown by high preference than low preference line colonies situated 7·2 km from a *M. sativa* field demonstrated that they were not merely selecting a bee that worked close to its hive. Unfortunately, they found that the handling qualities of the selected colonies had deteriorated but were confident that in a long range breeding programme this could be prevented and other desirable characteristics incorporated into the strain.

This pioneer work proved that the tendency to collect *M. sativa* pollen is heritable. Remarkably few colonies were needed to achieve the separation into high and low preference lines, and the procedure shows great promise for selecting special strains for pollinating many different crops.

Chapter 5

Using Bumblebees as Pollinators

The general biology of bumblebees has long been known (see Sladen, 1912; Plath, 1934; Free and Butler, 1959). They are social insects whose colonies are at a stage of organization which is in many ways more primitive than that of honeybees, but more advanced than that of solitary bees. Their colonies are annual and only queens fertilized the previous summer survive the winter in hibernation.

Annual Cycle of the Colony

The queens (Fig. 34) emerge from hibernation, sometime between early spring and early summer depending on the species, and during the next few weeks they consume nectar and pollen from flowers. As a result their ovaries, which during the winter were thin undifferentiated threads, begin to develop and the rudiments of the first eggs appear. At about this stage of their development the queens search in likely areas, such as hedgebottoms, banks and uncultivated ground, for locations in which to start their future colonies. Most often the site selected is a disused nest of a small mammal or bird, and consists of an accumulation of grass, moss or leaves. Whereas some species tend to choose sites that are underground, at the end of a tunnel which may be a few centimetres to several metres long, other species choose sites that are on the surface, although often well-concealed in a depression or under a thick tuft of grass.

Having located her nest material, the queen forms a cavity in its centre and during the next few days she makes a clump of pollen on the cavity floor and builds a wax cup on top of it in which she lays eggs (Fig. 35); she then seals the top of the egg cell with more wax. The larvae which hatch feed on their bed of pollen and on nectar and pollen which the queen regurgitates to them through temporary breaches that she makes in their wax covering. As a result they grow rapidly and the queen keeps adding more wax to their covering so they remain completely enclosed (Fig. 36). After about ten days the larvae spin their cocoons and pupate. The queen removes the wax from the outside of the cocoons and uses it to make egg cells on top of them. After about another ten days the adult bees emerge. Altogether, about three weeks

Fig. 34. Queen bumblebee (*Bombus lucorum*) drinking a drop of sugar syrup.

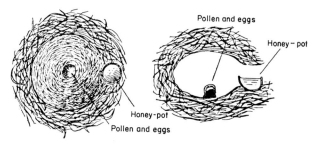

Fig. 35. Beginning of a bumblebee colony (after Sladen, 1912).

elapse between the time the egg was laid and the adult appears, but the length of all developmental stages varies with the environmental temperature and food supply. The cocoons vacated by emerging bees become storage cells for nectar and pollen (Fig. 37); wax cells may also be built for food storage only.

The first adult bees are all workers. When 2–3 days old, some of the workers begin to feed the larvae that have developed from the second batch of eggs laid by the queen and, when they are 3–4 days old, some of them start

Fig. 36. Queen bumblebee (*Bombus hortorum*) incubating her first batch of brood.

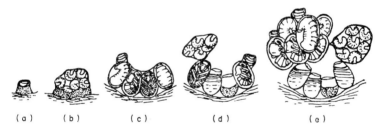

Fig. 37. Growth of bumblebee colony. a. The first batch of eggs. b. The eggs have hatched into larvae. c. The larvae have changed into prepupae (outside cocoons) and pupae (inside cocoons), and a second batch of eggs has been laid on the left hand cocoon. d. Adult bees have emerged from the central cocoons which are being used to store pollen and honey. e. All the first batch of adults have emerged—the comb grows upward and outwards (after Sladen, 1912).

to forage. When they are foraging regularly and collecting sufficient food the queen stops foraging and thereafter stays at home where she continues to lay eggs and incubate and feed the brood (Fig. 38).

As a colony becomes more populous, the rate of egg cell production increases and in general, the number of eggs a queen lays in an egg cell increases with the number of cocoons in the batch on which the egg cell is built. In this way, the number of eggs laid is adjusted to the number of workers that will be available to care for them during development. With the accumulation of more brood and bees the comb grows upward and outward and is approximately spherical when viewed from above, although the combs of underground nests have to conform to the shape of the nest cavity (Fig. 39 and 40).

Fig. 38. Part of a *Bombus agrorum* colony showing the mother queen incubating some of the cocoons.

The first workers produced in a bumblebee colony are, on average, smaller than those produced later, because during their larval stages they have only the queen to care for them and they receive less attention and food than those reared later. As a colony grows and there are relatively more workers to forage and care for the brood, on average larger workers are produced. However, particularly in some species, some small workers are nearly always produced, probably because for some reason they received less food than usual. The size of a bee is important in determining the duties it undertakes. Bumblebee workers tend first to do household duties only and later to forage as well, but larger workers begin foraging at an earlier age than smaller ones and, even when smaller bees begin foraging, they do so less often than the

Fig. 39. A colony of bumblebees (*Bombus lucorum*).

Fig. 40. A colony of bumblebees (*Bombus morrisoni*). (Photo: W. P. Nye.)

larger ones. This influence of a worker's size on its occupation is an advantage in that the smaller bees are able to move through the narrow complex passages of a bumblebee comb with greater ease than larger bees, and the larger bees can collect larger nectar and pollen loads and imbibe nectar more quickly than smaller bees. The type of food a bumblebee collects is primarily determined by its colony requirements which depend on the amount and type of food present and the amount of developing brood. However, large foragers collect pollen on proportionately more trips than small ones; because of this they are probably more efficient pollinators.

The size of a bumblebee colony at the climax of its development varies with the different species and there may be a considerable variation within a species; a large colony, whose comb is about 15–23 cm in diameter may have 150–200 bees, whereas a small colony may be only 8 cm in diameter and have only thirty to forty bees. Males and queens are produced at the climax of colony development. Males are produced from unfertilized eggs but queens, like workers, are produced from fertilized ones and externally resemble workers except they are larger. Males leave their nests when a few days old and never return but forage to satisfy their own needs, whereas young queens often forage for their maternal colony and continue to collect nectar and pollen for it even after they have mated. While the young queen is still attached to her maternal colony she develops large fat bodies which are her food reserve during the winter; eventually she fills her honeystomach, leaves her maternal nest and goes into hibernation.

Value as Pollinator

The bumblebee is regarded as one of the most efficient pollinators of many crops but is especially valuable in pollinating those flowers in which its large size facilitates pollen transfer while it is visiting the nectaries (e.g. *Gossypium* spp. and certain varieties of tree fruits) or in pollinating flowers with deep narrow corolla tubes from which only insects with long tongues can obtain nectar (e.g. *Trifolium pratense, Vicia faba*). However, various short-tongued species of bumblebees (e.g. *B. lapidarius, B. lucorum* and *B. terrestris* in Europe and *B. affinis* and *B. terricola* in North America) that are unable to reach the nectar by entering flowers with long corolla tubes do so through holes they bite near the bases of flowers, and so are much less valuable as pollinators of some crops than long-tongued species.

Although bumblebees are generally efficient pollinators, they are usually too few to pollinate large areas of agricultural crops; furthermore, their numbers show unpredictable fluctuations from place to place and from year to year so that even when they are relatively abundant one year they may be scarce the next. Moreover, it is supposed that the bumblebee population has declined over recent years because more intensive cultivation of the land has destroyed nest and hibernation sites, increased use of herbicides has destroyed wild flowers on which bumblebees rely for food supplies, especially in spring and use of insecticides has destroyed the bees themselves. Although it is probably true that the population of bumblebees and other wild pollinators has diminished, there is little real supporting evidence. Various suggestions have been made as to how the bumblebee population could be increased. Thus it has been suggested that farmers should grow small plots of nectar producing flowers to help provide for the colonies during times of scarcity and that they should leave small areas of uncultivated land for the bumblebees to hibernate and nest in (see Free and Butler, 1959; Holm, 1966a). However, I do not know of any attempts that have been made to test the value of such procedures.

Artificial Nest Sites

In contrast, many workers have provided artificial nest sites or "domiciles" for bumblebees, with the purpose of increasing the nesting population in an area where bees are needed as pollinators, or of obtaining colonies which can be moved to pollinate crops. Because different species of bumblebees often prefer different flower species, and because some species are more useful

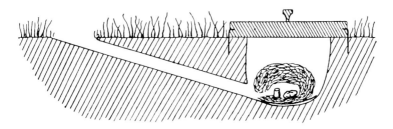

Fig. 41. Underground domicile for bumblebees (after Sladen, 1912).

pollinators of certain crops than others, the ultimate object is to domesticate selected species for different crops, even if not to select races that are specially amenable to domestication. The methods used to try to propagate bumblebees and the successes obtained have recently been reviewed in detail by Holm (1966a).

Sladen's (1912) domiciles consisted of a hole, about 30 cm deep, containing nest material and covered with a lid. A sloping tunnel, about 2·5 cm wide, connected the base of the hole with the surface of the ground (Fig. 41). In later models, a tin cylinder containing nesting material was put inside the hole. The nesting material was grass, moss or rope fibres. Although 26% of his domiciles were accepted by queens many of them were later abandoned probably because of increasing damp and invasion by various anthropods and rodents. Frison (1926) used a domicile which consisted basically of a metal can with a spout coming from its base as an entrance tunnel, the whole being buried in the ground with the lid of the can and the opening of the spout level with the surface. The nest of a mouse was put in each domicile and 47% were occupied by queens, but mature colonies were produced in only 17%.

The domiciles so far described attracted underground nesting species. Fye and Medler (1954) used domiciles on the surface of the ground. These consisted of wooden boxes each with an entrance hole in one side. Because bumblebees seemed to be particularly attracted to deserted mouse nests, Fye and Medler first induced mice to nest in the boxes by putting flax straw and a handful of grain inside each. After they had made their nests the mice were driven out, the size of the entrances was decreased so they could not re-enter, and the domiciles left to be discovered by searching queen bumblebees. Altogether 112 of the 130 domiciles were occupied by mice, thirty-seven by ants and fifty-two (40%) by bumblebees. This initial success was not maintained and the percentage occupied by bumblebees diminished to 3·7% over successive years (Medler, 1962). Discouraged, Medler supposed that the use of field domiciles was not a practical way to obtain bumblebees for pollination.

However, more consistent success has been obtained by Hobbs *et al.* (1960, 1962). They used wooden boxes containing upholsterer's cotton, which they found was readily accepted by bumblebees without prior occupation by mice. The boxes were converted for underground use by attaching a piece of plastic hosepipe between the entrance and the surface of the ground. In one year (1961), queens began their colonies in about 50% of the nest boxes; acceptance of underground nest boxes was particularly good and 72% of underground nest boxes in meadows were occupied. They pointed out that occupancy in locations on the prairies could be increased by using only underground nest boxes, as the four bumblebee species indigenous to the prairie (i.e. *B. borealis, B. fervidus, B. huntii* and *B. nevadensis*) prefer underground nesting sites. Later, false underground hives, which had their tunnels only beneath the surface of the ground, were used; they were as effective as underground hives and had the advantage that they were much easier to set in position (Hobbs, 1967a).

Laboratory Domiciles

Another approach to the problem of increasing local bumblebee populations has involved inducing queens to nest in especially prepared boxes in the laboratory (Figs. 42, 43 and 44), so that when colonies become large enough they can be taken to crops needing pollination. Unfortunately, colony initiation is often difficult to achieve, even when ample honey, pollen and nesting material are provided.

Frison (1927) was the first to obtain any marked success. He divided his

Fig. 42. Egg clumps made by a *Bombus agrorum* queen in captivity.

Fig. 43. A *Bombus hortorum* queen with her initial batch of brood and honeypot which she has produced in captivity.

nest boxes into two compartments, one of which was covered with a sheet of glass and contained a supply of diluted honey, and the other, the nest-chamber itself, was lined with several layers of wax coated muslin and contained an artificial wax cup and a small ball of pollen. By introducing only those queens that appeared ready to start colonies, Frison obtained eggs in 70% of his nest boxes but the first batch of larvae were successfully reared in only 32%. Hasselrot (1952, 1960) also used boxes with two compartments,

Fig. 44. A colony of *Bombus ternarius* started in captivity.

one for food and the other for the nest (Fig. 45). The nest compartment was filled with moss, except that in its centre there was a hollow sphere of cellulose wadding containing a lump of fresh pollen. Over several years 75% of 190 queens used started colonies. Some of the colonies grew very large; one *B. terrestris* colony produced 1,500 bees, 488 of which were queens. However, others (e.g. Valle, 1955; Holm, 1960; Holm and Haas, 1961; Horber, 1961) have been less successful with methods similar to Hasselrot's.

Both Frison and Hasselrot gave their queens well insulated nest boxes so it seemed that one reason for their success might have been the maintenance of a sufficiently high temperature inside. In their most successful method Plowright and Jay (1966) used well insulated nest boxes kept at 29°C (84°F) and in forty-five (79%) of fifty-seven nest boxes eggs were laid, in all but one of which adults were reared. They concluded that, provided these results could be maintained, bumblebee colonies cultured in captivity could play an important part in pollination. Their success is of the order which Holm (1966b) advocated as necessary to establish a practical method of domestication.

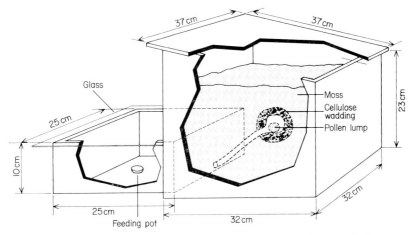

Fig. 45. Domicile used by Hasselrot (1952) for starting bumblebee colonies.

Artificial Hibernation

In order to have a readily available supply of disease-free queens in the spring, and especially ones not parasitized by the nematode *Sphaerularia bombi*, various workers have attempted to hibernate bumblebee queens artificially, and indeed have attempted to obtain complete control of the colonies by confining the queens and colonies in cages and glasshouses. If successful, this would also have the advantage that several generations could be reared in the same year and the production of bumblebee colonies would be independent of the natural population. Bumblebees readily mate and forage in captivity so these activities present no problems. Apart from colony initiation, hibernation was a major difficulty but this has now been overcome and the proportion of queens that can survive artificial hibernation is sufficiently large, to make it a practical proposition.

Horber (1961) successfully hibernated 80% of his bumblebee queens in small aluminium containers, filled with vermiculite and kept at 1°C (34°F). The queens that survived were allowed to fly in a glasshouse where nest boxes had been provided and although some queens started nest building, no real progress was made until the nests were put at temperatures between 25–35°C (77–95°F). Subsequently, the progeny of one of these queens was reared in captivity for five generations. Holm (1960) and Holm and Haas (1961) hibernated 84–90% of their queens in artificial mounds or boxes of sphagnum moss and steam sterilized soil inside a glasshouse. Over three years, a total of 37% of 245 queens started colonies and 19% of the queens produced colonies that reared sexual forms. Pouvreau (1965) introduced a total of 269 queens into a glasshouse containing forty empty nest boxes. The inner compartment of each box was lined with polystyrene and contained either dried moss, fine grass, cotton, kapok or horsehair. Of the queens, 19% occupied nest boxes, 11% founded colonies from which workers were produced and 4% founded colonies from which workers and sexual forms were produced. Pouvreau also hibernated queens in vermiculite at 7°C (45°F), and in soil and moss in an unheated glasshouse; 63% and 56% respectively of the queens subjected to these treatments successfully overwintered.

Some of the successes obtained in recent years on queen hibernation and colony initiation seem very hopeful, but for a method to be really useful it must be capable of repetition by other workers, often with far less skill than those who originated it. Perhaps the uncertainty which accompanies many colony-founding methods is basically because many of the essential factors associated with ovary development, nest searching and colony initiation are far from being fully understood, and the various problems will probably only be resolved when research has provided more information. Certainly, many of the procedures various workers have employed seem to be irrelevant and to have no influence on the final result. Too great a proportion of the colonies that are initiated in captivity fail to mature; probably, more fundamental work on colony growth and development must be done before this problem also can be solved.

Even when colonies are successfully established in domiciles or nest boxes, there are many difficulties to be overcome when they are transferred to sites where they are needed, and the process needs careful timing, otherwise the queens may desert. The boxes must be given distinguishing colours, otherwise some of the queens and workers become confused, enter the wrong nest and engage in mortal combat. Any foragers that have remained away from their nest overnight must be captured after the nest has been transferred. The boxes must be waterproof and prevented from becoming unduly hot. The colonies must be protected against enemies including parasitic bumblebees (*Psithyrus* spp.), small mammals and numerous parasites (see Hobbs *et al.*, 1962; Holm, 1966a; Hobbs, 1967a). In fact colonies in nest boxes seem more susceptible to attack than naturally occurring colonies, probably because they are more easily discovered.

Hence, although the prospect of producing bumblebee colonies for pollination on a commercial scale has improved during the last few years, much still needs to be discovered before it is feasible. The present, and for the most part, uncertain, methods involve too much effort relative to the number of pollinating insects produced to be economically justifiable. However, the use of domiciles in the field to attract nesting queens already seems to be worth while in Canada where, perhaps, the natural bumblebee population has not been depleted as much as in other countries; certainly the techniques developed there should be tested elsewhere.

Importing Beneficial Species

The introduction of bumblebee species, or for that matter other pollinating insects, into parts of the world where they do not already occur, awaits exploitation. In 1885 and 1886 bumblebee queens were sent from Britain to New Zealand where they became successfully established and apparently caused a large increase in *Trifolium pratense* seed production (see Thomson, 1922). Unfortunately, the three species at present in New Zealand, *B. subterraneus*, *B. ruderatus* and *B. terrestris* are not among the most useful pollinators of legumes, but the proposed introduction of more beneficial species was refused on the grounds that they might introduce the acarine disease of honeybees, although they have never been known to harbour honeybee acarine mites and even if they did so it is barely conceivable that they could transfer them to honeybees.

Theoretically, it seems desirable that long tongued species which are efficient pollinators should be introduced into areas even where other species already exist, as they might thrive in the absence of enemies specific to them, occupying ecological niches previously vacant; or, if chosen with care, might even compete successfully with less useful bumblebee species.

However, it is apparent that so long as there is any chance that they might carry disease or parasites with them, objections will be made to their importation. Perhaps in the distant future, when techniques of laboratory rearing of bumblebees have been developed, these objections could be overcome by importing beneficial species at immature stages. Even so, it is important that before any alien bumblebees are released in a country of importation they should be kept in enclosures and their behaviour studied while they are visiting species they are required to pollinate, in order to ensure they really are desirable.

Chapter 6

Using Solitary Bees as Pollinators

In many parts of the world solitary bees are valuable pollinators of certain crops, but their usefulness is limited because their numbers fluctuate greatly from year to year and from place to place so, in general, they cannot be relied upon.

For many years attempts have been made to induce solitary bees to occupy artificial nests. Fabre (1915) found that glass tubes and reeds he provided were readily occupied by *Osmia* spp. Balfour-Browne (1925) filled shallow wooden boxes with a mixture of wet clay and chopped straw and when it was dry he bored holes, 19 mm in diameter and 76 mm long, into it for bees to nest in. He also provided glass tubes rolled in black paper and pieces of *Sambucus* spp. stem with the pith bored out, either in bundles or pushed into the apertures of ventilation bricks. Some of each type of domicile were occupied by *Osmia* spp., *Megachile* spp. and *Anthidium* spp., but *Sambucus* spp. stems were the most successful. When a bee had chosen a tube, it rolled about inside it and he suggested that this behaviour conveyed the bee's body scent to the tube and so helped the bee to recognize her own nest. Frost (1943) also used glass tubes but found that the moisture from the food was unable to escape from them, whereas it was absorbed by the bramble stems. Kloet (1943) split his bramble stems lengthwise and bound the two halves together with string before using them, so that any nests built in them could be examined without injury.

Peck and Bolton (1946) found that *Megachile* spp. were the most important pollinators of *Medicago sativa* (alfalfa) in Saskatchewan and pointed out that, during the flowering of alfalfa, the bees had to search intermittently for nest sites, so the establishment of sufficient nests near the alfalfa field should result in the bees spending more time foraging. They put small bundles of *Heracleum lanatum* stems in sunny places but no bees were attracted to them; *Megachile nevalis* and *M. inermis* nested in holes which Peck and Bolton bored into logs but, although *M. frigida*, the most important, showed interest in the nesting sites, it did not rear brood in them. They suggested that an alfalfa field should be surrounded by a wide strip of uncultivated land in which *Megachile* spp. could nest, and that plants with hollow stems should be grown on the fringe of the field and old *Helianthus annuus* stems scattered along its edge. Medler and Koerber (1958) also investigated the possibility of attracting alfalfa pollinators to artificial domiciles. They used grooved boards bolted together so that the semi-circular grooves of adjacent boards fitted together to form a series of nesting holes; at the end of the season these boards were parted to examine the nests. However, they were not as effective as bundles of 20 cm long sumac stems with holes 6-8 mm drilled through their centres.

Megachile rotundata

Life history

The above attempts to induce solitary bees to occupy and multiply in artificial nests met with only limited success and were of doubtful economic value, but some of the techniques used were of considerable value when a solitary bee species, that was an efficient pollinator of alfalfa, was found to nest gregariously in artificial domiciles. This bee, the leafcutter, *Megachile rotundata* (Figs. 46 and 47), was inadvertently introduced from Eastern Europe or Western Asia to the east coast of North America in about 1930, and spread westwards reaching Utah in 1954 and Oregon in 1958. Once it was established in the western U.S.A. it became evident that it does not excavate its own burrows but would occupy a variety of nesting sites including beetle burrows, nail holes, holes bored in logs, hollow stems, drinking straws, metal and rubber tubes. Stephen (1961, 1962) and Bohart (1962) explored its life history and produced techniques that made possible its commercial exploitation.

Under natural conditions adult *M. rotundata* emerge over a 3-6 week period in about late May when the alfalfa hay crop is in flower, and the females are mated while they are basking in the sun, often in the vicinity of their maternal nests. Males can mate many times but females once only. The mated female makes a series of cells in the tunnels or tubes she selects for nesting; the walls and base of the cells are made with oblong leaf cuttings, mostly alfalfa, lightly glued together with a salivary secretion. Each cell is filled to between half and two-thirds its depth with a honey pollen mixture. When a female returns from foraging she enters her tunnel head first and deposits the nectar she has collected; she then goes to the tunnel entrance, turns round and backs into the cell to deposit her pollen. When sufficient food is present she lays an egg on it (Figs. 48 and 49). The cell is then capped with three to ten circular leaf cuttings and another one is started. The series of cells finishes a little below the tunnel entrance, and the end of the tunnel itself may be plugged with as many as 130 leaf cuttings. Foraging trips are

Fig. 46. Alfalfa leaf-cutter bee (*Megachile rotundata*) collecting nectar from *Borago officinalis*, Borage. (Photo: W. P. Nye.)

Fig. 47. Alfalfa leaf-cutter bee (*Megachile rotundata*) tripping a *Medicago sativa*, Alfalfa, flower. (Photo: W. P. Nye.)

USING SOLITARY BEES AS POLLINATORS

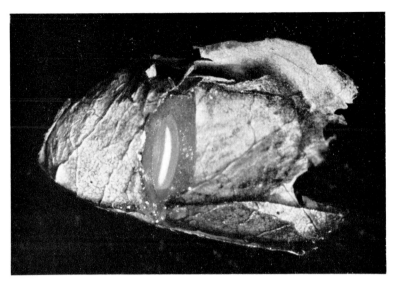

Fig. 48. Cell of alfalfa leaf-cutter bee (*Megachile rotundata*) opened to show egg laid on top of nectar and pollen provisions. (Photo: W. P. Nye.)

Fig. 49. Cell of leaf-cutter bee (*Megachile rotundata*) opened to show third instar larva feeding on pollen stores. (Photo: W. P. Nye.)

of extremely short duration; only 10–20 sec is spent collecting a leaf cutting and 90–150 sec a load of pollen. Whereas the females pollinate practically every alfalfa flower they visit, the males do not forage as consistently and collect nectar only, often without tripping the flowers. *M. rotundata* overwinters as a prepupa and during the warmth of spring changes to a pupa. The bee that develops from the last egg to be laid in the tunnel is the first to emerge.

In favourable locations there is a partial second generation each year and, when this occurs, 4–6 weeks separate the emergence of the overwintered generation and the second generation. The emergence of the second generation may even overlap that of the first. The first generation females produce about thirty to thirty-five cells, but the second generation females do not produce nearly as many, although in some areas the second generation bees produced are valuable in extending the working season.

Artificial nests and management

Bohart and Stephen tested corrugated paper, drinking straws and drilled wooden blocks as domiciles for *M. rotundata*. The corrugated paper was either provided as a roll about 15 cm in diameter, which had enough space for 600 nests, or cut into flat pieces which were stacked on top of one another with a piece of cardboard glued to the back; the attractiveness of the corrugated paper was increased when straws were interspersed in the rolls. When drinking straws (Fig. 50) were used they were either embedded in a thin layer of wax at the bottom of tin cans, or else presented in their cartons and protected from heat and rain by wooden or fibre-glass covers. Straws of diameters 4, 5 and 6 mm were used, cut to about 9 cm long; the cut, frayed ends were put in the bottom of the container. It was found that bees at the far end of tunnels were unable to emerge through uneaten pollen in a chamber containing a dead egg or larva and themselves died; the use of the short straws tended to diminish such losses as each cell in the short straws has, on average, fewer cells in front of it. The diameter of the straws used determined to some extent the amount of food provided, and the size of larvae and adults produced, but when smaller females occupied the straws of larger diameter they used several layers of leaf cuttings to decrease the aperture size. Early wooden domiciles merely consisted of blocks or logs in which 5 mm holes had been drilled; care was taken to ensure the holes were free of jagged fibres. By interspersing occupied blocks or straws among empty ones the acceptance of the empty ones was increased.

It was found that *M. rotundata* preferred nesting in wood to straws and corrugated paper, and use of the latter was abandoned because it lacked weather resistance. Stephen (1962) reported that bees reared in straws

Fig. 50. Drinking straws occupied by *Megachile rotundata*.

Fig. 51. Tunnels opened to show *Megachile rotundata* nests built inside. (Photo: W. P. Nye.)

preferred these when making their own nests, so perhaps conditioning to nest material occurs during the larval stage. The bulk and weight of the wooden domiciles made handling and storage difficult, and it was impossible to dismantle them to inspect and clean tunnels which contained refuse or dead immature stages. These objections were partially overcome by using layers of boards with semi-circular grooves (Fig. 51) which matched to produce linear series of circular tunnels (Hobbs, 1964; Nye and Bohart, 1964). These boards also had the advantage that at the end of the season the nests could be pushed out of the grooves, using a special apparatus to speed the operation (Hobbs, 1967b), and stored in a separate container, so the boards were clean and free of debris and ready for the next season. Recently, Hobbs (1967b) recommended grooved polystyrene boards, which have the advantage over wooden blocks that they weigh only a tenth as much, the leaf cells do not adhere as strongly to polystyrene as to wood, and climatic fluctuations have less influence on the size of polystyrene than wooden tunnels. Their disadvantages are that they are more easily damaged than wood and the bees can chew through them.

Hobbs (1964) reported that four times as many larvae died in tunnels 4 mm diameter as in tunnels 5·5 mm diameter; furthermore, three times as many males as females were produced in the smaller tunnels, but equal numbers of each sex in the larger ones. Consequently, he made his semi-circular grooves 3·0 mm deep so the completed tunnels were 6·0 mm diameter and 114 mm long. Stephen and Osgood (1965) found the ratio of male to female produced was between 5 : 1 and 11 : 1 in 4·0 mm diameter tunnels; 3 : 1 in 5·5 mm diameter tunnels; 2 : 1 in 6·0 mm diameter tunnels. A high proportion of males were also produced in tunnels less than 5 cm long.

A shelter to house domiciles should provide protection from sun, wind and rain, be screened to prevent birds attacking the nests, and face east or southeast so that the first rays of the sun stimulate the bees to activity. When many straws or tunnels are put together in the same container, a bee spends much time searching for its own tunnel, so it is advantageous to provide some form of orientation mark; the front of a shelter is either painted with a checkerboard pattern of different colours or has white and black identification letters stencilled on top (Figs. 52 and 53). At night the adults rest in their nests so, if the crop is likely to be sprayed with insecticide, there must be some means of confining the bees to their shelter until the danger is past. Bees will remain immobile in their straws for more than 48 h if the shelters are stored at 2–4°C (36–39°F).

Nests should only be moved to a new site either between the first and second generation, which requires careful timing, or at the end of October when all the adults are dead. If *M. rotundata* is scarce in a particular area where it is needed, "trap nests" may be put in a favourable area in the summer and

Fig. 52. "Shelter" provided at edge of *Medicago sativa* field for leaf-cutter bee (*Megachile rotundata*).

Fig. 53. Close-up of part of "shelter" showing tunnels occupied by nests.

moved to the desired site in the winter or early spring. Trap nests should contain some tunnels that have previously contained nests, as such tunnels are much more attractive than new ones, presumably because they still have an odour of the former occupants.

Sufficient empty tunnels should be provided in a shelter in the spring to allow for a five-fold expansion in population during the year, or about one tunnel for every cocoon present in the spring. However, when the tunnels are so short or so narrow, as to lead to the production of an exceptionally high ratio of male to female bees (see above) such an increase is not always realized. If the field beside which the shelters are situated is cut, some of the adult bees, and especially those of the second generation, may move to later flowering fields or to fields in which the second cutting is reaching its peak, and so a potential part of the nesting population may be lost. This may be overcome by leaving part of the first cutting for seed and taking part for hay. Stephen (1962) reported that late in the season the adults worked with diminished efficiency and he thought that this was possibly due to their senility and sparsity of forage.

The emergence of the bees can either be advanced by keeping the prepupae at 27–32°C (80–90°F), or delayed by keeping them at 4°C (40°F), and so timed to coincide with the flowering of alfalfa. It is even possible to store the prepupae for 2 years at low temperatures without heavy mortality (Stephen, 1962). Whereas in the western states of North America it is preferable to store the shelters and nests inside buildings for protection during the winter, in Canada this is essential as the prepupae are unable to survive at the outside winter temperatures. Although *M. rotundata* was introduced into North America without its natural enemies, several North American insects have now begun to attack it, encouraged no doubt by the large concentrations put near alfalfa fields; the major pests are the chalcid wasp, *Monodontomerus obscurus* and the carpet beetle, *Trogoderma glabrum* (Johansen and Eves, 1966). Because *M. rotundata* cannot survive in the wild in Canada, parasites and diseases can be controlled more easily than they can further south.

Hobbs (1964, 1967b) has given detailed instructions for the management of *M. rotundata* in Canada. In the autumn, all debris—dead, diseased and parasitized bees—should be removed from the nests and each cell rolled in turn between thumb and forefinger; if the larva is diseased or dead or the cocoon has not been completed, the cell will collapse; the healthy cells should be stored in closed jars or polythene bags at 4°C (40°F) during winter. The cocoons are removed from the jars about 15 days before the alfalfa comes into flower and put in shallow trays at 30°C (86°F) and 50–60% relative humidity. The first insects to emerge are any native bees and wasps which may be present as these develop much more quickly than *M. rotundata;* any parasites present will also emerge at this time. These emerging insects are attracted to light bulbs inside the incubator and drown in pans of water placed beneath them; a little detergent in the water helps the parasites to sink quickly to the bottom of the trays. The bees do not usually begin to emerge until the larvae have been incubated for 18 days, but as soon as the first males emerge lids are put on the trays and the lights in the incubator turned out. After another 3 days when about 40% of the bees will have emerged, the trays are taken to the field, put just beneath the roof of the shelter and the lids removed. In warm weather the bees leave the trays, make orientation flights and return to the shelter to mate and nest. As the remaining bees emerge much more slowly at outside temperatures than in the incubator it may be necessary to return them to the incubator for a day or so.

To avoid the emergence of a second generation, which would probably be destroyed by frost in Canada, Hobbs suggested that, ideally, all nests should be stored at 4°C (40°F) as soon as the first generation has finished nest building. However, the first generation larvae must have sufficient time to reach maturity, and by this time some of the second generation may have begun to emerge. Unfortunately, the factors causing the emergence of a second generation, and whether or not a diapause is involved, are incompletely understood. There is a suggestion that the tendency to produce a second generation is inherited, and this character is selected against in Canada where second generation bees die without nesting.

On the assumption that there should be one *M. rotundata* foraging per 4·2 sq. m of alfalfa, and that each female spends only half its time working the flowers, Bohart (1962) calculated that one shelter containing about 10,000 nesting females was needed for each 2 ha of crop. Although not oligolectic to alfalfa, *M. rotundata* shows a strong liking for it, but it will also visit other species including *Melilotus alba* and *Trifolium repens*. The constancy of bees nesting beside alfalfa fields to the alfalfa is probably mostly a reflection of the fact that, if possible, they restrict their foraging to within a hundred metres or so of their nest. Hence, in large crops, the shelters should be sited inside the field as well as around the edges to give an even distribution of the foragers.

Shipments of *M. rotundata* have been made from North America to other parts of the world including Chile and France. However, as it does not begin to forage at temperatures below 21°C (70°F), it can be used only in areas that are warm enough. Hobbs (1967b) calculated that 350 h above 21°C (70°F) during the flowering of alfalfa are necessary.

Nomia melanderi

Life history

Another solitary bee which is efficient at pollinating alfalfa and has been successfully exploited in western states of North America is *Nomia melanderi*

(Fig. 54), which nests in soil which is sub-irrigated. Menke (1952a,b), Bohart and Cross (1955), and Stephen (1959) described its life history. The adult, which is about two-thirds the size of the honeybee emerges from late June to mid-July, males about a week earlier than the females. The daily peak of emergence occurs between 09.00 and 11.00 h. The female becomes mated soon after it emerges, leaves the site and does not return until mid-afternoon

Fig. 54. Female alkali bee (*Nomia melanderi*) at the entrance to its nest. (Photo: W. P. Nye.)

when it begins to dig a tunnel (Fig. 55). It completes its main burrow during the night, the next day it prepares and provisions its first cell and the following day it lays an egg on the pollen and seals the cell entrance. This pattern of behaviour is repeated so that, during a typical day, a female lays an egg in a cell provisioned the day before, provisions a new cell already excavated and excavates a new cell. The branch burrows and new cells are built at night. Each completed nest (Fig. 56) consists of a vertical shaft 15–20 cm long leading from a mound of soil on the surface to a series of 3 or 4 branch burrows, each 5–7·5 cm long. A well-developed nest often has one of the side burrows much longer than the rest and it, in turn, is subdivided. About fifteen to twenty cells are built per nest. The rate of work is influenced by many factors including the distance between nest and forage, and the texture of the soil. Each female is capable of laying twenty-four to twenty-six eggs.

The egg stage lasts 2 days and the larval feeding period 6 days; the bee spends the winter as a prepupa and development of the pupa and adult

Fig. 55. Entrances to *Nomia melanderi* nests. (Photo: W. P. Nye.)

Fig. 56. Section of *Nomia melanderi* nest showing from left to right, top row: unfinished pollen ball; completed pollen ball with egg; feeding larva. Bottom row: adult female; feeding larva; mature larva; diseased larva. (Photo: W. P. Nye.)

awaits until the ground at brood-cell level has been warmed by the late spring sunshine. The time taken to develop from the prepupa to the adult depends upon the temperature, and adults do not emerge until the soil moisture has diminished to less than 25%. Although there is one main generation per year which is active for about 6 weeks, in parts of North America, there may be a partial second generation.

N. melanderi is very specific about the sites in which it will nest; alkaline flats or low mounds with sparse vegetation where the soil is constantly moist are usually favoured. In the right conditions there are extremely large concentrations of nests with as many as 540 entrance tunnels per square metre. Preliminary studies by Menke (1952a,b) and extensive ones by Bohart (1958a); Stephen (1959, 1960, 1965); Frick *et al.* (1960), Fronk and Painter (1960) and Bohart and Knowlton (1968) showed that the ideal nesting site has constant underground moisture extending up to the surface (which is provided by the underground movement of salt laden water over an impermeable layer), fine sandy loam soil with a low (less than 8%) clay size particle content, a well-drained surface (preferably with a slight slope), little or no surface vegetation, either no salty crust or one that is not too hard and no fluffy dry layer under the crust.

After a natural site has maintained a large bee population for a few years it often becomes unsuitable for a number of reasons and the population rapidly declines. The principal causes of the decline are ploughing, flooding, decreased moisture, development of a thick or hard crust, parasites and predators attacking the immature stages and encroachment by salt-tolerant vegetation. Alfalfa growers are encouraged to protect nest sites existing on their land from these hazards, and to maintain the site in its transitory state, either by regulating the water supply, reworking the soil surface or eradicating encroaching vegetation.

Artificial nests and management

Attempts have also been made to create new sites where they are needed (Stephen, 1960, 1965; Bohart, 1967; Bohart and Knowlton, 1968). These are modelled on natural nesting sites, but greater control can be exerted over the variables that lead to population changes (Fig. 57). Early artificial sites had an impermeable layer 60–90 cm below the surface, which was made by compacting the soil with heavy machinery or by depositing a layer of clay; its surface was covered by a series of graded parallel ditches to which water was supplied through pipes extending to the surface. This water kept the soil put on top of the impermeable layer moist and, if necessary, sodium or potassium salt was mixed with the top 5 cm of soil, at the rate of 2 to 5 kg per square metre, to draw moisture upwards and to provide a firm surface which minimized evaporation in summer and which was an effective seal against the penetration of surface moisture in winter.

More recent nesting sites are constructed so that excavation pits are lined with polyethylene sheets, or other suitable waterproof plastic, which are covered with layers of 5–15 cm of gravel and 5 cm of coarse sand. The pit is then filled with soil that has less than 10% clay sized particles and less than 40% sand sized particles. Water is provided to the gravel layer through down pipes, located every 37–56 sq. m, to maintain the surface in an optimum

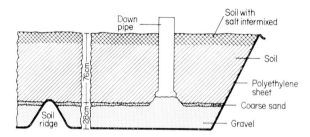

Fig. 57. Diagram of artificial bed for *Nomia melanderi*, alkali bee (after Stephen, 1965).

condition. The sizes of the beds used ranges from 1·2 by 1·2 m to 60 by 121 m, and in the larger beds a 30 cm high ridge of soil is made beneath the polyethylene every 7·6 m and so serves to isolate segments if drainage problems arise. Beds from 46–122 cm deep are found to be suitable, but the shallow ones dry out more quickly. As before, salt is added when necessary.

Menke (1952b) buried boxes of soil in existing nest sites, so they were occupied by nests, and in the winter moved them to establish nests in new areas. Stephen and Bohart seeded newly prepared nest sites with twenty-five to fifty cores of soil, each containing about 200 prepupae, obtained by driving steel cylinders (30 cm. in diameter) into existing sites in the spring and autumn. When the bees emerged they nested in their own soil cores and in the surrounding soil and their presence attracted other bees to the area. Another and faster method used is to cut a natural nesting site into blocks 30 cm^2, with a rotary tiller that has been modified by adding saw blades, and then prising the blocks out with a garden fork. However, when the blocks are transported long distances they are likely to fragment.

The combination of ample food supplies from alfalfa and adequate nesting sites results in large local populations, and Stephen (1960) found more than 2,680 burrows per square metre in a bed established for only 2 years in Oregon; this was eight times the concentration present in the best natural site in the state. However, in these populous conditions the efficiency of the females is curtailed and a female produces an average of only eight progeny compared to about twenty under ideal conditions; this decrease is probably partly because many females have to dig main shafts 50 cm long to avoid other nests, compared to 15–25 cm under normal conditions, and the entrances to the burrows are so close that the excavations from one burrow are deposited into the entrance of the adjacent one.

The number of beds needed depends upon the amount of alfalfa being grown and the cultural practices used. Stephen (1965) estimated that a well-populated bed (8 by 15 m) should provide enough bees to pollinate 16 ha of alfalfa. The time of emergence of the bees is important in determining whether the first or second cutting should be retained for seed. They begin working the nearest alfalfa field in flower when they emerge and, once they have become conditioned to visit a field that has plenty of forage, they rarely forsake it for fields nearer their nests. Any pollination which the males perform is incidental; indeed they prefer clover, corn and onion to alfalfa although they will collect nectar from alfalfa if there is no choice.

Search for other Species

Following the successful exploitation of *Megachile rotundata* and *Nomia melanderi* it is only natural that a search should be made for other suitable solitary bees, bearing in mind that a particular species may be more prolific when it is transferred to areas where it does not occur at present. Some species of *Osmia*, such as *O. excavata* and *O. cornifrons* which nest in bamboo or reed in Japan, and forage mostly on *Brassica campestris* and *Pyrus malus* respectively (Hirashima, 1959; Maeta and Kitamura, 1965), and *O. seclusa*

Fig. 58. Nests of *Osmia rufa* occupying tunnels in old piece of wood.

which is an efficient pollinator of alfalfa (Bohart, 1955) are likely candidates. Levin (1957) and Medler (1967) found that in some areas artificial wooden burrows were readily occupied by *O. lignaria*. Free and Williams (1970b) found that *O. rufa* (Figs. 58, 59 and 60) preferred to nest in straws 7 mm

Fig. 59. Can packed with drinking straws, many of which have been occupied by *Osmia rufa*.

Fig. 60. *Osmia rufa* larva feeding on bed of pollen.

rather than 5 mm in diameter; initial trials showed that bees had a strong tendency to nest gregariously and in some sites the artificial nest population increased considerably.

Nishida (1963) puts logs with holes drilled in them in fields of *Passiflora edulis* to attempt to increase the population of the carpenter bee *Xylocopa sonorina*, and reported that this was successful in some fields but not in others. Wójtowski (1964), in Poland, attempted to use *Anthophora parietina;* holes (7·5 mm diameter and about 20 mm deep) were drilled in blocks of earth (50 × 50 × 10 mm) which were suspended on earthen walls of old buildings in which *A. parietina* was nesting. In each block twenty-five to fifty-five nests were established and these were taken to pollinate a crop of *Vicia villosa*.

No doubt many other species will be found to occupy artificial nests but to be of use commercially a species must be gregarious, rapidly increase its population in man-made nests, visit a particular commercial crop in preference to other species, have its peak of activity coincide with that of the crop, be easily manipulated and managed, and not be subject to uncontrollable parasitism and disease. The encouraging results achieved during the last two decades give hope that other species of solitary bees may be successfully propagated to pollinate crops other than alfalfa.

Chapter 7

Pollination in Enclosures

The need for insect pollination of plants in enclosures arises either because the plants must be isolated to produce uncontaminated seed, or because attempts are being made to find whether the species gives increased seed or fruit following insect visits, or because the crop is being produced under artificial heat in a glasshouse. Whichever of these reasons applies similar problems are encountered.

During the early stages of plant breeding hand-pollination can produce sufficient seeds but with expanded programmes of breeding and testing this becomes too laborious and time consuming, and bumblebees, honeybees, blowflies, and more recently, solitary bees have been used to pollinate the flowers.

Bumblebees

Bumblebees work well in confinement and are especially valuable for use in small enclosures. They can usually be readily obtained from the flowers or by collecting their nests (see Free and Butler, 1959). If nests are difficult to find, local advertising usually produces the desired result. Lindhard (1911, 1921) was the first to use bumblebees in cages for pollinating. He enclosed colonies with *Trifolium pratense* plants after first caging them with *Lotus corniculatus* to free them of any viable pollen they might have been carrying. Williams (1925) used bumblebees to pollinate *T. pratense* in small compartments (90 × 90 × 90 cm) of a glasshouse. To ensure that they did not contaminate the selected strains with pollen from elsewhere he captured the bumblebees on flowers other than *T. pratense*, washed them in tepid water, and dried them in isolation for 3 to 4 h before use. Tests confirmed that this method was effective and, although pollination with pollen that had been immersed in water for only 5 min gave some seed, after it had been dried it was non-viable.

Usually only one bumblebee was kept in a cage; they remained active for 12–15 days in favourable weather and 4–6 days in cold weather. When more than one bumblebee was confined in a cage they became lethargic, probably because they were short of food. Some species (e.g. *B. agrorum*, *B. hortorum* and *B. humilis*) were more efficient in the cages than others (e.g. *B. lapidarius* and *B. sylvarum*). Essentially the same technique is still in use at the Welsh Plant Breeding Station, Aberystwyth. Williams also tried putting six to ten honeybees in each cage every 5 days but obtained poor seed sets from *Trifolium pratense* although *Trifolium repens* and *Medicago sativa* gave good sets.

Pedersen and Bohart (1950) used bumblebees to cross-pollinate *M. sativa* clones. They found that a single colony was ideal for pollination in a cage $3.5 \times 6.5 \times 1.8$ m high. Species successfully used included *B. morrisoni*, *B. mormonorum*, *B. fervidus* and *B. appositus*. The colonies were provided with a gravity feeder containing diluted honey. Because *M. sativa* was the only source of pollen available, a greater proportion than usual of bees of some species visited the flowers for pollen.

Although bumblebees are usually abundant enough in summer, there may be difficulty in finding sufficient to pollinate crops that flower in spring. Thus, Priestley (1954) reported that although he had used bumblebees to pollinate potted *Brassica oleracea* plants in glasshouses, the method was not entirely satisfactory as in cold or overcast weather the bumblebees were difficult to find, and furthermore, he supposed that it might diminish the bumblebee population as most of the bumblebees to be found when the *B. oleracea* flowered were queens. Dutch workers (e.g. Minderhoud, 1949; Kraai, 1958; Sneep, 1952) tried to overcome this last objection by using only queen bumblebees that harboured parasitic nematodes and therefore could never found colonies, but although such diseased queens have a characteristic flight pattern by which they can easily be recognized they are even more difficult to obtain in sufficient numbers.

Kraai (1958) reported that diseased bumblebee queens lived for 10–25 days when confined in small cages ($0.8 \times 0.8 \times 2.0$ m) with *B. oleracea*, *B. napus*, *Cichorium endivia*, *Raphanus sativus* or *Cichorium intybus* and visited flowers with the same frequency as healthy queens. He also used male bumblebees later in the season when few diseased queens were still available but found they died sooner and were less efficient than diseased queens. The parasitic bumblebees (*Psithyrus* spp.) were also very useful for pollination work. Strangely, he found that worker bumblebees were less efficient than either males or queens and lived only a short while, and thought that this was because they were less able to survive when isolated from their colonies. Kraai also shook all his bumblebees in lukewarm water to render any pollen inviable, before using them.

Honeybees

Because of the ease with which they can be obtained, honeybees are usually preferred to bumblebees for pollination in large cages or glasshouses and have been used for this purpose for many years. For example: Wilson (1929) mentioned that honeybees were being used to pollinate *Prunus persica* flowers in glasshouses; Murneek (1929) caged a *Pyrus malus* tree with a small honeybee colony; and Pearson (1932a) used small colonies each with about 5,000 bees to pollinate *Brassica oleracea* in cheesecloth cages.

Cages of many types of construction and material have been used for pollination work. Early ones often consisted of a wooden or metal framework, covered with muslin, wire mesh, or perforated zinc (Williams, 1925; Atkinson, and Constable, 1937). Many of these early cages were cumbersome and difficult to transport and the first field cage that was light, portable, easily assembled and stored was designed and used by Pedersen *et al.* (1950). Cages based on its design are now commonly used in pollination studies and have been adapted to cover large areas of crop (see Farrar, 1963). It consisted of a framework of 1·3 cm diameter electrical conduit, which supported a plastic screen cage with five meshes per centimetre. A zip at each corner of the cage allowed easy access.

Inevitably, a cage influences the light intensity, temperature, humidity and wind speed to which the plants inside are subjected; but the extent to which it does so varies with different weather and climatic conditions and with different types of cage; the effect of these differences on plant growth also depends on the plant species concerned. The cage designed by Pedersen *et al.* (1950) had little effect on the environment to which the plants were subjected, compared to cages of an earlier design with walls of metal screen or cloth (e.g. Weaver and Ford, 1953; Wafa and Ibrahim, 1960a); the relative humidity, light and wind speed inside were slightly reduced, but the temperature inside and outside was the same. Probably large cages have less effect than small ones. Palmer-Jones *et al.* (1962) found that large cages had little effect on the growth of *Trifolium repens*, but others have found that cages have a pronounced influence on plant growth. For example: Free and Spencer-Booth (1963b) found that *Brassica alba* and *B. nigra* plants caged with bees became etiolated and produced less seed than in the open; caging has also been found to diminish the yield of *Ribes nigrum* (Hughes, 1966) and *Carthamus tinctorius* (Rubis *et al.*, 1966). Free (1966b) obtained evidence that in some conditions the adverse effect of caging on *Vicia faba* seed production was so great that it tended to minimize the beneficial effect of honeybee pollination.

Because colonies confined without much forage inevitably dwindle in strength, arrangements have sometimes been made for them to fly alternately in the open and into cages on consecutive days (Butler and Haigh, 1956). However, there is a controversy as to the value of such a procedure; some workers (e.g. Scriven *et al.*, 1961 and National Agricultural Advisory Service, Derby, 1962) reported that under such alternating conditions, few bees appear to settle to forage in the cages and that a high mortality occurs on days the bees are confined, whereas Nye (1962) stated that the mortality rate was greater when colonies were confined continuously rather than intermittently. To preserve colony strength without providing artificial food, it is possible to obtain sufficient pollination by locating a colony outside a cage so it can forage freely, and then to direct one or two hundred foragers from the colony into it at the beginning of each day. However, this method is of no use when it is necessary to avoid contaminating the plants inside the cage with foreign pollen.

Although bees leaving their hives carry viable pollen on their bodies (Free and Durrant, 1966b) this pollen loses its viability after a few hours. When Butler and Haigh (1956) allowed bees to fly into a cage and outside the cage on alternate days, the change-over was made after flying had ceased for the day, and the *Brassica oleracea* and *Raphanus sativus* varieties in the cage produced pure seed although other varieties of each species were growing just outside. Kraii (1962) kept honeybees with varieties of *B. oleracea* (cabbage, kale, sprouts), *R. sativus*, *Begonia* spp. *Cheiranthus cheiri* or *Centaurea cyanus* that had dominant characters, and then either isolated the bees or confined them to their hives for a minimum of 10 h, and finally put them with varieties of the same species that had recessive characters. There was no contamination of the recessive varieties with pollen from the dominant. Hence, it would seem safe to transfer a honeybee colony between cages containing cross-compatible varieties without danger of contamination, after confining the bees to their hives overnight. Pankiw and Bolton (1965) and Pankiw and Goplen (1967) showed that when honey bee colonies had been isolated from a contaminating source of *Medicago sativa*, *Melilotus alba* or *Melilotus officinalis* for two days the bees no longer carried viable pollen from it.

Usually colonies occupying three or four combs are found to be sufficient for pollination in cages (e.g. Atkinson and Constable, 1937; Palmer-Jones, 1959; Scriven *et al.*, 1961; Steuckardt, 1963). Although in Holland and Belgium where colonies are used extensively for pollinating under glass, the bees must occupy seven combs, the area they are required to pollinate is much larger than is usual during breeding work or testing the pollination requirements of a species. It can be disadvantageous to use colonies that are too large for the areas needing pollination. Weaver (1956) put colonies of about 10,000 bees into cages of *Vicia villosa* and found that the anthers, stigmas, and corollas were badly damaged by such an excess of insects; he largely eliminated this trouble by replacing the colony with one containing less than 2,000 bees.

Experience has shown that using honeybees to pollinate early flowering crops in glasshouses is particularly likely to harm the colonies. Sometimes this is difficult to avoid, particularly when the flowering season is a long one; Killion (1939) reported that a *Cucumis sativus* grower in Iowa, who used seventy-five colonies in his glasshouse at the rate of one every 6,000 ft.^{2}, replaced the colonies several times during the flowering season which was from 15 December to 1 August. Consequently, efforts are often made to use colonies whose destruction will be no great loss. Pearson (1932a) prepared small colonies each with about 5,000 bees and an old queen in a small hive (15 × 15 × 22 cm) containing a piece of wax comb foundation. Each was fed 0·6 litre of sugar syrup per day while enclosed in the pollination cage and at the end of 5 weeks they had produced comb and about 100 cells of brood. Minderhoud (1949) claimed there was no difference in the pollination of *B. oleracea* by colonies, with and without queens. This is surprising as the presence of a queen stimulates foraging, and Goplen and Pankiw (1961) reported that queenless colonies were not as efficient as queenright colonies for pollinating *Melilotus alba* in cages. However, queenless colonies may provide sufficient foraging for some purposes.

Kraii (1954) found that small colonies of two or three combs, which were queenless but contained plenty of brood, were suitable for glasshouse compartments 3 × 5 × 2 m. For smaller compartments (0·8 × 0·8 × 2 m) he used very small colonies in miniature hives. These hives, which had walls 1·5 cm thick were of two compartments; the larger front compartment had two miniature combs and the smaller rear compartment contained sugar syrup and dry sugar. The two compartments were connected by a small round hole and the larger front compartment had two additional holes; one served as a hive entrance and the other covered with wire gauze as a ventilator. Each hive was given 400–500 bees initially and 100 newly emerged bees were added later. On average, a normal colony was sufficient to produce twenty to fifty of these small colonies during a season. They survived an average of 25 days in the glasshouse compartments and a few lived 112 days. Kraii successfully used his small colonies to self- and cross-pollinate selected strains of *B. oleracea* (cabbage, winter and summer cauliflower), *R. sativus*, *Brassica rapa*, *Cichorium endivia*, *C. intybus*, *Tragopogon porrifolius*, *Daucus carota*, *Apium graveolens*, *Asparagus officinalis* and *Fragaria* × *ananassa*. He was able to transfer his colonies from one glasshouse compartment to another but, to avoid possible contamination when the two compartments concerned contained closely related plants, he put his colonies with plants of another family for 2 days. Obviously honeybee colonies of such a small size are not viable and can only be used in heated glasshouses and to pollinate a few plants.

It is agreed that the colonies for pollination in cages and glasshouses should contain at least one comb of food stores but there is a controversy as to the best source of carbohydrate to provide. Thus, Atkinson and Constable (1937) stated that candy was better than sugar, Palmer-Jones (1959) and Scriven et al. (1961) preferred using sugar syrup and Steuckardt (1963) fed his colonies moistened sugar. The National Agricultural Advisory Service, Derby (1962) advocated using dry lump sugar unless a colony is short of food when a small amount of sugar syrup should be given, but not so much that egg laying space on the combs is diminished. To overcome the need for opening hives to determine the strength and food supplies of their colonies, Goplen and Pankiw (1961) suggested the use of observation hives. Ideally, sufficient combs of pollen should be provided either initially or progressively for brood rearing to occur, but if this is not possible efforts are sometimes made to provide either pollen or pollen substitutes (e.g. Neiswander, 1954; Nye, 1962; Palmer-Jones, et al., 1962; Miller and Amos, 1965; Cuypers, 1968). The pollen is obtained from pollen traps and the loads are squeezed together, if necessary with sugar syrup, to make a lump or "cake" which is placed on top of the combs. A pollen substitute can be made from soya bean flour, brewers yeast and skimmed milk; its attractiveness and value is enhanced if it is mixed with pollen (Haydak, 1945; Shemetkov, 1957a). Studies by McGregor (1952) and Rothenbuhler, et al. (1968) on keeping colonies in artificial conditions are very relevant; they found that when colonies were provided with water and sugar syrup and were able to collect dried, pulverized pollen from dishes in flight cages, the bees foraged vigorously and the colonies grew in size. Indeed, there are indications that when colonies have become adapted to cage or glasshouse conditions, there is no reason why they should not be kept confined indefinitely, provided they are properly maintained, although this may be expensive.

From the experience of the above workers and others it may be concluded that for most circumstances colonies containing three combs of brood and bees are adequate for cages in the field, and fewer bees may be a disadvantage. Should the colonies become weakened, combs of emerging bees can be introduced to them. When the combs do not contain sufficient pollen to maintain brood rearing a supply of pollen or pollen substitute should be made available. The bees should be provided with a source of sugar and, if dry sugar or candy is given, it is essential to give water also. Water may also be needed by the bees to regulate the temperature of their colonies. In addition to providing food for the colony, feeding sugar syrup stimulates pollen collection and hence probably results in more pollination; Free and Racey (1966) found that pollen-gatherers were more valuable pollinators of *Freesia refracta* in glasshouses than nectar-gatherers and their numbers could be increased by feeding their colonies with sugar syrup. This may well also apply to other glasshouse crops. The effect, if any, of feeding dry sugar on pollen collection in glasshouses is not known.

It is usually found that when colonies are first caged or put in glasshouses many of the bees spend much or all of their time trying to escape, fail to return to their hives and die. This seems to be particularly true in glasshouses, and is perhaps partly because the bees are stimulated to excess flight by the unaccustomed heat, and partly because the glass itself confuses them. Sources of artificial light may also act as traps and the bees fly round them until they become exhausted (Lecomte, 1955).

According to Thompson (1940), initial losses are likely to be minimized when colonies are put in the centre of a glasshouse and well away from the walls. Priestley (1954) found that shading a glasshouse roof helped to diminish the tendency of bees to fly up to it. Kraai (1954) noticed that "lost" bees congregated on the side of an enclosure facing the sun and found that putting hives near the north-west wall or corner facilitated their return in the evening. However, it is usually found that the proportion of bees trying to escape soon decreases, probably because the bees concerned are established foragers and soon die. Scriven *et al.* (1961) and Hawkins (1968) allowed the foragers of small colonies that they had prepared for pollination work to fly back to their original hives so that the small colonies consisted of young bees only, which quickly adapted to cage conditions. Hawkins pointed out that the absence of established foragers reduced the possibility of the bees carrying viable pollen into the glasshouse. Rothenbuhler *et al.* (1968) found that, when colonies composed entirely of young bees without previous flight experience were used in cages, the bees orientated when they left their hives and nearly all of them returned. However, it is not always possible to prepare colonies of young bees only, or colonies of bees without foraging experience, particularly in the winter or spring. Hitchings (1941) and Cuypers (1968) reported that releasing colonies in glasshouses after dark reduced the subsequent tendency of bees to fly against the glass; this tendency was also reduced when the sky was overcast during the first few days the bees were in the glasshouse, and the greatest loss was likely to occur on the first sunny day.

Free and Racey (1968b), who used honeybees to pollinate an early crop of *Phaseolus multiflorus* in a glasshouse, put the colonies just inside the open doorways so the bees could also fly outside; under these conditions few, if any, of the bees became lost in the glasshouse and the flowers were adequately pollinated. When conditions in a glasshouse are too unfavourable for honeybee colonies in the spring or summer, they may be kept outside and allowed access to the glasshouse through tunnels connected to their hive entrances. Sorokin (1958) and Shemetkov (1960) used hive entrances which allowed access to both a *Cucumis sativus* glasshouse and the open air; bees that started to work in the glasshouse continued to do so but in order to ensure sufficient recruits to maintain the glasshouse population, the part of the hive entrance leading to the outside was closed until 10.00 h each day.

Certainly, under the right conditions, when they receive adequate food, and are not subject to sudden temperature fluctuations, honeybees appear to forage as normally in a glasshouse as when visiting a crop in the open. For example, in a large glasshouse containing *C. sativus* Hitchings (1941) and Shemetkov (1960) found that the flight period of the bees was adjusted to the times of presentation of pollen by the flowers, and Lecomte (1955) observed that when a choice of species is present, individual honeybees tend to keep to one only.

Commercial glasshouse crops that honeybees are commonly used to pollinate include *Fragaria × ananassa, Lycopersicon esculentum, Prunus persica, P. domestica, Cucumis sativus, C. melo* and various ornamental flowers. A set of regulations governing the use of bees in glasshouses has been devised by Dutch beekeepers and growers to safeguard the interests of both (Cuypers, 1968). In particular, Dutch beekeepers prevent their colonies being exposed to cold when flowering in the glasshouse is finished, by keeping them for a few weeks at 5–11°C (41–52°F) until the brood nest has diminished to the size that is normal for the time of year. When a hive is moved from one glasshouse to another, it is placed in the same relative position in the second house as in the first and its hive entrance also faces in the same direction to avoid disorientation of the bees; it has been found that keeping hives clear of walls and sloping glass roofs also helps the bees to orientate.

The site at which colonies are located in a glasshouse is important for another reason. Shemetkov (1960) discovered that individual bees kept to one row only of *C. sativus* in a glasshouse. Free and Racey (1966) found that when colonies were located at one end of glasshouses 50 m long containing *Freesia refracta* the number of bees on the flowers became fewer as the distance from the hives increased and the bees tended to work along rather than across rows. D'Aguilar *et al.* (1967) reported that when a colony was placed at one end of a glasshouse containing *C. melo* comparatively few bees foraged on the plants most distant from their hive. Hence, to get an even distribution of honeybee foragers on a glasshouse crop, it would be better to have a single colony near the centre of a glasshouse than at one end, and with two or more colonies to have them at diagonally opposite corners or evenly distributed.

Solitary Bees

Recently, plant breeders have begun to use *Megachile rotundata* for pollination in cages and glasshouses. Bohart and Pedersen (1963) found that these bees were suitable pollinators of *Medicago sativa* in 6 m² cages in the field (about fifty female bees per cage), and in $1·2 \times 1·2 \times 0·9$ m cages enclosing plants on a glasshouse bench. The bees nested and foraged readily in the cages, although in the glasshouse natural light was supplemented with artificial light on overcast days, and the temperature was maintained above 28°C (82°F). They pointed out that either emerging bees or bees collected in the field can be put in the cages, although to avoid possible contamination in the latter circumstances they should be collected from species other than those they are required to pollinate. Because *M. rotundata* is apparently more susceptible to insecticides than the honeybee they recommended that use of insecticides should be terminated several days before the bees are introduced.

Holm (1964) reported that *M. rotundata* was an efficient pollinator of both diploid and tetraploid *Trifolium pratense*, as well as *M. sativa*, and worked more readily in glasshouses than honeybees. Heinrichs (1967) used 100 incubated cells, from which adults were beginning to emerge, per growth chamber (1·8 × 2·7 m) containing *Medicago sativa* plants. The chambers were subject to 16 h of continuous light per day (48,000 candela per m^2) and the temperature of 20°C (68°F) maintained during darkness was increased to 27°C (81°F) during the light period. The bees did not become active until the temperature exceeded 23°C (73°F). The females readily accepted the artificial domiciles provided and pollinated the flowers. In three successive tests thirty plants produced 324, 321 and 282 g seed. He calculated that, by carefully arranging for the emergence of the bees to coincide with flowering, a growth chamber could be used for the pollination of eight successive batches of *M. sativa* plants per year.

M. V. Smith and T. Szabo (personal communication) have found *M. rotundata* useful for pollinating *Lotus corniculatus* and *Cucumis sativus* in glasshouses and to be especially valuable in small cages in which honeybees will not forage. The bees readily collect artificial supplies of nectar and pollen to supplement the insufficient supplies from the flowers. However, they are extremely sensitive to light intensity in such conditions and because they do not accept *C. sativus* leaves for cell building, other plants have to be provided.

Blowflies

For an insect to be used as a pollinator either by a plant breeder or in the production of a commercial crop, it should be easily handled and readily available in large numbers; so far, apart from bees, only blowflies have filled these criteria. Attempts to breed other Diptera, including *Eristalis* spp., have not been successful (Phillips, 1933; Singh, 1954).

Blowflies were first used in California by Jones and Emsweller (1933) who found they were more suitable than honeybees for cross-pollinating selected lines of *Allium cepa* in cages (1 × 1 × 2 m high) and more efficient than hand pollination. Later, Jones *et al.* (1934) used blowflies to pollinate

A. cepa flower heads enclosed in pairs in small wire cages covered with muslin (Fig. 61). The flies were obtained by trapping or breeding. Trapping was done only when the flowers were unlikely to be contaminated with foreign pollen. Animal refuse of various kinds, such as fish heads and bullock lungs, were used to attract the flies into the base of the trap; once inside the trap they moved to its top section which was covered with fly-proof screen and from there were introduced to the pollination cages by means of small transfer cages (Jones and Mann, 1963).

Fig. 61. *Allium cepa*, onion, flowers enclosed in small cages with blowflies to pollinate them.

To ensure that the flies were free of undesirable pollen they were bred in captivity by the following technique (Jones and Emsweller, 1934). Refuse from abattoirs was put in open troughs and various species of blowfly, mostly *Phormia regina* and *Lucilia sericata* were attracted to it and laid eggs. The eggs hatched within 36 h and the larvae began to feed on the refuse; cattle lungs were especially suitable because of their porous nature and large feeding area. A roof was put over the refuse to protect the larvae from high temperatures and rain. After 5–7 days of feeding the larvae started searching for pupation sites, moved along the trough and fell into a bucket of sand; this was removed each day and replaced with a fresh bucket, so that all the larvae in a bucket were approximately the same age. The larvae pupated in the sand and the pupae were washed, dried and stored. When stored at 7°C (45°F) the adults emerged in about 2 weeks but at 3°C (37°F) the emergence was delayed for several months; by storing them at different temperatures a supply of adults could be made available for practically any time, although Jones and Emsweller reported that the flies appeared less active when the pupae were stored early in the pupation period. When the *A. cepa* began to flower a small handful of pupae were put in the cage and more added every 3 or 4 days during flowering to maintain a continuous supply of pollinating insects. In a refinement of this technique the pupae were put in small cages (15 × 15 cm), each with a cone-shaped top; the tip of the cone was inserted through an opening into the *A. cepa* cage so that adults entered it on their own accord when they emerged.

To avoid the disagreeable odour associated with breeding blowflies on meat, Wolz (quoted by Jones and Mann, 1963) developed a method of rearing blowflies on moist dog biscuits but, nowadays, it is unnecessary for growers to breed their own flies as firms supplying blowfly maggots for fishbait are also a source of pupae. The procedure is essentially the same as that outlined above except that adult flies are allowed to emerge in a room or cage provided with water and moistened sugar on which they feed, and small bowls of fish or meat in which they lay eggs. When eggs have been laid the contents of the bowls are emptied among the waste products of an abattoir so the larvae have plenty of food to complete their development. Whereas firms providing fishbait usually only allow enough larvae to pupate to maintain the stock, when given sufficient notice they can arrange for supplies of pupae to meet the requirements of local growers. Species of blowflies commonly used for fishbait in Britain include *Phormia terranovae* and *Calliphora vomitoria*. No doubt other species would be equally successful as pollinators (see Mackerras, 1933; Norris, 1965, for studies on life histories).

After Jones and Emsweller's (1933) original discovery blowflies were soon used as pollinators of other crops. Borthwick and Emsweller (1933) found that the hand pollination of *Daucus carota*, which was very laborious as each of the tiny flowers produced only two seeds, could be circumvented by using blowflies. They enclosed umbels with blowflies inside muslin cages with wire frameworks which were fastened to stakes driven in the ground. When there were ten flies per cage a good set was obtained, but umbels without flies gave little or no seed.

For many years it has been the practice for growers of brassica crops to produce their own seed from selected plants. These plants are caged or transplanted into large cages to prevent undesirable cross-pollination. According to Jones and Rosa (1928) in California, flies and other insects were enclosed within the cages to facilitate pollination. Blowflies are now satisfactorily used for this purpose and Faulkner (1962) found they were as effective as honeybees in pollinating the crop and more effective than hand pollination. In one test they gave 24 g of seed per plant compared to 9 g from hand pollination; probably this was because the flowers were being visited continuously by blowflies whereas hand pollination was done only every 3 or 4 days. He recommended growers to put about 500 adults, which are produced within a few days from half a pint of pupae kept at 15–21°C (59–70°F), in a cage containing 25 plants soon after flowering begins, and every 4 or 5 days thereafter during flowering, to put an additional handful of pupae on the floor of the cage and lightly cover with soil. He found that blowflies lived 2 to 3 weeks in the cages provided sufficient nectar was available; within limits the amount of pollination increased with the number of flies present.

Wiering (1964) made extensive tests of the efficiency of blowflies (*Phormia terranovae*), honeybees and bumblebees in pollinating *Brassica oleracea* (Brussels sprouts, kale and cabbage) in plant breeding work. He found that small isolation cages, which enclosed two plants only, were too small for honeybees and bumblebees soon died in them, but blowflies gave sufficient cross-pollination and seed. Even when caged with an isolated single plant, pollination by blowflies resulted in a small amount of seed which was sufficient to maintain inbred lines, whereas when plants were isolated without flies a much smaller and often insufficient amount of seed was produced. In large cages, blowflies, honeybees and bumblebees produced similar amounts of seed and cross-pollination. He pointed out that pollination was normally needed in early spring, when worker bumblebees are not available and the low temperature often encountered deterred foraging by honeybees but not by blowflies. However, honeybees moved more freely from plant to plant, especially when the plants were separated by more than 50 cm and he suggested that preference should be given to honeybees when uniform pollination of a larger group of plants is needed. Peto (1950) also recommended honeybees for pollination work in large cages as, in such circumstances, he found they were cheaper to use and involved less work. It seems that blowflies are especially suitable for use in small cages or glasshouse compartments; their other main advantages are that they are easy to handle, do not sting and can be used in small numbers. However, undoubtedly the choice between using blowflies or honeybees will often be determined by individual preferences.

There is little information on the behaviour of blowflies in cages; it seems probable that most pollination is achieved inadvertently while they walk over flowers. Faulkner (1962) reported that the cages should not be so high that there is much distance between the top of the plant and the cage roof, otherwise the flies spend much of their time on the roof and are less likely to alight on the flowers.

Other crops that blowflies have been used to pollinate include *Pastinaca sativa*, *Apium graveolens*, *Brassica napus*, *Rheum rhaponticum*, *Scorzonera* spp. and *Angelica* spp. (Gaag, 1955). Probably they could also be used with advantage to pollinate *Fragaria* × *ananassa* under glass. However, their short tongues will probably restrict their use to open flowers with readily accessible nectaries; they failed to pollinate *Phaseolus multiflorus* which have long corolla tubes (Free and Racey, 1968b). Perhaps smaller flies (e.g. *Musca domestica*, *Musca autumnalis* and *Fannia canicularis*), which could more easily enter long corolla tubes, could be used with advantage for some flower species but a necessary prerequisite is that the insects can be easily bred. Watts (1958) attempted to use hoverflies to pollinate *Lactuca sativa* in muslin bags but without success as the hoverflies made continuous efforts to escape and died within 36 h.

Part II

Crops Needing Insect Pollination

Chapter 8

Cruciferae

The flowers of Cruciferae are hermaphrodite and have four sepals, four petals, an inner whorl of four longer stamens, and an outer whorl of two shorter stamens (Fig. 62). The superior ovary of two united carpels, is surmounted by a style with a two-lobed stigma.

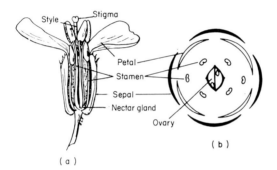

Fig. 62. Typical flower of Cruciferae (after Robbins, 1931, reproduced with permission of McGraw-Hill Book Company).

The genus *Brassica* includes many crop species of which the most important are *B. alba* Rabenh, white mustard; *B. campestris* L., field mustard; *B. carinata* L., Abyssinian cabbage; *B. chinensis* L., chinese cabbage, pak-choi; *B. juncea* Czern. and Coss, trowse mustard; *B. napobrassica* Mill, rutabaga, swede; *B. napus* L., rape; *B. nigra* Koch, black mustard; *B. oleracea* L., wild cabbage and its domestic forms; and *B. rapa* L., turnip. Other Cruciferae widely grown for food include *Armoracia rusticana* Gaertn, horse-radish and *Raphanus sativus* L., radish.

One of the major problems in producing *Brassica* seed is to prevent contamination as crops must be isolated from other varieties of the same species, or from other species with which they will readily cross. Occasionally severe contamination can ruin a seed crop (Davey, 1959).

Brassica species fall into three primary groups according to whether they have nine, ten or eleven chromosomes in the haploid state. There is free intercrossing between varieties within the same group so that when they are grown for a seed crop they must be planted far apart. Isolation between crops with different chromosome numbers is not very important as the chances of a fertile hybrid between them is extremely remote; however, in the rare circumstances in which chromosome doubling occurs after such a cross, the hybrids are fully fertile; in this way *B. napus* (nineteen chromosomes) has arisen from a cross between *B. oleracea* and *B. rapa* (nine and ten chromosomes), and *B. juncea* (eighteen chromosomes) has arisen from a cross between *B. nigra* and *B. rapa* (eight and ten chromosomes) (see Gill and Vear, 1958.) The situation can be expressed diagrammatically as follows.

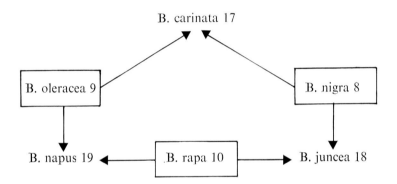

Estimates of the distances necessary to isolate *Brassica* seed crops that cross-pollinate, vary from 0·4–3·2 km depending on the purity required (e.g. Haskell, 1943). Haig (1956) reported that up to 0·6% crossing occurred between varieties planted about 1·6 km apart, and formed the opinion that, whereas such a little contamination would be insignificant in a commercial crop, it would be of considerable importance to a producer of pure seed. Gill and Vear (1958) pointed out that most contamination occurs within 90 m of a pollen source and regarded the small amount of cross-pollination between crops separated by 900 m as of no practical importance in a commercial seed crop.

Most of the *Brassica* species are self-incompatible to some extent but this varies with the species and variety concerned and even with the age of the plant (see Sampson, 1957). With increasing interest in the production of hybrid seed it is likely that perfectly self-incompatible varieties will find an important place in hybrid seed production. However, considering the economic importance of these crops remarkably little is known about their pollination requirements, or the likelihood of increasing pollination by using honeybee colonies.

Much more information is available on the various types of oilseed rape than other cruciferous crops so these will be considered first.

Brassica campestris L. and Brassica napus L.

Oilseed rape occurs as both annual and biennial varieties of the species *Brassica napus* var. *oleifera*, the swede rape, and *B. campestris* var. *oleifera* (the turnip rape). *B. napus* is the most popular in Europe; its production in Britain has increased greatly in the past few years, about ten times as much summer as winter rape being grown. *B. campestris* is only of local importance in Europe, but because of its earlier ripening it is important in countries with short growing periods, and the summer variety accounts for about 80% of the Canadian crop (Bunting, 1967). Toria (*B. campestris* var. *toria*) and sarson (*B. campestris* var. *sarson*) are important oilseed crops in India and Pakistan. Varieties of *B. juncea* (trowse mustard) are also grown extensively for oil in India, Pakistan and China; this species is considered later with other species of mustard.

B. napus is generally regarded as self-fertile, although in a normal population individual plants which are self-sterile or prefer foreign pollen do occur (Rives, 1957; Olsson and Persson, 1958). The rape flower (Fig. 63) is protogynic, and the anthers of the long stamens, which are about level with the

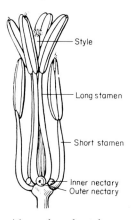

Fig. 63. *Brassica napus* flower with sepals and petals removed (after Ross-Craig, 1949).

stigma, dehisce outwards, whereas those of the short stamens, which are below the stigmas, dehisce inwards (Ewert, 1929). Hence, as any automatic self-pollination can only occur with pollen from the long stamens, insect visits would seem to favour the transfer of pollen and to be essential for the transfer of any pollen from the short stamens. Harle (1948) and Jenkinson and Glynne Jones (1953) found that plants kept in the relatively still air of glasshouses produced only a third to half as much seed as normal; probably this seed was set by autopollination or, less likely, by pollen falling from the earlier flowers on to ones lower on the stem, but clearly either wind or insect pollination is also needed to obtain a high set.

Fechner (1927) reported that cross-pollination gave a 14% increase in the weight of seed produced compared to self-pollination. Persson (1956) compared self- and cross-pollination with pollen from short stamens and from long stamens for both winter and summer rape. Her results confirmed and extended the preliminary experiments of Ewert (1929). She found that cross-pollination with pollen from short stamens was significantly superior to that from long stamens and gave a 14% greater weight of seed per pod. Cross-pollination with pollen from short stamens was also significantly superior to self-pollination from short or long stamens, producing 19% and 12% greater weight of seed per pod, but cross-pollination with pollen from long stamens was not. As in the field, pollen from both long and short stamens will be used in pollination, the benefit from cross-pollination will not be as effective as if pollen from short stamens only were present. Furthermore, Persson (1956) obtained evidence which suggested that with young flowers cross-pollination is inhibited and self-pollination favoured. Consequently, it is not surprising that any benefit of cross-pollination by honeybees in the field is so small that it is hardly measurable, although honeybees touch the stigmas in about three-quarters of the flowers they visit (Free and Nuttall, 1968a).

The amount that single plants with a recessive petal colour were cross-pollinated when grown in fields whose plants had a dominant petal colour varied from 20–42% with different varieties, with an average of 36%. The percentage distribution was bimodal, one mode occurring between 20–35% and the other at 100% indicating that the latter were completely self-sterile (Persson, 1956; Olsson, 1960). Even a five-fold difference in bee population between fields made no significant difference to the percentage of cross-pollination. Likewise, when dominant plants were grown in the centre of plots of recessive plants there was no correlation between the percentage of cross-pollination and the honeybee population. It was concluded that wind is the main pollinating agent of rape.

However, it has been claimed that the seed yield depends on the nearness to the crop of honeybee colonies (e.g. Koutensky, 1959; Belozerova, 1960) and various workers have attempted to discover the extent to which insect pollination can increase seed yield by caging some plots with bees and others without. Most of the experiments were unreplicated, so it is difficult to draw firm conclusions from them; and, although there was usually a slightly greater amount of seed produced in the plots caged with bees (e.g. Fujita, 1939; Zander, 1952; Pritsch, 1965) it is likely that any similar experiments that did not indicate an increase were not reported. Other experiments unfairly compared plots caged to exclude bees with those in the open field, and although it is to be expected that plants in the less favourable environment of a cage would have a lower yield (Louveaux, 1952; Rhein, 1952; Meyerhoff, 1954) this was not always so (Persson, 1956).

Free and Nuttall (1968a) had eight plots caged with bees, eight plots caged without bees and eight open plots. Although plants caged with bees produced 13% more seed than those without, the weight did not differ significantly between the different treatments. Moreover, the weight per seed was greater in plots caged without bees than in plots caged with bees, presumably reflecting the fact that these plots had fewer seeds to mature so the mean weight of seed produced per plant in the different treatments were even more similar (2·29, 2·18 and 2·17 g in cages with bees, without bees and not caged).

Perhaps the amount of automatic self-pollination from the long stamens or the amount of wind pollination depends on the variety, as suggested by the results of Sun (1937) and Olsson and Persson (1958), and on the local conditions including the amount of wind, but it seems that in general insect pollination only slightly increases the yield of *B. napus*, if at all; as at least some wild pollinating insects visit *B. napus* crops, taking honeybee colonies to pollinate them is of dubious economic value. However, rape flowers secrete nectar in abundance (e.g. see Belozerova, 1960; Jablonski, 1961; Demianowicz, 1965) and are most attractive to bees which will visit rape fields $3\frac{1}{2}$ to 4 km away from their hives and neglect fruit trees in favour of rape (Hammer, 1952). Koutensky (1958) and Hammer (1962) showed that the nearer honeybee colonies are to rape crops (maximum distance tested of 1,800 m) the more rape honey they produce, so beekeepers will probably benefit from taking their colonies to rape crops, although rape honey granulates rapidly and should be extracted from the combs within 6 weeks of collection (Palmer, 1959). In the Ukraine honeybees comprise 88% to 97% of for agerson rape, and the rest consist of Andrenidae and Bombycidae species (Radchenko, 1964).

The rape flower has four nectaries (Fig. 64); the two nectaries at the inner

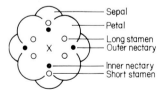

Fig. 64. Diagram of *Brassica napus*, rape, flower showing positions of nectaries after Meyerhoff, 1958).

bases of the short stamens secrete much more nectar than the two nectaries situated outside the ring of stamens (Hasler and Maurizio, 1950). The nectar contains glucose, fructose and ribose (Wanic and Mostowska, 1964). Hammer (1952) recorded that the sugar concentration of the nectar was low in the morning but increased during the day, and that even in cold and windy weather many bees visited rape and returned with 30–50 mg of nectar with 45–60% sugar content. Meyerhoff (1958) found that nectar production was high in the morning and mid-afternoon but low at midday, whereas in contrast Radchenko (1964) reported that both the rate of secretion and the sugar content increased toward the end of the day. Haragsimová-Neprašová (1960) and Petkov (1963a) reported that, on average, a rape flower secreted a total of 3·6 mg nectar of 29% sugar concentration and 0·6 mg nectar of 33% sugar concentration respectively. Maksymiuk (1958) found that in 1956 and 1957 rape flowers yielded means of 9·3 and 11·7 mg nectar of 32 and 39% sugar content and in 1957 flowers from which he removed nectar three times a day, instead of once only, yielded 26·7 mg nectar of 35% sugar content. He suggested that frequent bee visits might stimulate even greater nectar production. Meyerhoff (1958) observed that when plants had not been visited by bees for some hours all nectaries contained nectar, but after bee visits the inner nectaries were empty while the outer had the same amount of nectar as before; 5 min after a bee had visited the inner nectaries they had secreted a small amount of nectar, and 30 min later they were full again. During a single flower visit most honeybees visit both inner nectaries (Free and Nuttall, 1968a) but Meyerhoff (1958) found that when the bee population in a field was high some flowers were visited every 2 or 3 min so the bees must find many empty flowers, and more than half the bees he observed abandoned the flower they were visiting when they found that the first nectary they visited was empty. Although the outer nectaries are as easy for a bee to reach with its proboscis as the inner ones, and probably easier in older flowers whose sepals are wide apart, only one of fifty bees took nectar from the outer nectaries. Perhaps the outer nectaries have nectar with a lower sugar concentration, or less attractive sugar composition, or both.

Free and Nuttall (1968a) watched 187 individual honeybees for as long as possible while they were foraging and found that all of them collected nectar and that none collected pollen on its own (Fig. 65 and 66). Only rarely did a honeybee deliberately scrabble for pollen although many inadvertently became dusted with it as they brushed against the anthers. Some honeybees packed the pollen into their corbiculae whereas others discarded it; individual bees tended to keep constant to one or other type of behaviour. They spent a mean of 4·1 sec/flower. Belozerova (1960) and Petkov (1963a) found honeybees spent 5·9 and 5·0 sec/flower of winter rape and worked at about half the speed or less than *Andrena flavipes* and *A. tibialis*. However, Radchenko

(1964) observed that wild bees spent 12 sec/flower visit compared to 6 sec by honeybees which visited 300 flowers/trip.

Persson (1951) was able to correlate the honeybee population of different rape fields with the number of honeybee colonies in their vicinities. Bees began

Fig. 65. Honeybee collecting nectar and pollen from *Brassica napus*, rape, flower.

Fig. 66. Honeybee collecting nectar from *Brassica napus*, rape, flower.

working summer rape at about 06.30 h and their numbers increased in general with temperature until about 15.00 h when there was a concentration of about 5 bees/m^2; after 17.00 h the numbers decreased rapidly and at 19.00 h none were left. On average about twice as many bees visited summer as winter rape. Free and Nuttall (1968a) found a similar trend in the total bee population visiting a rape crop during the day, but most pollen was collected between 08.00 and 09.00 h, presumably because it was then most abundant. As the stage of flowering advanced fewer bumblebees and honeybees visited the crop but the percentage of honeybees with pollen loads increased; they suggested that this was either because toward the end of flowering the flowers contained less nectar, as found by Belozerova (1960), or because there was increased competition from other crops, especially *Trifolium repens*, but bees that collected pollen in addition to nectar from rape were sufficiently "satisfied" to remain constant to it.

In contrast to *B. napus*, there have been few experiments on the pollination requirements of *B. campestris* although it is generally accepted that this species needs cross-pollination. Canad. Dep. Agric. (1961) reported that the variety "Arlo" is 97% self-sterile. It was also found that tetraploid varieties were inferior to diploid in seed and oil production. When normal diploid and tetraploid strains of *B. campestris* were grown in alternate blocks the yield of the tetraploid strains decreased with the distance from the diploid, although the yield of the diploid was not influenced by proximity to the tetraploid. This was attributed to the greater pollen tube growth rate of the diploid pollen (Canad. Dip. Agric. 1963); these observations provide additional evidence of the role of cross-pollination in seed production. Koutensky (1958) showed that in Czechoslovakia fields of *B. campestris* well provided with honeybee colonies, had seed harvests which were 775, 830 and 820 kg/ha greater than control plots without colonies nearby. Fields with and without bees contained 198 and 229 seeds/g. Although Downey (1968, personal communication) was unable to show that honeybees increased the yield of *B. campestris* he found that when colonies were present the crop matured earlier and more evenly, and so resulted in a better quality harvest. Taking colonies to *B. campestris* crops should therefore be well worth while. It is unfortunate that compared to *B. napus*, for which insect pollination is not of great importance, so few experiments and observations have been made on the pollination by bees and other insects of *B. campestris*.

The pollination requirements of toria (*B. campestris* var. *toria*) and sarson (*B. campestris* var. *sarson*) were investigated by Mohammed (1935). Their floral structure and mechanism resemble those of rape. The stigmas of toria and sarson flowers remained fully receptive for the first two days after opening but, thereafter, the receptivity decreased and when a flower was not pollinated until the fifth day after opening very few seeds were formed. The petals and sepals were shed on the fourth and fifth day after opening but the pollen remained viable for 7 days. With the exception of a yellow-seeded variety of sarson, the anthers of the long stamens dehisce outwards but, in this variety, the anthers remain facing the stigmas in the fully opened flowers and dehisce inwards so pollen from them drops onto the stigmas. In a series of experiments over 6 years, Mohammed isolated plants by enclosing them in muslin bags and found that the average percentage of flowers that set pods were: toria, 12·3; normal, brown-seeded varieties of sarson, 20·3; and the yellow-seeded varieties of sarson, 91·0. A similar, earlier, less-extensive study (Akhter, 1932) had given sets of 0·3, 3·9 and 63·4% respectively. Thus the different structure of the flower of the yellow-seeded variety seems to be associated with autofertilization.

With the exception of the yellow-seeded variety, the pods produced inside the bags had many fewer seeds than on the plants not enclosed. When Mohammed tried self-pollinating the bagged toria flowers each day, it made little difference to the seed set. However, when he cross-pollinated bagged flowers, 100% of them formed pods, compared to 57% for self-pollinated bagged flowers and 37% for bagged flowers not hand pollinated; the mean numbers of seeds per 100 flowers of the different treatments were 1,863, 106 and 66. Cross-pollination gave much larger pods, and the loss of length of the pod formed from self-pollination was mostly confined to the seed-bearing portion. He obtained similar results with sarson; in fact, bagged sarson flowers that were cross-pollinated by hand gave more seed than for plants not enclosed, probably because some self-pollination of the flowers in the open occurs with a subsequent decrease in seed set. Latif et al. (1960) determined the seed yield on plots of sarson and toria that were either (a) caged to exclude insects; (b) caged with *Apis cerana* colonies; (c) left exposed to visits of *A. cerana* from nearby colonies; (d) left exposed but located three miles from *A. cerana* colonies. There were three replicates of each treatment. The average weight (g) of seed produced in two experiments in different years can be seen in Table 4.

TABLE 4

Treatment	Sarson	Toria
a	68	68
b	219	218
c	244	243
d	149	153

They supposed that the 60% greater yield of treatment (c) than (d) would represent the increase that might be expected from taking *A. cerana* colonies to crops of sarson and toria; this figure is, of course, based on the assumption that the growth conditions at the sites of treatments (c) and (d) were identical.

Mohammed (1935) recorded 117 species of insect belonging to seven orders visiting toria but, quantitatively and qualitatively, *Andrena ilerda*, *Apis florea* and *Halictus* spp. were the most important (in this order), and together formed 82% of all insect visitors. In general the same insects that visited toria also visited sarson, except that *Apis florea* was the most abundant visitor to sarson; this difference was probably because sarson flowers 6–8 weeks later than toria when the *A. florea* colonies are larger. Rahman (1940) confirmed these findings; in 1930 and 1931 *Andrena ilerda* comprised 44 and 40% of the visitors to toria and 3 and 5% to sarson, and *Apis florea* comprised 49 and 52% of the visitors to sarson and 16 and 17% to toria. He recorded 105 species of nine orders visiting the two crops; Diptera, especially Syrphidae were the most numerous pollinators present on a crop until 10.00 h and after 16.00 h each day, but between these times Hymenoptera were the most abundant. *Halictus* spp. began foraging at about 09.00 h, *Andrena ilerda* at about 10.00 h, and *Apis florea* at about 11.00 h; they visited about 3·5, 7·5 and 6·0 flowers/min. In an attempt to determine the pollinating efficiency of the more common species, Rahman (1940) enclosed flowering shoots in muslin bags and introduced six insects daily to them. The percentage sets obtained were: *Andrena ilerda*, 97; *Apis florea*, 94; *Halictus* spp., 86; *Eristalis* spp., 79; *Trichometallea pollinosa*, 46; *Sepsis* spp., 19.

Sarson and toria have average sugar concentrations of 49 and 45% (ranges 32–69, 31–64) (Sharma, 1958), and Mohammed reported that in good weather the frequency of insect visits to the flowers was so great that there was no chance of a flower escaping insect pollination and that on average an inflorescence was visited 71 times and a flower 26 times a day. Insect visits started at 10.00 h and reached a peak in numbers between 12.00 and 14.00 h; there was no pollen left in the flowers by 15.00 h and all the insects had deserted the crop by 16.00 h. He noticed that following a long spell of cloudy or rainy weather when few insects visited the flowers, podless gaps appeared on the branches, but in normal good weather there was an abundance of insect pollinators and pollination was no problem. In contrast, Rahman (1945) and Singh (1954) reported that five colonies of *Apis cerana* honeybees per hectare of either crop increased the yield by 30% and 10–25% respectively.

Cultivated species of *Brassica* other than rape are also very attractive to insects, especially honeybees, which obtain good honey crops (e.g. Pellet, 1923 and Ermakova, 1959) from the abundant nectar that is secreted; mustard and rape are recognized as serious competitors to *Trifolium pratense* for bee visits (Hammer, 1949). Flies (Syrphidae, Calliphorinae, Muscidae) and small beetles are occasional visitors and, in California, Pearson (1932b) suggested that solitary bees of the families Nomadidae, Megachilidae and Andrenidae were more important in cabbage seed production than honeybees because they worked at temperatures below 15°C (59°F) but bumblebees were not plentiful although a few collected pollen.

Pearson (1932a) reported that the inner nectaries of *B. oleracea* secrete 0·1 ml each day for 3 days, and Butler (1945a) found its nectar had an average sugar content of 39%. It has also been reported that a *B. alba* flower produces between 0·2 and 0·6 mg of nectar, whose sugar concentration may reach 60% (Ermakova, 1959; Haragsimová-Neprašová, 1960) and that *B. juncea* nectar has an average of 52% sugar content (range 22–65%) (Sharma, 1958). It seems to be general that the inner pair of nectaries secrete the most; bees can reach the outer nectaries without touching the stigma. An excellent description of the way bees visit the flowers was given by Howard *et al.* (1916a); when a bee lands on a *B. juncea* flower it pushes its tongue between the long and short stamens to reach the nectary on the side nearest to it and, in doing so, its head touches the stigma and the anthers of the long stamens and its thorax touches the anthers of the short stamens (Fig. 67a); it then passes over the top of the flower to reach the other inner nectary and, while pushing down between the short and long stamens, it touches the stigma with its pollen-covered thorax (Fig. 67b).

(a) (b)

Fig. 67. Behaviour of bees visiting *Brassica juncea* flowers (after Howard *et al.*, 1916a).

It seems true for most, if not all, species that the filaments of the long stamens curve so their anthers dehisce outwards and those of the short stamens curve so their anthers dehisce inwards, although some slight variations may occur. Knuth (1906) reported that when cabbage flowers are not visited by insects the upper part of their long stamens curve downward so their anthers touch the stigmas thus effecting auto-pollination.

Brassica oleracea L.

The flowers of *B. oleracea* are both self- and cross-pollinated, but cross-pollination is greatly favoured. Indeed, Darwin (1876) used different varieties of *B. oleracea* to exemplify this phenomenon; he planted two varieties of kohl-rabi, a broccoli, a Brussels sprout and a cabbage near together and found that about 83% of the kohl-rabi seeds were hybrids. Two varieties of cabbage planted together gave 49% hybrids. He pointed out that the large degree of cross-pollination occurred although pollen must have been carried much more often from flower to flower on the same plant than to different plants.

His observations have been confirmed by numerous workers. Roemer (1916) applied a mixture of equal amounts of self- and cross-pollen to emasculated cabbage flowers and found that 79% of the progeny were hybrids. A similar test by Pearson (1932b) gave 94% hybrids. Apparently, pollen tubes of self-pollen grow slowly (Kakizaki, 1930) and if 5 days or more is allowed for pollen tube growth self-fertilization can occur (Pearson, 1932b). The stigma is normally receptive for about 5 days after the flower opens but also for 4 days beforehand; plant breeders find that when flowers are self-pollinated in the bud stage the pollen tubes have sufficient time to reach the ovules before the styles begin to wither.

Although some selfing occurs in the absence of cross-pollination fewer ovules are fertilized. This also was first demonstrated by Darwin (1876). A cabbage plant he self-pollinated produced an average of 4·1 seeds/capsule compared to 16·3 seeds/capsule of a cross-pollinated plant. Darwin noticed that the seeds on self-pollinated plants were slightly heavier and attributed this to the extra nourishment available to them, but Sakharov (1958) claimed that the weight of 1,000 seeds produced in the absence of bee visits was 1·1 g, in the presence of a few bee visits was 2·2 g and with ample bee visits was 3·6 g. Radchenko (1966) reported that pollination increased the seed crop of cabbage by 300%. He found that honeybees comprised 85–100% of cabbage pollinators, and visited six cabbage flowers per min.

Pearson (1932b) found that some strains of broccoli were nearly or completely incompatible, and, although some were fairly self-compatible isolated plants in cages with honeybee colonies usually gave a poor seed set, but when inter-compatible plants were caged together with bees a good set was always obtained. An experiment by Jenkinson and Glynne-Jones (1953) was rather unsatisfactory because most of the plots developed bacterial rot before flowering began, but they stated that six broccoli plants caged with bees had about twice as many pods per unit length of stem as six plants caged without bees. The range of the number of seeds per pod was the same for each treatment, but the average number was much greater when bees were present (Fig. 68). These workers drew a current of air over broccoli flowers and found it contained some pollen, but they considered that the distribution of pollen by wind would be slight. Honeybees were the most common useful visitors and sometimes a bee obtained a load from a single plant; various beetles

(*Meligethes* spp.) were also common but they only moved from plant to plant in still air.

Recent work by Watts (1963) has done much to clarify the pollination requirements of cauliflower and explain the previously conflicting views about its self-compatibility. All the varieties of cauliflower he tested set some seed by autopollination, but the set was about 50% greater when the flowers were

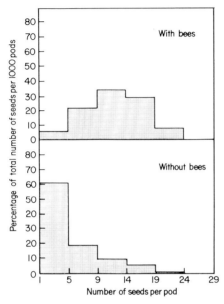

Fig. 68. Effect of presence or absence of bees on number of seeds per pod of *Brassica oleracea*, broccoli (after Jenkinson and Glynne-Jones, 1953).

self-pollinated by brush than when they were vigorously shaken; this indicates that insect pollination is necessary to ensure the maximum set possible by self-pollination. Cauliflower varieties in Britain belong to three main types—summer, autumn and winter. Although the different types set similar amounts of seed following artificial self-pollination, the tendency for autopollination to occur was greater in the summer and autumn than in the winter types. The different types also differed in their tendency to favour fertilization with foreign pollen as when pollinated with a mixture of self-pollen and Brussels sprout pollen the percentage hybrids in the progenies produced from summer, autumn and winter cauliflowers was 18, 57 and 72%. The average seed set per pod from mixed and self-pollination is given in Table 5.

The summer types were significantly more self-fertile than cross-fertile and the winter types were significantly more cross-fertile than self-fertile. The more thoroughly a plant was pollinated, i.e. the more pods that set, the fewer the seeds per pod suggesting that the bearing limit of the plant had been reached and that other factors, especially nutritional ones, may also influence the seed crop.

TABLE 5

	Summer types	Autumn types	Winter types
Mixed pollination	4·9	7·5	8·8
Self pollination	6·1	6·8	6·2

Haigh (1953) reported that Brussels sprouts are normally self-incompatible although a small amount of self-compatibility commonly occurs.

Brassica alba L.

By using mutants isolated from X-ray treated material as markers Olsson (1960) showed that an average of 99·6% cross-fertilization occurred in a field of *B. alba*, white mustard, and 31% crossing occurred between small plots 10 m long and 4 rows wide, and 45% between plots 2 m long and 3 rows wide. When 100 plants were isolated so only self-pollination could occur a total of only twenty seeds were obtained; the progeny from such seeds were smaller and shorter than their parents. In 2 years, Free and Spencer-Booth (1963) compared the seed production of four *B. alba* plots caged with bees, four plots caged without bees and four open plots. The plots caged with bees produced about twice as many seeds as the plants from which bees were excluded; they concluded that taking honeybee colonies to crops of *B. alba* might produce worth while increases in seed yield.

Brassica nigra L.

Akhter (1932) found that only 11% of bagged *B. nigra*, black mustard, flowers set seed, and Olsson (1960) reported that his preliminary observations on *B. nigra* had given results similar to those for *B. alba*.

Brassica juncea L.

Howard *et al.* (1916a) in India grew five varieties of *B. juncea*, frowse, in separate adjacent rows but found that only 24% cross-fertilization occurred between them. Furthermore, seed set under bags and nets without difficulty and when no insects were allowed to visit the flowers the anthers of the long stamens bent toward the stigmas so that any slight shaking of the flowers resulted in self-pollination. *B. juncea* plots caged without bees produced nearly the same amount of seed as those caged with bees, or not caged (Free and Spencer-Booth, 1963b), so there is no evidence that taking honeybee colonies to crops of *B. juncea* would increase the seed yield.

Other Brassica species

I can find little relevant information about other cultivated species of *Brassica*. According to Jones and Rosa (1928) *B. rapa*, turnip, sets less seed when self- than when cross-pollinated; Tedin (1931) found 18% contamination between adjacent plants which decreased to 1% for plants 25 m apart. Further increase in distance had little effect. Kakiziki (1922) (quoted by Jones and Rosa, 1928) found that only 5% of bagged *B. pekinensis* flowers set seed; when plants were self-pollinated they varied from being completely self-sterile to completely self-fertile but when they were cross-pollinated 96% of the flowers set seed. Cobley (1956) reported that *B. eruca* L., rocket cress, is self sterile.

Raphanus sativus L.

Most commercial varieties of *R. sativus*, radish, are self-incompatible (e.g. Akhter, 1932; Crane and Mather, 1943; Bateman, 1947a, 1947b), but self-fertilization of some varieties can be secured when the flowers are self-pollinated in the bud stage (Nilsson, 1927; Kakizaki and Kasai, 1933). Darwin (1876) found that covering a plant with a net halved the number of seeds per pod, and Radchenko (1966) reported that insect pollination increased the seed crop by 22%. The stamens do not rotate but bend away from the stigma when the flower opens although, when the flower fades, the anthers of the long stamens touch the stigmas (Knuth, 1906) any self pollination that occurs then is probably too late to effect self-fertilization.

Kremer (1945) reported that in Michigan, radish crops within 3·2 km of honeybee colonies yielded about 448–504 kg/ha, and one crop as much as 673 kg/ha whereas crops 3·2 km or more away yielded 224–336 kg/ha. Radchenko (1966) observed that in the Ukraine honeybees comprised 77–94% of the pollinators of radish, and visited eleven flowers per minute. The average amount of sugar secreted per flower was only 0·68–0·98 mg. Because bees sometimes neglect radish for other crops which secrete nectar more abundantly, Kremer suggested that either radish crops should be grown near permanent apiaries or colonies should be moved to them during flowering.

Cochlearia armoracia L.

According to Knuth (1906) nectar secretion from *C. armoracia*, horse radish, is also scanty; although the anthers of this species dehisce toward the stigma, the flowers are protogynous and self-pollination is almost entirely ineffective.

Eruca sativa Lam.

In India (Howard *et al.*, 1919) and in Poland (Jablónski, 1961) the flowers of *E. sativa* (rocket cress, taramira, or duan) open in the morning and remain open for about 3 days. Soon after a flower opens the anthers dehisce toward the style. Flowers have either long or short styles. Jablónski reported that nectar is not available until a flower has opened sufficiently for bees to reach it through gaps between the petals, but newly opened flowers are visited and pollinated by pollen-gathering honeybees. When honeybee colonies were taken to crops they produced surpluses of 25–117 kg each.

Howard *et al.* found that no flowers that were kept bagged produced seed, and only 7·1% of flowers bagged and self-pollinated produced seed compared to 98·2% of those that were cross-pollinated.

Because available evidence indicates that cross-pollination of most Cruciferae species and varieties is either necessary or gives more seed, and that when a variety is to some extent self-compatible little autopollination occurs, it can be concluded that, in general, insect pollination is essential to produce good crops.

Chapter 9

Malvaceae

Economically important species of this family are included in the genera *Gossypium* and *Hibiscus*.

Gossypium

The four most commonly cultivated species of cotton are *Gossypium hirsutum* L., (short staple or upland cotton); *G. barbadense* L., long staple or Egyptian cotton; *G. herbaceum* L., Asiatic cotton, and *G. arboreum* L., tree cotton. *G. hirsutum* and *G. barbadense* are the most important species, especially the former, and most of the experiments and observations described refer to *G. hirsutum*.

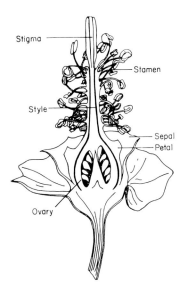

Fig. 69. Diagrammatic section of *Gossypium hirsutum*, cotton, flower (after Robbins 1931, reproduced with permission of McGraw-Hill Book Company).

The cotton flower (Fig. 69) is borne singly and usually there are not more than three open per plant per day. It is protected in the bud stage by an epicalyx of three or four large bracts which persist when the flower opens and the much reduced sepals are united to form a tube of five lobes. There are five large showy petals which are slightly united at their narrow bases; their colour varies with the species and variety concerned. They also open more widely in some varieties than others. Most varieties of *G. hirsutum* have flowers that open more widely than varieties of *G. barbadense*. The 100–150 stamens are united by their filaments to form a thick ridged tube which surrounds the style and is united to the base of the petals. The single celled anthers arise on short branches from the staminal column. The anthers dehisce just before (*G. barbadense*), or soon after (*G. hirsutum*), the flower opens and the pollen grains remain viable for about 12 h. The superior ovary has three to five united carpels each containing several ovules. The long style protrudes through the top of the staminal tube where it divides into three to five stigmas according to the number of carpels (see Cobley, 1956).

Pollination mechanism

The flower opens at dawn and withers by the evening of the same day. Thus, in Arizona, the corolla starts to expand at 06.30 h and has opened by 08.00–09.00 h; the flower begins to wilt in mid-afternoon and closes by sunset, never to reopen. The viability of the pollen decreases gradually from about 09.00 h but Kearney (1923a) found it was still able to fertilize 86% of the flowers to which it was applied at 17.00 h on the day of anthesis but only 38% when applied at 08.00 h the next morning. He also found that the stigmas were receptive at about the same time that the pollen was shed but by next morning only 8% of them remained so. Fertilization occurs about 30 h after pollination.

Kearney (1923a) noticed that in the fully open flower of "Pima" cotton (*G. barbadense*) the stigma projects 10 mm above the top of the staminal sheath and the highest stamens project 2·5 mm above it, so the lower quarter of the stigma is surrounded by anthers and is probably effectively screened by them from cross-pollination. When flowers were bagged so only selfing could occur, pollen was not deposited more than 2 mm above the anthers, whereas the stigmas of flowers in the open soon became covered with pollen over their entire lengths when insects were abundant. When the part of the stigma which projected above the anthers was excised no cross-pollination occurred; experiments demonstrated that in normal flowers the apical part of the stigma is a more favourable location for pollen germination and growth so that when self- and cross-pollination occur at about the same time the foreign pollen often reaches the ovules first; probably the amount of cross-fertilization partly depends on how soon the foreign pollen reaches the stigma. However, other experiments indicated that when foreign and self-pollen were applied simultaneously to emasculated flowers the self-pollen is favoured, so the matter is far from resolved. Kearney discovered that the structure of "sea island" cotton (*G. barbadense*) is similar except that the anthers do not form such a dense girdle, so that the lower part of the stigma is more accessible to foreign pollen. In contrast, most varieties of *G. hirsutum* have shorter stigmas and longer filaments, so all or most of the protruding stigma is surrounded by stamens whose erect position brings the anthers into contact with the stigma; presumably cross-pollination is more difficult with this species until the stamens later bend away from the stigma and leave the whole of it accessible to foreign pollen.

Nectar

Cotton has five sets of nectaries, one floral and four extrafloral. The floral nectary consists of a ring of closely aggregated secretory hairs on the inner side of the calyx. The five petals overlap except at their bases, where there are five small openings through which insects with long tongues can reach the nectar, although honeybees cannot readily do so until the nectar droplets coalesce to fill the calyx cup. These openings are usually guarded by interlacing hairs which exclude small insects until the petals begin to wither; moreover most small insects are unable to crawl between the closely fitting calyx and corolla to reach the floral nectar (see Tyler, 1908; Vansell, 1944a; Brown and Ware, 1958). The extrafloral nectaries consist of: (a) three irregular triangular shaped nectaries on the outside of the calyx near its base (inner involucral nectaries); (b) three nectaries on the pedicel of each flower just below each epicalyx bract (outer involucral nectaries); (c) nectaries occurring singly on the main veins on the underside of each foliage leaf, which vary in number from one to five per leaf; (d) minute unipapillate nectaries on the flower peduncles and young leaf petioles. Different leaf nectaries secrete at different times; Mound (1962) studied the behaviour of twenty leaf nectaries throughout the day and found that four secreted between 06.30 and 11.00 h, eight between 11.30 and 16.30 h, and eight between 16.30 and 06.00 h. Like the unipapillate nectaries they secreted most actively during the main growing season of the crop.

The amount and quality of floral nectar secreted varies greatly with different varieties; thus Vansell (1944a) found that the amounts and sugar concentrations of floral nectar produced per flower of three varieties of *G. barbadense* were 0–8 μl at 34%, 30–50 μl at 26% and 12 μl at 24%, and Min'kov (1956) found that one variety gave twice the amount of nectar of another of the same sugar concentration. The different concentrations may be partly related to differences in the amount of evaporation possible from flowers of different varieties.

Nectar secretion is favoured in calm clear weather with temperatures between 25 and 35°C (77 and 95°F) and ample soil moisture (e.g. see Kaziev, 1959). Although varietal differences again exist, in general, all the different nectaries of a plant produce nectar of about 20% sugar concentration but evaporation is much greater from the exposed extrafloral than from the floral nectaries, giving a more rapid increase in sugar concentration (e.g. Vansell,

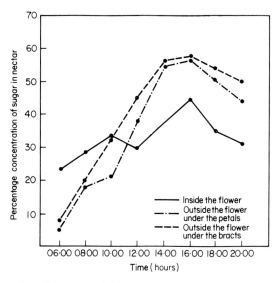

Fig. 70. Concentration of nectar at different times of day (after Glushkov and Skrebtsov, 1960).

1944a; Mound, 1962) which reaches a maximum of 60–82% compared to 20–54% of the floral nectar (e.g. Parks, 1921; Grout, 1955; Glushkov and Skrebtsov, 1965). Furthermore, whereas the floral nectaries secrete only on the one day the flower is open, the extrafloral secrete for several days, and any nectar that remains from one day to another has more time to become concentrated. Consequently honeybees do not visit the floral nectaries until the extrafloral have been emptied and it has been suggested that increasing the bee population by concentrating colonies in the area will help to achieve this (Grout, 1955; McGregor, 1959). Nectar removed by insects is partly compensated for by additional secretion and whereas the nectar concentration reaches its maximum between 14.00 and 16.00 h, it diminishes in the evening (Fig. 70) (Glushkov and Skrebtsov, 1965).

Pollinating agents

Allard (1910) believed that wind pollination of cotton might be important. He noticed much pollen on the surfaces of cotton leaves in the late afternoon and when he put pieces of glass covered with vaseline in a cotton field plenty of pollen collected on them. However, the period during which pollen becomes dry and windborne follows some hours after anthesis and such pollen may well be infertile; furthermore, later workers have been unable to corroborate his findings. For example, Afzal and Khan (1950) exposed forty sticky glass slides in a cotton field for 52 days and found thirteen of them each had one or two pollen grains on one day, but none on the other 51 days; Thies (1953) exposed six sticky slides for 30 min periods in and around cotton fields for three successive days but they caught a total of only six pollen grains, and none before 11.00 h; no pollen collected on the sticky slides Sidhu and Singh (1961) exposed. When cotton flowers were emasculated and exposed to wind, but not to insects, very few formed bolls compared with flowers not emasculated, or with emasculated flowers visited by bees (Afzal and Khan, 1950; Sidhu and Singh, 1961).

Cotton is visited by many different kinds of insects, especially in India and Pakistan. Thus, Sidhu and Singh (1961) observed forty-one species of twenty-three families of seven orders visiting cotton. However, many of the insect visitors did not visit the floral nectaries and many of those that did so failed to touch the flowers' sexual parts. Similarly, although hummingbirds were regular and persistent visitors to cotton fields in Georgia, U.S.A. (Allard, 1910), they thrust their long slender beaks between the epicalyx and corolla to obtain nectar and so did not contribute to pollination. The most important pollinators in various parts of the world are *Apis mellifera*, *A. dorsata*, *A. florea*, *A. cerana*, *Melissodes* spp., *Halictus* spp., *Bombus* spp., *Anthophora confusa*, *Elis thoracica* and *Scolia* spp.

In North America bumblebees are regarded as the most efficient pollinators of cotton (see Finkner, 1954). They rarely visit the extrafloral nectaries (Allard, 1910) and, because of their large size, they usually touch both the stamens and the stigmas when they enter flowers. Thies (1953) reported that, in Oklahoma, bumblebees of the species *B. americanorum*, *B. auricomus* and *B. fraternus* were the only insects that entered the cotton and they did so as soon as the flowers opened. During a single trip a bumblebee worked methodically from flower to flower in the immediate vicinity of the first flower it visited; Fig. 71 illustrates the behaviour of a *B. americanorum* worker that visited 193 flowers on 166 plants in 31 min. The average foraging statistics recorded for the three species can be seen in Table 6.

If it is assumed that, on average, bees were in the middle of their trips when observation of them began, the number of flowers visited during a complete trip should be about twice the observed number; this agrees quite well with the maximum number of flower visits recorded per observation. The peak period of bumblebee activity occurred between 09.00 and 10.30 h and they were so numerous that flowers received an average of forty-five visits each per day; the flowers that occurred on solid stands of plants and on the upper parts of stalks were visited more frequently than those on sparsely spaced plants and on lower branches.

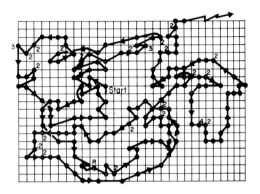

Fig. 71. Flight path of a bumblebee (*Bombus americanorum*) visiting 193 flowers on 166 cotton plants in 31 minutes. Horizontal lines represent cotton rows, and intersections with vertical lines represent plants (after Thies, 1953).

TABLE 6

	B. americanorum	*B. auricomus*	*B. fraternus*
No. flowers visited	111	74	62
Maximum no. flowers visited	193	146	156
No. flowers visited per minute	5·1	3·1	4·3

Melissodes spp. are also important pollinators of cotton in North America; they always visit the floral nectaries and are sometimes the predominant insects doing so. They are among the first visitors to squeeze into the flowers when they start to open at dawn. The carpenter bee, *Xylocopa* spp. also visits floral and extrafloral nectaries in Egypt and North America (e.g. Allard, 1910; Kearney, 1923b; McGregor *et al.*, 1955; Butler *et al.*, 1960).

Large wasps (e.g. *Elis plumipes*, Allard, 1910; *Campsomeris* spp., Kearney, 1923b and *Scolia* spp., Wafa and Ibrahim, 1959a) also frequent cotton fields in parts of North America and Egypt and always visit the floral nectaries and become covered with pollen. In India, Sidhu and Singh (1961) observed that an average of forty-seven insects visited each cotton flower per day.

The honeybees, *Apis dorsata, A. cerana* and *A. florea* were important pollinators and visited more flowers per minute and per trip than the wasps, *Scolia avreipennis* and *Elis thoracica*, but the abundance of the wasps and their greater tendency to visit the flowers in inclement weather more than compensated for this. The rate at which the principal pollinators visited the flowers can be seen in Table 7.

TABLE 7

Honeybees	No. flowers visited	
	Per minute	Per trip
Apis cerana	6·6	128
Apis dorsata	6·7	141
Apis florea	5·8	120
Various species of solitary bees	6·9–8·2	129–155
Wasps: *Scolia avreipennis*	3·4	91
Elis thoracica	3·6	76

Behaviour of honeybees

Although other insects may be more efficient pollinators, because of its abundance the honeybee, *Apis mellifera*, is the most important pollinator of cotton in many parts of the world and has been the most studied. Honeybees work cotton from 07.00–20.00 h daily but are most plentiful near midday (Kaziev, 1956, U.S.S.R.; Wafa and Ibrahim, 1957a, Egypt) when the amount and concentration of nectar is greatest, and it is generally supposed that they perform little pollination in the afternoon. Most honeybees prefer to visit the extrafloral nectaries which have the most concentrated nectar and as few as 6% visit the flowers (Min'kov, 1956). Extrafloral nectar is especially preferred in the early morning. Even when a honeybee does collect floral nectar it may do so by forcing its tongue down between the calyx and corolla (Wafa and Ibrahim, 1959a). A honeybee that enters a fully-opened flower usually crawls down the inside of the petals and either does not touch the stigma or stamens, or touches the stamens only and gets pollen on its back. Only occasionally, as when disturbed by another insect, does it touch both the stamens and stigmas of a fully-open flower. Probably most honeybee pollination is done in the early hours of the day when the bees have to squeeze past the stigmas of partly open flowers (Allard, 1910; Parks, 1921).

Parks (1921) stated that 1 ha of cotton produced about 31 kg of honey, and Min'kov (1957) and Avetisyan (1958) recorded that a hectare produced 19–36 kg and 15–20 kg honey respectively; Grout (1955) and Glushkov and

Skrebtsov (1965) found that in favourable circumstances colonies collected averages of 43 kg, and 10 kg of honey from cotton. Indeed, Kuliev (1958) who considered that cotton was one of the best nectar-producing crops in the U.S.S.R. stated that colonies collected between 0·8 and 2·5 kg/day. McGregor and Todd (1956) reported that *G. barbadense* gave excellent honey crops in Arizona and produced nectar at a time when few other plants did so. But whereas cotton provides a good honey crop on some occasions and in some localities, it does not always do so, and even when colonies are put beside cotton fields, sometimes only a few bees visit the flowers and the number that does so fluctuates at different times in the flowering season. No doubt this partly depends upon cultural conditions, especially water supply, and the species and variety concerned (see McGregor and Todd, 1956).

Although honeybees sometimes collect abundant nectar from cotton it is not a good source of pollen. Kaziev (1956) observed that only 15–25% of honeybees working cotton had pollen loads. Bees that collect pollen probably do so incidentally as they collect flower nectar, as it does not seem to be attractive enough for them to collect deliberately, although solitary bees do so. It would be interesting to know whether the same bees visit the floral and extrafloral nectaries at different times, and whether individuals collect pollen on some occasions but not on others. Because honeybees seem reluctant to collect cotton pollen, their colonies sometimes become very weak (e.g. see Grout, 1955; McGregor, 1959), and the possibility of using pollen substitutes to maintain colony strength needs investigating.

Benefits from insect pollination

Meade (1918) reported that in 1911 in California, and in 1912 in Texas, 22–52% and 35–82% of cotton flowers failed to develop into bolls. He thought that inadequate pollination, especially of varieties with long stigmas, might be the cause of this and found that hand-pollinating short stigma and long stigma varieties gave 5 and 11% greater set than natural pollination. He further reported that a minimum of twenty-five pollen grains is needed per stigma and that even then only one or two seeds mature in each carpel. Kearney (1923a) found that 10% fewer flowers enclosed in paper bags than flowers not enclosed produced bolls, and furthermore, the bolls produced from the bagged flowers contained fewer seeds; artificial pollination increased the set by 8% in an area where insects were scarce but failed to do so where they were abundant. Stanford (1934) stated that more than half the buds, flowers and immature bolls of cotton are shed before ripening. However, despite the evidence of inadequate pollination, at the time pollination by honeybees was regarded as beneficial in some areas only, but detrimental in other areas because of contamination of pedigree seed. Min'kov's (1953) results tended to confirm the earlier ones; in two experiments bagged flowers had 5·4 and 0·7% less set than controls.

Therefore, it seemed that although cotton is largely automatically self-pollinating and self-fertile insect pollination could increase the yield; experiments have been done to test this by caging plots of cotton and putting honeybee colonies in some cages but not in others. Thus, during each of 3 years in Arizona, McGregor and Todd (1956) found that *G. hirsutum* plants caged with bees gave 5–20% more cotton from the first picking than plants caged without bees and this increase resulted both from more bolls and more cotton per boll; however, the total amount of cotton produced per season was similar for plants caged with and without bees. Hence, honeybee pollination was primarily of value in producing larger earlier crops, and it was suggested that where the flowering season is of shorter duration bee pollination might also increase the total crop produced. The effect of honeybee pollination was more pronounced for *G. barbadense*, and the total crop of plants caged with bees was 16–24% greater than when bees were excluded.

There is now abundant evidence showing the value of bees in pollinating cotton. In the U.S.S.R. Shishikin (1952) stated that plots caged with bees set 41% more bolls and produced 26–43% more cotton than plots caged without bees. In two experiments Gubin and Verdieva (1956) discovered that bee pollination increased the number of seeds per boll by 5 and 6%, the weight per boll by 9 and 14%, and the amount of raw cotton per boll by 35 and 40%. Similarly, Kaziev (1960, 1961) found that bees increased the average amount of raw cotton produced by three varieties by 24%, the number of seeds per boll by 5%, the weight per boll by 12%, the weight of raw cotton per boll by 9%, the length of fibre by 2% and its strength by 6%; moreover, the bolls produced by bee pollination ripened more uniformly and 5–9 days earlier than the controls. Trushkin (1960) reported that bees increased the weight of cotton per plant by 13%.

In India, Mahadevan and Chandy (1959) caged about 350 plants to exclude insects, and found they produced 54% less weight of cotton in one test and 23% in another than plants not caged. Sidhu and Singh (1962) caged plots with either *Apis cerana* or *A. florea* or without bees. There were four replicates of each treatment. The results are given in Table 8.

The plots with bees yielded an average of 18% more cotton by weight than the plots caged to exclude insects, and the two species were equally effective. However, only *A. cerana* can be used for cotton pollination because *A. florea* cannot be kept in hives. It was suggested that apiaries of *A. cerana* should be established in large cotton growing areas to discover whether they increased the crops, but it was pointed out that because few other nectar-producing species are present it would be necessary to feed the colonies when the cotton is not in flower.

In Egypt, Wafa and Ibrahim (1960b) used cloth cages which they found seriously affected the plants by stimulating vegetative growth; nevertheless, plants caged with bees gave 15% more mature bolls per plant and 8% more cotton per boll than uncaged plants, and 31% and 15% more than plants caged without bees. They and other workers observed that insect-pollinated flowers wither at midday, or soon after, on the day of flowering, but flowers not visited by insects do not wither until the end of the day of opening.

TABLE 8

	Caged plots			Open plots
	With A. cerana	With A. florea	No insects	
% flowers set bolls	46	45	33	43
% bolls that matured	26	26	23	25

Insect visits to cotton flowers not only increase the amount of pollen on the stigmas, but also especially increase the amount of foreign pollen present. It is not apparent which of these factors is most concerned with producing greater yields. However, there is evidence that cross-pollination between different varieties gives better yields than cross-pollination within a variety. McGregor et al. (1955) caged a variety of G. barbadense (Pima S.I) with other varieties of G. barbadense and G. hirsutum. Honeybee colonies were put in some cages but not others. When bees were absent there was a negligible amount of hybridization but, when bees were present, considerable hybridization occurred and 24% more seed cotton was produced than in cages without bees. This increase in yield was caused by increases in the weight of bolls, seed and lint, and in the number of seeds per boll, together with a decrease in the number of bolls shed. Avetisyan (1958) claimed increases of 23 and 26% from intravarietal pollination and Skrebtsov (1964) reported that increased yields of between 6 and 13% were obtained from intervarietal cross-pollination but 33% from intravarietal pollination.

All the above results indicated that honeybee pollination inside cages increases both the quality and quantity of cotton produced. The pollinating value of honeybees when they are not restricted to cages was demonstrated by McGregor (1959). He put 200 honeybee colonies on the borders of a 16 ha field of Pima cotton (G. barbadense) and at weekly intervals during flowering he counted in designated plots throughout the field: (a) the number of flowers open; (b) the number of bees working them; (c) the number of stigmas that were well coated with pollen above the anthers. Very few wild bees were present and there was no correlation between their numbers and the amount of pollination that occurred, but in each of two years McGregor obtained a highly positive correlation between the number of honeybees and the number of pollen coated stigmas, and in many weeks more than enough bees were present to produce a maximum set (Fig. 72).

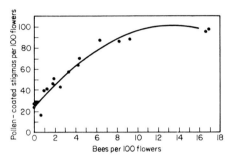

Fig. 72. Relationship between no. honeybees per 100 cotton flowers and no. stigmas coated with pollen (after McGregor, 1959).

Concentration of honeybee colonies needed

Because honeybees prefer to work the extrafloral rather than the floral nectaries, it is sometimes necessary to saturate an area with honeybee colonies to ensure that the floral nectaries are visited. In the U.S.S.R. Shishikin (1946) reported that fields saturated with honeybees had a 19% greater yield than fields with only local pollinators, and 43% more than plots caged to exclude insects, and Glushkov and Skrebtsov (1960) claimed that fields with 4·9 and 6·6 colonies per hectare produced 21 and 45% larger crops than a control field without bees. However, 1 colony/2 ha (Avetisyan, 1958) or 1 colony/ha (Radoev and Bozhinov, 1961) are sometimes regarded as sufficient.

McGregor (1959), who found that ten or more honeybees per 100 flowers was indicative of a good set in Arizona, suggested that the bee : flower ratio should be used as an indicator of the number of colonies required. By means of this technique any bees working the extrafloral nectaries, which are unlikely to pollinate the crop, would be ignored. He pointed out that the number of colonies needed to maintain the desired rate of flower visits may vary according to the number of hectares of cotton involved and, whereas even as many as 25 colonies/ha may be insufficient for small fields surrounded by competing plants, 2½ colonies/ha may be enough for fields of several hundred hectares. Artificial pollination would be a useful tool to determine whether the set in a particular area is adequate, or whether additional pollinating insects are needed. Sidhu and Singh (1961) noted that in India *A. cerana* ceased foraging in inclement weather and suggested putting colonies of this bee close to cotton fields so they could take full advantage of short spells of good weather.

Cross-pollination

The amount of cross-pollination of cotton and the distance over which it occurs is of interest both from the point of view of preventing intravarietal contamination and, more recently, of producing hybrid cotton. In order to determine the amount of cross-pollination that occurs, when conditions are most favourable for it, genetically marked plants of two different varieties have been planted: (a) in alternate rows; (b) as alternate plants in the same row; (c) as isolated plants of one variety spaced well apart in a block of another variety. Samples of typical results are given below.

Planting in alternate rows has been the most common method employed; the amount of cross-pollination has varied from 2–6% in India (e.g. Kottur, 1930; Afzal and Khan, 1950), 8% in China (Yu and Hsieh, 1937), 5–35% in Egypt (Balls, 1912), and 1–40% in U.S.A. (Allard, 1910; Kearney, 1923a; Ware, 1927; Simpson and Duncan, 1956). When varieties have been planted alternately in the same row, the percentage of cross-pollination was 2% in India (Afzal and Khan, 1950) and 2–19% in U.S.A. (Stroman and Mahoney, 1925; Brown, 1927). Most of these percentages must be underestimates as they do not include cross-pollination between plants of the same variety. When isolated plants have been grown in a block of another variety they have received from 10 to 48% cross-pollination (e.g. Kearney, 1923a; Simpson, 1954).

The amount of cross-pollination obviously depends on the abundance and concentration of pollinating insects, and Simpson (1954), who grew isolated plants of upland green leaf cotton at 3 m intervals in fields of red cotton in forty-eight locations, in twelve States during 4 years, found that cross-fertilization was low in areas of intensive cultivation, and high in small fields, presumably because the ratio of insect pollinators to cotton flowers was lower in the former circumstances. It is notable that the amount of crossing in India is much less than in N. America; this must be related to differences in the numbers and kinds of insects visiting the flowers.

Isolation distances

Indications of the isolation distances that are necessary between seed crops of different strains and varieties have been obtained by either: (a) growing a batch of genetically marked plants of one variety in the centre of the block of another and recording the amount of cross-pollination at different directions and distances from it; or (b) finding the amount of contamination that occurs at different distances from where two adjacent varieties join; or (c) finding the amount of cross-pollination between blocks of different varieties separated by different distances.

Trought (1930) in Egypt grew a clump of genetically marked red leaf cotton in the centre of a field of a green leaf variety; he marked out sixteen radii from the field centre corresponding to the sixteen main compass directions and sampled plants at different distances along the radii to find the amount of crossing that had occurred. On average it decreased from 0·9% at 2 m from the red leaf variety to 0·1% 22 m away (Fig. 73). A similar experiment was

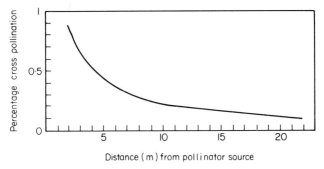

Fig. 73. The percentage of cross-pollination of cotton at different distances from a pollinator source (after Trought, 1930).

done by Afzal and Khan (1950) in Pakistan; they grew a patch of marker plants (5·3 m²) in the centre of a square field and found that most contamination occurred within 3·8 m of the marker plants, and pollination was very sporadic beyond this distance and none occurred beyond 30·5 m.

The first experiments to find the amount of crossing between two adjacent varieties were done by Brown (1927) in Mississippi. The average percentage cross-pollination at different distances from where the red leaf and green leaf varieties met can be seen in Table 9.

TABLE 9

	No. rows from where varieties met									
	1	2	3	4	5	10	20	40	80	119
Red leaf variety	4·3	0·7	0·4	0·5	0·8	0	—	—	—	—
Green leaf variety	14·8	6·5	6·9	4·0	3·2	1·9	0·5	0·2	0	0

Similar decreases with distance have been obtained by Balls (1929).

Fryxell (1956) investigated the crossing that occurred between two varieties planted in adjacent blocks in a field 192 m wide. Where the two varieties adjoined there was the same amount of crossing in the middle of the field as 46 m from either edge but, whereas crossing rapidly diminished with increasing distance separating the varieties along the centre line of the field, the dimunition was much less nearer the field's edge, and even where fifty rows separated the two varieties, 10 to 14% crossing still occurred. This must reflect a smaller population of pollinating insects near the middle of the field than near its periphery.

Attempts have been made to determine the distribution of pollen within a crop by using a technique involving methylene blue dye. At the beginning of the day anthers of selected flowers are dusted with methylene blue and the next day other withered flowers are re-opened and examined for its presence. The moisture from the flower causes the dye to stain the parts it contacts, and from the position of the stain it is possible to surmise whether the insect that transferred the methylene blue was collecting nectar or pollen. Thies (1953) dusted with methylene blue twenty-one flowers in the first row of a ten row block containing 195 open flowers; 30 min later 42% of the open flowers had received methylene blue and 2 h later every flower had. Others (e.g. Stephens and Finker, 1953; Simpson, 1954; Buffet, 1960) found that the dye spread to nearly all the flowers in the immediate vicinity of the few into which it was introduced and up to about 27 m distance. Stephens and Finker also obtained indications that the amount the dye spread tended to increase with a decrease in the number of flowers present, and was least during the period of maximum flowering and was greatest at the beginning and end of flowering; hence it is probable that unless the ratio of flowers to pollinating insects remains constant least cross-pollination occurs at the peak flowering time. Sidhu and Singh (1961) found that the percentage of flowers receiving the dye steadily decreased with distance as follows: 48% at 6·0 m, 18% at 12·2 m, 6% at 15·2 m, and 2% at 16·7 m. Hence the pattern of distribution of the dye is as expected and helps to confirm the findings based on the production of hybrid plants. However, although the results can be obtained more quickly by this method than by growing genetically marked plants, they are probably not as reliable and must be treated with caution. This is because methylene blue powder is lighter and drier than cotton pollen and has smaller and more numerous particles, and hence is more readily distributed (Thies, 1953). It would be most useful if a firm relationship could be established between dye distribution, cross-pollination and set.

In Uganda, Weatherley (1946) grew small plots (about 3·7 m square) of a genetically marked strain at various distances from other varieties and found that the percentage of hybrids formed was 26·2% at 0·9 m, 3·2% at 15·5 m, 2·7% at 64 m and 2·7% at 128 m. In Tennessee, Pope, et al. (1944) grew small plots of red leaf cotton at six sites which varied from 0·2–1·3 km from a large block of green leaf cotton; the percentage of crossing varied from 0–1·5%

and in general decreased with distance. They also grew barriers of corn between varieties in an attempt to reduce contamination, and although a reduction was achieved it was not great enough to give reasonable protection; thus the average amount of crossing during 2 years between two varieties when they were planted adjacent and separated by three, six and nine rows of corn was 23, 15, 12 and 10%. Hence, the idea of using barriers was abandoned in the U.S.A., and because the extensive cross-pollination that occurs, sometimes over considerable distances, makes production of uncontaminated seed difficult, it is recommended that varieties should be separated by 1·6 km or more (e.g. Pope *et al.*, 1944) and that one variety only is grown in each locality (see Brown and Ware, 1958).

Afzal and Khan (1950) pointed out that although one-variety cotton communities are practical for the U.S.A., they are not yet in Pakistan for educational reasons, and they re-investigated the value of barriers in Pakistan where natural crossing is much lower than in the U.S.A. When they used Sorghum as a barrier between varieties they found that some crossing still occurred, even when the barriers were up to 4·6 m wide. The amount of crossing between fields separated by different widths of fallow ground was also determined, and it was found that some crossing occurred when as much as 84 m separated fields. In fact, they found that cotton itself was the most effective barrier; they suggested that a belt 12·2 m wide on either side of the junction of two varieties should be discarded and the remainder of the fields could be picked for seed. Simpson and Duncan (1956) also reported that cotton itself was the most efficient form of isolation, and they found that cross-pollination between adjacent plantings of two varieties was reduced from 26 to 4% by a 7·6 m barrier of cotton.

Differences in the type and abundance of insects present could help to explain differences in the amount of crossing between different localities. It has long been known that the amount of crossing also varies with different species and different varieties (see above results of Brown, 1927). These differences may be caused by differences in the attractiveness of different varieties to insects; Stephens and Finker (1953), with the aid of the methylene blue dye technique, were able to show that bees preferred *G. herbaceum* to *G. hirsutum*, and Johannson (1959) similarly showed that bees visited one variety of *G. barbadense* twice as much as another when they were planted in alternate rows. Differences in the amount of crossing could also be caused by slight differences in the floral structure of the different species and varieties. Thus, Kearney (1923a) found that *G. hirsutum* was more susceptible to cross-pollination than *G. barbadense* and that two differences in their floral structures might account for this; the anthers of *G. barbadense* dehisce slightly earlier in the day than the anthers of *G. hirsutum* so that, when the flowers have unfolded just sufficiently for the bees to enter them, *G. barbadense* can be self-pollinated, but *G. hirsutum* can only be cross-pollinated; by the time the *G. hirsutum* anthers have dehisced, the flowers have opened sufficiently for the bees to obtain nectar without pollinating them. Varietal differences in the extent to which the stigma projects also occur (e.g. Min'kov, 1953). The less favourable location of the lower part of the stigma for pollen growth might explain why Meade (1918) found that hand pollination especially increased the set of varieties whose stigmas projected above the anthers.

Hybrid cotton

Recently, there has been much interest in using hybrid vigour to increase yield. Its practical use depends upon selecting and developing suitable lines for producing the hybrid and finding methods of producing the seed economically. It seems that suitable lines are, or soon will be, available. Thus, Peebles (1956) reported the results of preliminary trials with two strains of *G. barbadense* that, when crossed, gave seed which produced up to 25% increase in yield, and Kaziev (1958) found that cotton plants from hybrid seed were taller and healthier than control plants, and that third generation hybrids had 23–48% more bolls that, on average, weighed 20–27% more than self-pollinated plants in the same generation. Before hybrid seed can be produced commercially it will be necessary to produce a male and female line of the suitable varieties. The possibilities include: (a) lines in which male sterility is selected genetically or produced artificially (see Eaton, 1957); (b) lines whose anther dehiscence is delayed; (c) lines whose own pollen tubes grow so rapidly that immunity from cross-pollination is achieved (see Loden and Richmond, 1951). It has long been realized that the large amount of cross-pollination between plants should facilitate production of hybrid cotton (e.g. Cook, 1909), and the evidence reviewed above indicates that when varieties are planted in alternate rows and plenty of pollinating insects are present sufficient cross-pollination should occur. However, a greater bee population than normal, evenly distributed throughout the field, will be needed for such cross-pollination and suitable methods of determining this need formulating.

Hibiscus

The flower of *Hibiscus cannabinus* L., kenaf, is solitary, has five petals, a staminal column with numerous stamens and five to nine stigmas. Howard *et al.* (1910) observed that in India the flowers open before daylight, and soon after daybreak the anthers dehisce and bend back from the staminal column. At the same time the styles elongate beyond the staminal column so self-pollination rarely occurs at this stage, although sometimes the styles later bend outward and their stigmas touch the pollen. The flower closes at midday, and in so doing the corolla becomes increasingly twisted and becomes covered with pollen, some of which is transferred to the stigmas; hence, self-pollination always occurs (Fig. 74). However, they found that bagged flowers do not set seed when self-pollinated because the petals do not twist together in the normal way inside a bag, probably due to the unnaturally high humidity inside.

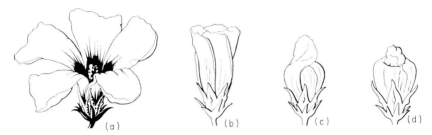

Fig. 74. Stages in closing of *Hibiscus cannabinus*, kenaf, flower (after Howard *et al.*, 1910).

Our knowledge of the insect pollinators is mostly due to the studies of Jones and Tamargo (1954) in Cuba. They exposed vaseline-coated slides in a kenaf field but very few pollen grains adhered to them, and those that did so were only badly shrivelled grains of doubtful viability, towards the end of the day's observation. They reported that honeybees were easily the most numerous insects present on kenaf crops but a wasp, *Campsomeris trifasciata*, was also quite abundant and a small wild bee, *Exomalopsis similis*, and a carpenter bee, *Xylocopa cubaecola*, were occasionally present. Some honeybees entered a flower to visit the nectaries at the flower base and in so doing became dusted with pollen; others collected nectar from five glands located on each capsule and never entered the flowers, although they noticed that bees entering the flowers sometimes visited the extrafloral nectaries.

Bees visited only 1·4 flowers/min but this was because more than two-thirds of their time in the field was spent between flower visits and an average of only 16·5 sec was spent in each flower. In contrast the wasp, *Campsomeris trifasciata* spent an approximately equal amount of time within and between flowers. It would be interesting to obtain comparative data for other crops and for other insects. Flowers under observation were visited 3–49 times/day with an average of 16·7, those flowers on the upper part of stems being visited slightly more frequently than those lower down. Honeybees were active on the kenaf crops before 07.30 h and foraged until 15.30 h with a peak of activity at 12.30 h. When flowers were abundant an individual bee kept to an area of the crop of 4·6 m^2 or less, but when flowers became scarce their foraging areas extended to 30·5 m^2. Presumably more cross-pollination occurred under these conditions.

Howard et al. (1910) mentioned that cross-pollination was fairly common and Tamargo and Jones (1954) found that the amount of cross-pollination between different varieties, grown in adjacent rows, varied from 1·8–23·8% (average of 7·2) in one year, and from 2·5–11·1% (average of 7·9) in another. Part of this difference reflected differences in the peak flowering periods of the different varieties, but they supposed that much of it could also have been due to incompatibility between some of the strains, or to differences in their attractiveness to insects. They supposed that the amount of cross-pollination could be greatly increased by increasing the honeybee population present.

Information on the pollination of *Hibiscus esculentus* L., okra, and *H. sabdariffa* L., roselle, is very sparse and it is supposed that they are normally self-pollinated although a little cross-pollination may result from visits of insects and hummingbirds (e.g. Howard and Howard, 1910; Robbins, 1931; Purseglove, 1968).

Chapter 10

Sterculiaceae

Theobroma cacao L.

The flowers of *Theobroma cacao*, cocoa, are hermaphrodite and arise directly from the main trunk as well as on the older branches. Each has five petals enclosed within five larger sepals which are its most conspicuous part (Fig. 75). The petals are white and each has two prominent purple guide lines. The superior ovary contains five carpels, each with numerous ovules. The single style divides into five stigmas; it is surrounded by an inner circle of five

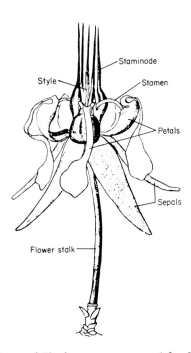

Fig. 75. Flower of *Theobroma cacao*, cocoa (after Hall, 1914).

stamens, each with four anthers, and an outer circle of five staminodes which are united at the base. The filaments of the stamens bend outwards so that the anthers lie within the pouched portions of the petals and the erect pointed staminodes form a barrier round the style.

Hand pollination has shown that a self-compatible flower can set with its own pollen from the time of opening, which is at daybreak, until late afternoon of the same day, but thereafter the receptivity of the stigma greatly decreases and Posnette (1938) found that only 20% of pollination done at 21.00 h resulted in cherelles being formed, and Entwistle (1956) discovered that pollination was only 45% successful 24 h after opening and only 13% successful 27 h after opening.

Abscission of an unpollinated flower occurs at a special constriction in the pedicel. Flower shedding is extensive, and only a very small proportion of flowers normally produce fruit with the consequence that the actual yield is much less than the potential and there is great variability from tree to tree. As trees mature they produce more flowers and usually a smaller percentage are pollinated. Most of the flowers that are not pollinated are shed within 3 days, but many of the pollinated flowers also fail to develop into mature pods.

Pound (1932) discovered that in Trinidad the wilting and dropping of cherelles might be caused by incompatibility. He self-pollinated some flowers and cross-pollinated others on the same trees and found that in seventeen of twenty comparisons crossing gave a greater set (means of 39 and 64%), and three of the eight clones tested gave consistently very poor results from selfing. The following year he demonstrated that pollen from a self-compatible tree can set flowers of self-compatible and self-incompatible trees but pollen from a self-incompatible tree set only self-compatible flowers. His findings that a large proportion of trees in Trinidad were self-incompatible were confirmed by Voelcker (1937, 1938) and Marshall (1934), who also demonstrated that when a self-incompatible tree was caged to exclude insects it set no fruit. Pound (1932) suggested that compatibility varied considerably according to the season of the year, but this suggestion has not been supported by subsequent studies although a small seasonal fluctuation does exist (e.g. Cope, 1939; Ostendorf, 1938).

Whereas self-incompatible trees of the Trinitario type studied by Pound set only if pollen comes from a self-compatible tree, some Upper Amazon types are fertile with any pollen but their own, and self-incompatibility is so uncommon in West Africa that cross-pollinating agents are of little importance in practical cocoa farming (Posnette, 1950).

There is no difference in the rate of pollen tube growth following compatible and incompatible pollination and the pollen tube penetrates the ovules, which are quite normal, in both circumstances (Cope, 1939, 1940a; Posnette, 1938; Knight and Rogers, 1955). Incompatibility is the result of the male nuclei failing to unite with the egg and polar nuclei and is genetically controlled. The site of incompatibility is the embryo sac; when pollen is introduced into a glucose-agar pollen-growing medium into which extracts of the gynecia of the pollen-producing flowers have been incorporated, the growth of the pollen tube is an indication of its compatibility, and when the tree is self-incompatible growth of the pollen tube is strongly inhibited. Another quick and easy method of determining the self-compatibility of a tree is by determining whether nuclei fusion occurs after self-pollination (Cope, 1958, 1962).

According to Posnette (1944), when an incompatible tree is self-pollinated, fertilization by compatible pollen applied simultaneously, or subsequently, is prevented, so any self-sterile pollination may even have a deleterious effect on setting. Knaap (1955) agreed that when the flower of an incompatible tree is selfed the developing fruit soon falls, but maintained that, if it is also cross-pollinated with compatible pollen within 24 h, some of the developing fruits are not shed. The sooner the compatible pollen was applied after self-pollination had occurred, the greater the percentage of pods that matured. His findings seem the more probable, as pollen is much more likely to be transferred between the flowers of a tree than between trees, and unless cross-pollination is favoured to some extent, it is difficult to understand how self-incompatible trees would set any fruit at all. Knaap also found that an average of 23% of the flowers of a self-incompatible variety were pollinated but of these 95% were shed; as 33% of the pollinated flowers of a self-compatible variety were also shed he concluded that selfing probably accounts for 62% of the pollination that occurs on a self-incompatible tree. Hence, there is a very considerable potential loss when a self-incompatible tree does not receive adequate cross-pollination.

Most self-sterile trees flower more profusely than self-fertile, so the percentage of their flowers that are pollinated is likely to be less, even without the detrimental effect of self-pollination, unless their more abundant blossom makes them more attractive to insects. In fact, it is generally accepted that self-compatible trees are high yielders and self-incompatible trees are low yielders. To increase the cross-pollination of self-incompatible clones, Posnette (1950) suggested that pollinizer trees should be placed not less frequently than every third tree in every third row. He pointed out that where cross-compatibility between self-incompatible clones is common (e.g. in Amazon type cocoa), the crop on the pollinizer trees would consist of the reciprocal cross and provided this gave a crop of similar market value, equal numbers of the two cross-compatible clones could be planted. But when self-incompatible clones do not cross (e.g. Trinitario type cocoa), the pollinizer trees would have to be self-compatible, and their crop would consist of both selfed and crossed pods.

Use has been made of the recessive seed character, "albinoism", to find to what extent clonal plots need to be isolated to produce pure seed, and successive experiments seem to indicate the need for increasingly greater isolation. Voelcher (1940) concluded that 25% of the flowers on self-compatible trees are cross-pollinated in Nigeria and that pollen is unlikely to be carried far beyond immediately neighbouring trees. Posnette (1950) found that cross-pollination varied from 18 to 43% in the Gold Coast and was much less frequent between trees two rows apart than between adjacent trees. He supposed that pollen is rarely carried more than the distance of two rows, and suggested that two rows of trees between plots should be adequate to act as guard rows.

Knaap (1955), in Indonesia, studied the pollination in a block of trees homozygous for white seed, against one side of which a row of trees of a clone with purple seed had been planted. About 20% of the trees in the twelve rows nearest the purple seeded clone had been cross-pollinated by them, and 1 to 2% cross-pollination occurred on trees as far distant as the fifteenth and sixteenth rows.

Glendinning (1958) in the Gold Coast used a plot of thirteen trees square of an Amelonado albino clone which when selfed gave an average of one albino to three normal seedlings, but when crossed to a normal clone gave only normal seedlings. He estimated that the percentage of pollination of the albino trees by normal trees outside the plot was as in Table 10.

TABLE 10

Row	1 (outer)	2	3	4	5, 6 and 7 (inner)
1956–7	47–56	46–53	35–49	32–57	30–60
1957–8	42–53	50–63	50–58	30–36	37–49

The pollination mechanism of cocoa is still imperfectly understood, and it is difficult to account for such cross-pollination that does occur. Although cocoa pollen is sticky and unlikely to be carried by wind, and the structure of the flower makes self-pollination difficult without insect intervention, early workers suggested that self-pollination occurs with the aid of wind, even before the flowers opened (e.g. Hall, 1932; Wellensiek, 1932). The evidence against wind-pollination now seems overwhelming. Jones (1912) and Billes (1941) enclosed flowers in cages in windy conditions but failed to obtain a set, whereas hand-pollination of such flowers sometimes produced one. Other workers (e.g. Harland, 1925; Stahel, 1928; Billes, 1941; Soetardi, 1950) shook flowers or blew on them to simulate wind action but failed to get any pollination, and concluded that the hairy staminodes surrounding the stigma prevented any pollen reaching it. Billes (1941) was unable to obtain any correlation between the amount of the set and the wind speed during flowering; in the dry season the set was minimal although the wind was maximal.

Soetardi (1950) in Java, put petri dishes containing 4% glucose-agar under clusters of newly opened flowers and counted the grains deposited in different weather conditions and at different times of the day, but the most pollen he caught in an hour was 21 grains, and only 5 of these were viable. He also drew a current of air from the flowers over the glucose-agar dishes but still only managed to catch a maximum of 39 pollen grains/h, 4 of which were viable. Finally, Posnette (1950) pointed out that the pollen clusters or smears on the styles and stigmas were directly opposite the staminodes and not between them as would be expected if the pollen were blown or shaken from the anthers onto the style.

Billes (1941) even investigated the possible effect of early morning dew or rain in causing transfer of pollen. He showed experimentally that the anthers of dew-laden flowers did not dehisce so no pollen transfer occurred under such conditions, and furthermore, there was no increased pollination in the early morning or after rain.

Most, if not all pollination is by insects, although the lack of nectar or detectable scent in the flowers does not favour insect pollination in general. In spite of efforts that were made to find what effects insecticidal measures against pests were having on pollinating insects, especially in areas where much cross-pollination is essential, it was many years before the principal insect pollinators were discovered.

Until 1941 various insects, especially ants, thrips and aphids were favoured as pollinators. However, no consistently satisfactory correlation has been produced between the abundance of any of these insects and the set obtained, but because of the large fluctuations in the number of flowers present and the percentage set at different times of the year, this is hardly surprising. Thus, Posnette (1942a), in the Gold Coast found that from April to July when flowering was greatest, about thirty times as many flowers were pollinated as from mid-August to the end of January which is the period of minimum flower production, but the actual percentage of flowers pollinated was only about half as great. It seems that the insect population determines the number rather than the percentage of flowers set, but the number of flowers present presumably influences the attractiveness of a tree, the foraging area of the insects and their tendency to move from one tree to another.

Cope (1940b) in Trinidad, found that the maximum set occurred in July and December and the minimum from January to May. The number of red ants and thrips also reached peaks in July and December but aphids, which were by far the most abundant insects, reached their maximum population during the time of minimum set.

Billes (1941) pointed out that although aphids were common on flower petals they seldom entered the flowers, and only the nymphs moved about much and ever carried pollen grains. However, when he enclosed flowers in tubes with aphids, some pollination did occur and so provided some support for the findings of Harland (1925) and Stahel (1928) that the percentage pollination of aphid infested flowers was slightly greater than that of aphid free flowers. Posnette (1942a,b) suggested that aphids may be indirectly beneficial by attracting an ant colony to the tree and ants to the flowers, although individual ants have never been found carrying more than a few pollen grains on their bodies (Billes, 1941).

TABLE 11

	Trees with ants nests	Trees without ants nests
Caged trees	6·1	0·1
Banded trees	19·0	9·7
Control trees	—	13·1

Posnette (1938), in the Gold Coast, found that two trees caged and sprayed with insecticide to kill all insects had 0·2% set, two uncaged trees sprayed with insecticide, and then isolated from crawling insects by putting sticky bands round their trunks, had 1·3% set and two uncaged control trees exposed to crawling and flying insects had 5·2% set. From this experiment he concluded that ants were the most likely pollinating agents. However, the extreme variability from tree to tree makes experiments, involving few trees only, unreliable and produces conflicting results. Thus, Billes (1941) found that banded and unbanded branches had the same average percentage of flowers form cherelles, but the percentage was greater from January to April on banded branches and from September to December on control branches, suggesting that crawling insects may be more effectual during the latter period.

By covering flowers for different periods of the day and night, Posnette (1942b) found that pollination occurred between 09.00 and 17.00 h on the day a flower opens (thus confirming Posnette's (1938) data on the duration of receptivity of the stigma) and most occurred in the afternoon. He also repeated his experiment on banded and caged trees but in addition nests of the ants *Crematogaster* spp. were put in the branches of half the caged and half the banded trees. There were three replicates of each treatment. The average percentage pollination is given in Table 11.

He interpreted these results as indicating that ants were normally responsible for about half of the pollination. Presumably any pollination by ants is self-pollination only. He had previously supposed (Posnette, 1942a) that the low set in March was due directly to the low humidity at the time, but he found that hand pollination in March was as effective as at other times of the year, and suggested that the low set was associated with a change in behaviour of the ants which were inactive and spent more time than usual inside the nests instead of crawling over the trunk and flowers. Because the set on banded trees without ants was not greatly dissimilar to that of the control trees he supposed that much of the pollination must be done by an insect that could gain access to the trees by flight, and he thought thrips were likely candidates.

Billes (1941) reported very large fluctuations in the population of thrips (*Frankliniella parvula*) at different times of the year in Trinidad, and they were most abundant in September, soon after the beginning of the wet season, when they occupied nearly every flower, but soon after the beginning of the dry season they were almost entirely absent. A greater percentage of the flowers with thrips than without were pollinated and the difference was most marked when the thrips were most numerous, probably because an increase in their numbers caused increased movement of individuals. However, another peak of pollination occurred in January, when the thrip population had decreased but midges belonging to the Ceratopogonidae (later identified as *Forcipomyia quasi-ingrami* and *Lasiohelea nana*; Macfie, 1944) were more common.

These midges were the only pollinators that he saw carry pollen from flower to flower and from tree to tree, and he obtained proof of their pollinating ability by enclosing them with flowers in glass tubes. Because they deposited pollen in neat bundles or, as characteristic smears on the stigmas, it was easy to determine when a flower had been pollinated by a midge, and the abundance of these typical pollen masses showed that the midges were important pollinators although they were rarely seen.

However, in any given area, pollination by midges was very uneven and varied greatly from tree to tree and even from branch to branch. Although abundant in the wet season, from July to January, the typical pollen masses were not seen in the dry season and self-incompatible trees did not then set fruit. In order that a self-incompatible tree may set fruit both self-compatible flowers and insect pollinators must be present when it flowers; when flowering is periodic and sparse, such as on old trees, setting is particularly likely to be low.

Billes (1941) pointed out that the structure of the cocoa flower seems to be perfectly adapted for pollination by ceratopogonid midges and thought that their efficiency at pollination compensated for their scarcity. When a female midge arrives at a flower she alights on one of the staminodes and walks along the inside of it probing the deeply pigmented tissue with her piercing mouthparts. She occasionally seems to feed, although the exact nature of the food she obtains is unknown. The midge often carries a dense mass of pollen in the curved bristles on the dorsal side of her thorax and this is automatically deposited in a characteristic cluster on the stigma or style, depending on the extent to which the staminodes are splayed away from the pistil. However, if the part of the staminode on which she is working is parallel to the style, the pollen may be deposited as a streak or smear on the pistil. Both types of pollen depositions are equally effective in pollinating the flowers (Posnette, 1938; Entwistle, 1956). Only rarely does the midge move along the outside of the staminode. After about 3–6 min the midge will have walked through the staminodes and into the petal hood where there are three ridges of raised tissue which is coloured like staminodes. As the midge walks along these ridges, her thorax becomes pressed against the anthers and once again gets covered with pollen. On emerging from the hood, a midge may rest for a while on the outside of the flower before flying to another one. Only rarely does a midge go from the hood to the staminodes of the same flower, and then apparently by accident, so she rarely causes self-pollination and her visits nearly always result in cross-pollination.

It is not known how often a midge flies from one tree to another although only such behaviour can result in the pollination of incompatible varieties; perhaps, once in flight it is readily carried by the wind from tree to tree. As a midge appears to deposit all her pollen each time she rubs against a stigma, only the first flower a midge visits when she arrives on a tree is pollinated with compatible pollen.

Knaap (1955) supposed that, because of the dense cover of leaves in the canopy midges tend to have more restricted foraging areas when visiting flowers there than on the trunk or open branches, and he suggested that, as a result, flowers in the canopy have less chance of being cross-pollinated and more chance of receiving undesirable self-pollination; as a remedy he suggested judicious pruning. However, there does not seem to be any evidence on this, and it would be interesting to sample the midge population in different parts of a tree and to determine the relative frequency with which midges land on the canopy and trunk, when they first arrive at a tree with compatible pollen. Pound (1932) demonstrated that the pods on the smaller branches contained slightly fewer seeds than those on the larger; this may have been because fewer insects visited the smaller branches.

Billes' (1941) initial discovery of the importance of ceratopogonid midges, was soon confirmed by other workers. Posnette (1944), in Trinidad, found they occurred throughout the dry season and at no time did the self-sterile trees entirely cease setting fruit. He found that the previous technique of sampling the insects present by cutting the flower stalks and immersing the flowers in alcohol was not efficient enough to trap most of the midges, and by carefully enclosing each flower and its insects in a tube before cutting the pedicel, he found that midges were quite abundant and not rare as previously supposed. He examined the styles of flowers that had set and were beginning to form pods, and found that 75% of them were pollinated by the characteristic ball of pollen left by the ceratopogonid midges, the remaining 25% might have been pollinated by other insects or by ceratopogonid midges leaving a pollen smear; bristles of the midges were found in some of the pollen smears.

Soetardi (1950) found 31 insect species in cocoa flowers in Java but only four occurred regularly. The number of visits made by these 4 species to 10 flowers during 30 min was: *Dolichoderus bituberculatus*, 4; Cecidomyidae spp., 11; *Drosophila* spp., 18; and *Forcipomyia* spp., 63 (4 males and 59 females). Only females of the midge *Forcipomyia* spp. entered the petal hoods and carried pollen; they were active between 07.30 and 10.30 h but seldom moved in the middle of the day. He investigated the possibility that night flying insects were pollinating the flowers by exposing caged flowers at sunset only, but none set fruit compared to 37 of 100 similarly treated flowers that were artificially pollinated.

Entwistle (1956) considered that in the Gold Coast also, cross-pollination is done entirely by three midges *Lasiohelea litoraurea*, *Forcipomyia ingrami*, and *Forcipomyia ashantii*, the former being the most common, although ants, aphids and other insects are responsible for a small amount of self-pollination. She examined 6,000 flowers at various times of the day in May and June and found that 87% of pollination occurred between dawn when the flowers opened, and about 09.00 h, and another 12% occurred between 14.00 and 15.00 h. She also observed that the midges rested on flowers in the middle of the day, so probably the same species are involved at the major and minor peak times of pollination.

Smit (1950) found that cocoa pollination in Costa Rica was done by midges, thrips, ants and also by honeybees, although it seems probable that the last would experience difficulty in obtaining the pollen.

In Indonesia, Knaap (1955) found an average of 10% of the cocoa flowers were pollinated and the percentage pollinated on 32 consecutive days varied from 1·6–13·5%. When he enclosed midges on a caged branch, a much greater than usual set was obtained, so he concluded that the midges were usually too few. Furthermore, he quoted Soetardi as finding that midges carried an average of only 8·6 pollen grains each and as a flower needs sufficient pollination to provide for 20–60 seeds/pod, he concluded that each flower needs to be visited several times to ensure adequate pollination. However, consideration of average numbers only can be very misleading and his conclusion could be verified by determining how commonly the pistils of cocoa flowers carry clumps or smears of only a few grains each.

More recently ceratopogonid midges have been reported as important pollinators of cocoa in the Congo (Dessart, 1961) where they develop in rotting pods, and in the Philippines (Fontanilla-Barroga, 1961).

The importance of ceratopogonid midges was unrecognized for so long because of their small size and difficulty in collecting them. Clearly much work still remains to be done to discover the species of midges and other insects that are responsible for cocoa pollination in different parts of the world, and to learn more about their life histories with the object of increasing their breeding facilities where necessary, or even culturing them artificially.

Chapter 11

Linaceae

The only species of agricultural importance in this family is *Linum usitatissimum* L., flax and linseed. The flower has five small sepals and five bright blue or white petals (Fig. 76). The ovary has five chambers each containing two

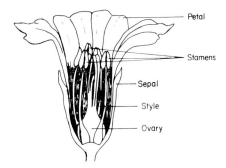

Fig. 76. Flower of *Linum usitatissimum*, flax (after Poehlman, 1959).

ovules, and is surmounted by five erect styles. The filaments of the five stamens are broadened at their bases. On warm clear mornings the flowers begin to open at sunrise and the anthers dehisce soon afterwards; the flowers are fully open by about 07.00 h and the petals drop between 10.30 and 12.00 h; on cool, cloudy days opening is delayed, and sometimes the flowers do not open fully until the next day (e.g. Howard *et al.*, 1919; Dillman, 1938; Gubin, 1945a). As a flower opens, the dehisced anthers touch the stigmas so self-pollination can occur. Soon afterwards the burst anthers often fall together to form a cap over the stigmas.

It is not surprising, therefore, that *L. usitatissimum* is almost entirely self-pollinated and that the two cultivated forms, grown for flax and linseed respectively, have remained distinct although crosses between them are possible. However, a small amount of cross-pollination, usually about 1% or 2%, does occur (Howard *et al.*, 1919). Henry and Tu (1928) in Minnesota and B. B. Robinson (1937) in Michigan and Oregon, grew white and blue flowered varieties in alternate rows and found that the percentage crossing decreased as the distance between the rows increased (Table 12).

TABLE 12

	Approximate distance (cm) between the rows					
	15	30	60	90	120	150
% hybrids from white flowered varieties						
Minnesota	—	1·3	0·9	0·5	0·5	0·3
Michigan	0·7	1·0	1·1	0·7	0·6	—
Oregon	0·7	0·4	0·2	0·2	0·2	—
% hybrids from blue flowered varieties						
Michigan	2·2	2·2	2·0	3·0	1·3	—
Oregon	0·9	0·4	0·4	0	0·6	—

Henry and Tu suggested that the presence of a tall leafy crop between rows 1·3 m apart would prevent any cross-pollination between them, but this has yet to be demonstrated. The reason why more cross-pollination occurred from white to blue varieties than vice versa is not known, but Robinson suggested that the greater amount of cross-pollination in Michigan than in Oregon might be because climatic differences between these States are such that the duration of flowering is about an hour greater in Michigan.

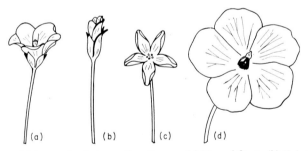

Fig. 77. Types of *Linum usitatissimum*, flax, flowers. (a) Funnel-form. (b) Tubular. (c) Star-shaped. (d) Disk-form (after Poehlman, 1959).

Dillman (1938, 1953) observed that four distinct types of flower occur in the different varieties which are most widely grown for seed flax (Fig. 77). The "funnel-form" type has overlapping and partly separate petals, the "tubular" type has petals which remain rolled in the form of a tube, the "star-shaped" type has narrow petals which are rolled at the margins and well separated and the "disk-form" type has large flat petals which are widely spread as a disc or saucer. Most varieties with funnel-form flowers have their five anthers completely surrounding and enveloping the stigmas, whereas other varieties, most often those with disk-form flowers, have stigmas which protrude through the top of the ring of anthers and so are more exposed to foreign pollen. Dillman (1938) grew different varieties in separate rows and found that most cross-pollination occurred in the varieties with disk-form flowers (about 2%) and least in the varieties with tubular flowers (about 0·3%). A variety with funnel-form flowers whose anthers entirely ensheathed the stigmas was completely self-pollinated. In hot dry weather the funnel-form types were more fertile than the disk-form, whose pollen and stigmas were more exposed to the drying effects of the sun, whereas the fertility of the tubular varieties was unimpaired.

Nectar glands located at the base of the filaments are visited by many insects including *Bombus* spp., various Diptera and butterflies, but honeybees are the most abundant (e.g. Howard *et al.*, 1919; Dillman, 1938; Kozin, 1954; Smirnov, 1954; Hassanein, 1955) although Gubin (1945a) found that honeybee colonies put beside a crop did not produce much honey. He also observed that, in order to reach the nectaries, about a third of the honeybees stood on the stalks outside the flowers and inserted their tongues between the petals and so did not touch the stamens or stigmas; he thought that such bees had learned to avoid relying for support upon the petals of the corolla which very easily drop later in the day. Smirnov (1954) observed that, although most honeybees collected nectar, a few collected pollen as well; his and Kozin's (1954) results taken together, indicate that honeybees spend an average of about 7 or 8 sec/flower visit. In contrast, Alex (1957a) recorded that bees visited 2·3 flowers per minute for pollen only and apparently collected no nectar. Henry and Tu (1928) noticed that many thrips entered flowers whose petals were just beginning to unfold but whose anthers had not dehisced, and supposed they might cross-pollinate the flowers before self-pollination could occur, but Dillman (1938) doubted whether this was so.

Opinions also differ as to whether insect visits increase seed production. Gubin (1945a) put glass slides, smeared with paraffin oil, at different heights in the centre of flax fields and found that pollen was deposited on them from some varieties of flax but not others, and most of the pollen deposited was on slides 50 cm or less above ground, indicating that it is not readily carried by air currents. He claimed that when varieties that had not deposited pollen on the slides were caged with bees, the number of seeds and seed weight was increased by 22 and 24%, compared to only 5 and 7% for varieties that had deposited pollen on the slides and were presumably wind-pollinated to some extent. Others (e.g. Kozin, 1954; Smirnov, 1954; Luttso, 1957) have reported that honeybee pollination of flax increases the number of seeds produced by 18–26% and the weight of seed produced from 22–49%.

Hassanein (1955) obtained less definite results. He used cloth cages to exclude wind and insects, metal screen cages to exclude insects or enclose honeybees, and uncaged plots. There were two replicates of each treatment on each of 2 years. The mean results are given in Table 13.

Table 13

	No. seeds/pod	Wt (g)/1,000 seeds
Cloth cages	5·0	6·3
Screen cages without bees	5·6	6·7
Screen cages with bees	6·8	7·0
Open plots	6·0	6·9

Alex (1957a) did a similar experiment and calculated the yield in kg/ha from plots caged without bees, plots caged with bees and plots not caged as 938, 954 and 947. Hence there is some doubt as to whether hiring colonies to pollinate flax would be economically justified.

Chapter 12

Rutaceae

Citrus

The commonly cultivated *Citrus* species are: *C. aurantifolia* Swing (lime), *C. aurantium* L. (sour or Seville Orange), *C. grandis* Osbeck (pummelo, shaddock) *C. limon* L. (lemon), *C. medica* L. (citron), *C. paradisi* Macf (grapefruit), *C. reticulata* Blanco (mandarin, tangerine) and *C. sinensis* Osbeck (sweet orange). The different species will first be considered together. Particular varieties that seem to benefit from insect pollination will then be discussed separately.

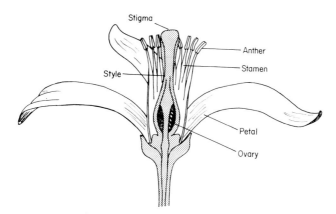

Fig. 78. Diagrammatic section of a citrus flower (after Hume, 1957).

The strongly perfumed citrus flower (Fig. 78) with its conspicuous white corolla of four to eight petals, but usually five, is 2·5–4 cm diameter. A ring of twenty to sixty stamens, partially united at their bases, surrounds the style; the superior ovary contains eight to fifteen united carpels, each containing two rows of ovules. In many varieties the stamens surround the stigma so closely that when the flower opens one or more anthers touch it.

Most species have hermaphrodite flowers only, but *C. limon* and *C. medica* sometimes have flowers with aborted ovaries and rudimentary pistils.

Abundant nectar is secreted by the part of the disc just inside the ring of stamens. Vansell *et al.* (1942) found that nectar in orange flower buds contained 13–17% sugar, and that the mean concentration steadily increased to 31% as the flower became older; bees preferred to visit the older flowers. Hassanein and Ibrahim (1959) reported that the average concentrations recorded for the different species were: *C. aurantium*, 15%; *C. limon*, 15–18%; *C. paradisi*, 16%; *C. reticulata*, 22%; *C. sinensis*, 11–18%. Fahn (1949) showed, with eleven species of *Citrus*, that there was a positive correlation between the size of the nectary and the amount of nectar secreted.

Citrus flowers are visited by numerous insects both for nectar and pollen, and citrus growers are able to support large concentrations of honeybee colonies which produce good honey crops. In N. America bumblebees, thrips and mites are also common visitors. Investigations of the types of insects foraging on citrus flowers in Egypt have been made by Hassanein and Ibrahim (1959) and Wafa and Ibrahim (1959a). The percentages of visitors belonging to different orders were Hymenoptera, 54; Diptera, 34; Lepidoptera, 8; Neuroptera, 4; and Coleoptera, 3. The Coleoptera were eating pollen and only effected pollination between flowers on the same branch, if at all. *Pieris rapae* was the most abundant member of the Lepidoptera; while collecting nectar it often touched stamens and stigma and its habit of moving between distant trees enhanced its value as a cross-pollinator. Most of the Diptera belonged to the genera *Eristalis*, *Musca* and *Syrphus*, which while collecting nectar inadvertently got pollen on their bodies. The Hymenoptera included *Andrena erincia*, *Xylocopa aestuans* and honeybees; the latter were the most abundant insects present and comprised 88 and 90% of visitors to the citrus flowers in 1955 and 1956.

Honeybees visiting citrus flowers collected either nectar only, or pollen only or both. Hassanein and El Bandy (1956) and Hassanein and Ibrahim (1959) made the important discovery, which may well be relevant to other crops, that bees that foraged for both nectar and pollen collected, on average, less of each commodity than bees that collected one or other only (Table 14).

Nectar-gatherers spent 15–20 sec per flower visit compared to a pollen-gatherer's 5–8 sec, and furthermore, tended to stand on the petals and approach the nectaries in such a way that they did not touch the stamens. However, in their favour, they began foraging earlier in the day than pollen-gatherers and their numbers on the crop quickly reached a peak and remained fairly constant from 09.00–15.00 h.

Although citrus flowers are very attractive to insects, most varieties do not need insect visits to set fruit. Certain varieties, e.g. "Tahiti" lime, produce parthenocarpic seedless fruits, with or without pollination, and some of these varieties, e.g. "Washington Navel" orange have no viable pollen (Vansell, 1944b). The tendency to parthenocarpy varies greatly with different varieties; in the absence of pollination some varieties set no fruit, and other varieties that normally have seeds produce seedless fruit but have a smaller set than when they are pollinated. Other varieties may be stimulated to produce seedless fruits by "foreign" pollen which is incapable of fertilizing them. Many varieties produce fertile pollen, are self-compatible, and auto-pollination occurs by the anthers touching the stigmas or by the pollen falling or being blown on to the stigmas; it is generally accepted that any insect pollination is of negligible importance (see Webber and Batchelor, 1948).

TABLE 14

The weight of loads collected by bees that foraged for nectar only, pollen only and for both from *Citrus* flowers. (After Hassanein and Ibrahim, 1959).

Year	Wt (mg) of nectar loads of bees that collected		Wt (mg) of pollen loads of bees that collected	
	Nectar only	Nectar and pollen	Pollen only	Nectar and pollen
1957	22·6 ± 2·4	15·3 ± 3·1	10·2 ± 2·5	6·7 ± 1·3
1958	20·2 ± 2·2	16·8 ± 3·9	9·9 ± 1·3	6·2 ± 1·9

However, recent evidence indicates that the set of some varieties of self-fertile oranges is increased by insect pollination. Fujita (1957) reported that "Satsuma" orange trees gave a bigger set when caged with bees; and Hassanein and Ibrahim (1959) found that six "Khalili" orange trees caged to exclude insects had an average of 3% flowers set fruit, compared to 10% of seven trees caged with honeybee colonies and 7% of seven trees not caged. These results were confirmed by Wafa and Ibrahim (1960c) who also reported that the bee-pollinated fruits were bigger and had more seeds and juice. Barbier (1964) found that three "Clementine" orange trees caged without bees gave only 37% of the set of uncaged trees and 40% of the set of a tree caged with bees; the presence of bees resulted in larger fruits with more pips and the diameter of the fruits increased with the number of pips present. Similar increases have been obtained from the insect pollination of "Washington Navel" orange and "Unshiu" tangerine but all the fruits suffer from the possible disadvantage that they contain seeds (Zavrashvili, 1967).

Citrus pollen is not adapted for wind pollination and cross-pollination is accomplished by insects. Self- and cross-pollination of varieties with abundant fertile pollen is equally effective. Although cross-pollination can benefit varieties with defective pollen, it could be detrimental when the variety concerned normally produces good crops of seedless fruits when self-pollinated, but has more seeds per fruit when cross-pollinated. Sometimes the choice is between more fruits with seeds following cross-pollination or fewer seedless fruits following self-pollination (see Webber and Batchelor, 1948). Coit (1915) reported that cross-pollination of incompatible varieties of orange may even result in a diminished set.

The varieties whose pollination requirements have received most attention lately are those that are wholly or partially self-incompatible. Thus, Oppenheimer (1948) discovered that the widespread unfruitfulness of "Clementine" mandarin in Palestine results from self-incompatibility but that pollen from other mandarin varieties was effective in giving a good set. This has been confirmed in California where three varieties of mandarin and one each of tangelo, orange and tangerine have been effectively used for cross-pollination (Soost, 1956). Oppenheimer (1948) suggested that pollinizer branches should be grafted on to every third tree in every third row of blocks of "Clementine" mandarin, and until these are sufficiently mature flowering branches of compatible varieties should be introduced to orchards.

Horn and Todd (1954) found that in Arizona two "Clementine" tangerine trees, caged to exclude insects, produced an average of only 30 fruits/tree compared to 295 fruits on 2 trees caged with honeybee colonies; 2 more trees, that were caged with bees and bouquets of a compatible orange or lemon variety produced an average of 510 fruits/tree, and although the fruits were of similar size to those produced by self-pollination they had a few more seeds each, were juicier and sweeter. Minessy (1959), in Egypt confirmed these results and, in addition, showed that the degree of self-incompatibility of "Clementine" tangerine varies from year to year, and that cross-pollination made no difference in a year when a good set occurred naturally whereas in a year of poor set it provided five or six times as much mature fruit as self-pollination. Furthermore, the average fruit size was greater and there were more seeds per fruit following suitable cross-pollination than self-pollination, and each year there was a positive relationship between the number of seeds per fruit and fruit size.

It is not surprising that the tangerine variety "Robinson" which is a hybrid of "Clementine" and "Orlando" is also self-sterile. When "Robinson" flowers were self-pollinated only 0·3% set fruit compared to 34·1% when they were cross-pollinated with Orlando. There were averages of 5 and 22 seeds/self-pollinated and cross-pollinated fruit respectively (Reece and Register, 1961).

The tangelo is a hybrid resulting from crossing grapefruit and tangerine. The "Minneola" tangelo sets little fruit with its own pollen, probably because the pollen tubes produced by its pollen grains are too short to transverse its long style, but gives a good set when crossed with some varieties of tangelo and orange (Mustard *et al.*, 1956). Robinson and Krezdorn (1962) found that 61% of 1,051 flowers of "Orlando" tangelo they cross-pollinated set fruit but only 2% of 1,051 self-pollinated flowers did so. Flowers not pollinated set no fruit. They pointed out that although "Orlando" was only very weakly parthenocarpic, erratically it was strongly so, and in some years normally unfruitful trees set large parthenocarpic crops. Krezdorn (1959) found that

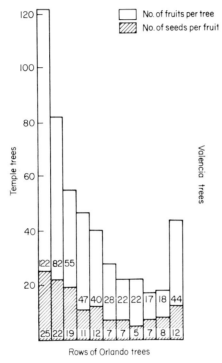

Fig. 79. Effect of proximity to "Temple" and "Valencia" orange trees on the yield of "Orlando" tangelos (after Robinson and Krezdorn, 1962).

pollen from different varieties of tangelo were equally effective when used to hand-pollinate "Orlando" tangelo, but many were ineffective or erratic under conditions of open pollination and he suggested this might have been because bees were showing a varietal preference.

These and similar findings help to explain why plantings of tangelos containing only one variety usually fail to produce commercial crops, and why when they do produce fruit it is mostly seedless. Other observations have shown that when solid blocks of "Minneola" or "Orlando" tangelos are adjacent to rows of suitable pollinizer varieties the percentage of flowers set and the number of fruits produced, on the "Minneola" and "Orlando" trees decreases sharply as the distance from the edge of the block increases (Fig. 79) and it has been suggested that they should be no further than two trees from a source of compatible pollen (Mustard *et al.*, 1956; Krezdorn, 1959; Robinson and Krezdorn, 1962).

Two and a half honeybee colonies per hectare has been suggested for varieties needing cross-pollination (Horn and Todd, 1954; Robinson and Krezdorn, 1962). Attempts to use pollen dispensers for citrus pollination have so far been unsuccessful because the viability of the pollen decreases too rapidly (Robinson, 1958).

Chapter 13

Vitaceae

There are two types of wild grape vine (*Vitis* spp.); one has staminate flowers only and the other has hermaphrodite flowers which are self-sterile. Cultivated vines have two types of hermaphrodite flowers, although intermediates occur; one type has upright stamens and fertile pollen and the other has reflexed stamens and mostly sterile pollen. There are no flowers without stamens. All European cultivated varieties have hermaphrodite flowers only. The principal cultivated species are: *Vitis vinifera* L., *V. rotundifolia* Michx. and *V. munsoniana* Simpson.

The flowers are in clusters; those at the base open first and all the flowers of a cluster open within 1–2 days. The hermaphrodite flower (Fig. 80) has a single slender-necked pistil and a two-chambered enlarged ovary, each with two ovules. There are usually five stamens, although the number may vary

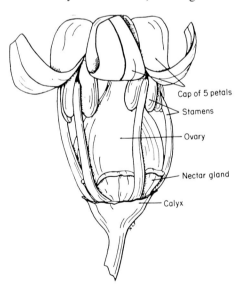

Fig. 80. Flower of *Vitis* spp., grape (after Robbins, 1931, reproduced with permission of McGraw-Hill Book Company).

from four to eight, even in the same cluster, and the same number of nectaries located between the bases of the stamens. The calyx, usually five sepals, is reduced to a ring at the base of the flower. The corolla is composed of five united petals which first open at the base of the flower and for a while remain united as a cap (calyptra) at the top of the flower. As soon as the calyptra is shed, the stamens move away from the pistil, although the anthers may dehisce beforehand. Opening can take a few minutes only or several hours. The male flowers have well-developed stamens, but rudimentary pistils and ovaries with undeveloped ovules.

The extent to which automatic self-pollination occurs with self-fertile varieties does not seem to be known. It is recommended that flowers grown in glasshouses should be hand-pollinated (Hughes, 1951) or the clusters tapped to facilitate self-pollination (Bewley, 1963).

Self-fertility differs greatly with different varieties. Whereas some are completely self-fertile and others completely self-infertile, most varieties are intermediate between these two extremes (Beach, 1898). Self-sterility seems to be mostly, if not always, associated with sterile pollen and does not seem to be caused by genetic incompatibility; according to Booth (1902) certain varieties produce mixtures of sterile and fertile pollen. A positive correlation exists between reflexed stamens and self-sterility and the pollen from such reflexed stamens always fails to germinate (Stout, 1921), so that vines whose flowers have reflexed stamens either fail to set fruit, or produce only loose clusters, unless they are cross-pollinated either naturally or artificially (Einset, 1930). However, some varieties with upright stamens are also self-sterile and it has been claimed that cross-pollination of self-fertile varieties increases the quantity of ripe berries produced (e.g. Sosunkov, 1953).

To obtain the maximum yield, self-sterile varieties are sometimes hand-pollinated, using pollen mixed with flour or lycopodium powder (Gladwin, 1937; Wellington, 1956). In Spain it is the practice to artificially pollinate "Almeria" grapes, which have completely reflexed stamens, by touching the flower clusters every day with small bouquets cut from a pollinizer vine (Olmo, 1943; Shoemaker, 1955); the first commercial plantations of this grape in California were failures because this practice was not followed.

Dunne (1943) attempted to compare hand pollinating the variety "Ohanez" with spraying the flowers with a suspension of pollen in water. Although about half of the pollen grains burst very quickly in water, the remainder seemed to be unaffected and gave a good set. However, other experiments (Hale and Jones, 1956) have shown that grape pollen very soon loses its viability in pollen-water sprays. It has also been claimed that improved sets are obtained by blowing an air current containing pollen onto the clusters (e.g. Cejtlin, 1956; Kovalev, 1958). More convincing data are necessary before these methods are acceptable.

When a variety is completely or partially self-sterile the interplanting of pollinizer varieties is recommended; sometimes every third vine in every third row or even all the vines of every other row are of the pollinizer variety (Lagassé, 1928; Auchter and Knapp, 1937). A greater than usual ratio of pollinizer plants has been recommended for muscadine grapes on which insect pollinators are said to be scarce (Reimer and Detjen, 1910). When partially self-sterile varieties are used as pollinizers of self-sterile varieties, the percentage set is about the same as that on the pollinizer plant itself (Beach, 1899), and self-fertile varieties usually give good results when used to pollinate either partially or completely self-sterile varieties.

Either wind or insects, or both must be responsible for transferring pollen to self-infertile varieties, but the relative value of wind and insects in self- and cross-pollination is a matter of some dispute, and few observations and experiments have been made.

Sartorius (1926) showed that few flower clusters whose stamens were removed set fruit unless they were within 7–10 cm of other clusters with intact flowers; he also found that when clusters with viable pollen were suspended in vines of a self-infertile variety they mostly influenced the fruit set only in their near vicinity. He suggested that self-pollination was the general rule, although the rapid dehiscence of the anthers which sometimes occurs could throw pollen on to adjacent flowers and, that to achieve cross-pollination between varieties, it is necessary to pollinate by hand or machine, a view that is also shared by Einset (1930). But later experiments by Gladwin (1937) showed that grape pollen can be carried up to 4·9 m by gentle to fresh air currents, but none was carried 7·3 m; hence, he recommended that pollen-bearing vines should not be planted more than 4·9 m from a self-infertile variety and, to allow for changes in wind direction, they should surround the self-infertile vines.

Olmo (1943) caged half of each of five "Almeria" grape vines to exclude large insects such as honeybees and syrphids, and found that 8·5% of the caged flowers set fruit compared to 10·5% of those in the uncaged halves and he concluded that wind was the most important pollinating agent. Wind, insect and cross-pollination were each probably greater outside than inside the cages, and this could account for the additional 2% set. Sharples *et al.* (1965) caged some "Cardinal" variety grape vines with honeybee colonies, caged others to exclude insects, and left others not caged. There were five replicates of each experiment. Although more bees foraged on the flowers of the vines caged with bees than on the open vines, there was no difference in the crop produced; perhaps the extra pollination inside the cages was balanced by the adverse effects of the cages themselves on plant growth and by lack of any cross-pollination. However, vines visited by pollinating insects, both in open plots and cages, had significantly more seeds per berry than those caged to exclude bees. Although berry size was significantly correlated with number of seeds present, the weight of berries on insect-pollinated vines, was not significantly greater than that on vines caged without bees.

It is assumed that the nectaries of most varieties either do not function, or the nectar is not attractive, as few insects visit them (Gladwin, 1937; Winkler, 1962). However, Armstrong (1936) believed that halictid bees were mainly responsible for distributing grape pollen in Georgia, U.S.A. and Olmo (1943) saw honeybees, sometimes numerous, and syrphid flies (*Lasiophthicus pyrastri*) visiting the flowers in many vineyards in California. Olmo observed three clusters of flowers continuously for 3 h, during which one was visited by six and another by five honeybees. Some honeybees collected large loads of grape pollen as well as nectar. He pointed out that although the abundance of honeybees varies greatly, at times they assist cross-pollination and he recommended introducing honeybee colonies to vineyards to supplement wind pollination. Christopher (1958) thought that because grapes flower relatively late in the season there are usually sufficient wild bees and other insects to pollinate them, provided the vineyards are planned so that the self-infertile varieties are within three to four rows of pollen sources. Steshenko (1958) found that, when other flowers were scarce, honeybee colonies obtained 68 to 84% of their pollen from grape flowers; he claimed that, as a result, the crop of one variety was increased by 38% and another by 48%. It has also been reported that bees can be trained to visit grape flowers by feeding them with sugar syrup in which grape flowers have been immersed, and that because of the increased number of bee visits the weight of crop was increased from between 23% and 54% for different varieties (Barskii, 1956).

Probably, the relative importance of wind and insects, in both self- and cross-pollination, varies under different circumstances and in different localities, but clearly there is much need for further more extensive and critical experiments on the lines attempted by Sharples *et al.* (1965), together with detailed observations on the behaviour and number of insect pollinators, before this can be elucidated.

Chapter 14

Anacardiaceae

Mangifera indica L.

The flower of *M. indica*, mango, (Fig. 81) has four or five ovate-lanceolate petals, which are red, pink or almost white, inserted in the base of an almost hemispherical disc. The disc of the perfect flower is surmounted by a greenish-yellow ovary, with a slender lateral style. The ovary has one chamber which contains one ovule. A single fertile stamen arises from the disc on one side of the ovary; sometimes there are two, and very rarely three, fertile stamens.

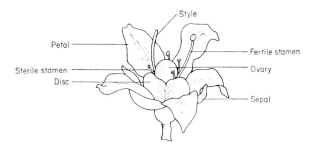

Fig. 81. Flower of *Mangifera indica*, mango (after Singh, 1960).

The remainder of the five stamens are much diminished and sterile. Male flowers are similar but have no ovary or style.

The flowers are in panicles which are up to 60 cm long. Each panicle contains from 200–6,000 flowers and a mature tree in full flower may have 600–1,000 panicles. A panicle contains both male and hermaphrodite flowers but the male flowers are always by far the more numerous and the percentage of hermaphrodite flowers varies from 1–35% (Cobley, 1956; Mukherjee, 1953) and is usually greatest in the terminal part of a panicle (Singh, 1960). The number of flowers per panicle, and the percentage of hermaphrodite flowers, is partly genetically determined and differs with different varieties, but is also influenced by environmental factors and especially the amount of light.

There is also a wide variation in the time at which varieties growing in identical conditions begin to flower, some varieties flowering more than once per year. Some of the flowers of a panicle open before it is fully grown. Opening starts early in the morning, but most flowers open between 09.00 and 10.00 h. They sometimes start opening during the night and so they are fully open by the morning (Singh, 1960).

According to Mukherjee (1953), anther dehiscence begins as soon as the flower is open and is continuous throughout the day but reaches its maximum between 08.00 and 12.00 h; however, Mallik (1957) stated that the flower is open for at least an hour before dehiscence. Anthesis and dehiscence can be delayed by relatively low temperatures or an appreciable rise in relative humidity. It is agreed that the stigma is receptive when, or before, the flower opens but estimates of the period for which it remains receptive vary from a few to 72 h (e.g. Spencer and Kennard, 1955; Mallik, 1957; Singh, 1960). Receptivity probably diminishes greatly after the first day.

When perfect flowers are emasculated and bagged, none set fruit (Naik and Rao, 1943; Spencer and Kennard, 1955). When panicles are bagged without emasculating the flowers, the set is decreased to about a quarter, so an external agent is normally responsible for pollen transfer (Mallik, 1957); the fertilization that occurs within the bags could result from pollen falling from the upper to the lower flowers of the panicles or from the self-pollination of individual flowers.

Naik and Rao (1943) obtained data which suggested that cross-pollination is more effective between some varieties than others, but there seems to be no barrier to self-fertilization. If Mallik (1957) is correct in his assertion that the stigma of a flower is receptive for at least an hour before anther dehiscence, cross-pollination will be initially favoured but, because of the great excess of male flowers in a panicle, this is likely to occur within the panicle.

Although mango trees produce an abundance of flowers relatively few mature fruits. This is probably due mostly to factors other than pollination. Naik and Rao (1943) examined 100 hermaphrodite flowers 7 h after they had opened and found as many as thirty-four had pollen grains on their stigmas but, when they hand pollinated 4,814 flowers, only 38% set fruit and only 2·5% of these matured. Other observers in India have recorded averages of 13–28% hermaphrodite flowers setting fruit, most of these dropping and only 0·1–0·25% of the flowers developing into mature fruits (see Mukherjee, 1953). As the percentage set would be adequate to produce a good crop if the fruits matured, it seems that the problem is largely nutritional and that additional pollination would only increase the crop very slightly, if at all.

Naik and Rao (1943) showed that varieties with the greatest percentage of perfect flowers produced the largest number of fruits and suggested that there is a positive correlation between the percentage of perfect flowers in a panicle and the number of fruits it produces. However, both the percentage and number of perfect flowers produced probably also depends on the nutritional state of the tree concerned.

Spencer and Kennard (1955) found that in Trinidad 16, 20 and 23% of stigmas had adhering pollen grains 6, 12 and 24 h after anthesis, and even 36 h afterwards less than a third had pollen. They also reported that 19% of the pollinated flowers had ten or fewer pollen grains on their stigmas, and 37% had less than fifty; they supposed that the presence of such comparatively few grains might explain why there was little correlation between pollination and fruit set, but they produced no evidence that these numbers were inadequate. They also thought that the average production of only 410 pollen grains/anther might be a factor limiting set, although it seems that the large number of male flowers could compensate for this.

Because the pollen grains soon lose their stickiness and become dry and dust-like it seems probable that wind pollination occurs to some extent although only a negligible amount has been found in the air. It has been claimed that, when freshly dehisced pollen is not taken by insects, it sticks to the anthers and cannot be removed by wind (Singh, 1962). Undoubtedly, most pollination is by insects that visit the flowers for pollen or for the nectar, which is secreted in large quantities by the disc and is readily available to short-tongued insects.

Many types of insect have been recorded visiting the flowers but their importance as pollinators is not always evident. In India the principal insects visiting mango belong to the Coleoptera, Lepidoptera, Hymenoptera and especially the Diptera; among the latter the genera *Syrphus, Musca,* and *Psychonosma* are the most important (Mukherjee, 1953). In Brazil, Simao and Maranhao (1959) found that the majority of insects visiting mango belonged to the Coleoptera (10%), Hemiptera (13%), Diptera (20%) and Hymenoptera (21%) but pointed out that most of them were not pollinating the flowers. In Trinidad, Spencer and Kennard (1955) found that Diptera, Hymenoptera and Thysanoptera comprised 49, 21 and 11% of the insect visitors in the early flowering period, and 37, 12 and 25% in the late flowering period. In the early period about two-thirds of the Diptera were midges, but in the later period there were relatively few midges and more larger flies. Very few of these insects carried pollen or were large enough to touch the stigma while collecting nectar and pollen. However, every bee had at least a few pollen grains and usually touched both stamens and stigma when visiting a flower. Only 1·5 and 2·8% of midges and thrips had pollen grains on their bodies; this finding agrees with the experience of Wolfenbarger (1957) who, having diminished the thrip population in parts of a mango orchard with an insecticide, found that it did not influence fruit set.

Local insect populations seem adequate to pollinate most mango orchards; when honeybee colonies are taken to them the bees visit the flowers (Wolfenbarger, 1957) but it seems that until cultural methods are improved their visits will give little increase in fruit production.

Anacardium occidentale L.

The panicle of *A. occidentale*, cashew nut also contains both male and hermaphrodite flowers (Fig. 82), the former being many times more numerous. The male flower usually has one long and nine short stamens, although some of intermediate size sometimes occur; the long stamen and most of the smaller ones produce pollen. The hermaphrodite flower also has one long and nine

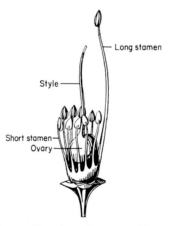

Fig. 82. Flower of *Anacardium occidentale*, cashew nut, with petals removed (after Nicholls and Holland, 1940).

short stamens. The ovary has a single chamber containing one ovule. The slender pistil is sometimes longer and sometimes about the same height as the long stamen; presumably in the latter circumstance, self-pollination is more likely to occur. Northwood (1966) reported a third type of flower on a tree of Jamaican origin; it is much diminished and has infertile reproductive organs and he suggested that its sole function is to attract insects to the panicles.

The flowers open between 06.00 and 18.00 h although the peak period is from 11.00–12.30 h. The stigmas are receptive when the flowers open but anther dehiscence does not occur until 1–5 h later, depending on the temperature, so cross-pollination is probably favoured soon after opening. However, the few experiments that have been made indicate that self-pollination is as effective as cross-pollination but, because bagged flowers do not produce nuts unless hand-pollinated, some agent is needed to transfer the pollen. As the pollen is sticky and adheres to the stigma, and no wind-blown pollen was caught on sticky traps, it seems that insect pollination is important (Northwood, 1966).

Numerous flies and ants visit cashew flowers in Tanzania, and Northwood (1966) found that 78% of open hermaphrodite flowers were pollinated with one or more pollen grains. As some of these flowers had presumably just opened, the final percentage that were pollinated was even greater; an average of 33% flowers produced mature fruits. Hence, in Tanzania it seems unlikely that there is insufficient pollination and it is probable that any lack of fruit is due to physiological causes. This was confirmed when it was shown that hand-pollination did not improve the set or crop produced by natural pollination.

However, in India only 3%, of hermaphrodite flowers produced mature fruit (Rao and Hasson, 1957). In India a few ants were the only insects that visited the flowers and the set was increased by artificial pollination and by spraying the flowers with water to simulate natural rain showers. The average percentage of perfect flowers found per panicle was about 4% in India (Rao and Hasson, 1957), 17% (Bigger, 1960), and 27% (Northwood, 1966) in Tanzania. Hence, varieties or localities with the smallest ratio of hermaphrodite flowers had the smallest set and *vice versa*. This suggests that either nutritional or genetical factors may be involved, but more work is needed to decide which.

Chapter 15

Papilionaceae : *Medicago*

A typical flower of Papilionaceae has a standard petal, two wing petals and two keel petals all of which are partially joined at their bases to form the corolla tube (Fig. 83). The keel petals enclose the staminal column of ten stamens and a single style. Nectar is secreted at the base of the corolla tube.

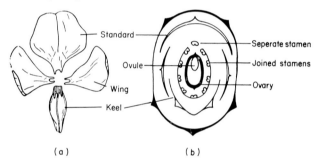

Fig. 83. Floral diagram and dissected flower of Papilionaceae (after Robbins, 1931, reproduced with permission of McGraw-Hill Book Company).

Medicago sativa L.

The flower of *M. sativa*, alfalfa, or lucerne, is constructed so that the staminal column is held under pressure within the keel by interlocking projections from the keel and wing petals. When the tension is released the staminal column snaps forward against the standard petal causing the pollen to be distributed. This process is known as tripping (Fig. 84); once it has occurred the staminal column does not return to the keel. Tripping is accomplished when the keel is pressed down by the weight of a visiting bee; at the same time the staminal column strikes the ventral side of the bee's head at the base of the proboscis and a small ball of pollen accumulates at this site.

Pollination process

During tripping the stigmatic membrane is ruptured. This must occur to provide a liquid medium for pollen germination and growth (e.g. Armstrong and White, 1935; Hadfield and Colder, 1936; Petersen, 1954; Umaerus and Åkerberg, 1959), and flowers that are not tripped do not set seed. About 7–9 h after tripping, the pollen tubes enter the ovary and, 24–27 h later, fertilization occurs. Viability of the pollen decreases only gradually with age, and even 8-day old pollen still shows 80% germination. The ovules are viable from opening to withering of the flower (see Pankiw and Bolton, 1965) but few of the ten to twelve ovules develop into seeds.

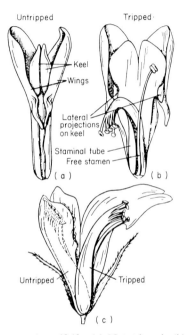

Fig. 84. Flower of *Medicago sativa*, alfalfa. (a) Not tripped; (b) tripped; (c) position of staminal column in tripped and untripped flower (after Robbins, 1931, reproduced with permission of McGraw-Hill Book Company).

Although it has been adequately demonstrated that high yields are possible, obtaining sufficient pollination is a complex problem. Tysdal (1940) focussed attention on the importance of tripping in relation to seed production and, since then, many studies on this process and the agents responsible for it have been made.

Individual plants range from being completely self-sterile to completely self-fertile (Bolton, 1948; Nielsen, 1958) but, on average, only 17–46% of flowers form pods after self-pollination (see Carlson *et al.*, 1950; Palmer-Jones and Forster, 1965) and more seeds are formed per pod from cross- than from self-pollination. Results obtained by different observers who artificially self-pollinated and cross-pollinated flowers are given in Table 15. Seeds from cross-pollination are also larger.

Pollen tube growth rate is slower following self-pollination than cross-pollination and fewer of the basal ovules are fertilized (see Bolton, 1962) and although self- and cross-fertilized seeds are intermingled in the same pod, self-fertilized seeds do not occur beyond the fourth position from the proximal end of the ovary (Pankiw and Bolton, 1965).

TABLE 15

	Average no. seeds per pod	
	From cross-pollination	From self-pollination
Bolton (1948)	5·4	1·6
Lesins (1950)	3·5	1·5
Pharis and Unrau (1953)	3·0 plus	1·7
Petersen (1954)	3·9	1·6
Palmer-Jones and Forster (1965)	3·9–4·9	1·5

TABLE 16

	Cross-pollination	Bee pollination	Self-pollination
% flowers that formed pods	77	67	31
No. seeds per pod	5·1	4·1	2·5

When a staminal column strikes a bee's body opportunity for cross-pollination can occur. Pankiw et al. (1956) compared the effectiveness of self-pollination by hand, cross-pollination by hand, and bee pollination. Cross-pollination was superior to self-pollination but, as expected, results from bee pollination, which sometimes cause self-pollination and sometimes cross-pollination, were between the two (Table 16).

The amount of natural cross-pollination in alfalfa fields varies from 84–94% (see Pedersen, 1953; Bolton, 1962) and, because alfalfa is mostly cross-pollinated, it is important to allow for sufficient isolation of breeding plots. Bolton (1962) suggested that 192 m is adequate separation for small plots, but Johansen (1968) found a relatively high percentage of crossing 150 m from a pollen source and Bradner et al. (1965) found still 6·5% crossing occurred at 1600 m distance.

Pedersen and McAllister (1956) cross-pollinated by hand a third of the flowers of some plants but all the flowers in others. In the former circumstances an average of 66% of the pollinated flowers formed pods of 12·7 mg weight, compared with 47% of 11·3 mg weight when all the flowers of each plant were pollinated. It seems therefore that the tendency of plants to benefit from additional cross-pollination decreases with the amount of set that has already occurred, presumably because the bearing limit of the plant is being approached.

Automatic tripping

A certain amount of tripping occurs without insect visits. This may be caused by a variety of factors including light frosts, high temperatures, hail and heavy rain. Bolton (1962) concluded that automatic tripping results from any force that destroys or weakens the turgidity of the restraining tissues of wings and keel. Dwyer (1931) and Dwyer and Allman (1932) found that in the laboratory, at various relative humidities, flowers tripped readily at 38–42°C (100–108°F) and they concluded that, in the field, dry hot conditions could remove the turgidity and result in automatic tripping of some of the flowers. Ufer (1933) and Izmaïlova (1934) also concluded that temperature was a most important factor inducing automatic tripping. Tysdal (1946) reported that warm bright days are conducive to tripping and whereas rain could cause up to 7% tripping, wind had little effect. Hobbs and Lilly (1955) subjected flowers to a series of temperatures and humidities in the laboratory and found that between 30 and 40% automatic tripping began to occur when temperatures reached 32°C (90°F).

The amount of automatic tripping recorded differs greatly. Only up to 0·9% was found by Pedersen and Stapel (1944) in Denmark, and 0·8% by Pharis and Unrau (1953) in North America. However, in Bulgaria, Stereva (1962) found that, over 5 years, 4·1–25·6% of flowers tripped automatically, and Lesins *et al.* (1954) in Sweden discovered that in favourable weather 13–60% of the flowers did so, the percentage differing with different varieties. Intermediate values of 5·4% and 3·5% were found by Hughes (1943) and Stephen (1955) in North America.

Although it has been possible to select self-fertile plants with a tendency to trip automatically and give heavy seed sets at normal temperatures, it has been found that the vigour, seed production and yields of their progeny were no better than from unselected self-pollinated plants (Kirk and White, 1933; Stevenson and Bolton, 1947; Lesins *et al.*, 1954). It seems more promising to select plants for their attractiveness to bees.

Mechanical tripping

From time to time various mechanical methods of tripping flowers have been used. These include beating the flowers with brooms, dragging ropes across the crop and using various sorts of rubber rollers (e.g. Koperzinskii, 1949; Zadražil, 1955). Although these methods cause a substantial proportion of flowers to trip, the inconvenience of the operations, the injury to the plants, and the low seed yields from selfing are generally considered to outweigh any possible advantages (e.g. Silversides and Olsen, 1941; Pedersen and Stapel, 1945; Pengelly, 1953; Pharis and Unrau, 1953). Tripping by mechanical means would be beneficial if the tripped flowers were then visited by bees which cross-pollinated them, but apparently honeybees rarely scrape pollen from the exposed anthers of flowers that have been tripped already, although, in general, the numbers that do so increase with the proportion of flowers tripped. Indeed, available evidence (see Pankiw and Bolton, 1965; Bohart et al., 1967) indicates that honeybees find tripped flowers unattractive, perhaps because nectar secretion soon ceases in tripped flowers, but further investigation is needed. Moreover, after tripping the stigmatic surface remains tightly pressed against the standard petal and it is doubtful how much cross-pollination could occur in such circumstances.

Pollen is sometimes found adhering to the standard petals of untripped flowers and it has been suggested that this may contribute toward cross-pollination when flowers are tripped (Dwyer and Allman, 1932; Hadfield and Calder, 1936). However, when pollen was artificially dusted onto standard petals and the flowers then artificially tripped their seed yield was not increased significantly (Lesins, 1950; Petersen, 1954; Pankiw and Bolton, 1965) indicating that any distribution of pollen by wind or machine is unlikely to have much influence.

Pollination by honeybees

Few insects other than bees trip alfalfa flowers (see Bohart, 1958; Bolton, 1962) but the relative importance of honeybees (Fig. 85) and wild bees (Fig. 47) differs greatly in different localities.

Cage experiments have demonstrated that honeybees are able to pollinate the flowers. Some of the results are summarized in Table 17.

There is also ample evidence that in some circumstances the seed yield of crops is related to the presence of honeybees. For example, Vansell and Todd (1947) reported that in 4 of the years between 1928 and 1934, when 4,000 or less colonies were moved annually into Utah from California, only between 75,000 and 363,000 kg of seed were produced compared to between 953,000 and 1,617,000 kg of seed in the years when from 10,000–14,200 colonies were imported. Sovoleva (1952) found that at 500 and 1,250 m from an apiary the seed crop was 310 kg/ha and 81 kg/ha respectively.

During 3 years Pankiw et al. (1956) in Saskatchewan compared the yield from 1·6 ha plots, separated by 4·8 km or more, which had $2\frac{1}{2}$, $7\frac{1}{2}$ or $12\frac{1}{2}$

colonies/ha. There were two replicates of each treatment each year. The average results are given in Table 18.

Kropáčová (1965) in S. Moravia obtained a positive correlation between the number of honeybees working on alfalfa crops and the number of tripped

Fig. 85. Honeybee visiting *Medicago sativa*, alfalfa, flower (Photo: W. P. Nye.)

TABLE 17

	Plots caged with bees	Plots caged without bees	Not caged
Hobbs and Lilly (1955)			
Wt (g) seed per cage, 1951	25	2	10
1952	202	22	92
Steuckardt (1961)			
Wt (g) seed per plant	4·2	0·5	2·9
Petkov and Simidchiev (1965)			
Wt (kg) seed per cage	119·0	0·6	223·1
Doull (1961) % flowers produced pods	57	10	35
Palmer-Jones and Forster (1965)			
% flowers produced pods	27·5	0	—

and pollinated flowers; the highest yields were obtained when 9–15 colonies/ha were used.

However, a high proportion of honeybees visiting alfalfa flowers fail to pollinate them. Vansell and Todd (1946) classified bees working alfalfa as "nectar-gatherers" entering the flowers, "side operators" and "pollen-gatherers". A pollen-gatherer visits mostly untripped flowers; it inserts its head into the centre of a flower or very slightly to one side, and braces itself with its middle legs on the wing petals to help force its head down the corolla tube. This releases the sexual column from the keel and it strikes the underside of the bee's head, momentarily trapping it and depositing pollen in the proboscidial fossa before glancing off onto the standard petal. The bee then uses its fore-legs to scrape pollen from the anthers. Rarely, a pollen-gatherer approaches from the standard petal and when it succeeds in tripping the flower, pollen is deposited by the staminal column onto the back of its head.

TABLE 18

	Colonies per hectare		
	$2\frac{1}{2}$	$7\frac{1}{2}$	$12\frac{1}{2}$
No. tripped flowers per 100 racemes	74	81	116
Seed yields (kg per ha)	50	81	131
Honey yields (kg per colony)	47	42	33

When a nectar-gatherer begins foraging on alfalfa it inserts its tongue into the throat of the flower but, after it has been momentarily trapped between the sexual column and standard a number of times, it learns to insert its tongue between the standard and wing petals so as to avoid the blow from the staminal column but it then fails to trip the flower on most visits. This change in behaviour is associated with an increase in working speed. Pollen-gatherers also learn to avoid becoming trapped by the flowers but, in contrast to nectar-gatherers their pollinating efficiency remains high and they trip most of the flowers they visit (Vansell and Todd, 1946; Reinhardt, 1952). A nectar-gatherer making a direct approach to a flower is often pinned down, and sometimes can only free itself with difficulty; very rarely it is unable to do so and dies in the flower. Until they learn the side approach, nectar-gatherers are apt to be very hesitant and tend to select withered or tripped flowers. They usually discard any of the pollen that gets on their bodies although some may have small pollen loads. Reinhardt (1952) observed that those nectar-gatherers that failed to learn the side approach within a few days usually deserted the crop. Kropáčová (1965) found that this learning process, if present at all, lasted for one day only.

Nearly all nectar-gatherers, as well as pollen-gatherers, working alfalfa, have alfalfa pollen in their proboscidial fossae (Grout, 1949; Levin, 1955; Vansell, 1955). However, Furgala et al. (1960) found that although 92–96% of bees on alfalfa had some alfalfa pollen in their fossae the predominant pollen was nearly always of another species; they supposed that the small amount of alfalfa pollen was probably deposited before the bees learnt the side approach to the flowers.

Pedersen and Bohart (1953) found that clonal differences existed in the force required to trip the flowers and in the force with which pollen is ejected, but increase in either of these factors was not related to any tendency of the bees to avoid them. Perhaps the change in behaviour of nectar-gatherers from entering flowers to working them from the side is related as much to increased speed of work as to avoiding blows from the staminal columns.

Nectar-trippers occur most abundantly when colonies are first moved to a field or when it begins to flower, but their numbers soon rapidly decline (Pedersen and Todd, 1949). It has been pointed out that colonies used for alfalfa pollination should contain plenty of brood which will provide a continuous supply of new foragers. The possibility of attempting to maintain the numbers of inexperienced bees by moving new colonies that have never worked alfalfa into the field at periodic intervals, has been suggested (see Bohart, 1957). However, Haragsin et al. (1965) reported that bees from colonies moved to an alfalfa field, and bees from colonies already present, behaved similarly and visited 15·1 and 15·3 flowers per min, and tripped 1·21 and 1·24% of the flowers they visited. Similar proportions of the bees had accumulated pollen in their fossae (89·7 : 84·0%).

A certain, but variable amount of tripping by experienced nectar-gatherers working flowers from the side does occur, but usually less than 2% of the flowers so visited are tripped. Unfortunately, it is not always clear in the literature whether the amount of tripping reported is by inexperienced nectar-gatherers, or accidentally by side workers while crawling over the flowers. Hobbs and Lilly (1955) observed that most accidental tripping of the flowers by nectar-gatherers occurred either when the bees alighted directly on the keels, or when they climbed over the keels, or stood on flowers below those from which they were gathering nectar. Occasionally, when a flower is accidentally tripped, a bee's head or tongue is temporarily trapped and, if the bee concerned also trips other flowers in the same way, cross-pollination is possible. However, nectar-gatherers that trip the flowers accidentally are usually not touched by the sexual columns and so cause only self-pollination which gives a low set; hence their value even when tripping the flowers is questionable. It seems possible that visits by bees that do not result in tripping may make the flowers easier to trip subsequently (Åkerberg and Lesins, 1949); this needs further investigation.

Tripping efficiency in different locations

Although only a small percentage of nectar-gatherers trip the flowers, their large numbers can help to compensate for this. Bohart et al. (1955) and

Bohart (1960) calculated that when 1% of nectar-gatherers trip the flowers sufficient set will be obtained when there are seven or more nectar-gatherers per square metre, but when less than 1% of the flowers visited are tripped it may be impossible for the crop to support enough foraging bees to produce sufficient seed. Unfortunately, in many parts of the world fewer than 1% of nectar-gatherers working alfalfa trip the flowers. Some examples are given in Table 19.

TABLE 19

Author	Location	% nectar-gatherers that tripped flowers
Hobbs (1950)	Alberta, Canada	0·8
Franklin (1951)	Kansas, U.S.A.	1·1
Fischer (1953)	Minnesota, U.S.A.	0·4
Menke (1954)	Washington, U.S.A.	0·4
Stephen (1955)	Manitoba, Canada	0·1
Stapel (1943)	Denmark	1·3
Stapel (1952)	Denmark	1·6
Petersen (1954)	Denmark	2·1–2·8
Åkerberg and Lesins (1949)	Sweden	0·8
Umaerus and Åkerberg (1959)	Sweden	0·8
Steuckardt (1962)	East Germany	1·3
Kropáčová (1964)	S. Moravia	1·2–2·2
Petkov and Simidchiev (1965)	Bulgaria	0·7–1·5

In North America, there is a tendency for the amount of tripping by nectar-gatherers to increase from the North East to the South West. It seems, in general, that nectar-gatherers are responsible for a greater set in arid than in moister regions (see Bohart, 1957). Probably as the critical temperature for automatic tripping is approached, nectar-collecting honeybees trip a greater percentage of the flowers they visit. Thus, McMahon (1954) in Saskatchewan found 99·8% of the flower visits were made by nectar-gatherers that approached the flowers from the side, and tripping occurred on only 0·3% of these. Hobbs and Lilly (1955) concluded that because honeybees trip so few flowers and cross-pollinate even fewer, and the pollinating period is so short, they are incapable of sufficient pollination to set a seed crop in Alberta. In contrast, Vansell and Todd (1946) found that in Utah nectar-gatherers tripped 17·5% of the flowers they visited.

Although the above information relates only to nectar-gatherers in many areas few honeybees collect pollen, and it seems that the tendency to do so is also greater in dry hot conditions and, like the percentage of tripping by nectar-gatherers, increases toward the south western parts of N. America (see Bohart, 1957). Thus, Hare and Vansell (1946) found that in California colonies collected about 32% of their pollen from alfalfa while it was in flower. Burkart (1947) reported that in the drier regions of Argentina where there are few competing flowers, honeybees are important pollinators of alfalfa from which they obtain much pollen. In Arizona, Butler et al. (1956) found that 15–50% of honeybees visiting alfalfa collected pollen.

Undoubtedly areas with high temperatures and constant sunshine favour pollen collection and give more days during alfalfa flowering on which foraging can occur, but the condition of the alfalfa plant is also important. Vansell and Todd (1947) investigated the behaviour of honeybees on adjacent fields with different irrigation levels. They found that the percentage of bees collecting pollen was much greater on fields that had received little irrigation, than on well-irrigated fields with lush growth (27–53% : 2–12%) and this was reflected in their seed production. In contrast, Móczár (1961), in Hungary found that an irrigated field received many more bee visits than a nearly dry field. He suggested that this was because the more abundant nectar in the irrigated field attracted more bees although it was of lower concentration. However, despite this, the set was greater in the dry field.

Probably, a main reason why pollen collection is greater in dry hot areas is that competing crops are largely absent, whereas in areas where irrigation of crops is not needed other pollen sources are usually within flight range. In fact, although where summer temperatures are high alfalfa is a fairly attractive source of nectar which has an average sugar content of 15–29% (e.g. Shaw, 1953; Montgomery, 1958; Haragsimová-Neprašová, 1960), most flowers are preferred to alfalfa as pollen sources. Vansell and Todd (1947) found that in one location, where little other pollen was available, honeybee colonies collected between 17 and 79% of their pollen from alfalfa; however, when they were transferred to alfalfa fields in other locations where competing crops occurred they ceased to collect alfalfa pollen. Linsley and MacSwain (1947), in California, found that cutting a nearby mustard field increased the percentage of bees working alfalfa for pollen, and McMahon (1954) reported that pollen collection from alfalfa increased during August in Saskatchewan as competition from other crops decreased. Indeed, when other crops are abundant honeybees may fail to collect any alfalfa pollen (e.g. Hobbs, 1950; Menke, 1954; Blagoveshchenskaya, 1955; Palmer-Jones and Forster, 1965). Major competitors vary with the area concerned but *Melilotus alba, Brassica nigra, Helianthus annuus*, and *Trifolium pratense* are among the most important.

Pollination by wild bees

The relative importance of wild bees increases in cooler humid areas where honeybees fail to trip the flowers or collect pollen. Bohart (1957) listed about seventy-five species of wild bees and indicated: (a) their importance as pollinators in different parts of the world; (b) their working speed; (c) the percentage of visited flowers that were tripped. Whereas honeybees visit averages of 7–17 flowers/min, *Bombus* spp. visit 10–30 and *Megachile* spp. 9–40. He pointed out that the relative importance of different species varied greatly from place to place and warned against generalizing about them. However, he concluded that wild bees more than 9 mm long are usually much more consistent alfalfa pollinators than honeybees, but bees 6 mm long never trip the flowers. Small halictids and adrenids often visit flowers already tripped; possibly, as suggested by Linsley (1946) and Pengelly (1953) they help to cross-pollinate them.

Most of the wild bees of widespread importance as pollinators in N. America and Europe belong to *Bombus* and *Megachile*, but other genera are important in more restricted regions. In U.S.S.R., Kostylev and Vinogradov (1934) found *Eucera clypeata*, *Andrena convexiuscula* and *A. labialis* were the most important pollinating agents. In Sweden, *Bombus terrestris* was found to be the most common pollinator although *Melitta leporina*, *Eucera longicornis*, *Megachile willughbiella* and honeybees were also important (Åkerberg and Lesins, 1949). In Denmark, apart from honeybees and bumblebees, *Melitta leporina* and *Eucera longicornis* are the most important pollinators (Stapel, 1952). Hobbs (1956) reported that *Megachile perihirta* is a particularly good pollinator in Alberta. He calculated that a single female provisions fifteen cells, and that it takes fifteen loads to provision a cell, 372 flower visits to obtain a load, and that five seeds are set on each flower visit; hence the visits from a single female are responsible for setting about 418,500 seeds. In Hungary, Móczáv (1961) found that *Melitta leporina*, *Eucera clypeata* and *Andrena ovatula* were among the most important pollinators. In Poland, Wójtowski (1965) reported that *Bombus terrestris* and *B. lucorum* were the most effective pollinators of alfalfa and pollinated 47 and 45% of the flowers they visited. Doull (1961) mentioned that in South Australia *Megachile quinquelineata* and *Nomia australiaca* were good pollinators.

The efficiency of different species of wild bees, even those belonging to the same genera may differ greatly, and the efficiency of the same species differs in different locations. Pengelly (1953), in southern Ontario found the tripping efficiency of bumblebees varied greatly with the different species from 35% (*B. borealis*) to 80% (*B. americanorum*); Stephen (1955) in Manitoba found *B. terricola*, *B. fervidus* and *B. vagans* tripped 87, 43 and 6% of the flowers they visited. Pengelly (1953) found that all *Megachile* spp., including *M. brevis*, were highly efficient (95–100%), in tripping alfalfa. Stephen (1955) found *M. frigida* and *M. latimanus* tripped 97 and 95% of the flowers they visited, whereas *M. brevis* tripped only 15% and obtained nectar like the honeybee by pushing its tongue into the base of the flower between standard and keel petals. In Kansas and California *M. brevis* is reported to be an efficient pollinator (Franklin, 1951; Linsley and MacSwain, 1947).

In contrast to the honeybee, most solitary bees visit alfalfa for pollen, and few collect nectar without pollen. Only bees with long slender tongues, such as *Anthophora* spp. and *Bombus* spp., can reach the nectaries through the throat of the flowers without tripping a large proportion of them. They leave their tongues extended for longer than pollen-gatherers and make no attempt to scrabble over the anthers. Holm (1966a) reviewed the data on the tripping efficiency of bumblebees and it was evident that nectar-gatherers trip few flowers compared to pollen-gatherers.

Although the pollinating efficiency of solitary bees is much greater per bee than that of honeybees, they are usually too few. Tysdal (1946) pointed out that pioneer seed areas with their associated high population of wild bees, give the best seed yields. In N. Saskatchewan (Peck and Bolton, 1946) and Manitoba (Stephen, 1955) it has been possible to associate a decrease in the production of alfalfa seed with increased clean cultivation of the land. When land was first cleared and alfalfa planted in small fields surrounded by bush the seed yield was high but, as the area planted increased, diluting the existing wild bee population, and the surrounding area was cleared, destroying nesting sites of wild bees, the yield became progressively lower, until growing alfalfa for seed was no longer worthwhile. Thus, seed production was limited to fields in recently settled areas.

In order to make use of the wild bee population to the best advantage, it has been suggested that: (a) the peak of flowering of the crop should be timed to coincide with the maximum wild bee population; (b) the fields should be kept to small size; (c) planting of competing crops should be avoided; (d) rough terrain and nest sites for wild bees should be provided; (e) existing nesting sites should be protected and extended; (f) other flowers should be provided early in the season to help the wild bee populations to build up (e.g. Bohart, 1957; Hobbs, 1958).

However, the number of wild pollinators fluctuates greatly and their peak of foraging does not always coincide with that of alfalfa flowering. It used to be generally thought that they were too unpredictable to be relied upon, and that there was no alternative but to use the less efficient honeybees. This situation has now changed dramatically in some areas by using as pollinators the alkali bee, *Nomia melanderi* and the leafcutter bee, *Megachile rotundata*, both of which are highly gregarious. *Nomia melanderi* shows a definite preference for alfalfa pollen and while collecting it pollinates about 95% of the flowers visited (Menke, 1952a,b). The bees work as close to their nests as possible, and usually do not move further afield until the flowers near at hand have been tripped. But they will, if necessary, work even 7 or 8 km away from their nests (Stephen, 1959). Local abundance of *N. melanderi* had often been correlated with high seed yields, and the introduction of irrigation into the N.W. States of U.S.A. at first greatly increased the area suitable for it to nest in. It has now been demonstrated that increasing the population of *N. melanderi* is a practical and worthwhile proposition (pages 111-116).

Bohart (1958b) discussed the possible advantages of introducing suitable wild bees into parts of the world where they do not already exist and mentioned Central Asia, the place of origin of alfalfa, as being particularly prolific in wild bee species that might be introduced into North America, to increase alfalfa pollination (see Popov, 1956; Ponomareva, 1959). Since then *Megachile rotundata*, which was accidentally introduced into North America from Eastern Europe or Western Asia, has become most important in pollinating alfalfa in North America, and by using artificial domiciles it has been found possible to produce large local populations (pages 103-111). *M. rotundata* prefers alfalfa to many other crops, and the bees' tendency to forage within about a hundred metres of their nests helps to restrict the visits of bees, housed in artificial domiciles on the field, to alfalfa only.

Efforts are being made to introduce both *Nomia melanderi* and *Megachile rotundata* to other countries, but because *M. rotundata* does not readily forage below 21°C (70°F), its use will inevitably be restricted. It is hoped that more introductions of this sort will be possible. As Bohart (1957) pointed out, a variety of species in any area helps to ensure that the short period of adult life of some, at least, is synchronized with the peak flowering of alfalfa.

Competition may occur between honeybees and other species of bees. Vansell and Todd (1947) observed that honeybees failed even to collect nectar from a field with a high population of *Nomia* spp. Menke (1952a) found that the rapid tripping of flowers by *N. melanderi* reduced the amount of nectar available to honeybees and hence their honey crop. Antsiferova (1957) reported that in the Soviet Union the larger the number of solitary bees present in an alfalfa field, the smaller the number of honeybees.

It has been suggested that, in areas where a very small percentage of nectar-gathering honeybees trip the flowers, their presence might be disadvantageous as they remove nectar and so decrease the attractiveness of the crop to wild bees which pollinate them (e.g. Peck and Bolton, 1946; Pengelly, 1953). But there was no information on this until Bohart *et al.* (1967) did a series of experiments to find the influence of wild bees and honeybees on each other. In general, they found that when the honeybee population on plots containing a wide variety of wild bees was increased, the number of wild bees decreased by between a third and a half. The population of *Nomia melanderi* was influenced by that of honeybees; it increased slightly when honeybees were removed and decreased when they were returned. However, the size of the honeybee population present did not affect that of *Megachile rotundata* but, when even moderate populations of *M. rotundata* were present, honeybees were no longer attracted to the field.

Even though honeybees sometimes trip only a small percentage of the flowers they visit, because of their relative abundance, they are more valuable than the few wild bees present. For example, Móczár (1961) in Hungary found that the tripping efficiency of honeybees was only a tenth of that of *Melitta leporina* but, because they were so much more numerous, they were the more valuable pollinating species.

Increasing the efficiency of honeybees

Various ways of increasing the pollinating efficiency of honeybees have been suggested. These include using sufficient colonies to saturate completely the nectar and pollen sources in an area. Increase in the number of colonies working a crop might increase the proportion of pollen-gatherers. Franklin (1951) reported that increasing the number of honeybee colonies present failed to increase the percentage of visited flowers that were tripped, but because McMahon (1954) found that the percentage of tripping by nectar-gatherers increased when the number of bees working the crop increased, this needs further investigation. Experiments to determine the value of periodic exchange of the colonies on crops by others from elsewhere are discussed on page 75.

Because honeybees are often reluctant to collect alfalfa pollen, colonies taken to pollinate a crop should have good pollen stores. If this is not possible, and no spare combs of pollen are available, feeding pollen substitutes might be advantageous. However, as pollen-gathering honeybees are so much more efficient than nectar-gatherers, ways of increasing pollen collection from alfalfa have been considered. Kropáčová (1965) found that increasing the pollen hunger of colonies was not effective, but reported that giving scented syrup to colonies, in an attempt to direct them to alfalfa, did give an increased set in one location. No attempt to increase alfalfa pollen collection by feeding sugar syrup to colonies has been reported (see page 86).

Breeding bees that are especially prone to collect pollen, and in particular alfalfa pollen, seems very important, and striking results have already been obtained (page 87). Åkerberg and Lesins (1949) found conspicuous differences between colonies of different races in their tendency to trip the flowers. Accidental tripping seems to occur more often with some strains of bees than with others (Petersen, 1954; Bieberdorf, 1949). If these characteristics are confirmed, perhaps they could also be selected by breeding.

Breeding plants that readily trip

Attempts have been made to select plants that are more readily tripped than average, or do not need tripping at all. Nielsen (1960) selected some plants whose stigmas and anthers projected through the tips of the keel petals and others whose keel petals were separated. Both types could be pollinated by bees without tripping and it was found that a greater percentage of bee visits to them than to normal flowers resulted in pollination. Further work on these is needed and, in particular, the possibility that under some conditions desiccation affects the exposed stigma needs investigating.

Honeybees would not be trapped by tripping flowers with vestigial standard petals. Pankiw (1967) found that honeybees tripped standardless mutant flowers on a greater percentage of visits than when they visited normal flowers, but transfer of pollen to the bee's bodies was not as effective as on normal flowers and seed production was no greater. The possibility of selecting an alfalfa flower from which bees cannot obtain nectar from the side does not seem to have been explored.

The sugar content of nectar of different varieties is similar (Kropáčová and Laitová, 1965) and contains about equal amounts of sucrose and fructose and slightly more glucose. The amount of nectar secreted per alfalfa flower and its sugar concentration has been found to vary from: 0·34–1·04 mg of 20–80% concentration (Pedersen and Bohart, 1953; Pedersen, 1956); 0·45–1·20 mg (Kropáčová, 1963); and 0·24–1·28 mg of 38–51% concentration (Petkov and Simidchiev, 1965).

There is good reason to hope that selecting clones for nectar secretion could increase bee visits and pollination. Ostaščenko-Kudrjavceva (1941) found that the amount and concentration of nectar varied with the variety concerned, and varieties that produced most nectar also produced the most concentrated. Abundant nectar appeared to be correlated with high sets. Bogoyavlenskii (1953) claimed that increases in either nectar secretion or sugar concentration led to more bee visits. Pedersen (1953) obtained positive correlations between nectar production per plant and seed production, and between seed production and honeybee visits regardless of whether the bees collected pollen or nectar only. Pedersen and Bohart (1953) found positive correlations between the number of visits pollen-collecting bumblebees made to different clones and their nectar sugar content, and Kropačová (1963) found, in three sites, a positive correlation between nectar sugar concentration and the number of bees present and, in one site obtained a positive correlation between nectar abundance and the number of bees present. Perhaps it may be possible to select some lines that are especially attractive to pollen-gatherers as well as to nectar-gatherers.

Increasing a crop's attractiveness

Various cultural practices can also increase pollination. Pedersen *et al.* (1959) in a series of tests showed that increasing the row spacing from 20 to 61 cm and reducing the seeding rate from 13 to 1 kg/ha doubled the seed yield. They obtained a positive correlation between the extent plots were thinned and nectar volume and concentration, number of bee visits and seed production. With more space surrounding plants, there was a reduction in relative humidity and an increase in light penetration, soil temperature and accessibility of all flowers to bee visits. Increased nectar secretion of a greater concentration attracted more bees resulting in more seed. Best seed yields were obtained in moderately dry plots. Adequate spacing seems to be equally, if not more, beneficial to pollen-gathering honeybees. However, Pedersen *et al.* pointed out that *Nomia melanderi* seems to prefer dense moist rather than dry sparse growth. Hence, depending on whether alfalfa in the area concerned is primarily pollinated by *N. melanderi* or honeybees, dense or sparse growth will give the best yields.

Staggering the cutting dates of the first bloom in a field so as to curtail the amount of second bloom present at any one time has also been suggested as a way of making increased use of the pollinator force available (Drake, 1949; Todd, 1951).

Estimating pollinator population needed

It is difficult for a grower to evaluate the pollination that is occurring in his crop. Although a low rate of tripping is sufficient for moderate seed production, the minimum required is difficult to determine accurately. Attempts at estimating the pollinator population present merely by counting bees is insufficient and, because many bees do not pollinate, their behaviour must also be observed. When alfalfa flowers have been extensively visited in the morning, there may be little forage left in the afternoon, so a low bee population late in the day is not necessarily indicative of poor pollination. To circumvent this difficulty counts must be made at regular intervals throughout the day.

Probably the best way to evaluate pollination is by the appearance of the field (Bohart, 1957). A field that is being adequately pollinated has a dull brown appearance because of the withered tripped flowers and developing pods. If many fresh flowers are apparent, insufficient pollination is probable.

Another possible way of determining the amount of pollination occurring was suggested by Nye and Pedersen (1962). The nectar-sugar concentration of an alfalfa flower varies in accordance with the prevailing relative humidity and the time available for evaporation. When nectar is removed from a flower additional secretion is stimulated. Nye and Pedersen argued, therefore, that hence the concentration of nectar collected by bees will depend upon the rapidity with which it is being removed, which by itself reflects the amount of bee visitation. They demonstrated that this was so, and suggested that in parts where the relative humidity is fairly constant and the relationship between nectar concentration and the number of trippers has been established, this method might be used to determine the pollination status of a crop. However, there is unlikely to be a firm relationship between the number of trippers and nectar concentration, and this method is probably more suitable for some species other than alfalfa.

The number of honeybee colonies needed depends largely on the proportion that are visiting flowers for pollen. Bohart (1951) suggested an average of six nectar-gatherers per square metre to get sufficient pollination, but if 5% of the bees present are pollen-gatherers 2–6 bees/m^2 would be sufficient. For profitable honey production not more than 2–4 bees/m^2 should be present; financial arrangements between grower and beekeeper should take this into consideration. Todd and Vansell (1952) recommended 15 or more colonies per hectare when nectar-gatherers are depended upon for pollination but 7½ or fewer when plenty of bees are collecting pollen. They also suggested that the colonies should be moved to crops at intervals to take full advantage of the higher rate of tripping that occurs at first. Lecomte (1959) suggested that when pollen-gatherers are abundant 6–7 colonies/ha are sufficient. In Australia where there are insufficient honeybee colonies to pollinate the large acreages of alfalfa harvested for seed, and wild bees are far too few, 12½ colonies/ha on irrigated fields and 5 colonies/ha on drier land are recommended (Doull and Purdie, 1960).

Chapter 16

Papilionaceae : *Trifolium*

Trifolium pratense L.

A flower head of *Trifolium pratense* L., red clover, is composed of 50–200 flowers (average of 140, Williams, 1930) which open in ascending order from its base to its top. The terminal heads have more flowers than those developing later and when two harvests are taken from the same crop, there are more flowers per head during the second than during the first flowering period (Pammel and King, 1911). It takes 6–10 days for all the flowers of a head to open and a single plant usually flowers for several weeks.

The flower (Fig. 86) is typical of the Papilionaceae. The slightly curved pistil is longer than the stamens so the stigma extends beyond the anthers. The anthers dehisce in the bud. Nectar, whose average sugar concentration is 29–66% (e.g. MacVicar *et al.*, 1952; Shaw, 1953; Maurizio and Pinter, 1961), is secreted at the base of the stamens and collects in the corolla tube. When sufficient pressure is exerted on the standard and wing petals, the stigma and anthers protrude from the keel petals, but return to their former position inside the keel when pressure is released. Hence the anthers and stigma are pressed against the underside of a visiting bee's head and pollination can occur. Fertilization occurs from 18–50 h after pollination, depending on the temperature. Although the two-celled ovary has two ovules, only one normally develops after fertilization.

Pollination requirements

The pollination of red clover has long been of considerable interest. In many parts of the world the seed yield of red clover is only a fraction of that theoretically possible and this is thought to be largely because of inadequate pollination (see Williams, 1925; Martin, 1941; Bohart, 1957; Åkerberg and Stapel, 1966). Much research has been done recently, particularly in Scandinavian countries where considerable emphasis has been put on red clover breeding.

Early work established that red clover is self-sterile. Darwin (1876) found that 100 flower heads, covered to exclude bees, failed to produce any seeds while 100 heads not covered gave 2,720 seeds. Williams (1925) covered 1,790 heads with muslin bags or wire cages and obtained only thirty-one seeds.

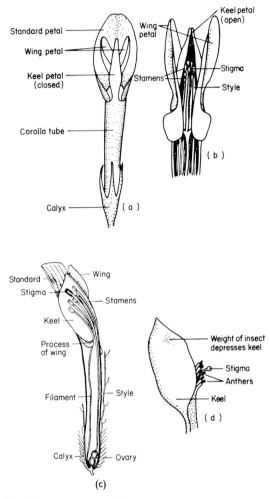

Fig. 86. Flower of *Trifolium pratense*, red clover: A. posterior view; B. anterior view with standard petal removed; C. diagrammatic longitudinal section; D. side view with keel depressed (after Pieters and Holowell, 1937; Poehlman, 1959).

He thought that because the anthers lie below the stigmas spontaneous self-pollination cannot occur unless the flowers are disturbed by wind or insects. Westgate and Coe (1915) found that when red clover flowers were self-pollinated, the pollen germinated normally on the stigmas, but the growth of the pollen tubes was slow and failed to reach the ovules before they degenerated. Williams (1925) self-pollinated flowers with dead bees and pieces of card, and by gently rolling heads between thumb and forefinger, but only 37 of 123 heads that were self-pollinated produced any seeds and then only a few. Hence, flowers need to be cross-pollinated by insects. The fertility of flowers rapidly decreases (Williams, 1925; Free, 1965a) so sufficient pollinators should be present to ensure that flowers are visited soon after they open. Woodrow (1952) caged plots of red clover to exclude pollinating insects during periods when no pollination was desired. He found that for maximum yields pollination should be provided throughout the entire flowering period.

Insect pollinators

Few insects other than bumblebees and honeybees (Fig. 87) play a part in red clover pollination. Although Lepidoptera are sometimes frequent visitors

Fig. 87. Honeybee entering the front of *Trifolium pratense*, red clover, flower.

they are unable to depress the keel (e.g. Pammel and King, 1911). Other insects that have been recorded and could pollinate the flowers include *Megachile* spp. (Jorgensen, 1921), *Osmia* spp. (Jorgensen, 1921; Maurizio and Pinter, 1961; Bohart, 1966), *Tetralonia* spp. and *Melissodes* spp. (Folsam, 1922), *Andrena wilkella* (Benoit *et al.*, 1948).

It has long been accepted that bumblebee species with long tongues, which can reach down to the base of the red clover corolla tubes, are ideal pollinators. Their effectiveness was dramatically demonstrated by the increase in red clover seed production soon after they had been introduced into New Zealand and had rapidly increased in number (e.g. Dunning, 1886; Montgomery, 1951), although the bumblebee population of New Zealand seems to have since declined. Hawkins (1956) found a close correlation between seed set and the number of long-tongued bumblebees present (e.g. *B. agrorum*, *B. hortorum*, *B. lapidarius* and *B. ruderatus*) in six single cut seed crops in 1955, but not between seed set and the number of honeybees or short-tongued bumblebees (e.g. *B. lucorum* and *B. terrestris*). Positive correlations were also found between seed set and the number of bumblebees present on eleven single cut and three double-cut crops in 1956 and nine single cut crops in 1957 (Hawkins, 1961). Richards (1953) suggested that red clover should be grown in small fields so that the limited bumblebee population would be more likely to pollinate sufficient of the flowers; and Hawkins (1958) found that in each of 3 years, during which he sampled the number of seeds per head on 29, 92 and 63 crops respectively, the set decreased with increase in size of the field. Another practical demonstration of the pollinating ability of bumblebees was provided by Vare (1960) in Finland, who found that the seed yield of flowers decreased as the distance between them and the nearest bumblebee colony increased from 0–200 m.

"Robber" bees

Unfortunately, some bumblebee species (e.g. *B. lucorum* and *B. terrestris* of Europe, and *B. affinis* and *B. terricola* of N. America) which have relatively short tongues (4·5–6·5 mm), bite holes in the bases of the corolla tubes and insert their tongues through the holes to obtain nectar (Fig. 88). These are known as "primary" robbers. Other bumblebee species and honeybees may obtain nectar through the holes that have been bitten but do not bite the holes themselves (Fig. 89); these are known as "secondary" robbers. Both primary and secondary robbers are sometimes referred to as "negative" bees in contrast to the "positive" bees which enter the flowers and pollinate them.

Mutilation of the corolla does not itself diminish the ability of the flower to produce seed (e.g. Williams, 1925; Hansen, 1934), and Skøvgaard (1936) in Denmark found that flowers that had been pierced were subsequently entered to the same extent as flowers not so pierced, so that pollination of them is not excluded. However, because of their habit of robbing the flowers, short-tongued bumblebees are, in general, less efficient pollinators than long-tongued bumblebees. When caged on small plots they gave only a third as much seed as long-tongued bumblebees or honeybees, and the set of a crop where they formed the majority of the bee population was relatively low (Palmer-Jones *et al.*, 1966). Besides failing to pollinate the flowers themselves, robber bumblebees enable honeybees to obtain nectar without doing so.

Fig. 88. Bumblebee (*Bombus lucorum*) biting a hole at the base of *Trifolium pratense*, red clover, flower.

Pedersen and Sørensen (1935) found there was a positive correlation between the number of robber bumblebees and the number of robber honeybees. Such a relationship appears to be general in Europe. However, although in New Zealand about 93% of short-tongued bumblebees rob the flowers when collecting nectar, honeybees and long-tongued bumblebees seldom collect nectar through holes bitten by them (Palmer-Jones *et al.*, 1966).

Williams (1925) pointed out that the effect, if any, of the behaviour of primary robbers on seed yield was not known but suggested that robber bumblebees might be attracted away from red clover by growing *Vicia sativa* nearby. Benoit *et al.* (1948) suggested growing *Phaseolus vulgaris* for the same purpose. Some authors (e.g. Stapel, 1934; Swart, 1960) have suggested that exterminating or suppressing primary robbers might cause an increase in the value of honeybees in red clover seed production, but others are not so sure. Thus, in Denmark, Pedersen (1945) found that *B. terrestris* was the most important bumblebee species visiting red clover and in New Zealand it has been found that although a large proportion of *B. terrestris* rob the flowers, the number that enter the flowers still exceeds that of other species (Forster and Hadfield, 1958). Hawkins (1961) obtained, in one year, a significant correlation between the seed set of crops and the number of robber bumblebees present. He pointed out that this supported the suggestion (Free and Butler, 1959) that honeybees which were secondary robbers might attract others that enter the flowers and pollinate them. However, he was unable to prove this as nearly all the honeybees he saw were robbing the flowers. It is

Fig. 89. Honeybee obtaining nectar through holes bitten by bumblebees in *Trifolium pratense*, red clover, flowers.

perhaps relevant that Schlecht (1921) observed that although two-thirds of the honeybees present on crops were robbing the flowers, the remaining third alone were more numerous than the bumblebees present. However, increasing the number of robber honeybees might also be disadvantageous as they reduce the amount of nectar available to bumblebees that enter and pollinate the flowers.

The extent to which robbing occurs varies with the number of short-tongued bumblebees, the length of the corolla tubes and the amount of nectar present (e.g. Schwan, 1953). Skøvgaard (1936) in Denmark and Bingefors *et al.*

(1960) in Sweden found that, although there was considerable variation from year to year, the proportion of negative *B. terrestris* was greater on early than late clover, and increased with the number of worker bees per unit area.

Pollinating ability of honeybees

Whereas at the beginning of this century it was generally agreed that bumblebees could pollinate red clover, it was recognized that there were often too few of them, and there was much controversy about the ability of honeybees to do so (see Westgate and Coe, 1915; Williams, 1925). These early doubts were based on the supposition that the corolla tube was too long for the honeybee's short tongue to reach the nectar (e.g. Martin, 1938). However, when honeybee colonies are enclosed in cages over plots of red clover satisfactory pollination occurs. Thus, Westgate and Coe (1915) found that two plots caged with honeybees and bumblebees respectively gave 32·7 and 30·4 seeds/head. Palmer-Jones *et al.* (1966) found that the set on small plots caged to exclude bees varied from 0–0·2%, whereas that in cages containing long-tongued bumblebees or honeybees was about the same as in the open field and varied from 58–79%. A summary of results from some other cage experiments with honeybees is given in Table 20.

TABLE 20

	Caged without honeybees	Caged with honeybees	Not caged
Richmond (1932) Seeds/head	—	61·5	67·3
Ševčenko (1939) % seed set	0·2	66·5	34·0
Crum (1941) Seeds/head	0	107	57
Dunham (1943) hl/ha	0·1	14·7	—
Anderson and Wood (1944) Seeds/head	1	56	37
Killinger and Haynie (1951) kg/ha	0	43	45
Laere and Martens (1962) kg/ha	39	540	585

Although honeybees can be forced to pollinate red clover when caged over it, they can be conspicuously absent from flowers in the open. Thus, Williams (1925) in Wales, failed to find a single honeybee working red clover in 1922, 1923 and 1924. But, in the right conditions nectar does rise sufficiently high in the corolla tube for the honeybee to reach it, and when this occurs the bee can suck up the rest by capillary attraction (Goetze, 1948). Hot, dry summers are especially favourable for nectar secretion and in such circumstances numerous honeybees may visit the crop (e.g. Stapel, 1934; Butler, 1941) and colonies may produce much red clover honey (Armstrong and Jamieson, 1940; Palmer-Jones et al., 1966). Goetze (1948) reported that honeybee colonies give considerable yields of honey from red clover only about once in every 12 years when the depth of nectar reaches 5 mm in hot and dry July weather. Furthermore, although long-tongued bumblebees usually work consistently on red clover crops for most of the day, the number of honeybees present are sometimes subject to large fluctuations (e.g. Stapel, 1934; Schwan, 1953; Free, 1955). This is believed to be due to the honeybee's inability to obtain nectar except when it is exceptionally abundant.

Honeybee colonies sometimes collect a large proportion of their pollen from red clover (e.g. Stapel, 1934; Synge, 1947). Thus, Braun *et al.* (1953) put fifteen colonies beside a red clover crop for 20 days and found an average of 83% of the pollen collected on the last day came from red clover; but, as expected, there was considerable colony variation and, over the whole period, one colony collected as much as 71% and another as little as 11% red clover.

Opinions on the relative importance of nectar and pollen in attracting bees to red clover differ greatly. Gubin (1936) suggested that nectar was more important than pollen and that honeybees did not visit red clover for pollen only, but pollen-gathering merely resulted automatically while a bee collects nectar. In contrast, Dunham (1939a) reported that few honeybees on red clover collected nectar but most deliberately collected pollen. He observed that nectar-gatherers kept their forelegs stationary on the outside of the keel, whereas pollen-gatherers combed their forelegs over the anthers and removed the pollen grains. He found averages of 148 and 1,153 pollen grains in florets that had and had not been visited by honeybees collecting pollen. Because pollen loads contain from 284,000–347,500 grains (Dunham, 1939a; Butler, 1941) honeybees visit about 300 flowers per trip. Forster and Hadfield (1958) found that all bumblebees and honeybees entering flowers collected pollen but very little, if any, nectar. Free (1958d) was unable to show any correlation between the size of nectar loads of bees entering flowers and the weight of their pollen loads; some had large nectar loads but small pollen loads, and others had large pollen loads but little or no nectar in their honeystomachs. This confirmed that honeybees visit clover flowers even when they can only collect pollen from them. Undoubtedly, many factors govern the proportion of bees collecting pollen; these include its availability relative to nectar and the needs of the colony at the time. Skøvgaard (1936) in Denmark reported that, although pollen-gathering honeybees make a considerable contribution toward pollinating early clover, honeybees are particularly inclined to collect red clover pollen when there is a scarcity of other forage plants, as commonly happens during the second bloom.

Free (1965a) determined the ability of honeybees to pollinate red clover in the field by exposing previously bagged heads until they were visited by insects and then re-bagging them. Visits by honeybees and bumblebees were almost equally effective and the set of newly opened flowers was between 60 and 80%; this compared well with that obtained by hand pollination. Honeybees and bumblebees that entered flowers were more efficient when they collected pollen than when they collected nectar only. Perhaps nectar-gatherers discard any pollen they collect inadvertently and so keep their fossae more free of pollen and this, in turn, affects pollination. The number of florets visited per head varied with the amount of nectar and pollen available. Presumably the tendency of pollen-gatherers to visit more florets per head than nectar-gatherers, reflected their greater satisfaction with it. However, there was no evidence that the percentage of seed set decreased with the number of florets visited per head. It is difficult to understand why this was so, because the compatible pollen a bee carries when it arrives at a plant would soon become diluted with the plant's own pollen. Further experiments to test this are needed. Although honeybees visit only about 3·0 florets per head compared to the 4·5 visited by bumblebees (Umaerus and Åkerberg, 1959; Dennis and Haas, 1967), there is no evidence that this makes them better pollinators.

Attempts have been made to compare the seed yield of fields with and without honeybee colonies. Gubin (1936) reported that early experiments by Klingen showed red clover seed production could be increased three-fold by introducing honeybees of a Caucasian strain to the field. Beard, *et al.* (1948) found that fields with few honeybees, and which therefore relied upon wild insects for pollination, had calculated yields of 0·4–0·6 hl/ha compared with 3·7–11·1 hl/ha when honeybee colonies were present. Thomas (1951) found that in 1949, 9 crops with honeybee colonies gave 77 kg seed/ha compared to 40 kg/ha of 9 crops without colonies, and in 1950, 13 crops with colonies gave 87 kg/ha compared to 50 kg/ha of 13 crops without colonies. Peterson *et al.* (1960) reported that when honeybee colonies were adjacent to red clover fields, seed yields were good, but that they decreased greatly when they were more than 0·8 km away. Finally, Pritsch (1966) discovered that in five districts of East Germany where additional honeybee colonies were imported for pollination, red clover seed yields were 30–220% greater than in districts where no additional colonies were imported.

Hawkins (1961) was unable to correlate the honeybee population on crops with seed set but, as he pointed out, because the honeybee population on a crop fluctuates more than the bumblebee population it is more difficult to estimate, and the fact that no correlation between seed set and honeybee population is found does not necessarily mean that none exists.

In New Zealand, Palmer-Jones *et al.* (1966) compared the seed set of several fields with the population of honeybees, long-tongued bumblebees and short-tongued bumblebees present. On fields that had few long-tongued bumblebees but many honeybees, a good set was obtained, and they concluded that red clover can be pollinated as effectively by honeybees as by the principal species of long-tongued bumblebees. These findings are difficult to reconcile with the supposed low seed yields of red clover in New Zealand before bumblebees were introduced, unless the honeybee population then was inadequate. Perhaps the best evidence of all that honeybees can pollinate red clover in the right conditions is provided by the favourable seed yields in Australia (Hills, 1941) where there are no bumblebees.

Braun *et al.* (1953) showed that seed yields decreased as distance from colonies on the edge of a field increased (Table 21).

TABLE 21

	Distance (m) from colonies					
	0–122	122–244	244–366	366–488	488–610	610–732
Average no. honeybees/1·7 m²	16·6	15·9	9·7	9·7	4·2	5·2
No. kg seed/ha	57·3	46·0	29·4	15·6	9·6	11·4

In a similar experiment Walstrom (1958) obtained similar results.

Length of corolla tube

There is, therefore, ample evidence that, in some circumstances at least, honeybees do increase the seed yield of red clover. However, unfortunately they cannot always be relied upon to visit it. Red clover pollen is readily accessible to honeybees and bumblebees and it is not clear why it is not always sought. Stimulating colonies to collect red clover pollen by feeding them with sugar syrup has proved to be successful (page 86) but needs to be tested more extensively. Red clover nectar is often inaccessible to honeybees and to some species of bumblebees, and many investigations have been made on the relationship between the tongue length of bees and the length of corolla tubes and the height of their nectar. An illustration of the problem is provided by Dunham (1939b). He calculated that a honeybee would be able to reach 7·9 mm into a corolla tube; this distance comprised: the length of its tongue, 6·3 mm; additional reaching distance of the tongue, 0·5 mm; depth of the corolla tube into which a bee can push its head, 1·1 mm. He found that the the average corolla tube length was 9·5 mm indicating that the nectar must rise 1·6 mm in the corolla tube before the bee can reach it. Similar conclusions have been reached by others (e.g. Martin, 1938; MacVicar et al., 1952). Apparently small differences in any one of these factors can make pronounced differences to the number of honeybees visiting a crop.

Several attempts have been made to increase visitation by using or breeding honeybees with longer tongues, and flowers with shorter corolla tubes or greater nectar secretion. Although there may be a wide range in the corolla tube length of a variety, even on different parts of the same plant, a positive correlation has been demonstrated between the corolla tube length of parent and offspring (Starling et al., 1950) and consistent differences occur between varieties (Åkerberg, 1953) which can sometimes be correlated with the number of honeybees visiting them (Julén, 1953a).

TABLE 22

Variety of red clover	Mean length (mm) corolla tube	Calculated yield kg seed/ha	No. seeds/head	No. bumblebees Entering flowers	Robbing
Cornish Marl	10·4	310	51	432	129
Aberystwyth S 123	10·3	301	64	560	97
New Zealand Montgomery	10·2	254	53	493	160
Montgomery late	10·2	322	52	401	86
English singlecut (Cotswold)	9·9	511	62	936	96
Dorset Marl	9·4	379	72	398	47
Essex Broad Red	8·9	436	84	1054	26
Altaswede	8·7	397	65	919	17

Differences in the corolla tube lengths of different varieties may also influence which bumblebee visitors they receive. Many authors have measured the tongue lengths of worker bumblebees and although, because of the different techniques used, the actual measurements given by different authors differ greatly, the comparative average sizes for the different species are relatively constant (see reviews by Brian, 1954; Bohart, 1957; Holm, 1966a). Hawkins (1965) investigated during 3 years, the corolla tube length, seed yield and pollinating insects of eight varieties of red clover (Table 22). The same relative differences in corolla tube length between varieties were found each year. In general varieties with the shortest corolla tubes had a greater set than those with the longest and were visited by more bumblebees. Hawkins grouped his bumblebees in to subgenera and demonstrated that *Lapidariobombus*, which has a shorter tongue than *Agrobombus* and *Hortobombus*, showed a greater preference for varieties with shorter corolla tubes. *Terrestribombus* that entered flowers behaved in a similar way to *Lapidariobombus* but *Terrestribombus* that robbed the flowers preferred varieties with longer corolla tubes, presumably because less of their nectar had been removed.

Although positive correlations between shorter corolla tubes, increased bee visits and seed set are sometimes found, this is not invariably so. Armstrong and Jamieson (1940) and Wilsie and Gilbert (1940) were unable to demonstrate that the lengths of the corolla tubes of different varieties influenced their seed yield when pollinated by honeybees. Moreover, attempts to breed clones with shorter corolla tubes have been unsuccessful because short corollae seem to be genetically linked to lack of vigour and other undesirable characteristics (e.g. Wexelsen, 1940; Starling *et al.*, 1950; Stählin and Bommer, 1958). Indeed, Damisch (1963) found that the shorter the corolla tube, the less was the percentage seed set.

Tongue length

There seems to be similar uncertainty regarding the value of honeybee strains with different tongue lengths. Stapel and Erikson (1936) found the tongue length of Italian bees was slightly greater than that of Danish, and that an average of 33% of the bees of fifteen Italian colonies visited red clover compared to 13% of the bees of sixteen Danish colonies. Others (e.g. Pedersen, 1945; Schwan, 1953; Stapel, 1935) have also reported that long-tongued Italian bees are more consistent visitors to red clover than short-tongued Northern bees, which are more inclined to rob the flowers. Alpatov (1948) found that the seed set on red clover fields increased with nearness to Ukrainian and Caucasian strains of bees with long tongues, but not with nearness to North European strains with short tongues; the long-tongued bees also stored more red clover honey. Lehmann (1952) supposed that the abundance of Caucasian bees on red clover was associated with its wedge-shaped head which allowed it to penetrate more deeply into the flowers. Laere and Martens (1962) found that in Belgium honeybees from colonies with average tongue lengths of 6·8 and 6·4 mm made 40 and 28% visits in which they did not rob the flowers. However, in contrast to the findings of the above workers, others (e.g. Gubin, 1936; Skøvgaard, 1956; Åkerberg and Umaerus, 1960; Valle *et al.*, 1962) have concluded that there are no consistent differences in the tendencies of different races to choose red clover, or to obtain nectar without robbing, and it does not seem to have been adequately demonstrated that long-tongued strains of honeybees collect relatively more pollen or work faster than short-tongued strains.

The importance of the amount of nectar secreted by a flower in determining whether it is visited by bees was first demonstrated by Gubin (1936), who obtained correlations between the percentage of flowers with nectar reaching a certain minimum height in the corolla tubes and the number of bees on the crop. About 0·22 mg of nectar of 30% sugar concentration is secreted per flower in 24 h (Maurizio and Pinter, 1961), but due to evaporation its sugar concentration increases during the day to about 50–60%, while its volume diminishes. In association with this, nectar collection by honeybees entering flowers is more common in the morning than in the afternoon, and pollen collection and robbing are more common in the afternoon (e.g. Braun *et al.*, 1953; Schwan, 1953; Johansen, 1966). Genetic variability in nectar secretion occurs between different clones (Shuel, 1952; Ryle, 1954) and different varieties (Anderson and Wood, 1944) of red clover, but the actual selection of lines giving increased nectar production does not seem to have been undertaken.

Indeed, as suggested by Åkerberg (1952) and Stählin and Bommer (1958), selecting plants for attractiveness to bees and for the amount of seed produced seems more promising than breeding for selected characters. Thus, in an attempt to select a strain of tetraploid red clover suitable for pollination by honeybees, Bingefors and Eskilsson (1962) caged honeybees over a plot and found that plants from the seed produced had a greater set when visited by honeybees than plants from seed produced in uncaged plots. In a similar preliminary experiment, Hawkins (1966) showed that seed from honeybee pollination gave plants whose flowers contained more nectar, which was more easily reached than in the original stock. However, further selection failed to bring an increase in nectar availability.

Tetraploid red clover

In recent years use of tetraploid red clover varieties in Scandinavian countries has given increased fodder yield, good persistency and increased disease resistance compared to diploid varieties. However, they also give less seed. There are a number of reasons for this including contamination with diploid pollen. Diploid pollen tubes grow twice as quickly as tetraploid and, when tetraploid red clover is fertilized with diploid pollen, the resultant embryos abort; however, this detrimental effect of pollination from adjacent diploid fields is limited to a narrow strip along the tetraploid field edge and is considered of no great practical importance (e.g. Wexelsen and Vestad, 1954).

One of the most important factors limiting seed yield in the tetraploid varieties is inadequate pollination which is linked with the deeper corolla tube of the tetraploid flower. Wexelsen and Vestad (1954) in Norway found that, in 1950 and 1953, tetraploid corollae were 0·8–1·1 mm longer than diploid. In Germany, Skirde (1961) reported that, although tetraploid corollae were 1·0–1·5 mm longer, this was more than compensated for by their extra width which allowed bees to reach further into the flowers. Similar measurements have been recorded by others (e.g. Åkerberg, 1953; Julén, 1956; Bingefors and Eskilsson, 1962). Although more nectar is produced by tetraploid than by diploid flowers (e.g. Maurizio, 1954; Valle et al., 1962; Skirde, 1963) because the corollae of the tetraploid are wider, the height of nectar in both types was similar.

Dennis and Haas (1967) found the average lengths of the corollas of diploid and tetraploid forms were 9·5 and 10·0 mm respectively, but each showed considerable variation. There was little difference in the nectar height in the diploid and tetraploid (1·3 and 1·4 mm) so the depth a bee would need to reach to obtain it would be 8·2 and 8·6 mm respectively. However, the average diameters of the diploid and tetraploid flowers were 1·9 and 2·2 mm, and they calculated that a bee would be able to push into the tetraploid flowers slightly further (1·6: 1·4 mm). Even so, nectar would be slightly more accessible (about 0·2 mm) in the diploids than tetraploids.

Therefore, on the basis of the above findings, any difference in the nectar accessibility of diploids and tetraploids is very small and it is difficult to believe that it could by itself make much difference to bee visitation, especially in view of the large overlap in the width and depth of the flowers of the two forms. However, considerable differences in the type of bees visiting diploid and tetraploid clover do occur. Several workers (e.g. Julén, 1953b, 1954; Umaerus and Åkerberg, 1959; Valle et al., 1960; Bohart, 1966; Dennis and Haas, 1967) have demonstrated that relatively more honeybees visit diploid, and relatively more bumblebees visit tetraploid flowers, and that long-tongued species of bumblebees are relatively more abundant than the shorter-tongued species on the tetraploid. Hence, long-tongued bumblebees are especially important for pollinating tetraploid red clover.

Behaviour of bees visiting diploid and tetraploid clovers

The behaviour of the bees is also influenced by the type of flowers they are visiting. Friden et al. (1962) observed that honeybees only entered tetraploid flowers when they were collecting pollen, and Vestad (1962) found the percentage of robber bumblebees and honeybees was greater in tetraploid than diploid fields. Haas (1966) made similar findings.

Hence, despite such small differences in accessibility of nectar, there are large differences in the type and behaviour of bees visiting tetraploid and diploid varieties, and it seems possible that the different species are merely exhibiting preferences for flowers with corolla tubes of different depths (page 25) irrespective of their ability to obtain nectar.

Relative importance of honeybees and bumblebees

Stapel (1933) pointed out that the efficiency of a species depends upon its working speed, its numbers and whether or not it robs the flowers. Most observers find that, whereas bumblebees visit 25–35 flowers/min, honeybees visit only 10–15 (e.g. Westgate and Coe, 1915; Gubin, 1936; Pedersen and Sørensen, 1936; Dunham, 1939b; Jamieson, 1950; Forster and Hadfield, 1958; Åkerberg and Umaerus, 1960). There is also some evidence that pollen-gatherers work slightly faster than nectar-gatherers (Skøvgaard, 1952; Dennis and Haas, 1967). The working speed of different species is closely correlated with, and increases with the length of their tongues (Holm, 1966a; Dennis and Haas, 1967). Therefore, the longer a bee's tongue the faster it works and the greater its value as a pollinator. The term "bee unit" has been devised to indicate pollinating efficiency. Based on their working speeds, Stapel (1933) gave the relative value of honeybees, short-tongued bumblebees and long-tongued bumblebees, when visiting red clover, as 1·0, 1·5 and 2·5 bee units. Dennis and Haas (1967) confirmed that a similar relationship existed between bees visiting tetraploid clover.

In many countries attempts have been made to determine the bee population on crops and relate it to requirements. For example, Williams (1925) observed that bumblebees visited about 30 flowers/min and calculated that 0·4 ha of

TABLE 23

	1960		1961		1962	
	Uncultivated	Cultivated	Uncultivated	Cultivated	Uncultivated	Cultivated
Bumblebees	1327	969	1911	723	1670	1400
Honeybees	0	805	143	1318	70	2230

crop, with 160 million flowers in bloom per day, and a 14-day flowering period, would need 640 bumblebees working 10 h/day to pollinate it. Pedersen (1935) estimated that 500 bee units are needed per 1,000 m² during a 40-day flowering period to produce satisfactory seed yields. In an extensive survey of 191 crops he found honeybees were twice as numerous as bumblebees and a third of each were robbers. Assuming that positive bumblebees worked twice as fast as positive honeybees, he calculated that they did 68% and 43% pollinating respectively, on early and late red clover. A similar ratio of honeybees to bumblebees was recorded by Pritsch (1966) in Germany. The proportion of honeybees is sometimes considerably greater; they formed 82% of the pollinating insects in Ohio (Beard *et al.*, 1948) and from 76–89% in New Zealand (Forster and Hadfield, 1958).

The number of bumblebees visiting a crop varies greatly with the local population (e.g. Pedersen, 1945; Schwan, 1953; Åkerberg and Hahlin, 1953) and accounts for differences in seed set in different parts of a county. Thus,

Umaerus and Åkerberg (1967) found more bumblebees but fewer honeybees in uncultivated than in cultivated areas (Table 23).

The relative importance of honeybees and bumblebees varies at different times in the flowering season. It has long been known that corolla tube lengths are shorter on the second than the first flowering period (Westgate and Coe, 1915), and Pedersen (1935) found that late red clover, whose corolla length is about 1 mm less than early red clover, had more positive honeybees and fewer negative bumblebees visiting it.

Williams (1925, 1930) pointed out that in Wales the peak period of flowering of the second crop of red clover is planned to coincide with the maximum bumblebee population which occurs in the second and third week of August, and several workers (e.g. Valle, 1938 in Finland; Swart, 1960 in Holland; Bawolski, 1961 in Poland) have found that crops flowering late in the season, when bumblebees are more abundant give more seed than those flowering earlier. Valle (1959) found that, in Finland, honeybees were responsible for about 70% of the pollination during the early stages of flowering but only about 17% during the entire flowering period. However, because of the wet climate there, Valle et al. (1960) advised against management designed to postpone the flowering of red clover to coincide with the maximum bumblebee population.

The numbers of bumblebees and honeybees visiting a red clover crop will also of course depend upon local competition and in Europe *Trifolium repens*, *Brassica alba*, *Daucus carota*, *Medicago sativa* and *Brassica napus* are serious competitors for honey bee visits (e.g. Stapel, 1935; Hammer, 1949, 1963; Damisch, 1963). In Canada, Hobbs (1957) reported that, although honeybees collected pollen from red clover, they preferred *Medicago sativa* as a nectar source. Different species of bumblebees behaved differently; *B. nevadensis*, *B. fervidus* and *B. borealis* preferred red clover to *Melilotus* spp., *Trifolium hybridum* and *Medicago sativa*, whereas *B. huntii*, *B. rufocinctus* and *B. occidentalis* preferred *Melilotus* spp. to the other three legumes; the former three bumblebee species had tongues 7 mm or more in length and the latter three had tongues 7 mm or less in length (Hobbs et al., 1961). According to Holm (1962) bumblebees prefer *Lotus* spp. and *Vicia* spp. to red clover.

Hence the presence or absence of competing species is most important. Undoubtedly one of the reasons for greater seed production from the second crop is that there tends to be a scarcity of other forage plants, and consequently the flowers are visited more frequently by both honeybees and bumblebees than the first crop. Braun et al. (1953) found the amount of red clover pollen collected by colonies steadily increased from 4 to 83% as competing sources, particularly *Fagopyrum esculentum* and *Solidago virgaurea* ceased flowering.

After reviewing the available evidence Åkerberg and Stapel (1966) reached the interesting conclusion that in Europe the relative importance of bumblebees increases from south to north, and that whereas the length of the corolla tube appears to decrease in general from north to south the tongue length of honeybees seems to increase.

Thus the corolla is shorter in varieties in southern than northern Sweden (Umaerus and Åkerberg, 1959) and it is suggested that natural selection resulting in a shorter corolla tube has occurred in S. Sweden. The limited information available indicates that this may apply to tetraploid varieties also. Whereas honeybees are important for pollination in S. Sweden, bumblebees are far more important in central and N. Sweden (Schwan, 1953; Åkerberg and Hahlin, 1953) where there are more nest sites and more wild flowers to provide food in spring. In central Finland, where good seed sets can be obtained, it has been reported that bumblebees are the most abundant pollinators of red clover, sometimes practically the only ones, and that in particular *B. distinguendus* occurs almost exclusively on red clover fields (e.g. Pohjakallio, 1938; Valle, 1946; Sigfrids, 1947). In contrast, in S. Finland, few bumblebees are present and the more numerous honeybees, which occur equally on diploid and tetraploid clover are still too few to give a good set (Valle *et al.*, 1960). As a result of these findings it has been suggested that tetraploid seed production should be attempted in those areas of central Sweden and Finland where bumblebees are common (e.g. Bingefors, 1958; Bingefors and Eskilsson, 1962, Valle *et al.*, 1960), and that other nectar and pollen plants should be provided before red clover is in flower to help increase the bumblebee population.

Whereas bumblebees seem more valuable in areas inclined to be relatively cool and wet, honeybees are especially useful pollinators in hot dry conditions which result in shorter corolla tubes (Åkerberg, 1952). There is some indication that in the southern part of Europe, the honeybee is so abundant on red clover crops that no pollination problem exists, or at least none as great as in N. Europe (Åkerberg and Stapel, 1966). Probably this applies to other parts of the world also, including much of N. America, New Zealand and Australia, and may help to explain some of the apparently contradictory statements regarding the behaviour and value of bumblebees and honeybees.

Because the number of bumblebees is often inadequate, and moreover, is subject to large variations from year to year (e.g. Bird, 1944; Valle, 1948; Schwan, 1962) and from place to place (e.g. Schwan, 1953; Åkerberg and Hahlin, 1953), it is suggested that, where practicable, honeybees should be introduced for pollination. Although sometimes only 3–5 colonies/ha are regarded as adequate (e.g. Thomas, 1951; Johansen, 1960a), often a greater concentration is recommended. Stapel and Erikson (1936) thought that 10–12 colonies/ha of Danish bees, or 2–3 colonies/ha of more efficient Italian bees are needed; Beard *et al.* (1948) supposed 10–15 colonies/ha are necessary. Hammer (1950) suggested 3 colonies/ha for late flowering red clover and

6–8 colonies/ha for early flowering red clover, but when competing crops are within 1 km more colonies should be used.

Conclusions

Although the pollination of red clover appears to be a complex problem, some general conclusions can be drawn. Whereas bumblebees are more efficient pollinators than honeybees, the situation is complicated by the robbing behaviour of some species, and also by fluctuations in local populations. When there is no alternative but to use bumblebees as pollinators, the crops, particularly tetraploid ones, should be grown in areas where bumblebees are usually abundant, and also in small fields so as to facilitate an adequate and even distribution of the pollinating force. Attempts should be made to increase the bumblebee population. Cultural practices should aim to achieve coincidence between peaks of flowering and bumblebee populations. Red clover is less attractive to honeybees than to bumblebees and, when honeybees are used as the main pollinating agents, efforts should be made to try to avoid competition with other flower species, or to overcome it by using high concentrations of colonies and inducing bees to collect red clover pollen when nectar is inaccessible. Although there are many strong indications that clover with shorter corolla tubes, or bees with longer tongues, facilitate pollination, breeding to produce these characteristics has not been generally successful, whereas breeding on the basis of selection and pollination by bees is probably easier and shows more promise. It seems important that attempts to select a strain of honeybee that prefers red clover to other pollens should begin as soon as possible.

Trifolium repens L.

The corolla tube of *T. repens*, white clover, is only 3 mm long, so even short tongued bees are able to reach the nectar. When the flowers are pollinated they soon become reflexed, turn brown and cease producing nectar.

Darwin (1876) found that twenty heads covered to prevent insect visits produced only a single aborted seed, whereas twenty heads visited by bees produced 2,290 seeds. In each of two years Oertel (1934) caged white clover plants in three locations in Louisiana; a few shrivelled seeds were produced in one cage but no seeds were produced in the other five cages. However, Atwood (1943) found that different clones differed in the extent to which they are self-fertile, and hence in the amount they are cross-pollinated by bees. Thus only 19% cross-pollination of one self-fertile clone occurred. Because honeybees usually visit only three to eight flowers per head (Oertel, 1961; Weaver, 1965a) cross-pollination is usually favoured.

Green (1956) in Mississippi found the average yield during three years of plots caged with bees, without bees and not caged was respectively 278, 0 and 282 calculated kg/ha of seed. Additional plots from which all except small halictid and andrenid bees were excluded by cages, with walls of 8 mesh screen, gave yields equivalent to 56 kg/ha, indicating that wild pollinators made some contribution to set. Weaver (1957) reported that plots caged with bees, without bees and not caged (five replicates of each treatment) produced respectively the calculated equivalent of 92, 13 and 145 kg/ha of seed. He pointed out that insects other than bees might have caused the set that occurred in the plots caged to exclude insects, but he also suggested that the seed might have come from a few plants that were self-fertile. Palmer-Jones et al. (1962) obtained 90 and 97% set in cages with bees, but none in cages without bees.

E. E. Edwards (quoted by Pryce-Jones, 1948) imported colonies to Ramsey Island, off the Welsh coast, and increased the yield of white clover seed from 4 or 5 to about 280 kg/ha. Langridge (1956) reported that taking colonies to seed crops in Australia increased yields from 39 to 112–118 kg/ha. Oertel (1961) reported that a survey in Louisiana showed that white clover crops with and without honeybee colonies had averages of 434 and 299 kg/ha of seed respectively.

Shaw et al. (1954) found that during 1 day 52% of honeybees visiting white clover collected nectar only, 8% collected pollen only, 28% collected both and 12% had collected neither when examined. Because bees collecting nectar only contact the anthers and stigma (Oertel, 1961), their pollinating efficiency is probably not greatly dissimilar from that of pollen-gatherers, but this has yet to be determined.

Weaver (1965a) observed that although inexperienced bees probed their tongues against the upper side of flowers or between flowers, they quickly learned to forage efficiently. He also reported that whereas many bees collected nectar only, or nectar and pollen, few collected pollen only, and most bees that collected pollen did so without specifically foraging for it. He could find no correlation between the amounts of nectar and pollen carried by bees collecting both. On one day bees with pollen had averages of 5·2 mg pollen and 37·3 mg nectar, and bees without pollen had 37·9 mg nectar when they had completed their trips.

White clover nectar contains predominantly sucrose; equal quantities of fructose and glucose are also present (Bailey et al., 1954). Average sugar concentrations of 37–44% have been reported (e.g. Shaw, 1953; Montgomery, 1958). Oertel (1961) found that although there was a considerable variation in nectar yield between glasshouse and outdoor samples and between samples taken on the different days, the amount of nectar produced differed with the strain concerned (from 4·2–9·9 μ/head in outdoor plots), and he suggested that better nectar producing strains that might be more attractive to honeybees could be selected.

Weaver (1965a) found that a flower contained 0·02–0·08 μl of nectar containing 42–65% sugar. He observed honeybees visit about 18–19 flowers/min and calculated that when 0·08 μl/flower is present a bee would take 26 min to collect an average size load.

Oertel (1954) observed that an average of 2,980 pollen grains are contained in a single flower and 385,250 in a pollen pellet. Hence assuming that a bee collects all the pollen from each flower it visits, he calculated it would need to visit 520 to obtain a load (two pellets). However, as some flowers would have been visited already and a bee is unlikely to collect all the pollen present in a flower this is probably somewhat of an underestimate. Weaver (1965a) saw two pollen-gatherers, with very small pollen loads, visit 502 and 454 flowers before they flew home.

Wild bees do not seem to be important for white clover seed production and are generally relatively few in number compared to honeybees (e.g. Oertel, 1934; Green, 1957). However, Lecomte and Tirgari (1965) found that *Anthidium punctatum* and *Melitta leporina* were very abundant in white clover fields in France and Palmer-Jones *et al.* (1962) suggested that bumblebees are important in assisting reseeding of white clover in permanent pasture in New Zealand.

Probably because of the ease with which honeybees can work the flowers, the high pollinating efficiency of their visits, and the attractiveness of the crop which tends to eliminate other competition, a large population of honeybees is not needed.

Green (1956) calculated that there was no advantage in having more than 1·2 honeybees/m^2 of white clover, and Green (1957) reported that excellent seed yields occurred in a field in which wild bees were scarce, and the estimated population of one honeybee colony per 4–8 ha gave an average of 0·39 honeybees/m^2 at peak flowering time. However, the number of honeybees and seeds per pod decreased with increase in distance from known honeybee colonies; at 32 and 892 m from the colonies there was an average of 0·63 and 0·52 bees/100 flower heads and 3·10 and 2·59 seeds/pod were produced.

Weaver (1957b) suggested a maximum of one colony per 1·2 ha. He pointed out that when bands of withered flowers occur at the bottom of flower heads and about twenty or fewer flowers are open on each, sufficient pollination is occurring, but when fresh looking flowers occur from top to bottom of many of the heads, pollination is inadequate.

Oertel (1960) suggested one colony per 0·4–1·2 ha depending on competing floral sources, and indicated that experiments are needed to determine the relationship between colony concentration and seed crop.

Palmer-Jones *et al.* (1962) were unable to find a correlation between the concentration of bees per hectare, or per 10,000 flowers and the percentage of florets that set seed. In one district thirty or more honeybees per 10,000 flowers produced 90% or more seed set. In another district where honeybees were fewer, four fields that had twenty-five or more honeybees per 10,000 flowers gave 85% or more set, and the seed sets of six other fields with 12 bees (bumblebees and honeybees) or fewer per 10,000 flowers ranged from 5-85% (average 54%). Although they pointed out that the data were inadequate for any definite conclusions to be drawn they suggested that twenty-five bees per 10,000 flowers, or about 1 colony/3 ha was ample.

From the information given by Oertel and Weaver it is possible to calculate the number of bees needed to pollinate 1 ha of crop. Assuming there are 60 flowers/head and 717 heads/m², there are 430,200,000 flowers/ha. Assuming each was visited once only, it would take 917,269 bee trips, in which 520 flowers were visited per trip to visit them all. Assuming 30 min for a complete trip, including time spent in the hive, and that foraging was continuous between 09.00 and 17.00 h, each bee would make 16 trips/day. Therefore, in 1 day 57,329 bees would pollinate all the flowers in a hectare. This is an underestimate because many flowers would be visited more than once. However, dividing this figure by the number of flowering days of the crop, and multiplying it by the number of hectares should give an indication of the number of honeybees needed.

Trifolium repens Latum L.

T. repens Latum, Ladino clover, also needs insect pollination. Vansell (1951) found that bagged flower heads produced little or no seed, whereas most exposed heads had two or three seeds per flower. Scullen (1952) obtained 2·7 seeds/head from plants caged to exclude insects and wind, 16·3 when caged to exclude insects but subject to strong winds, and 213·2 when not caged. Dunavan (1953) obtained 89·9, 2·7 and 72·6 seeds/head from plants caged with bees, without bees and not caged respectively. In the open field an average of 2·5 seeds were produced per flower, although a few contained as many as seven seeds.

Vansell in California found an average of eight flowers opened per head per day, but few opened before 11.00 h and after this time flowers opened intermittently until evening. No flowers had nectar when they opened and it only accumulated slowly during the next 24 h. Consequently bees obtained little nectar from the crop and all those observed collected pollen. At peak periods of visitation two to three bees were present per square metre.

Scullen (1956), in Oregon, confirmed that the flowers produced little nectar although it contained an average of 45% sugar. He found that honeybees visited an average of 5·4 flowers/min (2,592/working day), and set an average of 1·2 seeds/visit or 3,110/day. He calculated that 2·4 bees/m² are necessary to provide sufficient pollination to set the approximately 7,170 ovules available for pollination per square metre per day of flowering, and that this concentration of bees should be produced by 5 colonies/ha.

Bumblebees appear to be numerous enough to make an important contribution to set in some areas (e.g. Oregon, Scullen, 1952; S. Carolina, Dunavan, 1953). Miller *et al.* (1951) reported that, in California 2½–3½ honeybee colonies/ha provide sufficient pollination.

Trifolium incarnatum L.

T. incarnatum, crimson clover, can set some seed through self-pollination but only a limited amount of automatic self-pollination occurs, and insect pollination is necessary to achieve maximum seed yields.

Darwin (1876) found plants to which bees had access produced between five and six times the weight of seed of plants bees were unable to visit, and many seeds of the latter were aborted. Results of workers who have caged plants of crimson clover and compared their yield with plants in the open field are summarized in Table 24.

TABLE 24

	Caged plots	Open field
Amos (1951) Tennessee, g seed/50 heads	2·6	6·4
Johnson and Nettles (1953) S. Carolina, kg seed/ha	46	238
Beckham and Girardeau (1954) Georgia, kg seed/ha	101	527
Blake (1958) Alabama, kg seed/ha	100	658
Girardeau (1958) Georgia, kg seed/ha	47	184
Scullen (1956) Oregon, no. seeds/head	5·1	69·2
Girardeau (1958) Georgia, no. seeds/head	7·1	38·7

Experiments in which plants were exposed for various stages of the flowering period, showed that the set increased fairly regularly with duration of exposure (Girardeau, 1958). In other experiments plants caged with bees, without bees and not caged have given average yields (kg/ha) of 53, 3, and 64 in Florida (Killinger and Haynie, 1951) and 261, 66, and 333 in Texas (Weaver and Ford, 1953).

Few wild bees visit crimson clover in Texas or Georgia and it is a good source of nectar and pollen for honeybee colonies. Girardeau (1958) found that, in Georgia, bees collected about 50% (maximum of 62%) of their pollen from crimson clover while it was flowering. Nectar-gatherers were more abundant on the fields in the morning and pollen-gatherers in the afternoon.

An average of 25 kg/of honey per colony was collected during flowering. Weaver and Ford (1953) found that crimson clover nectar from the honey-stomachs of foragers contained 38% concentration of sugar, but when flowers were not visited their nectar concentration increased to 58%.

They stated that most bees collected both nectar and pollen, visited an average of 13·5 flowers/min and 2·8 flowers/flower head. One bee was observed throughout the 25·5 min it spent collecting a load of pollen from 526 flowers. The crop contained 4,755,392 flower heads or 456,517,685 flowers/ha (96 flowers/head). It was calculated that the 10,292/ha of foraging honeybees present would have been able to visit the flowers that were opened on any one day in slightly less than 2 h, and they estimated that, provided competition from other sources was not intense, $2\frac{1}{2}$ colonies/ha would be sufficient.

Weaver and Ford (1953) noticed that whereas the flowers all remained open on flower heads isolated from pollinating insects, when sufficient pollination occurred there was only a narrow whorl of open flowers with buds above and withered flowers below. Hence, as for *T. repens* the appearance of flower heads is a reliable indication of the amount of pollination that has occurred. Blake (1958) found no advantage in using more than 5 colonies/ha.

Trifolium alexandrinum L.

Hassanein (1953) proved that *T. alexandrinum*, Egyptian clover, is practically self-sterile and needs cross-pollinating by bees. The number of seeds per head of plants enclosed in screen cages with bees, without bees and not caged (four replicates of each treatment) was 38·9, 1·9 and 23·5 at one location and 42·7, 1·3 and 22·5 at another. In addition, at each location there were four cages covered with cloth to help exclude wind currents as well as insects; plants in these cages had averages of 0·4 and 0·1 seeds/head at the two locations. The number of seeds per head in the screen cages without bees was not sufficiently greater to indicate that wind pollination is of any consequence.

Tobgy and Said (1956) found that single plants caged with bees, so they could only be self-pollinated, gave 7·0 seeds/head, compared to 1·5 when no bees were introduced and 28·0 in the open field. If it is assumed that the bees introduced into the cages did not carry compatible pollen on their bodies, this experiment indicates that Egyptian clover is slightly self-fertile.

Latif *et al.* (1956), in Pakistan, found that plants caged with *Apis cerana* gave 17·0 developed and 9·1 undeveloped seeds per head, compared to 0·5 and 0·6 when caged without bees. Flower heads in the open field, visited by *A. cerana* and other insects, gave averages of 46·4 developed and 6·8 undeveloped seeds, whereas in a field where *A. cerana* was absent 17·1 and 6·1 seeds were formed per head. In India, Narayanan *et al.* (1961) found 100 flower heads caged to exclude bees and 100 flower heads in the open gave averages of 83 and 2314 seeds in one year and 140 and 5707 seeds in another.

A comprehensive study by Wafa and Ibrahim (1960b), in Egypt, using five replicates of each treatment, confirmed and extended these results (Table 25).

TABLE 25

		Caged with bees	Caged without bees	Open plots	Cloth cages without bees
No. seeds/head	1958	43·8	1·3	22·7	0·4
	1959	43·7	1·1	21·8	0·4
% flowers set seed	1958	66·0	1·9	34·5	0·6
	1959	66·0	1·7	33·6	0·7
Wt (g) seeds/100 heads	1958	13·8	0·2	6·1	0·1
	1959	13·0	0·2	5·6	0·1

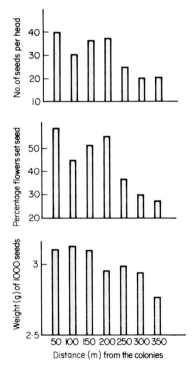

Fig. 90. Effect of distance from honeybee colonies on the yield of *Trifolium alexandrinum*, Egyptian clover (after Wafa and Ibrahim, 1960b).

Flowers that were not visited by bees remained open for 10–12 days. When bees were present flowers began to wither within a few hours, and all the flowers of a head withered within 3–4 days. During flowering such heads had a band of withered flowers at the bottom, a zone of 15–20 open flowers and a band of buds at the top.

Wafa and Ibrahim (1960b) also demonstrated that the percentage of seeds set, the number of seeds per head, and the weight per 1,000 seeds diminished with increasing distance from honeybee colonies (Fig. 90).

Hassanein (1953), in Egypt, found that honeybees comprised 60% of insect visitors to Egyptian clover, and butterflies 34%, but the value of the latter as pollinators has not been ascertained and is doubtful. Wafa and Ibrahim (1959b) collected four species of Lepidoptera, 14 of Diptera, 16 of Coleoptera and 35 of Hymenoptera (*Anthidium*, *Andrena*, *Halictus*, *Osmia* and *Xylocopa*) visiting the flowers. Narayanan *et al.* (1961) found in India that in 2 years honeybees constituted 84 and 97% of the insects pollinating Egyptian clover. *Apis cerana* was the most abundant each year. *A. dorsata* was also present in both years but *A. florea* was only present during one year when nectar was secreted freely and was sufficiently high in the corolla tubes for this short tongued bee to collect it. Bee visitation began at 07.00 h, reached a peak between 08.00 and 11.00 h and had ceased by 13.00 h.

The behaviour of insects while visiting the flowers does not seem to have been studied.

Trifolium hybridum L.

It is generally accepted that *T. hybridum*, alsike clover, is self-sterile. Williams (1951) obtained only 0·63% set of flowers he self-pollinated. Dunham (1939c) found that the average numbers of seeds per head in cages with bees, without bees and in the open field were 122, 2 and 39 respectively. In a similar experiment Crum (1941) obtained 107, 0·4 and 57 seeds/head. Scullen (1956) found that plants caged to exclude bees and plants in the open field produced averages of 0·4 and 126 seeds/head respectively.

Gooderham (1948) reported that the yield (kg/ha) in plots caged with bees, without bees and not caged was 259, 6 and 416 in 1 year and 138, 9 and 84 in another. Fischer (1954) found that plots caged without bees gave yields equivalent to only 22 kg/ha, compared to a range of 355 to 501 kg/ha in the open field. He demonstrated that in Minnesota there was a positive correlation between the honeybee population and seed yield per square metre of crop; and that fields with 10 colonies/ha had larger seed yields than fields with 5 colonies/ha.

Dunham (1957) suggested that 7,413 bees/ha are needed during the major portion of the flowering period. He compared the percentage of seed set in

Ohio in: (a) areas of Ohio where there were no honeybees and all pollination was by wild bees, (b) in areas where there were few honeybees and (c) in areas 0·5–1·2 km from commercial apiaries. The set during four years was: (a) 1·5–6·5%, (b) 1·5–2·3% and (c) 31·1–64·1%. Fischer (1954) and Dunham (1957) recommended using at least 5 colonies/ha for pollination.

The results of Pankiw and Elliot (1959) can be seen in Table 26.

TABLE 26

Treatment:	Percentage flowers that formed pods			Open field with 7·5 honeybee colonies per hectare
	Caged with bees	Caged without bees	Open field without honeybee colonies	
1955	77	1	47	75
1957	81	10	14	62

During other observations they found that 82% of the flowers set seed, and 420 kg/ha of seed was produced when an average of 8,401 honeybees/ha were present, but when honeybees were absent and wild bees (95% bumblebees) were responsible for pollination 32–328 kg/ha were produced depending on the number of wild bees present. The number of flowers visited per minute by honeybees, bumblebees and *Megachile* spp. was 18·7, 28·6 and 20·0.

Dunham (1939) remarked on the scarcity of wild pollinating insects. Valle (1947) found the pollinators of alsike clover in Finland were 93% honeybees and 7% bumblebees. Fischer (1954) found that in two years honeybees comprised 93% and 97% of insects pollinating alsike clover in Minnesota. However, Scullen (1956) found that *Nomadopsis* spp. were abundant in parts of Oregon; and honeybees formed only about 75% of the pollinators of alsike clover in Egypt (Wafa and Ibrahim, 1959b).

In a preliminary observation, Shaw *et al.* (1954) found that 24% of honeybees visiting alsike clover collected pollen only, 28% collected nectar only, 44% collected both nectar and pollen, and 4% had collected neither. The nectar collected had a 42% sugar concentration. A thorough study of the behaviour of honeybees and bumblebees on the crop is needed.

Trifolium fragiferum L.

Most plants of *T. fragiferum*, strawberry clover, are self-fertile but a small proportion produce a few seeds when selfed. Thus, in Australia, Morley (1963) found that only 5 of 20 heads he self-pollinated produced seed (average of 2·0 each) whereas 20 of 22 heads cross-pollinated produced an average of 13·0 seeds. In one area where honeybees were plentiful 70·3% of flowers set seed compared to 24·8% where bees were scarce. The flowers became reflexed when pollinated.

Other clovers

Table 27 shows the pollination requirements of a number of other clovers have been determined by Weaver and Weihing (1960) as in Table 8.

TABLE 27

	Microlitres Nectar per floret	Average wt. (g) seed per m of row		
		Caged with bees	Caged without bees	Not caged
T. resupinatum, Persian clover	0·006	16·6	8·4	25·4
T. nigrescens, ball clover	0·09	10·0	0·1	6·2
T. xerocephalum	0·06	3·1	0	4·3
T. michelianum	0·12	19·2	0·4	21·4
T. isthmocarpum	0·14	17·8	0·2	18·4

All species except *T. resupinatum* set little or no seed without honeybee pollination; it was not determined whether this was because they are self-fertile but not automatically self-pollinating, or because they are self-sterile and need cross-pollination. Although *T. resupinatum* produced a considerable amount of seed without bees, their presence increased the yield.

All the different species produced nectar of about 50% sugar concentration but in different amounts (see above). Weaver and Weihing observed that *T. michelianum* and *T. isthmocarpum* were very attractive to bees; *T. resupinatum* and *T. nigrescens* were fairly attractive. *T. xerocephalum* was not attractive; they thought this was because it has a long corolla tube and the bees have to struggle to reach the nectar.

According to Knuth (1906) and Robinson, D. H. (1937) the flowers of *T. subterraneum*, subterranean clover, are self-fertile.

I can find no information on the pollination requirements of *T. ambiguum*, Kura clover; *T. dubium*, small hop clover; *T. hirtum*, rose clover; *T. lappacaeum*, lappa clover; *T. medium*, zig zag clover; *T. procumbens*, large hop clover; *T. arvense*, hare's foot clover.

Chapter 17

Papilionaceae: *Vicia*

The *Vicia* species to be considered are *Vicia faba* L., broad bean and field bean, *V. villosa* Roth, hairy vetch and *V. sativa* L., common vetch.

Vicia faba L.

A normal field bean plant has fifty to eighty flowers (Fig. 91) but a large proportion of flowers and young pods are shed during the season. Soper

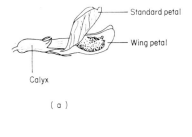

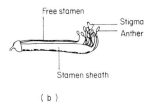

Fig. 91. Flower of *Vicia faba*, field bean: a, complete flower; b, petals and sepals removed (after Robinson, 1937).

(1952) found that in three seasons a mean of only 6·5, 11·1 and 8·0 pods/plant were harvested; he found it was exceptional for the earliest flowers, which generally appear on the sixth node, to form pods, and the first pods are usually found on the seventh and eighth nodes. Of the five or more flowers at each node, three or more dropped off; sometimes there were podless nodes above and below, presumably due to temporarily unfavourable conditions for fertilization. Soper suggested that no one factor alone is responsible for the insufficient setting in a normal crop, but that adequate nutrition, light intensity and cross-fertilization are all important, their relative importance changing from year to year. He observed that isolated plants on the headlands often had more pods than usual, sometimes over sixty, and thought this was partly because they were more frequently visited by bees than those in the centre of a dense crop. A commercial crop of field beans consists of about one-third hybrids and two-thirds inbred plants (Sirks, 1923; Hua, 1943; Fyfe and Bailey, 1951; Rowlands, 1958) and Soper thought that many of the pods that reach maturity are from cross-pollinated flowers.

Drayner (1959) found there was great variation in the set of field bean flowers left undisturbed in a glasshouse and, whereas hybrid beans were fairly fertile under these conditions, inbred plants were practically sterile. As was to be expected, plants from commercial plantations were on average intermediate in behaviour. She manipulated flowers of inbred plants in the glasshouse in various ways to facilitate pollination; she depressed the keels of some so that self-pollination could occur, and she depressed the keels of others and put pollen from another plant on their stigmas so cross-pollination could occur. However, the set from these two treatments was similar, and about twice that of flowers not so manipulated. This suggests that the undisturbed inbred flowers were self-fertile but there was some mechanical barrier to pollination which was absent in hybrid flowers. She was unable to find a completely satisfactory explanation for this but noticed in X-ray photographs that hybrid flowers had more pollen than inbred flowers, and that the plug of pollen above the stylar branch was denser. In further experiments she emasculated flowers and either self- or cross-pollinated them; the pollen tube growth rate from their own and from foreign pollen was similar and equally effective in fertilizing the flowers.

Thus it seemed that in a crop of field beans the inbred plants would not set seed unless visited by insects, whereas the hybrids would be able to set seed by autofertilization. Drayner confirmed that in commercial conditions hybrids set a high proportion of seed by this means. An additional factor favouring self-pollination of hybrids could be that their more abundant pollen decreases the chance of any foreign pollen reaching their stigmas, as an insect visitor would push the dense plug of the plant's own pollen against the stigma first.

Drayner pointed out that as the progeny of the hybrid plants (approximately a third of the field population) is predominantly self-fertilized, whereas that of many of the inbred plants (approximately two thirds of the field population) is cross-fertilized, an equilibrium between inbreeding and outbreeding in the population is maintained. Whereas, under normal conditions, cross-pollination occurs giving flexibility to the population, under conditions of poor pollination the autofertility of the hybrid plants will ensure survival of the population. Hence, one would expect a fair crop without insect pollination, although insect pollination would improve it. This, in general, is what experiments have shown, although the results depend, of course, on the proportion of plants concerned that have been grown from self-fertilized and cross-fertilized seeds.

Oschmann (1957) found that field bean plants, enclosed in cages to exclude bees, produced 71% as many pods and 77% as many seeds as plants in open plots. However, these differences could have resulted from the detrimental effects of the cages themselves.

Riedel and Wort (1960) compared the yield of three plots enclosed in cages with honeybee colonies, three plots caged without bees and six plots not caged, and found that plants enclosed with honeybees produced about half as many beans as those in open plots (17·6 : 28·3), but not significantly more than in cages without bees (17·6 : 13·1); their plants did not produce mature pods from more than about a third of their flowers, even when more were pollinated. However, in the open plots and the cages with bees a greater percentage of the inflorescences lower on the stem developed pods; a greater setting of the upper inflorescences of plants caged without bees partially compensated for this and so ensured a moderate yield without insects. Riedel and Wort suggested that the degree of concentration of the pods at the bottom of the stems may be a useful indication of the adequacy of pollination.

Scriven *et al.* (1961) thought that the supposedly greater yield of beans in the past was because there were then more honeybees and wild bees present; they claimed that in Lincolnshire honeybees and bumblebees were numerous in districts that habitually had good bean crops, but few in districts where beans were regarded as a poor or unprofitable crop. They stated that the introduction of honeybees into these low yielding areas gave better and increased yields. From three years of unreplicated experiments they calculated that plots caged with bees and plots not caged gave average yields equivalent to 4,520 kg/ha compared to only 2,378 kg/ha for plots caged without bees. Further experiments in Lincolnshire (Watts and Marshall, 1961) using four replicates of each treatment showed less difference between plots caged with and without bees (3,377 : 2,687 kg/ha); however, whereas plants caged with bees and plants not caged had averages of about 9, 9, and 2 pods respectively in the bottom, middle and top thirds of their stems, those caged without bees had 3, 5 and 4 pods in the equivalent positions. Scriven *et al.* (1961) also stated that plants from cross-pollinated seeds were more resistant to frost and better able to survive the winter than plants from self-pollinated seed.

Free (1966b) had six replicates each of plots caged with bees, without bees and not caged. Although the plants in the cages with bees had fewer formed seeds per pod than plants in open plots (2·7 : 3·1), they had significantly more than plants caged without bees (2·7 : 2·5). Many of the pods contained small unformed seeds and, when these were included with the properly developed seeds, so giving the potential number of seeds per pod, the difference between plants in cages with bees and in open plots ceased to be significant but that between plants in cages with and without bees remained so (3·2 : 2·9). Significantly more seeds per plant matured in the cages with bees than in those without bees (18·1 : 14·5) but they weighed less, so that the total weight of seeds per plant in cages with bees was not significantly greater than that in cages without bees. However, plots without cages had more seed per plant than the plots caged with bees (31·8 : 18·1) and they weighed more; as the plants in both types of plots were visited by many bees, the large differences in the yield between them probably resulted from the adverse effect of the cages themselves on the growth of the bean plant, which decreased the number of pods and seeds each plant could mature. It seems that the plants in the cages had reached the level of production possible in the circumstances concerned; and although those caged with bees produced slightly more seeds, these did not grow as large as seeds of plants isolated from bees.

The detrimental effect of the cages is likely, therefore, to minimize any differences in seed production, and the beneficial effect of bee pollination is probably greater than the results implied. Furthermore, because insect pollination sets a greater proportion of earlier flowers, it will give an earlier and more compact set and more simultaneous ripening, and so facilitate harvesting. Although in any one year seed production by the autofertilization of hybrids might result in a moderate seed harvest without insect pollination, any subsequent crop produced from the seed would contain only inbred plants whose flowers, as Drayner (1959) showed, set little or no seed when they are not disturbed.

The pollination requirements of broad beans seem to be similar to those of field beans but the few experiments that have been done have given more definite results. Darwin (1876) put netting over 17 plants and found they produced only 40 seeds compared to 135 produced by 17 plants left exposed. Brandenburg (1961) put insect proof cages over two groups of 20 plants, each in different localities and recorded only 0·03 pods/plant in one cage, and 0·28 in the other, compared with 4·4 and 7·4 on plants growing just outside the cages. However, much of this difference could be due to the effect of the cages themselves on plant growth.

Wafa and Ibrahim (1960a) had cages enclosing honeybee colonies as well as cages without bees and open plots; they used three replicates of each treatment in 1958 and five replicates of each treatment in 1959. In both years the plots caged with bees had significantly more mature pods per plant than plots caged without bees (9·1 : 8·1 in 1958 and 9·3 : 8·1 in 1959). The plots without bees also had fewer seeds per pod, many having no seeds at all. About three times as many of the seeds in the plots without bees were abortive. Free (1966b) found more seed was produced per plant in eight cages with bees than in eight cages without bees (24 : 15), and the seeds were heavier; but about twice as much seed was produced per plant in eight

Fig. 92. Honeybee entering *Vicia faba*, field bean, flower and collecting pollen.

plots not caged as in the plots caged with bees. At the first harvest plants caged with bees had more mature pods than those caged without bees (4·9 : 2·6) but fewer than those not caged (8·5). Nevertheless, the total number produced during both harvests was about the same for plants caged with and without bees (7·7 and 7·9) and less than on plants in the open (11·5). The greater number of pods produced in cages with bees than in those without bees during the first harvest must reflect a greater setting of the early than the late flowers, as happened in the experiments with field beans.

Wafa and Ibrahim (1960a) found that at harvest time the cages without bees contained a greater number of green pods than in cages where bees were present, and they noted that pollination by bees accelerated setting and pod growth so that such pods reached maturity sooner; usually there were more open flowers in the cages without bees and the plants were still flowering when those caged with bees had finished doing so. This was confirmed by Free (1966b) who found that when only 1% of the plants caged with bees still had flowers, 55% of the plants caged without bees were still flowering.

It seems likely, therefore, that inadequate pollination limits seed production in broad beans more than in field beans. However, this apparent difference may merely reflect differences in the growth conditions inside the cages in different years, rather than any difference in the capacity of broad beans and field beans to set seed by autofertilization. More experiments are needed to investigate this.

Fig. 93. Honeybee obtaining nectar through hole bitten by bumblebee at base of *Vicia faba*, field bean, flower.

Only insects with long tongues are able to reach the nectar in a bean flower through the mouth of the corolla tube, and honeybees and short-tongued bumblebees that enter the flowers probably obtain only pollen on many of their visits although they may attempt to obtain nectar as well (Fig. 92). However, some insects (e.g. some bumblebee species with short tongues, e.g. Knuth, 1906; male carpenter bees, *Xylocopa aestuans*, and ants, *Cataglyphis bicolor*, e.g. Wafa and Ibrahim, 1959b), circumvent this difficulty by obtaining nectar through holes they have bitten near the bases of the flowers.

Whereas bumblebee species with short tongues (e.g. *Bombus lucorum* and *B. terrestris*) "rob" the flowers, bumblebee species with long tongues (e.g. *B. agrorum* and *B. hortorum*) enter the mouths of the flowers and pollinate (Free, 1962b). Honeybees also rob the flowers by using the holes already bitten (Fig. 93) but it is generally accepted that they are unable to bite holes themselves. Although Scriven *et al.* (1961) recorded that when plants were caged with honeybee colonies every flower was pierced, their observations have not been confirmed; furthermore, any such behaviour in cages might result from abnormal conditions; for example, Zander (1951) found that when honeybees are confined to a glasshouse they will pierce lilac flowers but they do not do so in the field.

Honeybees and other insects also visit the extrafloral nectaries located on the undersides of the stipules of field beans (Fig. 94). Insects visiting the

Fig. 94. Honeybee obtaining nectar from extrafloral nectary of *Vicia faba* field bean.

extrafloral nectaries obviously will not pollinate and it seems unlikely that those robbing the flowers do so to any extent, although it is possible that they may do so by shaking pollen from the anthers onto the stigmas, as suggested by Soper (1952), provided the flowers are not self-sterile. Only insects entering bean flowers touch the anthers and stigmas and so can pollinate them.

Free (1962b) observed honeybees foraging on a crop of field beans and found they spent only 4·7 sec/extrafloral nectary, compared to 8·0 sec when "robbing" flowers and 11·9 sec when gathering pollen, but there was no indication that the size of the foraging area differed according to the type of flower visit being made.

During a single trip, or part of a trip, 86% of the bees were constant to one type of behaviour and most of the bees that were inconstant both robbed the flowers and visited the extrafloral nectaries on the same trip. Bees visiting either type of nectary often ran along the plant from one nectary to another, and it is easy to envisage how they could discover nectaries of a different kind to those they had been visiting. Bees that made both visits for nectar without entering the flowers and visits for pollen, or changed from one type of visit to the other, were very infrequent, particularly when the nectar visits were to extrafloral nectaries.

Bees visiting extrafloral nectaries tended to be most numerous at midday, while those visiting the flowers, especially for pollen, were most numerous from 14.00–16.00 h. This later time of day is undoubtedly associated with pollen availability, because although bean pollen is presented from 10.00–17.00 h, some 91% of it is presented in the peak period which occurs from 12.00–15.00 h (Percival, 1955). Furthermore, all new flower buds first open in the afternoon, 74% between 12.00 and 14.00 h (Percival, 1965). In fact, many of the pollen-gatherers did not arrive on the crop until the afternoon; they may have stayed at home in the morning or they may have visited another crop, but they did not collect nectar only from field beans. Bond and Hawkins (1967) observed that only 18% of the honeybees on one crop of beans and 24% on another entered the flowers.

Free (1967b) found that honeybees visited the extrafloral nectaries of field beans before any flowers opened, and the proportion so doing increased during the flowering period at the expense of the floral visits; presumably the extrafloral nectaries continued to secrete after the floral nectaries had finished doing so and there was little pollen left. Most bees making floral visits deserted the crop rather than change their behaviour. It is difficult to see what value the extrafloral nectaries have in propagating the species, but bees gathering nectar from them may recruit other bees that enter the flowers and pollinate. Knuth (1906) pointed out that the secretion from extrafloral nectaries is eagerly taken by ants which may protect the plants against various harmful phytophagous insects. According to Darwin (1876), secretion from the extrafloral nectaries is very dependent upon sunshine; it would be interesting to compare the concentration and abundance of the nectar secreted by floral and extrafloral nectaries before, during and after flowering.

Although honeybees can pollinate field beans, the extent to which colonies forage on them varies greatly; thus, the mean amounts of bean pollen collected daily by colonies located beside a bean crop in full flower was found to be approximately 63% (Free, 1959); 81% (Edwards, 1961); 1% (Free, 1965d); 70% (Steuckardt, 1965); 3% and 87% in different years (Bond and Hawkins, 1967), 35% and 4% for colonies beside different crops in the same year (Free *et al.*, 1967). This variation probably partly reflects different weather conditions during different years, as the lowest temperature at which anthesis occurs is 12°C (54°F), free anthesis occurring at 13°C (55°F) or more (Percival, 1955), and partly reflects competition from other sources. The amount and concentration of field bean nectar is probably relatively low; when possible, beans should be planted where there are few competing nectar and pollen sources as bees prefer to visit many wild flowers, and *Trifolium* spp. and *Brassica* spp. crops to field beans.

The main object of breeding beans is to increase the yield, and some authorities (e.g. Bond and Fyfe, 1962) believe that the only way to do this is to produce a hybrid form. Hybrid seed production depends upon the transfer of pollen from a male fertile plant to a male sterile plant of the female parent. If hybrid beans are to be produced on a field scale they probably will be planted for convenience in blocks of 3-6 rows of 76 cm apart. The width of these blocks is well within the foraging area of a bee during a single trip as Free (1962b) found that the average distance travelled by honeybees was 132 cm for up to 10 flower visits, 256 cm for 11-20 visits, 305 cm for 21-30 visits and 335 for an average of 62 flower visits. However, Bond and Hawkins (1967) studied the foraging behaviour of honeybees and bumblebees in fields planted in blocks of male sterile and male fertile lines, separated by an empty row, and they found that most of the small proportion of honeybees that entered the mouths of flowers visited the male fertile plants; the few honeybees that entered the male sterile flowers soon deserted them and probably learned by trial and error to avoid the male sterile flowers. Whereas *B. lucorum* and *B. terrestris* foragers, nearly all of which robbed the flowers, were much more frequent on the male sterile than on the male fertile blocks, *B. hortorum* foragers, which were entering the flowers, showed the reverse tendency, if any, although many *B. hortorum* foragers entered the male sterile flowers and were probably mostly responsible for such pollination of them as did occur. Furthermore, bumblebees were less readily deterred than honeybees from moving across the empty row between blocks.

Bond and Hawkins suggested that to increase the pollination of male sterile beans the crops should be grown in areas where beneficial species of bumble-bees are normally abundant, and the male fertile and male sterile lines should be planted in alternate single rows, or the seed of the two lines even mixed at random during sowing so as to increase the frequency of chance visits to the male sterile lines by bees entering the flowers. However, these latter suggestions are dependant upon there being no colour or scent differences between the lines, so that the bees could not easily distinguish between them, and upon either adequate methods of segregating the resultant seed, or using the minimum proportion of male fertile plants that are necessary to allow adequate pollination of the male steriles without unduly diluting the hybrid with inbred seed. This dilution reached 30% without a measurable loss of yield of one hybrid tested (Bond *et al.*, 1966).

There is little information on the behaviour of insect foragers on broad beans, but Wafa and Ibrahim (1957a) recorded that in Egypt, honeybee nectar-gatherers did not start foraging until 11.00 h and honeybee pollen-gatherers were most numerous between 13.00 and 14.00 h which was also the time of maximum temperature. In 1955 and 1956 honeybees formed 77 and 83% of the insects visiting broad bean flowers. The carpenter bee (*Xylocopa aestuans*) was also numerous and, although the males bit holes in the calyx and corolla to reach the nectar, the females entered the flowers and so helped in pollination. To find whether the hole-biting activities of male carpenter bees diminishes a flower's ability to set seed, some broad bean plants were caged with honeybees and male carpenter bees, and other plants were caged with honeybees only; the presence of the male carpenter bees made no difference to the amount of seed produced so their activities are probably not harmful. Brandenburg (1961), in New Zealand, examined fifty broad bean flowers punctured by bumblebees (*B. terrestris*) and found that the sexual parts of the flowers had not been damaged.

In New Zealand the broad bean crop flowers in early spring and unfortunately the only bumblebee species that is available is *B. terrestris* whose colonies tend to be perennial and which is useless as a pollinator. The useful species of bumblebees (*B. ruderatus* and *B. subterraneus*) whose colonies are not perennial, are not abundant until flowering has ceased. Serious consideration should be given to the introduction into New Zealand of other bumblebee species that do not bite holes (e.g. *B. agrorum* and *B. hortorum*) and whose colonies might become perennial. Brandenburg (1961) suggested that honeybee colonies taken to broad bean crops should be changed every 1 or 2 weeks to prevent the bees finding the nectar through holes bitten by *B. terrestris*, but it seems more probable that as soon as a sufficiently large proportion of the flowers have been bitten honeybees will "rob" them and will not spend several days visiting the flowers first.

There have been few attempts to demonstrate the value of taking honeybee colonies to crops of field beans or broad beans. Brandenburg (1961) put four colonies beside one crop, but not beside another 4·8 km away, and stated that the former had twice as many pods as the crop without colonies but he gave no numerical data. It is supposed that there are enough wild bees in most parts of England to pollinate field bean crops of up to 2 ha in size but that larger crops may need $2\frac{1}{2}$–5 colonies/ha to provide sufficient pollination (National Agricultural Advisory Service 1964). Because bees visit the extrafloral nectaries of field beans before any flowers open it is especially important that honeybee colonies should not be moved to field beans for pollination before they begin to flower, otherwise the bees might become conditioned to work the extrafloral nectaries and never visit the flowers.

Vicia villosa Roth

Experiments have established that *V. villosa*, hairy vetch, is self-sterile and needs cross-pollination to produce seed. Schelhorn (1942, 1946) caged vetch plants with and without bees; when bees were absent only a few flowers (up to 4%) set seed whereas the set on plants caged with bees was reported to be as great as in the open field. In further experiments, he grew plants with recessive white-coloured flowers in the midst of normal plants with dominant red coloured flowers, and found that 80 to 100% of the progeny of the white flowers had red flowers indicating that cross-pollination generally occurs.

Bieberdorf (1954) caged seven plots in a field of vetch to exclude insects and found that they produced a mean amount of seed equivalent to 149 kg/ha compared to 490 kg/ha on similar uncaged plots. Unfortunately the cages were not put over the vetch plants until honeybees had been visiting the flowers for 4–5 days so the effect of insect pollination is probably greater than the data imply, but as no control cages with honeybees were used, it is impossible to determine what effect the cages themselves had in decreasing the yield.

Much more satisfactory experiments were done by Weaver (1956b) who, in each of three years, determined the yield of five plots of hairy vetch caged with bees, five plots caged without bees and five plots not caged. The mean yields of seed per hectare in the above types of plot was calculated as equivalent to 491, 26 and 464 kg respectively. In one year in which fewer than usual thrips and moths were present inside the cages, the plot caged without bees yielded only 1 kg/ha of seed. The pods that did develop in the plots caged without bees had only half as many seeds as pods on plants caged with bees, but they weighed about 50% more, probably because more nourishment for development was available per seed.

Although the above results had established that vetch needs insect pollination to produce seed little information was available on the relative value of self-pollination and cross-pollination by insects until Młyniec and Wójtowski (1962) in Poland, did an extensive series of experiments in which they caged plants singly and in pairs, and either self- or cross-pollinated them with insects or by mechanical means, or left them uncaged. They found that although in an undisturbed vetch flower pollen from the anthers falls onto the nearby brush-like stigma, this automatic self-pollination failed to effect fertilization unless the stigma was stimulated mechanically, as happens when a bee visits the flower. The exact effect of this mechanical intervention is unknown, but Młyniec (1962) noted that squeezing a stigma caused it to exude a sticky liquid which, he suggested, might facilitate retention and germination of the pollen grains on its surface. Different species of bumblebees differed in their efficiency as pollinators and he supposed this was probably partly because they differed in their ability to perform the necessary mechanical operation. *Bombus agrorum* was the best pollinator; *B. lapidarius*, *B. subterraneus*, *B. lucorum* and *B. terrestris* also induced a certain amount of pod setting but the latter two species punctured the bases of the corolla tubes and in contrast to the findings for *Vicia faba*, most of the damaged vetch flowers fell off and failed to set pods. Although Młyniec (1962) expressed doubt about the ability of honeybees to pollinate vetch, his data indicate that sometimes, at least, honeybees are as efficient as *B. agrorum*. Cross-fertilization was much more effective than self-fertilization both in setting pods and in producing mature pods with seeds. Probably the pollen tubes of cross-pollen grains grow more rapidly than those of self-pollen grains and are more likely to reach the ovaries first; this could explain why Schelhorn (1946) found that cross-pollination was the general rule in the field. The set obtained depends not only upon the type of insect visitor, and on whether self- or cross-fertilization is occurring but also on the variety of vetch concerned, as Młyniec found that the variety "Euvillosa" set twice as many pods as variety "Glabrescens" both when self-pollinated and when cross-pollinated.

We are indebted to Weaver (1954, 1956a, 1956b, 1957a, 1965b) for a detailed study of the behaviour of honeybees while visiting hairy vetch flowers. He observed that it was difficult for a honeybee to reach the nectar through the mouth of a vetch flower, and to help push its tongue between the standard and keel petal it used its head as a wedge and lever and pressed its legs downward on the keel. As a result the keel was depressed at least to a right angle with the corolla tube, and the sexual column freed. When a flower tripped, the bee's tongue was sometimes temporarily trapped between the sexual column and the keel; some bees got their tongues trapped in nearly every flower they visited but others rarely did so.

Some of the bees that collected nectar through the mouths of flowers had pollen loads while others did not. A sample of nectar-gatherers with pollen loads had 14·7 mg of nectar and 4·5 mg of pollen and a sample of nectar-gatherers without pollen had 19·8 mg nectar. Whereas the head and thorax of a nectar-gatherer made slight contact with the sexual column when it tripped a flower, when it visited a flower that had already been tripped a nectar-gatherer often did not even touch the sexual column. However, other bees gathered pollen only, and as such bees contacted both stamens and stigmas on every flower visit they were better pollinators than nectar-gatherers; unfortunately there were relatively few of them.

A bee could also reach the nectaries without tripping a flower by pushing its tongue between the standard and keel petals at the base of the corolla tube. Such a so-called "base worker", alighted on the ventral side of a flower. When a bee first began to forage on vetch flowers it attempted to insert its tongue into any part of a flower, apparently at random but, as soon as it had managed to work one or a few successive flowers successfully, it usually became conditioned in its behaviour and, henceforth, methodically collected nectar by whatever approach had proved successful. However, when attempting to reach the nectary, a bee's tongue often slipped down the side of a corolla tube between the standard and keel petals, and although most bees immediately withdrew their tongues and attempted to enter the mouth of the flower again, a few accidentally learned to forage as "base workers".

Weaver (1956a) recorded the behaviour of individual bees for up to ten consecutive flower visits; 17% collected nectar from the bases of flowers, 35% collected nectar only through the mouths of flowers, 42% collected nectar and pollen, 3% collected pollen only, and only 3% visited both the bases and mouths of flowers for nectar. Late in the season, some "base workers" entered the mouths of previously tripped flowers but most of the bees that made these two types of visits attempted to enter the mouths of all flowers and, when one of them did not trip easily, the bee concerned changed its approach to the base of the flower.

When the nectar flow deteriorated, the proportion of "base workers" increased and most of the nectar-gatherers that did not change to become "base workers" began collecting pollen in addition to nectar, or else ceased to forage on vetch. The proportion of "base workers" to nectar-gatherers that entered the flowers was also greater when the vetch flowers were few and scattered than when they were numerous, and also when the nectar supply was diminished by a heavy aphid infestation. Under such conditions of poor nectar secretion, the "base workers" collected larger nectar loads than bees entering the flowers, and so were less likely to abandon the crop. There seem to be two reasons why they could collect larger loads. Firstly, they used less energy in collecting nectar from the flowers; secondly, because they did not actually enter the flowers their vision was unimpaired throughout foraging, and consequently they were more sensitive to the presence of other foragers and more likely to be forced into areas of the field where competition was less and foraging more rewarding.

On average, "base workers" spent less time per flower than those entering flowers for nectar only (8·5 : 10·5 sec); but the latter worked at about the same speed as those that entered flowers and collected pollen in addition to nectar, probably because few bees that collected pollen foraged specifically for it, but merely packed into their corbiculae the pollen that accidentally clung to them. However, bees deliberately gathering pollen, but not nectar, spent only an average of 4·4 sec/flower visit. There was great variation in the foraging speed of individual bees belonging to the same category. Weaver (1957a) suggested that these differences were due in part to variation in the ease with which tripping occurred because flowers that developed during cool weather were narrower and required more time and effort to trip them, than those that developed during warm weather. Weaver (1965b) also found differences in the speed at which bees foraged in two different fields; it seems that this was partly because the flowers of one field could be more quickly and more easily tripped than those of the other, and partly because they contained less nectar and it was much more common for a bee to insert its tongue into a flower and then withdraw it immediately. However, even when differences due to location, date and method of entering a flower were excluded there were still differences between the speed of work of individual bees.

On most fields honeybees were the only flower-visiting insects (Weaver, 1956b); but on one field carpenter bees (*Xylocopa* spp.) were numerous and bit holes near the flower bases (Weaver, 1965b); about a third of the honeybees collected nectar through these holes and spent much of their time searching for flowers that had been bitten. In contrast Schwanwitsch (1956) in the U.S.S.R. reported that few honeybees entered the flowers and most robbed them and had either no pollen loads or loads of red clover pollen. Although he demonstrated that the opened mandibles of a honeybee exactly fitted the holes pierced in vetch flowers, there is no evidence that honeybees make them; indeed the absence of robber honeybees when hole biting insects did not occur in Weaver's fields is evidence to the contrary.

The need for importing honeybee colonies to pollinate vetch crops and the number of colonies required will depend upon the population of wild pollinating insects already present. Competition for bee visits from other plants will also be of importance in deciding the number of honeybee colonies needed. Weaver (1954 and 1956b) found that the amount and concentration of nectar present varied greatly, and vetch that had little vegetative growth and that had just started to flower yielded the most nectar (up to $1 \cdot 2\ \mu l$/flower) with the greatest sugar concentration (up to 55%) and, in good conditions, honeybee colonies beside vetch crops gained about 82 kg/day of honey. When no other competitive crop was in the area, 90% of the pollen collected was from vetch. However, the difficulty bees experience in obtaining nectar and pollen from vetch makes it very susceptible to competition from other crops; Alex *et al.* (1950) and Weaver (1965b) mentioned *Melilotus alba* as a serious competitor and noted that vetch is erratic in its nectar production and attractiveness to bees and, as a consequence, farmers find it necessary to hire colonies for pollination. They attempted to correlate the number of honeybee colonies in the vicinity of various vetch fields, in Texas, with seed production and their data suggest that whereas a farmer might harvest about 224 kg/ha of seed from fields with no honeybee colonies nearby, the presence of $3\frac{1}{2}-7\frac{1}{2}$ colonies/ha might treble this yield.

From the data he obtained on the speed of foraging Weaver (1956a) calculated that one bee should be able to visit all the flowers in 0·8 m² of vetch duing a 10 h working day, so that $2\frac{1}{2}$ colonies of bees should provide more than enough bees to pollinate a hectare provided competition from other crops was not severe. However, whereas some flowers would be visited more than once, others would not be visited at all; furthermore we have no idea what proportion of flower visits result in self- or cross-pollination, and of the average number of visits necessary to achieve fertilization. There should be sufficient pollinators to pollinate flowers on the day they open as in hot weather the flowers begin to wither within two days. Therefore, to achieve maximum pollination, probably $2\frac{1}{2}$ colonies/ha are needed, but as Weaver (1965b) pointed out, an increase in the number of pollinators beyond that required to pollinate the majority of flowers would probably not be economically worth while because of the decrease in honey production per colony and the increased expenditure in hiring colonies.

Vicia sativa L.

There is no evidence that *V. sativa*, common vetch or tare, benefits from bee visits. The anthers dehisce in the bud and self-pollination and self-fertilization occurs (Knuth, 1906). Darwin (1876) found that plants covered with a net produced as many seeds as plants not so covered.

Chapter 18

Papilionaceae : *Phaseolus*

Although several species of *Phaseolus* are grown for food, there is little information on the pollination requirements and insect visitors to species other than *Phaseolus multiflorus* Willd (runner bean, scarlet runner bean), *P. lunatus* L. (butter, sieva or lima bean) and *P. vulgaris* L. (common, French, Kidney, haricot, white pea bean). Purseglove (1968) presumed that *P. acutifolius* var. *latifolius* Gray (tepary bean) is self-pollinated, but pointed out that this is not known for certain. He stated that self-fertilization generally occurred in *P. aconitifolius* Jacq (mat or moth bean), *P. calcaratus* Roxb. (rice bean), *P. aureus* Roxb. (green or golden gram, mung), *P. mungo* L. (black gram or urd), and *P. angularis* Willd (adzuki bean) although cross-pollination is frequent in the last species (see also Hector, 1938). Pollen is shed in the bud of *P. aureus* and *P. mungo* in the evening; only about half the flowers open and these do so the following morning. Experiments are needed to confirm the pollination requirements of the above species, especially because the species *P. multiflorus* differs greatly from *P. vulgaris* and *P. lunatus* in its ability to produce seed without insect visits.

Phaseolus vulgaris L.

Automatic self-pollination of flowers of *P. vulgaris* occurs before or at about the time they open. However, fertilization is not accomplished until 8–9 h later (Weinstein, 1926) and honeybees and bumblebees sometimes visit the flowers and cross-pollinate them (e.g. Jones and Rosa, 1928; Cruchet, 1953). Probably the pollen tubes of foreign pollen grow more quickly than those of self-pollen otherwise self-pollination only would occur. Mackie and Smith (1935) suggested that thrips might also cross-pollinate the flowers in California; they sometimes found numerous western grass thrips (*Frankliniella occidentalis*) in a single flower and noted that when they left a flower they carried several pollen grains with them. Furthermore, in contrast to bees, thrips are able to enter flowers before they open and so effect early pollination.

The amount of crossing is normally only about 1% (e.g. Kristofferson, 1921; Mackie and Smith, 1935; Vieira, 1960) but may sometimes reach 8–10% between adjacent rows, although it decreases rapidly with distance from the pollen source (Barrons, 1939). To reduce occasional hybridization when producing commercial seed, it is recommended in the U.S.A. that varieties should be isolated by 1·8–3·7 m and a tall, dense barrier. In Canada, however, the recommended distance between varieties is 46 m when producing "registered" seed and 0·5 km when producing "elite" seed (Webster, 1944; Hawthorn and Pollard, 1954).

Although insect visits may cause cross-pollination, they do not increase yield; Darwin (1876) found that *P. vulgaris* plants, covered with a net, appeared to produce as many pods as exposed plants and Free (1966c) found that *P. vulgaris* plants, caged to exclude insects, had as many pods and seeds as those caged with a honeybee colony, and the average weight per seed in the different cages was also similar. Hence, taking honeybees to crops of *P. vulgaris* is unlikely to influence seed yield.

Phaseolus lunatus L.

Lambeth (1951) stated that *P. lunatus* is almost entirely self-pollinated, and that mechanical tripping is unnecessary to ensure adequate pollination, but no adequately controlled experiments have been made to investigate the effect of insect pollination in increasing yield. Amos (1943) and Vansell and Reinhardth (1948) found that *P. lunatus* plants caged to exclude pollinating insects gave a good yield of seed, but not quite as great as that of plants not caged; because no cages with bees were used, it is impossible to differentiate between the lack of pollinating insects and the effect of the cages themselves on reducing yield, but even if the reduction were caused by lack of pollinating insects, it was probably too small to make it worthwhile importing bees to pollinate a crop.

The amount of cross-pollination in field crops of *P. lunatus* ranges from 1–89% depending on the variety and conditions (Barrons, 1939; Magruder and Wester, 1940; Welch and Grimball, 1951). Lambeth (1952) found that, when weather favours pod-setting by *P. lunatus*, pollen and stigmatic fluid are both plentiful, whereas drought limits their production and fewer pods are produced. Differences in the local abundance of insect visitors undoubtedly also partly explain differences in cross-pollination. Vansell and Reinhardth (1948) found that *P. lunatus* secreted abundant nectar with an average sugar concentration of 42–59% and that it was very attractive to honeybees; several other Hymenoptera, including five species of Halictidae, two species of Sphecidae and one species of *Polistes* visited the extra-floral nectaries.

Mackie and Smith (1935) observed that honeybees and bumblebees never visited *P. lunatus* flowers that had not opened naturally and never attempted to force an entry into a flower bud, but all the flowers they examined contained numerous thrips (*Frankliniella occidentalis*). In contrast to bees, the thrips usually entered a flower before it opened, and Mackie and Smith supposed that overcrowding in a flower caused some thrips to leave and, when they did so, they carried pollen with them to another flower. Because thrips could pollinate a flower soon after its stigma became receptive, Mackie and Smith suggested that thrips were more likely than bees to be responsible for such pollination as does occur.

Phaseolus multiflorus Willd

Knuth (1906) observed that the *P. multiflorus*, runner bean, flower is so constructed that a bee can reach the nectary with its proboscis only by standing on the left wing petal (Figs. 95 and 96); he supposed that the floral mechanism prevented automatic pollination and that only large bumblebees were heavy enough to cause the style to protrude from the spirally wound keel.

Blackwall (1964) thought that although several types of insect can pollinate runner bean, bumblebees are the most reliable, and she pointed out that the small marks made on the left wing petal by bumblebee's feet indicate that the flower has been visited and pollinated. She supposed that the failure of the early flowers to set, as sometimes occurs in England, is associated with a small bumblebee population. Blackwall (1969) found that runner bean plants grown in an insect free glasshouse set ten times as many pods on hand pollinated trusses as on control trusses. The average yield of runner bean in England is related to rainfall and Blackwall (1969) showed that it is important to irrigate before the first flower buds turn red; thus, difficulties in getting early pickings may also be associated with lack of irrigation.

Darwin (1876) covered runner bean plants with a net and found they produced only an eighth as many pods as exposed plants in one experiment and only a third as many in another. He thought that these pods might have been from flowers fertilized by thrips. However, more variable results were obtained by Tedoradze (1959). In one year, plants from which honeybees and other insects were isolated had a lower set than plants visited by honeybees (6 : 72%) but this was not so in another year (63 : 76%). Pollination by bees was more effective than hand pollination. No cross-pollination occurred in cages without insects but 8% of the seeds produced in cages with honeybees were hybrids. In contrast, Kristofferson (1921) found that only about 1% cross-pollination occurred when a small batch of plants were grown in the middle of a field of another variety.

Free's (1965c) results with "Kelvedon Marvel" variety of runner bean confirmed those of Darwin. Seven plots, each containing 9 plants, were caged

260 INSECT POLLINATION OF CROPS

Fig. 95. Bumblebee, *Bombus hortorum*, entering *Phaseolus multiflorus*, runner bean, flower and obtaining pollen.

Fig. 96. Honeybee entering front of *Phaseolus multiflorus*, runner bean, flower.

with honeybees, 7 plots were caged without bees, and 7 plots were not caged. The ripe pods from 6 plants in each plot were harvested every few days. Plants caged with bees produced an average of 96 pods weighing 1,479 g compared with an average of 11 pods weighing 180 g produced per plant caged without bees. In the earlier harvests the open plots had a much greater yield than plots caged with bees. This may have been because the open plots were visited by bumblebees, particularly *B. agrorum*, throughout the flowering period, whereas the honeybees were not seen visiting plants in the open or in cages until about 3 weeks after flowering had begun. The reason for this difference is not known; perhaps it merely reflected changes in the competing local flora. If it is a more general phenomenon, perhaps the nectar produced from the first flowers is less attractive than that produced later. For the remainder of the flowering period, the plants in the open were visited by many honeybees and bumblebees, whereas those in the cages were visited by honeybees only. At most harvests, plants caged with honeybees yielded less than plants in the open, but the adverse effect of the cages themselves on plant growth and capacity yield could easily account for this.

Most of the pods that did mature in cages without bees were harvested later in the season; perhaps auto-fertilization occurs more in flowers produced at the end of the season, or perhaps the cages then contained more small insects that pollinated the flowers. Plants caged without bees had obviously not reached the limit of their bearing capacity at the end of the season as the average number of flowers still open per plant that were producing pods was 1·5 in the open plots, 8·2 in cages with bees, but 42·4 in cages without bees.

The pods on the remaining three plants in each cage were allowed to produce mature seeds which were harvested at the end of the season. Plants caged without bees had an average of only 30 seeds compared with 206 per plant caged with bees and 260 per plant in the open; the average number of seeds per pod was 2·7, 3·9 and 3·9 respectively.

The pod bearing capacity of a plant is an important factor limiting the crop it produces, and Blackwall (1969) showed that only about a third of the flowers produce pods irrespective of cultural and other treatments. However, supplementary hand pollination sometimes increased the set in the field, indicating that natural pollination was inadequate. As honeybees are able to pollinate runner bean, their use should increase yield in circumstances where pollination is a limiting factor. Such a circumstance occurs when runner beans are grown in glasshouses. Free and Racey (1968b) attempted to compare the ability of honeybees, bumblebees (*B. agrorum*, *B. hortorum* and *B. pratorum*) and blowflies to pollinate runner beans enclosed in cages in a glasshouse. Blowflies were ineffective as pollinators, probably because their tongues were too short for them to obtain nectar, and plants caged with them produced only a few more pods than plants from which insects were excluded. Pod production on the plants enclosed with bumblebees was no better than on plants enclosed with honeybees. Thus, there seems to be no disadvantage in using honeybees to pollinate runner beans in glasshouses, so as to facilitate the production of an earlier and hence more profitable crop.

However, the pollinating efficiency of honeybees may be much greater in a glasshouse or cage than in the field, as bumblebees with short tongues (e.g. *B. lucorum* and *B. terrestris*) obtain nectar through holes they bite in the bases of the corolla tubes (Fig. 97) and honeybees also obtain nectar through these holes although they cannot make them (Fig. 98) (Darwin, 1858; Henslow, 1878; Jany, 1950; Free, 1968d). Bumblebees and honeybees that rob the flowers are probably useless as pollinators unless their movements sometimes lead to self-pollination; this needs investigation. However, their behaviour does not appear to injure the formation of the seed or pod or cause the flower to shed prematurely.

Honeybee colonies put beside runner bean plantations collect very little runner bean pollen (Tedoradze, 1959; Free, 1968d). Even when honeybees entered runner bean flowers very few of them collected pollen loads (Free, 1968d); perhaps most honeybees failed to become dusted with pollen and, if so, their visits would mostly effect self-pollination only; they may however have discarded any runner bean pollen that collected on their bodies. *B. agrorum* foragers were consistently the best pollinators; they always entered the flowers and about half of them collected pollen as well as nectar. Many of the *B. agrorum* pollen-gatherers deliberately collected pollen by scrabbling over the anthers, but honeybees never did this.

In general the number of bees gathering pollen was greatest between 08.00 and 10.00 h each day, presumably because most was then available. The number of bees robbing was greatest during the late afternoon; probably this was partly because as nectar was collected its level in the corolla tube fell so it was inaccessible through the mouth of the flower, and partly because more of the newly opened flowers were pierced as the day progressed.

All bumblebees that entered flowers, and collected either nectar, pollen or both, stood on top of the left wing petals only, as described by Darwin (1858) and Knuth (1906), but only 78% of the honeybees observed did this, and the remainder stood on the standard petals to reach the nectaries. Contrary to the common supposition that bumblebees are much more erratic foragers than honeybees, it was found that bumblebees visited slightly more flowers per plant than honeybees, but because they also visited more flowers per trip, the foraging areas of bumblebees and honeybees were similar and it was calculated that, on an average trip, they both moved about 12 m along rows and visited about three rows.

A single bumblebee can pierce 2,000 or more runner bean flowers per day and, as a result, the number of pierced flowers can increase rapidly when *B. lucorum* and *B. terrestris* foragers visit a crop. The number of honeybees robbing the flowers depends on the number of robber bumblebees present (Jany, 1950; Free, 1968d) and presumably most honeybees visit the robbed

Fig. 97. Bumblebee *Bombus pratorum* biting hole in base of *Phaseolus multiflorus*, runner bean, flower.

Fig. 98. Honeybee collecting nectar through hole bitten by bumblebee at base of *Phaseolus multiflorus*, runner bean, flower.

flowers directly they begin foraging on a crop and have not entered the flowers earlier, but Free (1968d) found that many honeybees, that were entering the flowers at a time when robbing was prevalent, changed to robbing. It is difficult to understand how bees conditioned to alighting on the wings discover the perforations. However, although the increase in the honeybee population on a runner bean crop was initiated by the appearance of robber bumblebees, its maintenance was independent of the population of robber bumblebees, and after the robber bumblebees had disappeared for the season and no further holes were being bitten, many robber honeybees changed to collecting nectar through the mouths of the flowers.

In fact, honeybees changed from one type of nectar-collecting behaviour to another more readily on runner bean than on other leguminous crops; perhaps this was because the proportion of flowers with pierced corollas was small when the robber bumblebee population was decreasing, and under these conditions would-be robber honeybees spent much more time searching for pierced flowers and, while doing so, found and reached the nectaries through the mouths of the flowers. The piercing of flowers by bumblebees might even be considered advantageous in that it attracts to the crop honeybees that eventually enter the flowers and pollinate. Therefore, taking honeybee colonies to runner bean crops is probably worthwhile.

Chapter 19

Other Papilionaceae

Arachis hypogaea L.

The flowers of *A. hypogaea*, groundnut or peanut, are of characteristic papilionaceous structure but with a very long slender calyx tube (Fig. 99). Each flower has 10 stamens whose filaments are fused for more than half

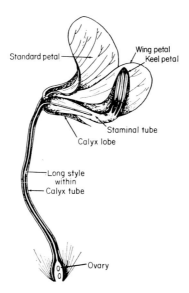

Fig. 99. Flower of *Arachis hypogaea*, peanut (after Robbins, 1931, with permission of McGraw-Hill Book Company).

their length. The 2 stamens opposite the standard are sterile; 4 of the remaining stamens have round anthers with a single locule and 4 have oblong bilocular anthers. The ovary, which contains 2–6 ovules, is surmounted by a long slender style whose club-shaped stigma is enclosed within the anthers of a mature flower. The flowers open at sunrise and at about the same time the anthers dehisce; automatic self-pollination usually occurs within the enclosed keel and self-fertilization seems to be the general rule.

It has been suggested (Hammons *et al.*, 1963; Hammons and Leuck, 1966) that visits by thrips and bees may facilitate self-pollination. Girardeau and Leuck (1967) found that in a glasshouse mechanical tripping increased the yield of "Early runner" but not of "Argentine Spanish". Plants in the field, enclosed by 8 mesh screen cages, produced an average of 19 kernels/plant, compared to 24/plant in cages of 2 mesh screens, and 30/uncaged plant.

The abundance and types of insects visiting peanut flowers seems to depend very much upon local conditions. Heide (1923) in Java, reported that peanuts were not visited by insects during the east (dry) monsoon when there were many competing sources, but during the west (wet) monsoon when there was little competition, many types of insects foraged on the flowers; these included butterflies of the family Lycaenidae, and bees of the families Xylocopidae, Ceratinidae and Apidae, the latter including *Melipona* spp. and *Apis cerana*. Carmin (1959) in Israel mentioned the small wasp *Ceratina bispinosa* as the most important visitor to peanut flowers.

In Georgia, U.S.A., a very few syrphid flies and butterflies were seen visiting the flowers but 6 species of Halictidae, 7 of Megachilidae, 3 of Apidae and 2 of Bombidae, were recorded doing so. Whereas the Halictidae and Megachilidae were most abundant during the cool morning hours, when most pollination occurred, and had practically disappeared from fields by midday, the numbers of Apidae and Bombidae tended to keep constant throughout the day (Hammons *et al.*, 1963; Girardeau and Leuck, 1967).

Leuck and Hammons (1965a) have recorded the behaviour of bees of different species when visiting the flowers; their studies might well be repeated with advantage on other leguminous crops. When the wing petals of the peanut flower are depressed by a suitable insect, the keel is lowered and the stigma, covered with pollen from the dehisced anthers, is forced out of the opening at the keel apex and against the insect's body. When pressure on the wing and keel petals is released they return to their former positions. A flower releases most of its pollen the first time it is tripped, but it may continue to release small amounts of pollen during several subsequent trippings.

It was found that honeybees effectively tripped the flowers while collecting pollen; however, some honeybees that were probably searching for nectar approached the flowers from the side between the standard and wing petals, and others stood on the standard petal and probed downward. On average, honeybees spent 6 sec/flower visit. *Melissodes* spp. were also efficient trippers and by pumping the wing petals sometimes tripped a flower more than once during a single visit and so collected more than one ejection of pollen. Worker bumblebees were among the most persistent trippers; large workers and queens depressed the keel to such an extent that it split open permanently exposing the stamens and stigma; however, there is no evidence that this damaged the reproductive abilities of the flower. In contrast the Halictidae observed were not heavy enough to depress the wing petals and keel but they managed to gather large loads of pollen from just inside the keel tips which protruded from between the wing petals. They even entered between the wing petals to collect pollen from an unprotruding keel. As a result of the difficulties they experienced, a single visit often lasted 3 min or more.

Stokes and Hull (1960) in Florida reported that only very rarely do insects other than thrips visit peanut flowers, and Pelerents (1957) claimed that thrips were the principal agents causing natural hybridization of peanuts in the Congo. Hammons and Leuck (1966) did an experiment to test this later assertion. They caged 10 plots, each containing 4 plants of one strain and 3 of another, to exclude bees and other large insects, and left 20 similar plots uncaged. Although there were as many thrips (*Frankliniella fusca*) in the caged as in the uncaged plots (average of 14·9 and 12·6 thrips/10 flowers respectively) no flowers in the cages were cross-pollinated compared to 0·22% of flowers in the uncaged plots which were visited by bees, mostly *Halictus*. An earlier experiment (Leuck and Hammons, 1965b) had given similar results (0% and 0·54% cross-pollination in caged and uncaged plots respectively) but the thrips in the two types of plot had not been counted. Thus, it seems unlikely that thrips contribute toward the cross-pollination of peanut flowers; their tendency to remain in a flower until it had completely withered and collapsed would account for this.

Observations and experiments by Kushman and Beattie (1946), Bolhuis (1951) and others have indicated that natural hybridization is greater than had previously been supposed and may even reach 6%. According to Srinivasalu and Chandrasekaran (1958) the stigma and stamens protrude more from the keel of some varieties than others; it seems probable that the extent to which they protrude may influence the likelihood of cross-pollination, but no experiments have been done to test this. Because cross-fertilization occurs at all, various physiological factors must act in delaying the growth of pollen following self-pollination. Probably cross-fertilization occurs only when cross-pollination has taken place very soon after a flower opens, and this will depend upon the abundance of suitable pollinating insects.

Cajanus indicus Spreng

The flowers of *C. indicus*, pigeon pea, open at any time between 09.00 and 17.00 h and remain open for 1–2 days. The anthers, which surround the stigma, dehisce in the bud, and the stigma is covered with pollen when the flower opens. Howard *et al.* (1919) in India, found that self-fertilization readily occurred when flowers were isolated from insect visits. However, in the field the flowers were constantly visited by bees and natural cross-pollination was common. Mahta and Dave (1931) also in India, found that *Megachile lanata* is heavy enough to depress the keel and release the staminal column which contacts the abdomen of the bee while it is collecting nectar; this bee also deliberately collects pollen from the anthers before it leaves a flower. In contrast, *Apis dorsata* was found to be too light to depress the keel, and had to force the keel petals apart with its head and legs to reach the pollen.

Howard *et al.* (1919) reported 65% crossing occurred, and Mahta and Dave (1931) found that when in three experiments, genetically marked varieties were grown in adjacent rows, there was 14, 13 and 13% crossing.

Cicer arietinum L.

In India, Howard *et al.* (1916b) found that the flowers of *C. arietinum*, chick pea or gram, opened between 09.00 and 16.00 h, closed at sunset, opened again the next morning and finally closed in late afternoon. Anther dehiscence occurred in the bud, and before the flower opened the filaments lengthened to carry the anthers above the stigmas so self-pollination could occur. When plots were covered to exclude insects, they produced a normal amount of seed. However, they noted that numerous bees visited the flowers and suggested that they were responsible for the small amount of natural cross-pollination that occurred.

Coronilla varia L.

The only studies on the pollination requirements of *C. varia*, crown vetch, have been by Anderson (1958, 1959) who obtained no seed from plots caged to exclude bees, 10·1–21·1 seeds/head from plots caged with bees and 17·8–24·6 seeds/head in the open field.

According to Knuth (1906) nectar is secreted in the fleshy calyx. Bees collecting nectar push their tongues down the corolla and between the narrow bases of the petals to reach the calyx. Anderson observed that bees appeared to experience considerable difficulty in tripping the flowers to obtain pollen but could obtain nectar without tripping them; however, most flowers contained no nectar. Honeybees and bumblebees visited 10·4 and 25·2 flowers/min respectively. He found an average of only 2·6 honeybees and 0·5 bumblebees on 37 m² plots with 1 honeybee colony/1·8 ha, and 14·0 honeybees and 1·6 bumblebees with 1 honeybee colony/0·4 ha. Many more honeybees were attracted to nearby crops of *Melilotus alba* and *Lotus corniculatus*.

Crotalaria juncea L.

The large yellow flowers of *C. juncea*, sann or sunn hemp, occur in inflorescences. They need insect visits to set seed. Each flower has five stamens with short filaments and long narrow anthers, and five with long filaments and small round anthers. The long anthers dehisce in the bud and the filaments of the round anthers then elongate and push the pollen to the orifice of the keel (Fig. 100). The stigma is located at the orifice above the pollen mass.

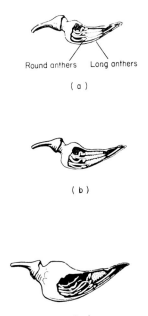

Fig. 100. Stages in the flower opening of *Crotalaria juncea*, sann or sunn hemp: a, before anther dehiscence; b, long anthers have dehisced; c, filaments of round anthers elongate (after Howard *et al.*, 1919).

Howard *et al.* (1919) observed that when a heavy insect, such as *Megachile anthracine* or *Xylocopa amethystina* alights on the wings, the stigma and pollen mass are forced against the insect's abdomen. *Apis cerana* and *A. florea* are not heavy enough to work the pollinating mechanism, but collect pollen left by other insects.

When flowers are not visited by insects, the continued elongation of the filaments presses the pollen mass onto the stigma so that self-pollination is possible. However, self-fertilization does not occur unless the stigmatic surface is rubbed against an insect's body, and lack of pollinators probably helps to explain why in parts of India few flowers set seed.

Glycine max L.

While the small flowers (Fig. 101) of *G. max*, soyabean, are still in the bud stage the ten anthers begin to dehisce, and the stamens elongate, so that when the flowers open they are nearly as long as the pistil, and the pistil and stamens are covered with pollen grains. Sometimes further elongation of the stamens pushes pollen out of the end of the keel (Woodhouse and Taylor, 1913). Experiments by Piper and Morse (1910, 1923) in Virginia, and Woodhouse and Taylor (1913) in India, and Milum (1940) in Illinois, showed that bagged or caged plants set pods and seeds as perfectly as plants in the open. Although these experiments were not properly controlled, and there is no evidence that the plants in the open were visited by pollinating insects, it seems that automatic self-pollination is followed within 24 h by self-fertilization (Oganjan, 1938). As this happens at about the time a flower opens or earlier, there is little opportunity for cross-pollination and very little occurs.

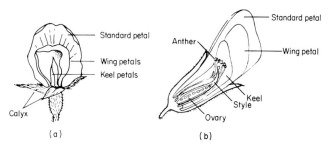

Fig. 101. Flower of *Glycine max*, soyabean: a, front view (after Poehlman, 1959); b, sectional side view (after Purseglove, 1968).

Woodworth (1922) found that only 0·16% crossing occurred when plants of different varieties were alternated; Piper and Morse (1923) stated that the amount of crossing exceeded 1% in occasional seasons only, and Woodhouse and Taylor (1913) concluded there was a lower percentage of cross-pollination of soyabean in the plains of India than in N. America. Probably the amount of cross-pollination depends upon the population of pollinating insects, and on how soon the flowers are visited after they have been opened. Gordienko (1960a,b) caged two varieties of soyabean with honeybee colonies and found that an average of 35% of the seeds produced were hybrids.

Morse and Cartter (1937) thought that thrips were mostly responsible for cross-pollination, but that bees also visit the flowers, especially for pollen. Woodhouse and Taylor (1913) found that in India the peak period of insect visits was between 10.00 and 12.00 h; *Nomia cognata* was the most common, followed by *Apis cerana* and *Apis dorsata;* muscid flies were also abundant.

The amount of nectar secreted by soyabean can be small and its attractiveness as a nectar crop varies greatly; honeybees may ignore soyabean in some years, despite a dearth of nectar from other sources, and gain 14 to more than 45 kg/colony of honey from it in others (Milum, 1940; Johnson, 1944). There is some evidence suggesting that nectar production by soyabean is very dependent on soil fertility, although no experiments have been done on this (Davis, 1952).

Indiofera sumatrana Gaertn

The anthers of *I. sumatrana*, Sumatrana indigo, burst in the bud just before the flowers open, and visits by bees (*Apis florea* and *A. cerana*) trip the flowers and cause an "explosive" discharge of pollen. Howard *et al.* (1919) found that flowers did not set seed unless tripped, and when they were artificially tripped, resulting in self-pollination only, setting was less than when they were visited by bees.

Indiofera arrecta Hochst

Howard and Howard (1915) found that *I. arrecta*, Java indigo, needs to be visited by bees to set seed, and there was little or no self-fertilization of isolated flowers. They also reported that extensive cross-pollination occurred.

Lotus corniculatus L.

Darwin (1876) covered several plants of *L. corniculatus*, birdsfoot-trefoil, and found they produced only two empty pods and no good seeds, but the first thorough investigation of the pollination requirements of this species was made by Silow (1931). He grew plants in bee-proof glasshouses and either: (a) did not touch the flowers so that any seed produced was from auto-pollination; (b) self-pollinated the flowers by depressing the keel; or (c) cross-pollinated them. As a result of these three treatments 14, 64 and 100% of the plants produced seed with averages of 2, 40 and 481 seeds/100 florets respectively. He concluded that birdsfoot-trefoil is practically self-sterile, although an occasional isolated plant does set a few seeds. This conclusion has since been confirmed by others (e.g. Seaney, 1962).

It has long been known (e.g. Knuth, 1906) that the only effective pollinators of the flowers are members of the Hymenoptera whose weight is sufficient to depress the wing and keel petals, so forcing the stamens and pistil to contact the underside of the petiole region of the insect where pollen is deposited (see Fig. 102; Schwanwitsch, 1956). Honeybees comprised 96% of the insects visiting the flowers in New York (Morse, 1958), whereas Bohart (1960) reported that in Utah wild bees (*Megachile*, *Osmia*, *Anthidium* and *Bombus*) were more abundant than honeybees.

Each flower can be tripped many times, and Morse (1958) found that, although a flower produced a large number of seeds as a result of a single honeybee visit, from 12–25 visits were necessary to achieve maximum pollination. Flowers that were caged with bees and received eight times this number of visits gave no more seed. When flowers received many visits they remained open for about 3 days only, but when they received few visits they remained open for up to 10 days. He observed that bees visited an average of 12 flowers/min, and calculated that an average of one honeybee present per 0·8 m² , containing 1,000–3,000 flowers, was adequate to ensure that each

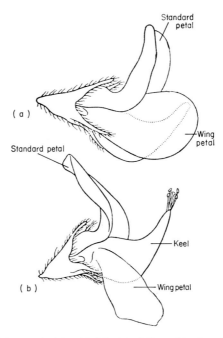

Fig. 102. Flower of *Lotus corniculatus*, birdsfoot-trefoil: a, side view; b, side view with keel and wing petals depressed (see Schwanwitsch, 1956).

flower was visited sufficiently often to give maximum seed production; he pointed out that attempts to increase the number of bees beyond one per 0·8 m² would have little effect on pollination. The results of Shchibrya and Mart'yanova (1960) in general confirmed those of Morse and indicated that with increase in the number of visits to a flower up to a total of 10 visits the number of seeds per pod increased as follows: 1–2 visits, 2·5 seeds; 3–5 visits, 7·3 seeds and 6–10 visits, 10·8 seeds.

Bader and Anderson (1962), in Ohio, covered 1·2 m² plots of a birdsfoot-trefoil crop with small cages which they removed at intervals during flowering to allow bees to visit the flowers. They recorded the number of times each flower was visited, the duration of each visit and whether the bee concerned collected nectar or pollen. Data on flowers that were visited by both nectar and pollen-gatherers were discarded. Although flowers visited by pollen-gatherers received an average of fewer visits than those visited by nectar-gatherers (1·4 : 1·6), and pollen-gatherers spent less time per visit (5·4 : 7·1 sec), they produced more seed per pod (7·0 : 4·9). Bader and Anderson concluded that nectar was more difficult and time-consuming to collect than pollen, but exactly how the behaviour of pollen-gatherers is more effective in pollinating the flowers than that of nectar-gatherers does not seem to have been elucidated. It seems that honeybees gather either nectar or pollen, but not both together, from the flowers, although this too needs clarifying. However, Bader and Anderson found that an increase in the number of both types of visit increased the number of seeds per pod, and they also obtained a positive correlation between the time pollen-gatherers spent visiting a flower and the number of seeds it set. Butler (1945a) and Montgomery (1958) reported that birdsfoot-trefoil nectar had average sugar concentrations of 26 and 15% respectively; Morse (1958) discovered that both its abundance and sugar concentration compared favourably with that of other possible competing species, and on a typical day the sugar concentration was 19% at 09.00 h, 31% at 12.00 h, 39% at 15.00 h and 34% at 18.00 h. Birdsfoot-trefoil is regarded as a valuable honey plant in N. America (e.g. Pellett, 1948 and Lovell, 1955) and in the U.S.S.R. (e.g. Shchibrya and Mart'yanova, 1960), although apparently not in Poland (Demianowicz and Jablonski, 1966).

Lupinus

There is little or no cross-pollination of *Lupinus angustifolius* L., blue lupin, but about 20% cross-pollination occurs when plants of different varieties of *Lupinus luteus* L., yellow lupin, are grown intermingled, and 1–4% when they are grown in discrete rows; the amount of cross-pollination of yellow lupin also differs with the variety concerned (Hagberg, 1952, Wallace *et al.*, 1954, Leuck *et al.*, 1968).

Kozin (1967) reported that honeybee visits to many varieties of lupin increased the number and length of pods, and the number and weight of seeds produced. He found that 83% of the pollinating insects were honeybees and 10% bumblebees, particularly *Bombus agrorum*. Some honeybees collected pollen only (33%), others nectar only (10%), but most (57%) collected both nectar and pollen. They spent an average of 6·4 sec/flower visit. As a result of his studies, Kozin advocated that honeybees should be used to increase pollination of lupin seed crops, 1 ha of which yielded an average of 26 kg honey. Leuck *et al.* (1968) surveyed the insects visiting blue lupin in Georgia and Florida and found that species of Hymenoptera, especially honeybees, but also *Bombus* spp., *Xylocopa virginica*, *Anthophora ursina*, *Tetralonia dubitata*, *Lasioglossum* spp. and *Andrena* spp. were the most common visitors. Bumblebee queens, because of their size, sometimes split the keel petal and permanently exposed the stigma; such flowers were then pollinated by smaller insects not strong enough to depress the keel.

Melilotus

The calyces of *Melilotus alba* Desr., white melilot or sweet clover, and *Melilotus officinalis* L., yellow or common melilot, are short and moderately wide (Fig. 103) so the nectar is readily reached by the numerous honeybees and other insects with short tongues that visit the flowers.

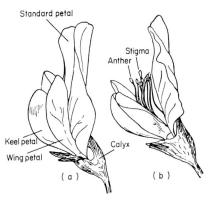

Fig. 103. Flower of *Melilotus alba*, sweet clover: a, side view; b, side view with wing and keel depressed (after Coe and Martin, 1920).

Darwin (1876) reported that a yellow melilot plant visited by bees produced thirty times as many seeds as a plant protected from insect visits. Coe and Martin (1920) found that sweet clover plants protected from insect visits and plants exposed to insect visits produced averages of 0·4 and 31·0 pods/raceme. Other early experiments (e.g. Kirk, 1925; Elders, 1926; Dann, 1930) confirmed that whereas yellow melilot is generally self-infertile and sets few seeds without insect pollination, sweet clover is generally self-fertile and automatic pollination occurs to some extent. Experiments by Ufer (1930) and Kirk and Stevenson (1931) showed there were great differences in the self-fertility of different strains, but sweet clover was more inclined to be self-fertile. Whereas some strains of sweet clover produced seeds without artificial manipulation or insect visits, others were self-fertile but not self-pollinating and others were completely self-sterile. The tendency of plants to produce seed by automatic self-pollination was influenced by the length of stamens, the amount and condition of pollen, the stage of flowering at which pollen was liberated, the distribution of free pollen within the flowers and the receptivity of the stigmas.

However, Hartwig (1942) found that selfed plants of sweet clover and yellow melilot gave inferior progeny, and he thought it likely that less vigorous plants grown from selfed seed would be crowded out as a result of competition so that most commercial strains of both species would be highly cross-pollinated.

Killinger and Haynie (1951) in Florida found that plots of "Hubam" sweet clover caged with honeybees, without bees and not caged gave yields of seed equivalent to 224, 171 and 180 kg/ha. The high yield in the cage without honeybees may have occurred because the strain they used was self-pollinating. In contrast to the above results, Weaver *et al.* (1953) obtained yields of 146, 18 and 176 kg/ha of seed from the same treatments, each of which was replicated five times. Haws and Holdaway (1957) in Minnesota, found that a field with 5·5 honeybees/m² gave 510 kg/ha whereas a field with no honeybees, that depended on the few bumblebees present for pollination, gave 71 kg/ha of seed. Goplen (1960) reported that plots caged to exclude bees produced yields of 56 kg/ha of seed, compared to about 560 kg/ha from the open field where there were 3·6 honeybees/m². From these results it appears that most strains commonly grown require self-pollination, if not cross-pollination.

An additional illustration of the extent to which cross-pollination occurs was provided by Goplen and Cooke (1965). They found that pollen from a block of "Arctic" sweet clover was responsible for 21, 15 and 6% of the pollination of three 0·8 ha plots of "Cumino" sweet clover 0·4, 0·8 and 1·6 km away respectively.

Many types of insect, including Coleoptera, Lepidoptera, Diptera, as well as Hymenoptera, visit sweet clover (e.g. Coe and Martin, 1920) but the honeybee is the most important pollinator. Sweet clover is a good source of nectar and pollen for honeybees and readily attracts them from other crops. The nectar has an average sugar concentration of 22–52% (Shaw, 1953; Montgomery, 1958; Velichkov, 1961). In U.S.S.R. nectar production varies from 130–500 kg/h of sugar according to the locality (Kopel'kievskii, 1959) and it has been calculated that one hectare of sweet clover yields 120 kg honey in Rumania (Copaitici, 1955), 500 kg in Yugoslavia (Kulinčevic, 1959) and 180–218 kg in Bulgaria (Velichkov, 1961).

Weaver *et al.* (1953) observed that both nectar- and pollen-gatherers pollinated the flowers; they visited an average of 37 flowers/min and 5·5 flowers/flower head. To collect a complete load of pollen one bee visited 494 flowers on 84 flower heads.

Pollinated flowers quickly wither, whereas ones that have not been pollinated remain open for days, so the appearance of flower heads can be used as an index of the amount of pollination that has occurred. Haws and Holdaway (1957) recommended 5 colonies/ha for pollination although they stated that experiments had shown that 15–25 colonies/ha gave a progressively greater bee population on the crop and greater seed yields.

There is little information about *Melilotus dentata*, *M. sauveolens*, *M. taurica* and *M. indica*. According to Ufer (1930) the first two species need insect pollination and Hartwig (1942) thought cross-pollination was important in the seed production of *M. taurica* and *M. sauveolens*.

Onobrychis viciifolia Scop

The flower of *O. viciifolia*, sainfoin, has small wings and the keel bears the full weight of a visiting insect (Fig. 104). Thomson (1938) noticed that as the keel is depressed the style projects 1–1·5 mm from the keel cleft so automatic self-pollination is largely prevented. When the keel is released it returns

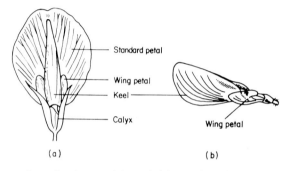

Fig. 104. Flower of *Onobrychis viciifolia*, sainfoin: a, from below (after Knuth, 1906); b, from side, standard petal and calyx lobes removed (after Ross-Craig, 1954).

to its original position. Because the petals are free, and the calyx tube short the nectar is readily accessible to many insect species but the honeybee is the most frequent visitor.

Thomson (1938) reported that an average of 54% of the flowers of a crop set seed. He did an experiment in which only 1·0% of flowers kept bagged set seed, compared to 5·1% of flowers bagged and self-pollinated by tripping, and 51·6% of flowers not bagged. However, Bogoyavlenskii (1955) found that never more than 50% of the flowers of a raceme set seed; although single flowers were visited up to 21 times, no further increase in set was obtained after the eighth visit. Kropacova (1969) calculated that 2–3 colonies/ha are needed to enable 5–6 visits to be made to each flower.

It has been found (Rozov, 1952) that sainfoin secretes nectar at temperatures between 14 and 30°C (57 and 86°F) and maximum production occurs at 22–25°C (72–77°F). The amount secreted has also been shown to depend on meteorological factors, agricultural practices and the variety concerned. Various workers (e.g. Rozov, 1952; Petkov, 1958; Haragsimová-Neprašová,

1960; Kropacova, 1969) have reported that, on average, sainfoin nectar contains 39–50% sugar; that a single flower produces 0·06–0·89 mg nectar and 0·01 to 0·15 mg sugar; and that 33–174 kg nectar/ha of crop are produced.

Pisum sativum L.

The anthers of the *P. sativum*, field or garden pea, dehisce in the bud and the flower is both self-pollinated and self-fertile. It is seldom visited by insects and cross-pollination probably rarely occurs, although Renard (1930) reported that as much as 3% crossing occurred in some varieties, and Haskell (1943) reported that studies in the Argentine showed that cross-pollination varied from 0·2–6·5% for most pea varieties, but was 32% and 24% for the varieties "Bountiful" and "Kentish Invicta"; he suggested that seed crops should be isolated by 200 m or more.

Vigna unguiculata (L.) Walp.

Ants, flies, bumblebees and honeybees are attracted to the extrafloral nectaries of *V. unguiculata*, cowpea, but only heavy insects can trip the flowers. The flowers are said to be self-fertile and self-pollinating in dry climates but in humid conditions insect pollination is necessary and some cross-pollination occurs (see Robbins, 1931; Purseglove, 1968).

The small amount of information I have been able to obtain (see Sampson, 1936; Cobley, 1956; Purseglove, 1968) about other species of Papilionaceae, especially tropical ones, that are of economic importance is given below. Much of the information needs confirmation.

The following are reported as being insect pollinated: *Dipteryx odorata* Willd (Tonka bean); *Lablab niger* Medik (hyacinth bean).

The following are reported as self-fertile and mostly self-pollinating but the flowers are visited by bees and some cross-pollination occurs: *Canavalia ensiformis* L. (Jack bean, horse bean); *Canavalia gladiata* DC (sword bean); *Lens esculenta* Moench (lentil); *Acacia mearnsii* de Wild.

The following is self-fertile and self-pollinating: *Voandzeia subterranea* L. (Bambara groundnut).

I can find no information about the pollination requirements of: *Canavalia plagiosperma* Piper; *Cassia angustifolia* Vahl (Indian or Tinnevelly senna); *Cassia auriculata* L. (avaram); *Cassia senna* L. (Alexandrian senna); *Ceratonia siliqua* (Carab bean, locust bean); *Cyamopsis tetragonoloba* L. (cluster bean); *Derris elliptica* Benth (derris, tuba root); *Dolichos uniflorus* Lam (horse gram); *Lathyrus sativus* L. (grass pea; chickling pea); *Mucuna deeringiana* (velvet bean); *Pachyrrhizus erosus* L. (yam bean); *Pachyrrhizus tuberosus* Lam (yam bean, potato bean); *Prosopis juliflora* (mesquit bean, algaroba, honey locust); *Psophocarpus tetragonolobus* L. (goa bean).

Chapter 20

Grossulariaceae

Ribes nigrum L.

The calyx of a *R. nigrum*, blackcurrant, flower is bell-shaped, only 5 mm deep, and the pendulous flower with its small white petals is rather inconspicuous (Fig. 105).

About a month after flowering, that is at the end of May or the beginning of June, many berries begin to fall. The potential crop may be diminished by

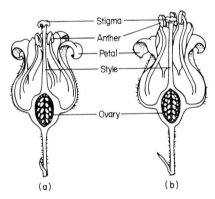

Fig. 105. Flower of *Ribes nigrum*, blackcurrant: a, variety with style above anthers; b, variety with style level with anthers (after Wellington *et al.*, 1920).

as much as half, and the quality of the rest is impaired by the presence of bare strigs and stalks. This phenomenon is known to fruit growers as "running off" (Table 28).

Wellington *et al.* (1921) investigated the cause of fruit drop and concluded that it was due to lack of pollination. They found that about half the dropped fruits had no developed ovules and the other half had very few developed; when they hand-pollinated flowers the fruit drop was negligible, irrespective of whether self- or cross-pollination had occurred.

They also enclosed numerous trusses of flowers in grease-proof paper bags to exclude insects, and found that although the variety "Baldwin" set the same amount of fruit in the bags as outside, all the other varieties they tested set much less fruit when bagged. A high percentage (82%) of Baldwin flowers have their styles just below or level with the stamens, and Wellington *et al.* (1921) suggested that the sepals press the anthers against the stigmas, or that pollen is transferred to the stigmas as the flowers are knocked against each other by the wind. This has recently been confirmed by Wilson (1964), who found there was very little set on bagged "Baldwin" branches firmly tied to a stake to prevent movement, whereas a normal set occurred on bagged

TABLE 28

The percentage fruit drop of black currant varieties recorded by Teaotia and Luckwill (1956) during different periods in 1951

Variety	May 15–June 5	June 6–June 19	June 20–July 2	July 2–harvest	Date of harvest
Seabrook's Black	42·2	6·7	1·8	6·1	July 10
Westwick Choice	39·9	9·7	4·6	6·0	July 23
Baldwin	25·4	1·7	3·0	12·9	July 18
Mendip Cross	21·2	13·5	2·8	3·5	July 12
Cotswold Cross	21·6	5·2	1·0	3·9	July 16

branches which were briefly and gently shaken each afternoon during flowering. However, in two of the other varieties that Wellington *et al.* (1921) tested, ("French" and "Boskoop Giant") more than half the flowers have their styles above the level of the stamens so that no autopollination is possible and insects are needed to transfer the pollen. There is a general tendency for the style to be larger in successive flowers of the raceme; so that whereas the basal flowers of "Boskoop Giant" tend to have the style at the same level as the stamens and to be self-pollinating, those at the apex have the style well above the level of the stamens (Fig. 105). Wellington *et al.* (1921) correlated these structural differences with the general tendency of the basal flowers to set and those at the apex to "run off".

Teaotia and Luckwill (1956) confirmed that fruit drop increases from the base toward the apex of the raceme, but showed that the fruit weight also decreases (Fig. 106). This decrease was much more marked in some varieties than others, and especially occurred in varieties such as "Seabrook's Black" and "Mendip Cross", which exhibit a large range of berry size. The number

280 INSECT POLLINATION OF CROPS

of seeds per berry also decreased toward the apex, and they concluded that differences in seed number are the principal cause of variation in fruit size, and in the amount of drop at different positions in the raceme (Fig. 107).

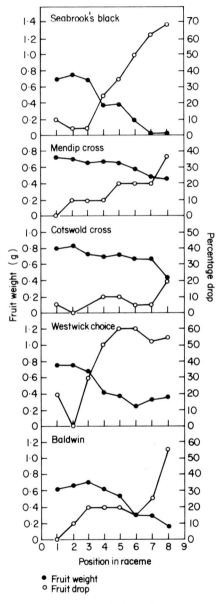

Fig. 106. Relationship of fruit weight and fruit drop of *Ribes nigrum*, blackcurrant, to position in raceme (after Teaotia and Luckwill, 1956).

Although the terminal flowers of a raceme had fewer ovules than the basal ones, there were still more than enough (range 106–149 ovules/ovary) present to provide the optimum content of 30 seeds/fruit. In the absence of pollination the drop increased, but when natural pollination was supplemented by hand pollination no decrease in drop occurred. Although they concluded from this that lack of pollination cannot be considered the primary cause of the drop, it would seem more correct to have concluded that although it was an important cause of lack of set, it was not the only limiting factor under their conditions.

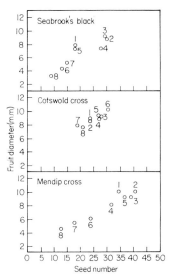

Fig. 107. Correlation between seed number and fruit diameter of *Ribes nigrum*, blackcurrant, at different positions in raceme. Position in raceme indicated by numbers, number 1 being the lowest (after Teaotia and Luckwill, 1956).

Teaotia and Luckwill (1956) further found that, in both attached and dropped fruits, most ovules that are initially fertilized never develop into seed, but abort soon after starting their development in an apparently normal manner. They concluded that the smaller seed content of dropped berries was because a larger proportion of their fertilized ovules aborted. They did not demonstrate the cause of this abortion, but because it was more frequent in berries at the apex of the raceme, inadequate nutrition seems likely. Perhaps when nutritional requirements are no longer the limiting factor the drop would be more directly dependent on pollination.

The experiments of Wellington *et al.* (1921) on excluding insects from flowers have been repeated by other workers. Schander (1956a,b) found that bushes of two varieties exposed to bee pollination had 3·8 and 2·5 times the fruit set of bushes he caged without bees. Zakharov (1958) bagged branches of two varieties to exclude bees and found that, compared to unbagged branches a smaller percentage of flowers produced mature fruits (62 : 49% and 60 : 37%), the fruits weighed less (0·42 : 0·36 g and 0·57 : 0·36 g) and had fewer seeds per berry (14·5 : 6·6 and 24·0 : 8·0). Pollination by bees was superior to hand pollination. Later experiments (Zakharov, 1960a) with seven varieties gave similar results. Glushkov (1958) reported that "Laxton" blackcurrant bushes isolated from insects and others pollinated by bees, had 2% and 46% set respectively and gave 0·1 and 1·9 kg/bush of fruit. Hughes (1962, 1963) found that three bushes of "Baldwin" (one a year) caged to exclude bees had an average crop of 1·2 kg compared to 4·1 kg for uncaged bushes.

There have been few experiments that demonstrate the actual ability of honeybees to pollinate blackcurrants. Schander (1956a,b) found that when honeybees were enclosed with caged bushes the yield of two different varieties was increased 2·5 and 3·8 times in one year, and 10·9 and 17·3 times in another. Hughes (1966) found that in one series of experiments "Baldwin" bushes caged with bees yielded an average of 2·9 kg of fruit compared to 1·6 kg for a bush caged without bees and 3·6 kg for a bush not caged, and in another series of experiments bushes caged with bees gave as much fruit as uncaged bushes.

Attempts to trap wind carried pollen in blackcurrant plantations have been unsuccessful (Baldini and Pisani, 1961).

Experiments comparing self- and cross-pollination of blackcurrants have given conflicting results. Wellington et al. (1921) reported that all the varieties they tested were self-fertile and self-pollination was as effective as cross-pollination. Ledeboer and Rietsema (1940) found that some varieties were partially self-unfruitful, and that although following self-pollination the pollen tubes grew the entire length of the style and entered the ovaries, the male and female nuclei sometimes did not fuse. Various Russian workers (e.g. Simonov, 1949; Neumann, 1955a; Zakharov, 1960a) have since found that the average set and weight of mature fruit from cross-pollinated flowers is greater than that from self-pollinated ones. Furthermore, when pollen from more than one variety is used the set is greater. Thus the average set from pollinating a variety with pollen from 1, 2 and 3 other varieties was 22, 31 and 41% respectively (Zakharov, 1960a). This presumably indicates that the cross-compatibility of the pollen differs with its source. Neumann (1955a) also reported increased crops when pollen mixtures of different varieties were used, compared to pollen from one variety or with self-pollination.

Baldini and Pisani (1961) showed that cross-pollination of flowers of bagged branches gave a greater percentage set than self-pollination. Potter (1963) compared the set from self- and cross-pollinated bushes grown in a glasshouse. Emasculated flowers of all five varieties tested (i.e. "Baldwin",

"Boskoop Giant", "Seabrook's Black", "Victoria" and "Wellington XXX") set more fruit when cross-pollinated than when self-pollinated, but some varieties benefited from cross-pollination more than others, and varieties differed in their cross-compatibility. Cross-pollination also resulted in more seeds per berry and larger berries.

It has also been reported that artificial pollination with pollen of another species (i.e. *Prunus avium*, *Ribes grossularia*) is as effective as pollen from some blackcurrant varieties. For example, in each of 2 years the weight of fruit harvested from a row of blackcurrant bushes directly adjacent to a row of gooseberry was greater than that from a second row of blackcurrant, which in turn was greater than that from a third row (Zakharov, 1958, 1960a).

The mechanism by which the differences resulting from self- and cross-pollination might arise was suggested by observations of Klämbt (1958) and Williams and Child (1963). They showed that, although varieties were perfectly self-compatible, the pollen tube growth-rate was greater following cross-pollination than self-pollination. Hughes (1966), failed to obtain an increased crop by enclosing bouquets or bushes of "Mendip Cross" with "Baldwin" bushes in cages with honeybee colonies. She pointed out that although differences in pollen tube growth-rate could be responsible for differences in the set and yield of the above experiments, particularly as some were done in artificial conditions, it was unlikely that they would give rise to economic differences in yield.

Free (1968e) argued that if yields were increased by cross-pollination between different varieties, the fruit set in plantations should be greatest on those bushes growing next to another variety, and should decrease with increase in distance from other varieties. In each of 3 years he determined the initial and final set on bushes adjacent to another variety, and at various distances from it in blackcurrant plantations containing blocks of two or more varieties. However, the distances separating bushes of two different varieties had no measurable effect on the quantity of fruit, indicating that the varieties concerned, which include those most commonly grown in Britain (i.e. "Baldwin", "Cotswold Cross", "Davison", "Mendip Cross", "Seabrooks", "Wellington XXX", "Westwick Choice"), do not benefit from cross-pollination.

The fruit set on all the varieties Free (1968e) used was as great or greater than the Russian workers obtained from cross-pollination. Varietal differences in self-fertility have been reported (e.g. Neumann, 1955a; Hilkenbaumer and Klámbt, 1958) so perhaps lack of any favourable effects of cross-pollination in Free's observations reflects the greater ability of the varieties he used to set fruit when self-pollinated, or their more favourable environment, or both.

It is easy to envisage that under sub-optimum conditions (as occurs when branches are bagged) the faster growth-rate of the pollen-tubes in cross-pollinated than in self-pollinated flowers is more likely to influence set even with varieties that are usually completely self-compatible, for example "Baldwin", "Wellington XXX" and "Victoria" (Williams and Child, 1963). Free (1968e) concluded that there was no reason to interplant rows of the different varieties he tested, particularly as planting them in separate blocks facilitates cultivation and picking, and is essential if close planting and fully mechanized harvesting is to succeed. However, possibly with less self-fertile varieties, or less favourable conditions, planting different varieties near each other may give better crops. If this is demonstrated the tendency of honeybees and bumblebees to keep to one row per foraging trip (Free, 1968b) must be considered when planning the arrangement of the different varieties.

Zakharov (1960b) found that honeybees spent between 7 and 18 sec/flower visit depending on the variety, and Free (1968b) found that honeybees spent an average of 14 sec/flower, but three species of bumblebees (*B. lapidarius*, *B. lucorum* and *B. terrestris*) spent an average of only 5–6 sec/flower. Whereas honeybees visited an average of only 5 flowers/bush the different species of bumblebees visited between 9 and 13; bumblebees also visited more flowers per cluster. This is surprising as it is often supposed that bumblebees have larger foraging areas than honeybees, although few direct comparisons have been made (page 262). However, because bumblebees work faster than honeybees, and have a larger nectar carrying capacity, they visit many more flowers and spend longer per trip. Consequently they are at least as valuable as honeybees in cross-pollination.

When visiting a blackcurrant flower, a bumblebee or honeybee grips the corolla, or more rarely a nearby leaf, and pushes its tongue and the front of its head between the stigmas and stamens in order to reach the nectaries; thus, one side of its head touches the anthers and the other the stigmas (Hooper, 1919; Free, 1968b). Very few honeybees and bumblebees foraging on blackcurrants collect pollen (0–10% and 0–22% on different days) and when colonies were put in blackcurrant plantations, usually less than 1% of the total pollen they collected came from blackcurrant (Free, 1968b). Probably any blackcurrant pollen that is dusted onto the heads of bees as they enter the flowers is so scanty that pollen loads rarely accumulate. However, as nectar-gatherers are probably as efficient as pollen-gatherers in pollinating the flowers, there seems no reason to encourage pollen collection by colonies taken to blackcurrant plantations.

Butler (1945a) found that blackcurrant nectar had 25% average sugar concentration. Sazykin (1953) reported that a blackcurrant flower secretes from 1·4–2·7 mg/day of nectar with an average sugar concentration of 22%, and

Rymashevskii (1957) found that a flower secretes 4·4 mm³/day of nectar of 23% average sugar concentration. The concentration and abundance of nectar secreted differs for the different blackcurrant varieties. Differences in the amount of nectar present had no apparent effect on the number of bee visits, but varieties secreting the most concentrated nectar (about 25% sugar concentration) had more than twice as many honeybees visit them as varieties with the most dilute nectar (about 15% sugar concentration) (Zakharov, 1958, 1960b). Similar but less pronounced results were obtained by Rymashevskii and Rymashevskaya (1960). Zakharov (1960b) found that *Ribes grossularia* and *Prunus avium* flowers, whose nectar concentration is greater than that of blackcurrant, are preferred by most bees, and when they come into flower the number of visits made to blackcurrant decreases greatly.

Although, at the time blackcurrant flowers most or all of the bumblebees present are queens, they are in general more abundant on the crop than honeybees (Wellington *et al.*, 1921; Hughes, 1965); their numbers fluctuate less and they are plentiful on cool days and at the beginning and end of the day when honeybees are few or absent (Free, 1968b). As free anthesis occurs at relatively low temperatures (8–14°C, 46–57°F, Percival, 1955) these bumblebees may pollinate the flowers. On days when many honeybees visit the crop their population reaches a peak by about midday and rapidly decreases in the early afternoon. Similar rapid fluctuations in the honeybee population occur on other crops, and presumably reflect the honeybees' ability to communicate the existence and location of a favourable source of food to other members of their colony (page 26) and rapidly exploit it.

A correlation has been claimed (Skrebtsova, 1959) between the number of honeybee colonies present and the fruit crop produced. With 0·5 colonies/ha only 53% of the flowers set fruit, but with 3 colonies/ha an 88% set was obtained; an increase to 6 colonies/ha did not result in a further increase in crop, presumably because the plants were incapable of bearing more fruit. However, the value derived from taking honeybee colonies to blackcurrant will be more dependent on the prevailing temperature and weather than with most other crops, and it is of course important that the nutritional status of the soil is adequate for the berries to mature.

Ribes grossularia L.

Colby (1926) bagged small flowering branches of *R. grossularia*, gooseberry, to exclude insects. He found that in nine out of ten varieties the number of seeds per berry was less on bagged than on control branches, and sometimes much less; however, the seeds were of similar size and maturity irrespective of the treatment. He noticed a wide variation in the relative lengths of the style and stamen of the different varieties, and suggested that the ease with which self-pollination could occur as a result of movement of the bagged branches by the wind might account for varietal differences. The gooseberry flower (Fig. 108) is protandrous and is normally inverted; both these factors usually favour insect pollination. Hence, it seems probable that although gooseberry is self-fertile, at least some varieties need insects to transfer the pollen.

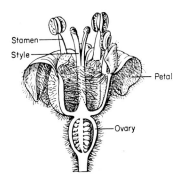

Fig. 108. Median section of flower of *Ribes grossularia*, gooseberry (after Ross-Craig, 1957).

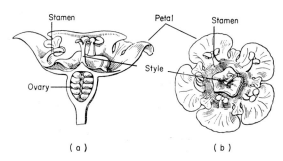

Fig. 109. Flower of *Ribes rubrum*, red currant: a, sectional side view; b, from top (after Ross-Craig, 1957).

Glushkov (1958) reported that a gooseberry variety "Finik" isolated from insects gave only 12% set and 1·8 kg of fruit/bush compared to 43% set and 7·1 kg/bush when pollinated by bees. Preliminary cage experiments by Hughes (1961, 1962) suggested that the gooseberry variety "Careless" produces better crops with insect pollination. She also recorded that caged bushes produced smaller berries with fewer seeds than uncaged controls, and thought that some parthenocarpic development might occur.

Hooper (1919) suggested that because gooseberry and red currant flowers are more open than blackcurrants and their nectar is more accessible, they are visited by more insects. However, the gooseberry flowers even earlier than the blackcurrant, and cold weather during flowering often deters insect visitors. Eaton and Smith (1962) reported that gooseberry flowers are often visited by honeybees in Ontario where their pollination does not present a problem, as the plants are self-pollinating and self-fruitful.

Sazykin (1953) investigated nectar secretion from four varieties of gooseberry for 4 years and found that an average 1·1 1·7 mg of sugar was secreted per flower per day. Rymashevskii (1957) found that a gooseberry flower secreted 1·4 mm^3 of nectar of 34% sugar concentration per day.

Ribes rubrum L.

Very little is known about the pollination of *R. rubrum*, red currant (Fig. 109), although there is an indication that the variety "Earliest of Fairlands" produces better crops with insect pollination than without (Hughes, 1961, 1962).

Chapter 21

Myrtaceae

Species of economic importance are present in the genera *Eucalyptus, Eugenia, Feijoa, Psidium* and *Pimenta*.

Eucalyptus

Species of Eucalyptus have been introduced into most tropical countries but only about twenty have been exploited commercially. The principal commercial species are: *E. citriodora*, Hook; *E. dives* Schau; *E. fruticetorum* F. Muell; *E. globulus* Labill; *E. radiata* Sieb ex D.C., *E. smithii* R. T. Baker, and *E. staigeriana* F. Muell ex F. M. Bail.

Eucalyptus flowers are important sources of nectar and pollen for honeybees and it is generally supposed that they are insect pollinated, although bird pollination may be contributory in some areas (Pryor, 1951; Penfold and Willis, 1961). In Egypt honeybees have two daily peaks (at 07.00 and 15.00 h) of pollen collection from *Eucalyptus* spp., least pollen being collected at midday (Ibrahim and Selim, 1962).

The calyx and corolla of a *Eucalyptus* flower are united to form an operculum or cap. Soon after this is shed the anthers of the numerous stamens dehisce. The ovary has three to six cells, each with numerous ovules. The stigma does not become receptive until a few days after the operculum has fallen so cross-pollination is favoured. However, some of the innermost anthers retain their pollen until the stigma is receptive, and as they unfold they brush against it and so self-pollinate the flower if cross-pollination has failed to occur (Pryor, 1951). Furthermore, pollen may be transferred from younger to older flowers on the same plant. Probably most species of *Eucalyptus* are self-fertile; Pryor (1951) found that flowers of the three species he tested (*E. bicostata, E. blakelyi* and *E. maidenii*) set fruit when bagged, and when bagged emasculated and self-pollinated by hand, but not when bagged and emasculated only. However, one tree of *E. bicostata* failed to set fruit with its own pollen, so perhaps it was self-sterile or produced defective pollen.

The amount of cross-pollination that usually occurs is not known but a preliminary experiment by Foot Guimarães and Kerr (1959) indicated that cross-pollination may be more effective than self-pollination. They determined the set on a branch of an *E. alba* tree caged with a honeybee colony and on a branch not caged. The branch not caged had more fruits per umbel (4·5 : 2·5) and more seeds per fruit (2·1 : 0·8), but they stated that the branch caged with bees gave a better set than hand pollination. This experiment needs repeating with more replication.

Eugenia

I can find no information about the pollination requirements of *E. jambos* L., rose apple; some authorities state that *E. caryophyllus* Bullock and Harrison, clove (Fig. 110), is almost entirely self-pollinated whereas others think it is

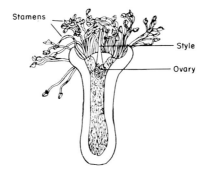

Fig. 110. Sectional side view of *Eugenia caryophyllus*, clove, flower (after Purseglove, 1968).

probably cross-pollinated by bees which visit it. The little evidence that is available is conflicting as bagged clove flowers set fruit in Tanganyika but not in Zanzibar (see Tidbury, 1949). Because the results may have been influenced by the extremely high temperatures and humidities inside the bags, experiments, in which entire trees or branches are caged, are necessary.

Feijoa

Experiments in California and Russia have shown that bagged flowers of *F. sellowiana* Berg, feijoa, set little or no fruit and hand pollination experiments have demonstrated that, in general, cross-pollination gives a better set and bigger fruits than self-pollination although some varieties benefit more from cross-pollination than others. Bees are frequent visitors to the flowers and it is assumed they are responsible for most of the pollination (Clark, 1926; Korolev, 1936; Schroeder, 1947).

Psidium

Purseglove (1968) reported that *Psidium guajava* L., guava, is visited by bees and other insects and although it is self-fertile about 35% cross-pollination occurs.

Pimenta

The important species is *Pimenta dioica* L., allspice or pimento. The inflorescences contain several dozen small flowers. Each has four spreading calyx lobes and four round white petals (Fig. 111); there are numerous stamens and a single style with a yellow stigma; the inferior ovary has two carpels each with one ovule. Although the flowers are hermaphrodite they

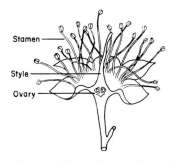

Fig. 111. Median section of *Pimenta dioica*, pimento, flower (after Purseglove, 1968).

are self-sterile (Ward, 1961) and functionally dioecious; male and female trees occur in approximately equal proportions. The flowers on the male tree are barren and their stigmas tend to disintegrate a few hours after the flowers have opened; each has a broad receptacle with about 100 stamens which produce abundant viable pollen; they usually drop after shedding pollen and the fruit, which is very rare, is usually one-seeded. Flowers of the female trees have about 50 stamens whose pollen is non-viable (Chapman and Glasgow, 1961).

Ward (1961) thought pimento was wind-pollinated because bees and other insects were absent during flowering, but Chapman (1965) saw solitary bees belonging to the genera *Halictus*, *Exomalopsis* and *Ceratina* and honeybees visiting allspice flowers for pollen; he suggested that honeybee colonies might with advantage be taken to allspice plantations during flowering. Bees seeking pollen might prefer to keep to male trees only, but it seems that sufficient pollen is produced by the female tree to ensure adequate visits under natural conditions; indeed the only function of the pollen of female flowers is to attract visitors.

Most of the crop of allspice in Jamaica is produced from semi-wild trees, among which male and female trees occur in approximately equal numbers, but such a high proportion of non-bearing trees is obviously not economically desirable in plantations and Chapman suggested two planting arrangements to provide the minimum number of male trees that were sufficient to ensure adequate pollination of the female trees. In the first, every third tree in every third row is a male (ratio of 8 females : 1 male) so each female tree is adjacent to one male; in the second arrangement there are 3 female trees to each male as follows:

F	F	F	M	F
F	M	F	F	F
F	F	F	M	F
F	M	F	F	F
F	F	F	M	F

The female trees in the mixed rows are adjacent to two males and the female trees in the female rows are adjacent to three males.

Presumably the correct proportion of male trees will depend upon many factors including the insect population and the frequency with which insects move from male to female trees. This can only be discovered when the plantations now being prepared have matured.

Chapter 22

Passifloraceae

The two main cultivated species of the Passifloraceae are *Passiflora edulis* Sims, passion fruit, and *Passiflora quadrangularis* L., giant granadilla.

The large (7·5–10 cm diameter) showy flowers (Fig. 112) of *P. edulis* have five spreading spongy sepals which are white above and yellowish-green below, and five white petals which alternate in position with the sepals. The

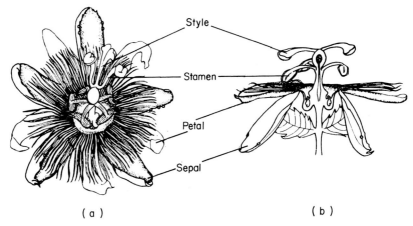

Fig. 112. Flower of *Passiflora edulis*, passion fruit. a, from top; b, side section (after Purseglove, 1968).

five stamens are united into a tube for half their length, but spread out at the top and their large anthers hang downwards. The unilocular ovary is on a stalk, or gynophore, above the level of the anthers, and it is surmounted by three styles. Abundant nectar is secreted at the base of the gynophore.

In Hawaii, the passion fruit is in flower for 8–9 months a year with two flowering peaks, one in April and May and the other in July and August. The flowers of the purple passion fruit open at dawn and close at 12.00 h and the flowers of the yellow passion fruit open at about 12.00 h and close at 22.00 h. Shortly after the flower opens the erect styles curve downwards, and when the process is completed, they are more likely to be touched by insects collecting nectar or pollen. Shortly before the flower closes the styles become erect again, the process of recurvature occupying about an hour. However, the styles of some flowers do not curve downwards as much as others, and because there is a greater distance between anther and stigma for a pollinator to bridge they are less likely to get pollinated. This applies particularly to those flowers whose styles always remain upright, many of which are infertile when acting as the female part of a cross. In natural conditions, the stigma remains receptive on the day of opening only, and the pollen loses its viability after 24 h (Akamine and Girolami, 1959). In New South Wales the flower starts to open in the night or early morning and starts to close at about midday the following day, but the stigmas are fully receptive on the morning of the first day only; the anthers of most flowers do not dehisce until the afternoon of the first day (Cox, 1957).

A large percentage of passion fruit flowers (e.g. 21–64% in four different localities in Hawaii; Nishida, 1958) are not fertilized and drop, and it was thought that the protogynous habit of the flower contributed to this lack of pollination (e.g. Pope, 1935; Cox, 1957) but recent studies have shown that it is nearly completely self-sterile (Akamine and Girolami, 1957). Only 1% of 2,600 flowers self-pollinated by hand set fruit; individual flowers bagged and caged with honeybees or carpenter bees also failed to produce fruit (Nishida, 1958).

Because the flowers are nearly self-sterile any action of wind in transferring pollen from the stamens to the stigmas of the same flower is unimportant, and only a very few of the large sticky grains are carried in air currents. Nor is wind important in cross-pollination. This was confirmed when it was shown that two plots caged to prevent access by insects set no fruit during several months although they flowered profusely (Akamine and Girolami, 1957).

Nishida (1958) found that in two localities flowers, bagged and cross-pollinated by hand, had about the same percentage set as flowers not bagged but in two other localities they had a much greater set, indicating that the local insect pollinators were too few (Fig. 113). Akamine and Girolami (1959) also found that natural fruit set was less than that achieved by hand pollination. Nishida suggested that either the acreage of passion fruit should not exceed the pollinating capacity of the insects present, or else the number of pollinators should be increased. His results also show that the set from cross-pollinating bagged flowers varied from 50–100% depending on the locality, so the maximum set in some localities was limited by factors other than pollination (Fig. 113). The percentage set from hand pollination and the difference in set between hand and natural pollination also varied at different times of the season in the same locality. Pope (1935) also stated that hand-pollinated flowers failed to set in certain localities.

Akamine and Girolami (1959) found that a minimum of 190 pollen grains/flower was needed to achieve any fruit set and, whereas one of the plants they used needed 600 pollen grains to achieve 100% set, another needed more than 1,800. The fruit size, weight and number of seeds all increased with the number of pollen grains used (Akamine and Girolami, 1959). Hence, low fruit set,

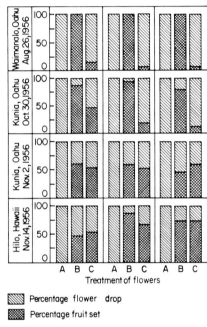

Fig. 113. An experiment to determine the efficiency of natural pollination of *Passiflora edulis*, passion fruit, in three sites in each of four localities: a, flowers bagged; b, flowers bagged and hand pollinated; c, flowers not bagged nor hand pollinated (after Nishida, 1958, with permission of the Entomological Society of America).

and small fruit size in natural conditions, may be because insufficient pollen is transferred to the stigma; the former condition could, of course, also be caused by some flowers receiving no visits. It would be interesting to discover the number of pollen grains deposited during a single bee visit.

The percentage set following cross-pollination depends on the varieties concerned and indicates that while some crosses are compatible others are incompatible. Thus, Akamine and Girolami (1959) found that any cross that involved variety "C 39" mentioned in Table 29 was compatible, but crosses between any of the other three varieties were nearly completely incompatible.

Sometimes the degree of compatibility varied according to whether a variety was acting as the male or female in the cross. They suggested that

Chapter 23

Cucurbitaceae

Cucurbita L.

The genus *Cucurbita* contains numerous species of pumpkin, squash, gourd and vegetable marrow and includes *C. maxima* Duch, *C. mixta* Pang., *C. moschata* Duch and *C. pepo* L.

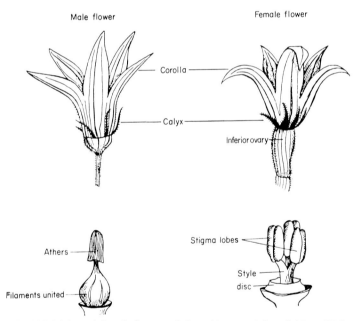

Fig. 115. Male and female flowers of *Cucurbita pepo* (after Cobley, 1956).

The yellow flowers are monoecious and occur singly in the axils of the leaves (Fig. 115). The corolla is divided into five pointed lobes. The pistillate flower surmounts an easily recognizable undeveloped fruit; the style usually has three stigmatic lobes corresponding with three chambers in the ovary;

a ring-like nectary surrounds the base of the style; the stamens are rudimentary. The staminate flowers have five stamens, with united filaments and anthers; they are always more numerous than the pistillate flowers and sometimes there are even twelve times as many of them (Robbins, 1931). The ratio of staminate to pistillate flowers is influenced by the season and the numbers of developing fruits already present (Whitaker and Jagger, 1937), and relatively low temperatures and short day length tend to increase the proportion of pistillate flowers (Nitsch *et al.*, 1952). Shaw (1953) found that the nectar sugar content of *C. maxima* ranged from 18–38% with an average of 30%.

Seaton and Kremer (1939) found that pumpkin and squash flowers opened and the anthers dehisced at 9–10°C (48–50°F) so they were always open at daylight when the temperature was above 10°C (50°F). When the temperature was relatively low and the humidity high, they remained open until noon or later, but under conditions of high temperature and low humidity the corollas started to wither as early as 08.00 h. However, the response to temperature and humidity probably differs with different species because under the same environmental conditions flowers of some species open early in the morning and close at about midday while others remain open for about 24 h. Bhambure (1958a) listed the approximate time of day at which flowers of seven species of gourd and pumpkin opened and closed near Bombay; of the five that opened at 06.00 h, one closed at 12.00 h, one at 14.00 h, one at 17.00 h and two at 18.00 h; another species opened at 08.00 h and closed at 13.00 h. The remaining species did not open until 17.00 h and closed sometime during the night.

Different varieties of the same species of *Cucurbita* will, of course, cross freely and so should not be grown near each other. Crossing can also occur between some species of *Cucurbita* but not others; according to Whitaker and Davis (1962) *C. moschata* can cross with *C. pepo* and *C. mixta*, so isolation of these species is necessary to produce pure seed; but, *C. pepo* and *C. mixta* do not cross with each other and neither of these nor *C. moschata*, crosses with *C. maxima*. However, pollen from one of these species can stimulate the production of parthenocarpic fruit by others, and although this will not result in contaminated seed, it may reduce the total seed yield. Therefore, to obtain the greatest yield it is probably best to separate all species from each other.

Because the male and female organs do not occur in the same flower, and the pollen grains are too large and sticky to be carried by wind, mechanical transfer of pollen is necessary. In glasshouses this is usually done by hand (e.g. vegetable marrow, Bewley, 1963) and in the field by insects. Verdieva and Ismailova (1960) found that pollination of squash by honeybees is as effective as hand pollination, or even more so. The staminate flowers have long peduncles and are at approximately the same level as the upper leaves of the plant but the pistillate flowers have short peduncles and lie under the leaves; consequently, bees are more inclined to visit the more obvious male flowers when they arrived on a field and so cross-pollination is facilitated.

Wolfenbarger (1962) demonstrated in three ways the value of honeybees in pollinating squash plants. Firstly, in each of 3 years he caged squash plants to exclude insects and found that their average yield was only 19% of that of uncaged plants; secondly, he found that the yield of fruit decreased with distance from a group of twenty honeybee colonies put at one end of the field; and thirdly, he found there was a positive correlation between the number of honeybee colonies per hectare of field and the number of baskets of fruit obtained, i.e. an average of 0, 1·2, 2·5, 5 and 7·5 colonies/ha gave an average of 366, 383, 398, 415 and 427 baskets of fruit/ha. Battaglini (1968) found that only 7% of caged female *C. pepo* flowers set fruit compared to 61% of female flowers of 38 uncaged plants.

Sănduleac (1959) suggested that one or two honeybee colonies are necessary to pollinate 10 ha of a cucurbit crop. He found that varieties of *Cucurbita maxima*, *C. pepo* and *C. moschata* were worked intensively by bees from 06.00 h to 12.00 h daily, and the numbers of bees reached a peak between 08.00 and 09.00 h. The male flowers were preferred to female indicating that they were collecting pollen deliberately. However, Linsley (1960) found that honeybees only managed to scrape pollen from the anthers with great difficulty and their pollen loads were very small compared with those of solitary bees of the genera *Peponapis* and *Xenoglossa* which visited the flowers earlier in the day when the pollen was first available. Because of the difficulty honeybees have in collecting pollen, their numbers are strongly influenced by the presence of competing crops. This also applies to the Indian honeybee, *Apis cerana*, which Bhambure (1958) found sometimes collects pollen from *Cocus nucifera* and grasses in preference to gourds.

Numerous insects, most of which belong to the Hymenoptera, Diptera and Coleoptera, have been recorded visiting cucurbit flowers. Durham (1928) suggested that the striped cucumber beetle, *Acalymma vittata*, pollinated summer squash, but too many beetles may perhaps be detrimental to pollination as Fronk and Slater (1956) found an inverse relationship between the number of beetles per flower and the number of times it was visited by bees; although bees alighted on all flowers indiscriminately, they avoided entering those that contained many beetles. In the Ukraine, Nevkryta (1937) recorded 63 species of Hymenoptera, 16 Diptera, 7 Lepidoptera, 1 Hemiptera, and 1 Coleoptera on cucurbit flowers. In Iowa, Fronk and Slater (1956) found that 98% of all insects visiting *Cucurbita pepo*, *Cucurbita maxima* and *Lagenaria siceraria* were Hymenoptera and Coleoptera; only 2% of the Hymenoptera were honeybees, and the most important Hymenoptera were *Xenoglossa strenua* and *Peponapis pruinosa*. They suggested that the superficial resemblance of these solitary bees to honeybees, especially when their bodies are dusted with pollen, might explain reports of many honeybees visiting cucurbit flowers (e.g. Jones and Rosa, 1928; Pammel and King, 1930).

It has long been known (see Hurd and Linsley, 1964) that species belonging to the genera *Peponapsis* and *Xenoglossa* (Anthophoridae: Eucerinae) obtain their pollen solely from the indigenous and domestic cucurbit species, although they may obtain nectar from several other sources. Because of their close association with squash, pumpkin and gourd, they are known as "squash bees". The available data suggest that certain species of bees exhibit a preference for pollen of certain species of *Cucurbita* including domestic species.

The genera *Peponapis* and *Xenoglossa* share in common a number of features which are adapted to the *Cucurbita* flowers (see Hurd and Linsley, 1964). Unlike many insect-pollinated plants, *Cucurbita* pollen is available very early in the day, sometimes before daylight, and the bees of both genera are able to fly at low temperatures and at low light intensities to collect it, with the result that their activity is synchronized with the opening of the flowers and competition with insects that arrive later is avoided. "Squash bees" are ideally equipped to gather and manipulate the large pollen grains, and females of both genera have a narrow band of dense hairs along the anterior margin of the hind basitarsi which may be a special adaptation for this purpose. The male bees habitually spend much of the day and night in closed, mostly staminate flowers of *Cucurbita*; they are usually covered with pollen which they carry with them when they visit newly opened flowers the following morning; however, the viability of such pollen has yet to be determined. Female bees that have not yet begun to nest, also spend the night in flowers.

The majority of species of *Peponapis* and *Xenoglossa* are found in Mexico and Central America, probably because the genus *Cucurbita* reached its maximum development there, and there is some evidence that the distribution of the different species of bees in America was altered after aboriginal man developed and introduced domestic species of *Cucurbita* (Hurd and Linsley, 1964, 1966a,b). Today, these domestic species have been introduced into many parts of the world but the "squash bees" have remained in America. Because of their value in pollinating these crops, it has been suggested that "squash bees" should also be introduced, due care being taken to select species of bees suited both to the particular *Cucurbita* species concerned and to the climatic and topographical characteristics of the area where they are being grown (Michelbacher *et al.*, 1968).

Despite the great importance of "squash bees" in pollinating cucurbits, Hurd (1964) emphasized that the value of other Hymenoptera should not be minimized because in some places at certain times of year "squash bees" may be few or absent, and bumblebees, carpenter bees, halictid bees or stingless bees may account for most of the pollination.

Cucumis sativus L.

Most varieties of *C. sativus*, cucumber, are monoecious and bear staminate and pistillate yellow five-lobed flowers (Fig. 116). There are usually many more staminate than pistillate ones. The staminate flowers each have three

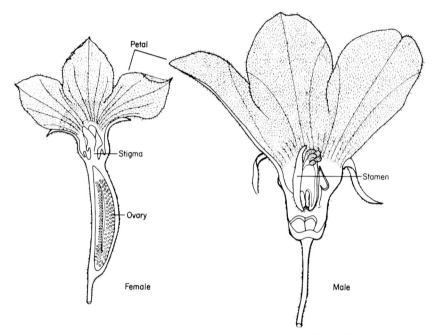

Fig. 116. Male and female flowers of *Cucumis sativus*, cucumber (after Purseglove, 1968).

stamens, two of which have two anthers and one of which has one anther only; dehiscence begins at about 17°C (63°F) and reaches its optimum at 18–21°C (64–70°F) (Seaton and Kremer, 1939). The pistillate flowers have rudimentary stamens but each has a well-developed, three-chambered ovary, each with several rows of ovules and a short thick style.

Maturing fruits of fertilized flowers have an inhibitory effect on the further development of the plant and on the formation of pistillate flowers, but maturing parthenocarpic fruits have no such effect (Tiedjens, 1928; McCollum, 1934). The ratio of pistillate to staminate flowers also decreases with decrease in the availability of nitrogen.

Whereas most varieties of cucumber grown commercially need to be pollinated to produce satisfactory fruits, certain varieties used in some countries (e.g. Britain and Holland) develop parthenocarpic fruits. Should these varieties become pollinated, their fruits become misshapen with greater development near the stalk, instead of being straight and columnar, and their commercial value is considerably diminished. Banks (1951) recorded the presence in a glasshouse of numerous syrphids (*Tubifera pertinax*) which had probably been introduced as larvae in horse dung; they pollinated the flowers, and the resultant fruits became unmarketable due to the development of large seeds, swollen distorted ends and a bitter taste. Great efforts are therefore made to prevent cucumbers becoming pollinated and frequently ventilators of glasshouses are screened to prevent bees and other insects entering them and as an additional precaution, male flowers are removed whenever possible. However, it is not practical to protect completely cucumbers grown under Dutch lights from bee visits. In Holland, honeybees and bumblebees are responsible for most of the pollination that occurs and three weeks after a crop has been visited by bees the first so-called "bull-necked" cucumbers are harvested. These misshapen cucumbers are especially likely to occur in July when, as a result of the increased hours of daylight and high night temperature, the proportion of male to female flowers increases and the ability of the female flowers to respond parthogenetically decreases. Honeybees usually visit cucumber flowers only in the absence of more attractive crops such as rape or clover, but they are especially troublesome in areas where cucumbers are grown extensively. Laws are in force to prevent honeybee colonies being kept in certain areas of the Dutch Glasshouse District from the end of May to the beginning of August, and this restriction has successfully reduced the number of malformed fruits (Vriend, 1953; Berkel and Vriend, 1957; Berkel, 1960; Koot, 1960).

In contrast, attempts are made to encourage and not discourage pollination of the cucumber varieties favoured in most countries, whether they are grown in the field or the glasshouse. Shemetkov (1957) reported that the number of visits a flower received, influenced the number of seeds and weight of fruit produced; thus 2–8 visits/flower gave fruits of 221 g average weight and 60 seeds, and 50 visits/flower gave 500 g fruits with 140 seeds. Alex (1957) reported that the calculated average yield of three plots in a cucumber field caged without honeybees, three plots caged with honeybees and three plots not caged was 80, 409 and 472 hl/ha respectively. He attributed the set in the cages without honeybees to pollination by small ground nesting solitary bees. Similar results were obtained by the Canadian Department of Agriculture (1961); the mean yield, calculated in kg/ha, during 5 years of a plot caged without honeybees, a plot caged with honeybees and a plot not caged was 1,754, 4,683 and 5,787 respectively.

Pollination by honeybees is as effective as hand-pollination; Peto (1950) found that 1,300 plants pollinated by hand gave a mean of 1·3 fruits each compared to a mean of 8·5 fruits for 1,200 plants pollinated by bees. Hand pollination and bee pollination gave no advantage over bee pollination alone. Markov and Romanchuk (1959) found that the average yield of cucumbers per frame from 6 frames isolated from bees, 3 frames pollinated by hand and 24 frames pollinated by bees were 0·7, 7·3 and 11·8 kg; Edgecombe (1946) found that hand and bee pollination were equally effective in producing hybrid cucumber seed in the field. Ankinovič and Ljubimov (1954) suggested that the efficiency of honeybee colonies in cross-pollinating cucumbers in frames would be improved if foragers leaving their hives were forced to walk through a tube of pollen of the desired variety.

Few insects other than honeybees visit cucumber flowers. Alex (1957b) noted that in Texas, halictid bees and ants obtained nectar from the cucumber flowers without pollinating them, and that only a few solitary bees of the species *Melissodes communis* seemed to be effective pollinators. Attempts have been made to correlate the abundance of honeybee colonies in an area with the size of the crop produced, and Gubin (1945) succeeded in doing this for both cucumbers grown in hot beds and those grown in the field near Moscow. He also concluded that there were insufficient wild pollinating insects present to give a full crop. Thus the yield, in kg/ha, was 1,420 from three collective farms with no honeybee colonies, 4,075 from four collective farms with honeybee colonies located 200–800 m from the crop, 7,570 from seven collective farms with honeybee colonies located 100–200 m from the crop, and 16,111 from eight collective farms with honeybee colonies located 50–100 m from the crop. Kaziev and Seidova (1965) stated that taking bees to cucumber fields increased the weight of the crop by an average of 44%. However, Warren (1961) was unable to show that cucumber fields with $2\frac{1}{2}$ honeybee colonies/ha gave a better yield than those to which no honeybee colonies had been taken. Peto (1950) found that the weight of fruits and seeds did not decrease with increasing distance from honeybee colonies up to the maximum distance sampled of 256 m, but this result was probably obtained because his fields had $12\frac{1}{2}$ colonies/ha, and so were saturated with bees.

Although honeybees and wasps are the most common visitors to cucumber flowers (Peto, 1950), taking colonies to a crop is no guarantee that the bees will work it and the extent to which a crop of cucumber flowers attracts honeybees needs more investigation. Nemirovich-Danchenko (1964) reported that the average daily nectar yield of female and male flowers was 1·29 and 0·69 mg respectively, and was greatest 3–4 h after opening; pollen was also most abundant at this time and bee visitation greatest. Kaziev and Seidova (1965) found that a female cucumber flower secreted between 1·1 and 2·4 mg of nectar compared to between 0·9 and 1·6 mg by a male flower, the amount secreted depending to some extent upon the variety and environmental conditions. Edgecombe (1946) mentioned that honeybee colonies put beside a cucumber field gave a low yield of honey. In a glasshouse honeybees soon collect all the nectar and pollen available for the day (Shemetkov, 1960). Bhambure (1958a) observed *Apis florea* and *Melipona* spp. but not *Apis cerana* working cucumber fields near Bombay.

Pollination is, of course, important in producing a seed crop. Zobel and Davis (1949) allowed plants to develop either all the fruits that set, or 3, 6 and 9 fruits only. Although when the number of fruits was restricted the seeds were larger and heavier, their germination was no better than that of seeds produced by the former treatment which gave a total of more seeds.

There is little information on the behaviour of bees when working cucumbers, although studies, especially of their foraging areas, could give information which would be most useful to plant breeders. When producing hybrid seeds, plants of the "female" type have their male flowers removed, and plants of the "male" type have their female flowers removed; it is sometimes necessary to plant the two parents of the hybrid at different dates to ensure that plenty of pollen is available when the female flowers open. In preliminary tests Edgecombe (1946) grew rows of "female" and "male" plants in ratios of 1 : 1, 2 : 1, 3 : 1, and 4 : 1 in different fields, but found that the different ratios made no difference to the number of seeds produced per fruit. He suggested that the higher ratio might be used with safety.

Bees do not appear to differentiate between different varieties as Skrebtsova (1964) found that about 70% of the bees he observed moved from one variety to another when they were planted in alternate rows. In the field, there is 65–70% natural cross-pollination (Jenkins, 1942) and when two varieties are grown in adjacent blocks about 25% cross-pollination may occur between them (Knysh, 1958). Thus, it is essential to separate different varieties being grown for seed production by 0·4 km or more and when stock seed is being grown a distance of 1·6 km or more has been suggested (Hawthorn and Pollard, 1954).

Cucumis melo L.

The species *C. melo* includes muskmelons, canteloupes and sweetmelons among others, but as all varieties of *C. melo* are self-fertile and interfertile they will be considered together.

Most European varieties are monoecious and have separate staminate and pistillate flowers whereas most American varieties are andromonoecious and have staminate and hermaphrodite flowers (Fig. 117), although certain varieties may produce perfect or pistillate flowers depending on local conditions. Hermaphrodite and pistillate flowers are solitary, and occur on the first and second axils of the fruiting branches, whereas the more numerous staminate flowers occur in small axillary clusters of three to five on all the remaining branches. The staminate flowers are present both earlier and later in the season than the hermaphrodite and pistillate. Griffin (1901), in Colorado, kept a record of the staminate and hermaphrodite flowers on six vines from 27 June to 13 July, by which time the vines had become so interwoven that individual vines could not be distinguished. He found that the percentage of the total number of flowers that were hermaphrodite was: 27 June, 0·5;

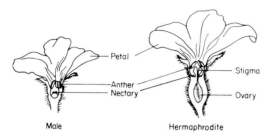

Fig. 117. Male and hermaphrodite flowers of *Cucumis melo* (after University of Arizona, 1961).

30 June, 3·1; 3 July, 5·6; 7 July, 11·2; 10 July, 11·6; 13 July, 4·6. Perhaps the increased scarcity of hermaphrodite flowers later in the season is associated with the presence of developing fruits, as McGregor and Todd (1952) found that vines isolated from pollinating insects had a greater proportion of hermaphrodite flowers than vines whose flowers had been pollinated.

In a female flower the ovary is divided into three or five chambers, depending on the variety, each of which contains several longitudinal rows of ovules, and the divisions of the lobed stigma correspond in number to the chambers in the ovary. Seed counts from mature melons indicate that usually 400–600 ovules are fertilized in each flower that sets a fruit; melons containing fewer than 400 seeds are, in general, unmarketable (McGregor and Todd, 1952). In the hermaphrodite and staminate flower, there is a whorl of stamens whose anthers contain viable pollen. Mann (1953) found that, prior to anthesis, both types of flower with stamens contained 8,000–13,000 pollen grains, but that there were usually only about 3,000–4,000 left by 13.00 h; thereafter the number decreased less rapidly and there were still 2,000 grains by 17.00 h, although they were so well scattered that the flowers seemed to be devoid of pollen.

Seaton and Kremer (1939) found the minimum temperature for anther dehiscence of canteloupe was 18°C (64°F). Rosa (1924) found that in California, the flowers usually open between 07.00–08.00 h and at about 09.30 h the anthers dehisce; the stigma of the hermaphrodite flower is receptive when the flower opens; the flowers begin to close by 18.00 h and the corolla is withered by the next morning. In contrast, Nandpuri and Brar (1966), in India, found that anthesis occurred between 05.30 and 06.30 h and dehiscence occurred 30 min earlier.

Although hermaphrodite flowers are self-fertile, the stickiness of the pollen makes it tend to adhere to the anthers after dehiscence so that automatic self-pollination is unlikely to occur. Furthermore, because the anthers dehisce outward, any pollen that falls from them is deposited at the base of the corolla; consequently, when flowers are bagged they fail to set fruit (Rosa, 1924). It has long been recognized that hand pollination is essential for fruit production of plants in glasshouses and is a wise precaution for frame-grown plants in England (e.g. Sanders and Lansdell, 1924; Lansdell and Macself, 1949; Bewley, 1963) and that insects, particularly honeybees, are responsible for pollination of the flowers in the field.

Nandpuri and Brar (1966) found that stigmas were very receptive from 2 h before anthesis to 2–3 h afterwards. Pollen fertility decreased during the day, and pollination was more successful early than late in the day.

Self-pollination of an hermaphrodite flower is the same genetically as the transfer of pollen from staminate to hermaphrodite or pistillate flowers of the same plant, and is equally as effective in fruit production. No difference in set has been obtained between pollinating flowers with their own pollen and that from other plants (Mann, 1953), and early experiments of Rosa (1926) failed to show any difference in the time of ripening or the shape of fruit from cross- or self-pollination.

Rosa (1924), in California, found that 42% of the flowers located on the first spur of each branch set and developed into fruits, whereas only about half as many flowers on the second and third spurs did so. But if the flower on the first spur failed to set a fruit, and produce an inhibitory effect, the flower on the second and third spur usually did so. The set usually decreased on successive spurs until it was practically nil, presumably because the bearing capacity of the plant had been reached. It increased again to about 16 and 17% on the tenth and twelfth spur and then finally decreased; probably this secondary increase was associated with the maturing of some of the earlier fruit. To avoid the inhibiting effect of developing fruit, breeders remove all of them from a vine before hand-pollinating the selected flowers.

Rosa's observations were confirmed by Mann and Robinson (1950) and Mann (1953). They hand-pollinated vines that had not been thinned, and about 10% of the perfect flowers set fruit, but when they removed all the flowers except one from each vine and hand-pollinated it a mean of 60–70% of the flowers set fruit, indicating that nutritional factors were again involved.

Whitaker and Pryor (1946) had previously noted that only about a third of the flowers they artificially pollinated set fruit, and Mann and Robinson

(1950) found that the set on emasculated and hand-pollinated flowers was only about half as much in general as on flowers left to be naturally pollinated, probably because many of the flowers were injured during the process; the fruit from naturally pollinated flowers was on average 0·1 kg heavier and had 180 more seeds/fruit. Three applications of pollen by hand were no more effective than one, indicating that the number of pollen grains transferred was not the limiting factor.

Nectar is produced in a cup-like gland in the centre of the receptacle in a staminate flower and in a ring gland around the base of the style in an hermaphrodite flower. Bees collecting nectar spend about 9 sec/flower visit (Mann, 1953). The staminate flower has the shorter corolla tube and honeybees can easily reach the nectaries; the hermaphrodite flower has a deeper corolla tube and the entrance is more constricted, so that honeybees have to squeeze between the anthers and stigmas to reach the nectary, and in so doing transfer pollen to the stigmas. The narrow entrance and depth of the flower must greatly restrict the type of insect able to obtain nectar and probably helps to explain why the honeybee is the most abundant pollinator recorded (e.g. Rosa, 1924; McGregor, 1950; Peto, 1950; McGregor *et al.*, 1965). McGregor and Todd (1952) observed other insects (e.g. native bees, thrips, and ladybirds) on the flowers, but found their activity was not conducive to pollination and obtained no evidence that they contributed to fruit set.

More than one bee visit per flower is probably necessary to transfer the large number of pollen grains needed to pollinate it satisfactorily. Experiments in which canteloupe flowers have been bagged to exclude visits entirely, or bagged and then exposed for a predetermined number of bee visits, have demonstrated that the percentage of marketable melons increased with the number of visits a flower received up to about ten visits, but additional visits then made little difference. It was calculated that to ensure that each flower receives sufficient visits there should be one foraging bee present for every ten hermaphrodite flowers (McGregor *et al.*, 1965).

McGregor and Todd (1952) found that the sugar concentration of the nectar of staminate canteloupe flowers was greater than that of pistillate ones (56 : 27%). Kaziev and Seidova (1965) reported that a mixed sample of male and female flowers had nectar whose sugar concentrations varied from 20–40% and was greater after, than before, midday; whereas hermaphrodite flowers secreted 1·3–1·6 mg of nectar, male flowers secreted much less and the further a male flower was from the main stem the less nectar it secreted. At least part of the difference in the amount of nectar produced by staminate and hermaphrodite flowers reflects the different lengths of time for which it is secreted as, although nectar secretion began at between 08.00 and 09.00 h in flowers of both types it ceased in staminate flowers at 11.00 h, but continued until late afternoon in hermaphrodite ones.

Although bees passed readily from one type of flower to the other, there was a slight preference for hermaphrodite flowers when equal numbers of each were presented together (Mann, 1953). Such a preference probably also exists in the field after midday as pollen collection was found to reach a peak at about 11.00 h and then rapidly diminished whereas nectar collection declined much less abruptly and a few bees were still collecting nectar after 15.00 h (Fig. 118). Hence, the prolonged nectar secretion by the fewer

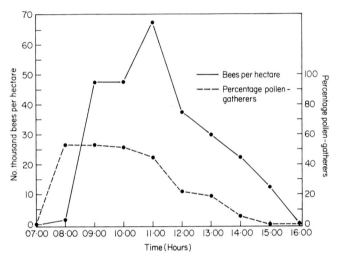

Fig. 118. Number and behaviour of bees visiting *Cucumis melo*, cantaloup, flowers at different times of the day in Arizona (after McGregor and Todd, 1952 with permission of the Entomological Society of America).

hermaphrodite flowers can be seen as a floral adaptation encouraging visits to them after the bees have become dusted with pollen from the staminate flowers.

Rane (1898) suggested there was a connection between the presence of wet weather and little sunshine during flowering with a lack of insect pollinators and a poor fruit crop, but the value of honeybees in pollinating canteloupes was first demonstrated by McGregor and Todd (1952) in Arizona. They demarcated plots, 3 by 6 m, in a canteloupe field so that each plot included 40 plants, 10 in each of 4 rows. The plots were either: not caged, caged to exclude bees, caged with bees throughout the flowering period, or caged but bees not introduced until after the crown flowers had withered; there were four replicates of each treatment. In the last treatment no melons set until the bees were introduced but, as soon as this was done, there was a rapid set resulting in a total production of 184 marketable melons compared to 180 melons of plots caged with bees throughout flowering. When bees were present throughout flowering, melons tended to develop mostly from the early crown flowers and were sweeter than those from flowers produced later. The open plots had an average of 145 melons, and the plot caged without bees only 4. Hence, to obtain both the maximum quantity and quality of melons plenty of honeybees should be present from the beginning of flowering.

These results were confirmed by Alex (1957b) who had 3 plots caged with bees, 3 caged without bees and 3 open plots. An average of only 3·9 melons were produced per plant caged with bees and in the open, but only 0·4 per plant caged without bees (equivalent to 805 and 94 crates/ha respectively). He pointed out that some pollination in the cages without bees may have been done by ladybird beetles which had been introduced to control the number of aphids.

Studies have been made of the behaviour of bees when visiting a nectarless strain of canteloupe. Staminate flowers of this strain have small flat-topped nectaries, and hermaphrodite flowers have less nectar tissue than usual; little or no nectar was present in either type (Bohn and Mann, 1960; Bohn and Davis, 1964). It was found that when a few nectarless muskmelons were planted among normal plants, bees that visited them appeared to be discouraged and soon moved to plants farther away but, despite this, the flowers were visited sufficiently often to give a good set of fruit; when the situation was reversed, so there were more nectarless than normal plants, the nectarless plants set fruit late and erratically; when nectarless plants alone were present they were avoided by bees and few set fruit. In addition to providing extra evidence of the dependence of high yields of muskmelon on honeybee pollination, these experiments were particularly interesting because when nectarless and normal varieties were flowering together, pollen was still present in nectarless flowers after it had been removed from normal ones, indicating that the nectar was of prime importance to the foraging bees and that they were reluctant to treat it as a crop providing pollen only.

Taylor (1955), in Arizona, compared the fruit production of 17 fields with no honeybee colonies in their immediate vicinity, with that of 20 fields with honeybee colonies beside the fields or within a mile of them, at the average rate of 1 colony/0·8 ha. Fields with, and without, honeybee colonies in their vicinity gave 1·1 and 0·7 melons/plant respectively (equivalent to 598 and 398 crates/ha).

Although the normal individual canteloupe flower secretes abundant nectar there are so few flowers per hectare that a canteloupe crop does not provide much nectar, and McGregor and Todd (1952) recorded that even when there was only 1 colony/5·7 ha of canteloupes the colonies lost weight.

Kennerley (1961) proposed that one colony was adequate to pollinate 2–2½ ha, as he supposed that when more bees are provided only enough visit the crop to get the available nectar and pollen. In support of this contention he reported work of A. H. Alex of Texas who found that a field with only 1 honeybee colony/2·4 ha had the same number of bee visits per flower, and the same percentage of fruit set as a field with 1 colony/0·6 ha.

However, 2½ colonies/ha is recommended by Pew *et al.* (1956) and the University of Arizona (1961), although they point out that honeybees are present in all Arizona melon fields and it may not be necessary to supply additional colonies.

There has been considerable interest in the extent to which flowers are self- and cross-pollinated, and seed growers have been especially interested in the distances over which cross-pollination can occur. The hybrids resulting from cross-pollination can often be recognized by the characters of their fruit in the following year. Rosa (1927) reported that when varieties were planted in alternate rows the amount of cross-pollination for different varieties ranged from 5–73%. Ivanoff (1947) planted four vines of a strain of canteloupe, whose fruit had a recessive colour, 9–12 m apart in a field containing plants of a dominant strain. The seed produced was sown the following season and he found that the percentage of cross-pollination of the four vines had been 1, 24, 100 and 100% respectively. He was unable to offer any satisfactory explanation for these large differences, especially as there was no reason to suspect differences in self-incompatibility.

The methods of cultivating melons are largely governed by the necessity that the developing fruit shall rest on dry soil and, in addition to being planted in continuous rows, they are often planted on "hills" about 2 m², the furrows between the hills carrying water which seeps through the soil to the roots. Whitaker and Bohn (1952), in California, planted each of ten such hills, well separated from each other, with two muskmelon plants with a recessive colour, and found that there was again a considerable variation in the extent to which the plants had been cross-pollinated, both from one hill to another (mean of 11–46%) and within the same hill (0–91%). More of the early than of the later flowers had been self-pollinated, and they supposed that this was because fewer bees were on the crop when the first flowers were open, and these bees were more likely to have smaller foraging areas than later in the season when competitive crops had ceased flowering and more bees were visiting the muskmelons. If this proves to be generally true, it might be beneficial to select for breeding melons developed from flowers that were pollinated when comparatively few bees were on the crop.

Because of the preponderance of male to female flowers it seems likely that a bee will usually visit a staminate flower before it visits a female one on the same plant, and so is likely to cause self-pollination even when it has visited another plant first. This would especially be so when relatively few female flowers are open, as happens near the beginning of flowering, and may well help to explain why more self-pollination occurs then.

In Georgia, James et al. (1960) attempted to discover a planting arrangement that would either greatly diminish or eliminate cross-pollination of muskmelons. The centre hill of each of two rows of nine hills was planted with a strain of melons whose long stem was genetically dominant to the short stems of the variety on the remaining hills. In each of 2 years this was done, the plants on hills adjacent to the centre ones received most cross-pollination from it (average of 1·5%), and the plants on the hills at the end of the rows received least cross-pollination (average of 0%). But, paradoxically, plants on the second and seventh hills in the rows were cross-pollinated more than those on the third and sixth hills. These results represent the amount of cross-pollination a plant received from one hill in one direction only, whereas in a normal crop it would probably receive pollen from several surrounding hills; hence, the larger amount of cross-pollination found by Whitaker and Bohn (1952).

Foster and Levin (1967) studied the transfer of pollen between two varieties planted in different arrangements. Cross-pollination was greatest where the varieties adjoined (up to 30%) and still occurred (up to 5%) 10·7 m from the pollen source. The amount of cross-pollination was also influenced by cultivation practices; it was greater *across* but less *along* high bed rows than flat bed rows, probably reflecting the greater influence of wind in causing erratic bee flight on the high beds.

On the basis of their findings, James et al. (1960) suggested that when labour is limited but plenty of space is available, single isolated rows of eight to ten hills of a given strain should be planted and only the fruits of the middle rows, which are unlikely to be cross-pollinated, should be harvested for seed. They also advised that when breeding programmes involve two or more species (e.g. cucumber, muskmelon, watermelon and squash) the different strains of the same species should be isolated from one another with rows of a different species. They noted that when different species grew side by side, bees kept constant to one species only. Contrasting results were obtained by Foster and Levin (1967) who watched the behaviour of individual honeybees on a field containing two strains of muskmelons. They found that, although bees spent about twice as long per visit to one strain as to another (6·6 : 3·1 sec) and visited 8·9 and 6·4 flowers/min respectively, the bee population as a whole showed no preference for either strain and, although during short periods, at least, individual bees tended to prefer one strain to another, most visited both during the same trip.

Colocynthis citrullus L.

The flowers of *C. citrullus*, watermelon, are similar to those of muskmelon except they are slightly smaller and have a greenish-yellow corolla, and the different sexes are not segregated on separate branches. Some varieties of watermelon bear hermaphrodite and staminate flowers, but most are monoecious and have staminate and pistillate flowers only (e.g. Jones and Rosa, 1928; Goff, 1937). The mechanism of pollination is similar to that of muskmelons, but few experiments have been done to determine their pollination requirements. Although the hermaphrodite flowers are self-fertile, none set fruit when they are bagged unless they are hand-pollinated so some agent is necessary to transfer the pollen; the results from hand and open pollination are similar (Rosa, 1925; Adlerz, 1966). There is no information on the amount of cross-pollination that occurs, but it seems likely that it would be less in varieties with hermaphrodite than pistillate flowers.

Mann (1943) tried to discover whether an uneven distribution of pollen grains on the stigma would alter the fruit shape; he found that in normal conditions, when each of the three lobes of the stigma received enough pollen, the pollen tubes grew down to their own carpels; when one lobe only received pollen grains some of the pollen tubes crossed to adjacent carpels, but insufficient to prevent the fruit developing assymetrically.

According to Seaton and Kremer (1938) watermelon flowers open at 14.5–15.5°C (57–59°F) and the anthers begin to dehisce when the temperature reaches 17°C (63°F). Both staminate and pistillate flowers open between 06.30 and 08.00 h, in California (Rosa, 1925), but Bhambure (1958a) recorded that staminate flowers open at 08.30 h and the pistillate at 09.30 h near Bombay, and during the next 3 or 4 h, as the temperature rises, the flowers change from a cup-shape to a saucer-shape and finally to an inverted umbrella shape in which the anthers and stigma are fully exposed. *Apis cerana*, *Apis florea*, and *Melipona* spp. collected watermelon pollen from 08.30 h and reached a peak in numbers on the crop at 10.30 h; *Apis cerana* deserted the crop by 12.00 h each day, but *A. florea* continued working it until sunset so presumably the flowers must fail to close. In California honeybees began to visit the flowers soon after they opened and were most numerous on fields between 08.00 and 10.00 h (Goff, 1937; Adlerz, 1966); thereafter they became steadily fewer until the flowers had closed. It seems that flower visits occurring between 09.00 and 10.00 h are more likely to lead to pollination than earlier ones (Adlerz, 1966).

Rosa (1925) thought that the cucumber beetle (*Diabrotica* spp.) might play a part in watermelon pollination, and Goff (1937) observed eight species of solitary bees, which he listed in order of their abundance, visiting the flowers but it is agreed that honeybees are the principal pollinators. Adlerz (1966) found that the fruit set and yield of flowers that received eight or more bee visits was better than those that received four or fewer visits. As bees seldom changed position after landing during visits of normal duration, he suggested that distribution of pollen on a flower depended more on multiple visits than on movement of the bee while visiting it. The necessity for each lobe of the stigma to receive adequate pollen undoubtedly helps to explain why numerous visits per flower are necessary to obtain the best result. Adlerz (1966) calculated that when $2\frac{1}{2}$ colonies/ha were present, there would be enough bees to visit each flower eight or more times in slightly more than one hour, and that this would ensure adequate pollination.

However, there is little actual evidence that the presence of honeybee colonies in a watermelon field increases yield. Goff (1937) reported a better set along the margins of a 405 ha crop than in its centre and supposed this was because there were fewer bees in the centre. He mentioned that one field, that had only a tenth of the bee population of another, had a poor set, but gave no data. More observations and experiments are needed.

Other Cucurbitaceae

There have been few attempts to discover the pollination requirements of economically important species belonging to the genera *Lagenaria* and *Luffa*.

According to Fronk and Slater (1956) the flowers of *Lagenaria siceraria* Standl, bottle gourd, open in the early evening and pollination occurs during the evening or night.

Bhambure (1958a) reported that near Bombay the flowers of *Luffa acutangula* L., angled loofah, ridge gourd, open from 09.00 to 11.30 h during which time they were worked vigorously by *Apis cerana*, *A. florea* and *Melipona* spp. The same species, together with *Bombus* spp. and *A. dorsata* visit *Luffa aegyptiaca*, smooth loofah; the male flowers of this species are more numerous than the female and open about a week earlier. When female flowers are bagged they do not set fruit (Bhambure, 1957).

I can find no information about the pollination of members of the genera *Momordica*, *Sechium* and *Trichosanthes*.

Chapter 24

Umbelliferae

The typical species of this family have small flowers massed together in conspicuous umbels. The individual flower has five sepals, which are much reduced or absent, five petals and five stamens. The inferior ovary consists of two chambers, each containing one ovule; it is surmounted by a fleshy disc (stylopodium) around which nectar is secreted; two short styles arise from the centre of the disc. Protandry seems to be the general rule so self-pollination is largely prevented although the flowers are self-fertile.

Daucus carota L.

The only species whose pollination has been extensively studied is *D. carota*, carrot (Fig. 119), Borthwick and Emsweller (1933) caged carrot plants individually and put blowflies in some cages but not others; they

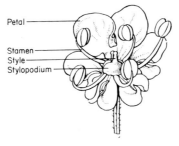

Fig. 119. Flower of *Daucus carota*, carrot (after Ross-Craig, 1959).

reported that plants caged with flies produced a good seed crop, whereas plants without flies had little or no seed. However, when they caged only a single umbel with flies, no seed was obtained; they discovered that the pollen lost its viability soon after it was shed and although all the flowers of an umbel did not open at exactly the same time, the anthers of those that were last to open dehisced and fell before the stigmas of the earliest flowers became receptive. It was necessary, therefore, to enclose umbels at different stages of development within the same bag for the flies to accomplish self-pollination. Although all the varieties they tested were self-fertile, they found that by enclosing an umbel of each of two varieties together in the same bag with blowflies they could obtain suitable cross-pollinations without the likelihood of self-pollination occurring.

Hawthorn et al. (1956) had plots (6·4 × 3·3 m) which were either: (a) enclosed by nylon screen cages with honeybee colonies; (b) enclosed by nylon screen cages without bees; (c) enclosed by muslin cages without bees; or (d) not caged. There were four replicates of each treatment. The average yield of clean seed from the different treatments was: 864, 367, 112 and 674 calculated kg/ha of 96, 88, 67 and 94% viability respectively. The relatively small set inside the muslin cages was thought to result from pollen that had either been transferred between umbels that touched one another, or had fallen from one umbel onto another. Only a few small halictid bees, tiny gnats and midges penetrated the screen cage walls but they set a substantial amount of seed; presumably they were also responsible for a part of the set in the cages with bees. Although the uncaged plots were visited by large numbers of insects of many different species, their yield was no greater than that of plots caged with honeybees. This experiment was repeated on 3 subsequent years with essentially similar results (Hawthorn et al., 1960) except that in one year several small sweat bees (*Halictus confusus arapahonum*) entered the plots caged without honeybees and increased the set. During another year, insects were scarce on the open plots which, as a result, had much smaller yields than on plots caged with bees, thus indicating that there are not always sufficient natural pollinators. In plots that had received little pollination the plants remained in flower for longer periods than normal and produced unusually high proportions of both abnormally large and abnormally small seeds. In contrast the presence of abundant pollinators was associated with early seed maturation.

Further evidence that carrot depends upon insect pollination was provided by Pankratova (1958), who found that plots visited by insects gave 15 times as many seeds and 10 times the weight of seed as plants covered with muslin.

The frequency of cross-pollination, and the rarity of self-pollination help to explain why it is necessary to isolate different varieties by 0·8 m or more (see Jones and Rosa, 1928; Haskell, 1943; Hawthorn and Pollard, 1954). Thompson (1962) distributed isolated plants of an orange-rooted variety in a block of a white-rooted variety, and found that only 1·1% of their progeny resulted from self-pollination. He suggested that hybrid carrot seed could be obtained from a few plants of one variety distributed in a block of another, but it is doubtful whether using such a relatively few seed-producing plants would be an economically sound proposition.

Carrot flowers are visited by many different types of insects but especially honeybees, Coccinellidae, Hemiptera, Syrphidae, solitary bees and solitary wasps (Treherne, 1923). Bohart and Nye (1960) pointed out that the pollen is abundant and both it and nectar are readily accessible to nearly all insects; and although the individual flowers are small the flat umbels are able to support the weight of large insects. Bohart and Nye collected 334 species belonging to 71 families on carrot flowers at Logan, Utah. The proportions of the different species present varied from place to place, in different years, and at different times of the 4-week flowering period. In general, however, bees were most numerous toward the middle of the flowering period when pollen was most abundant, whereas Syrphidae were most numerous toward the end of flowering when flowers with fallen anthers were still secreting nectar. Attempts were made to rate the pollinating efficiency of the more common species on the basis of their abundance, size, tendency to touch stamens and stigmas and the amount of loose pollen on their bodies, and it was concluded that most of the species did little pollination although they comprised a substantial proportion of the insect population. Because these inefficient pollinators collected nectar and pollen, their presence was detrimental as they removed forage which could have attracted the beneficial visitors. The most efficient pollinators belonged to the families Apoidea (genera *Apis*, *Andrena*, *Halictus*, *Chloralictus* and *Colletes*), Sphecidae (genera *Cerceris*, *Lindenius*, *Philanthus*, *Nysson* and *Sceliphron*), Syrphidae (genera *Syritta* and *Tubifera*) and Stratiomyidae (genera *Eulalia* and *Stratiomys*). The pollen-gathering honeybees were especially valuable as they scraped the pollen from the anthers with their forelegs and moved to and fro over the flowers more than honeybees collecting nectar did. The females of other bee species behaved like pollen-gathering honeybees. Pankratova (1958) found that Diptera comprised 90% and honeybees 9% of the insects visiting carrots near Moscow but, whereas honeybees collected both nectar and pollen, the Diptera collected nectar only.

Bohart and Nye (1960) suggested that attempts should be made to propagate some of the wild pollinators, especially Diptera species, that breed in decaying vegetation and the alkali bee, *Nomia melanderi*. Although honeybees are effective pollinators of carrots and their population can readily be increased, they do not seem to be especially attracted to carrot flowers, so taking their colonies to carrot crops may not be particularly useful when there are more attractive crops nearby. If possible, fields for carrot seed should not be located near competing crops, but should be in areas with varied habitats capable of supporting many kinds of pollinators; to avoid diluting the pollinator population the amount of carrot grown in any one area should be restricted.

Pastinaca sativa L.

Protandry is also very evident in *P. sativa*, parsnip, (Fig. 120) as the stigmas do not become receptive until about 5 days after the anthers of the same flower have dehisced. The flowers open in whorls from the outside toward the centre of an umbel so that a flower usually receives pollen from one of

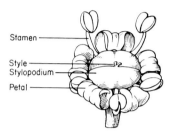

Fig. 120. Flower of *Pastinaca sativa*, parsnip (after Ross-Craig, 1959).

the flowers of an inner whorl with the result that the centre flowers are often not pollinated and produce no seed (Beghtel, 1925). The flowers are visited by many types of insect, especially solitary bees (Treherne, 1923) and although pollination within the umbel appears to be the rule, it is customary to isolate parsnip seed crops (Haskell, 1943).

Foeniculum vulgare Mill.

Narayana *et al.* (1960) reported that insect activity on *F. vulgare*, fennel or saunf, was at its maximum between 11.00 and 13.00 h and that *Apis florea* was the most abundant pollinator present and comprised 82% of the population compared to *Apis cerana* 6%, *A. dorsata* 1%, Syrphidae 6% and other Diptera 5%. They pointed out that, unfortunately, *A. florea* cannot be kept in hives so little can be done to ensure that fennel is adequately pollinated. However, there are no data to indicate the extent to which seed production might be related to bee visits and it would be interesting to know whether the number of *A. cerana* bees visiting a crop could be increased by importing colonies of this species.

It is often assumed that other Umbelliferae also need cross-pollination (see Hawthorn and Pollard, 1954) but there have been no investigations on the pollination of most species although many economically important ones are concerned, including: *Anethum graveolens* L., dill; *Apium graveolens* L., celery; *Arracacia xanthorrhiza* Bancr, arracacha; *Carum carvi* L., caraway; *Coriandrum sativum* L., coriander; *Cuminum cyminum* L., cumin; *Petroselinum crispum* Mill, parsley; and *Pimpinella anisum* L., anise.

Chapter 25

Rubiaceae

Coffea

The four most commonly grown species of coffee in order of importance are *Coffea arabica* L. which is cultivated throughout the tropics, *C. canephora* Froehn, *C. liberica* Hiern and *C. excelsa* A. Chev.

The flowers are in groups of two to twenty in the axils of the leaves. Each flower (Fig. 121) has a five-segmented small calyx, five white petals, the lower

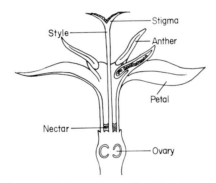

Fig. 121. Flower of *Coffea arabica*, coffee (after McDonald, 1930).

half of which are united to form a cylindrical corolla tube which is sometimes quite elongated, five stamens with long anthers and short filaments which are inserted into the corolla tube, a long thin style with a two-branched stigma and an inferior ovary of two chambers each containing one ovule. The stigma is receptive when a flower opens at dawn and the anthers dehisce soon afterwards. The disc surrounding the base of the style secretes nectar.

In regions where the change from dry to wet monsoons is rapid, as in the main coffee growing areas of Central and East Java, flowering occupies a few days only at the end of the dry season. When there is no such seasonal contrast coffee flowers throughout the year, but flowering maxima occur periodically (e.g. Ferwerda, 1948).

McDonald (1930) found that nearly every *C. arabica* flower was pollinated within 2 h of opening, and within 8 h, pollen tube growth was completed and fertilization had occurred. The flowers usually withered within 48 h of opening but, when they were not pollinated, withering was delayed. Cloudy days sometimes inhibited opening, but maturation continued to proceed and pollen was shed within the bud.

The nectar of *C. arabica* contains about 38% sugar (Nogueira-Neto *et al.*, 1959), and McDonald (1930) observed that the inflorescences received numerous visits by bees, wasps and flies; he suggested that growers in E. Africa should keep honeybee colonies in their plantations and grow nectar-producing flowers in spare areas of ground to provide forage for the bees when the coffee was not in flower.

Nogueira-Neto *et al.* (1959) found that in Brazil the honeybee was the most common visitor to *C. arabica* flowers and it was especially abundant at midday; in wet weather about half the honeybees collected pollen only. The stingless bees, *Nannotrigona testaceicornis* and *Plebeia* spp., were also numerous and mostly collected pollen, but some of these bees and *Tetragona jaty* collected nectar only from the almost dried corollas of older flowers. They observed *Trigona hyalinata*, and the wasp, *Brachygastra augustii*, biting holes in the bases of the flowers, and *Plebeia* spp., *N. testaceicorniss* and *Tr. ruficrus* were seen to use them. They considered that the large bees, such as *A. mellifera* and *Melipona quadrifasciata anthidioides*, to be more efficient at pollination than the smaller ones, although both these species preferred the larger and more scented flowers of *Coffea dewevrei* hybrids which were growing further away.

C. arabica is tetraploid and self-fertile. To try to find the extent to which self-pollination occurred without insect visitors Amaral (1952) caged five bushes and compared their crop with that of eight uncaged bushes which were visited by numerous bees from a nearby apiary. The uncaged bushes produced significantly more seeds than the caged ones (4704: 3379; 39% difference). Sein (1959) found that bushes caged to exclude bees, and bushes not caged, had sets of up to 60% and 70% respectively; bees visited the exposed bushes only but various other insects, including flies and ants were present in the cages. In another experiment, Amaral (1960) covered 77 branches with muslin and found they had an average fruit set of 62% compared to 75% of 77 exposed branches which were visited mostly by honeybees. Noqueira-Neto *et al.* (1959) caged 26 bushes during flowering and compared the fruit set with that of 26 control bushes left exposed to visits by honeybees from 15 colonies about 700 m away, as well as by solitary bees from artificial nesting sites about 20 m away. In 4 of the 6 years this was done the set was slightly, but not significantly, less on the caged bushes and they supposed that this slight difference was partly because a greater amount of pollen was liberated by insects in the uncaged bushes and partly because the diminished light intensity in the cages had an adverse effect on growth. However, the covered bushes had a significantly greater proportion of berries with only one developed ovule ("peaberries") and, although the cause of this abnormality is not understood, it could be due to lack of pollination. As the flowers in the open were visited by numerous bees, they concluded that insect pollination had little effect and was of only secondary importance compared to gravity and wind pollination in *C. arabica*. They suggested that the beneficial effect of bee visits is restricted to diploid self-sterile species. Unfortunately, proper controls, in the form of bushes caged with bees, were not provided in any of the above four experiments, so there is no means of deciding whether the better crop on the uncaged bushes merely reflected better growing conditions; the production of fewer flowers on the caged bushes (Sein, 1959) indicates this might have been so.

It seems to be agreed that because the pollen of *C. arabica* is not carried far by wind, insects are largely responsible for such cross-pollination as occurs. Depending on the recessive variety used, estimates of the amount of cross-pollination have varied from 39–93%, 40–50%, 4–5% and 7–9% (Taschdjian, 1932; Krug and Costa, 1947; Carvalho and Krug, 1949); the last two estimates seem the more reliable, and it is supposed by Wellman (1961) that cross-pollination is less still among protected, shade-grown coffee bushes. Haarer (1956) stated that in the middle of a block of *C. arabica* almost 100% self-fertilization occurs.

Most flowers of the diploid species of *Coffea* are self-sterile and depend on cross-pollination by insects and wind. Different clones can give different results when self-pollinated. Hall (1938) found that some clones of *C. canephora* produced no seed when self-pollinated, but the majority gave a small amount, although always much less than that produced by cross-pollination.

In contrast to the heavy sticky pollen grains of *C. arabica*, those of *C. canephora* and *C. liberica* are light and dry and easily carried by light breezes. Ferwerda (1948) trapped large amounts of coffee on sticky glass slides 100 m from coffee plantations and 20 cm above the ground. Moist calm weather hindered pollen transfer. Only few insects have been reported visiting plantations of diploid species and it has been supposed that wind is the most important pollinating agent (McDonald, 1930; van Hall, 1938); if insects really are scarce on crops of these diploid species it would be interesting to know why; perhaps insects find their dry powdery pollen more difficult to collect.

Under natural conditions, only 20–25% of the flowers develop into mature beans, and sometimes only 10–15%. During the first 4 or 5 months of their development many fruits fall; the majority of which are not fertilized.

During the final period of development the fall is much diminished, but any that are shed have well-developed ovaries and presumably exceed the bearing capacity of the plant (Ferwerda, 1948). In commercial plantations it is essential to interplant different clones to provide for cross-pollination, and in favourable circumstances 30–40% fruit set is obtained (Purseglove, 1968).

Cinchona

Various species of *Cinchona* (e.g. *C. calisaya* Wedd., *C. ledgeriana* Moens, *C. officinalis* L., and *C. succirubra* Pav.) are the source of quinine. The small flower has a united calyx, a tubular corolla with five spreading lobes, five stamens which alternate with the corolla lobes, a bilocular inferior ovary with numerous ovules, and a single style with a bilobed stigma. There are two types of flower; in one the anthers are at the mouth of the corolla and the stigma is half way along it, and in the other the situation is reversed.

The flowers are reported to be cross-pollinated by insects (McIlroy, 1963; Purseglove, 1968).

Chapter 26

Compositae

The economically important species of Compositae whose pollination is discussed are *Helianthus annuus* L., sunflower; *Carthamus tinctorius* L., safflower; *Guizotia abyssinica* Cass, niger seed, which are grown for oil, and *Chrysanthemum cinerariaefolium* (Trev.) Bocc, the principal source of pyrethrins, *Parthenium argentatum* A. Gray, guayale and *Lactuca sativa* L. lettuce. I can find no reliable information about the pollination of *Helianthus tuberosus*, Jerusalem artichoke; *Tragopogon porrifolius* L. salsify or oyster plant; *Scorzonera hispanica* L., black salsify; *Cichorium endivia* L. endive; *C. intybus* L., chicory; and *Cynara scolymus* L., globe artichoke.

Helianthus annuus L.

In common with other Compositae the flower head, or capitulum, of *H. annuus*, the sunflower, is surrounded by an involucre of green bracts which protects the unopened head (Figs. 122 and 123). The head contains a single

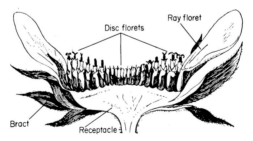

Fig. 122. Median section of capitulum of *Helianthus annuus*, sunflower (after Groom, 1906).

outer row of ray florets, whose large yellow ligules help to make the head conspicuous, followed by many concentric rings of tubular disc florets (Fig. 124). The corolla is of five united petals, the sterile ray florets being

Fig. 123. Honeybee foraging on *Helianthus annuus*, sunflower.

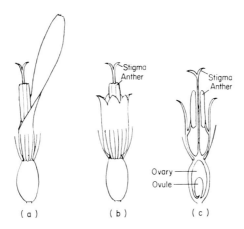

Fig. 124. *Helianthus annuus*, sunflower: a, side view of ray floret; b, side view of disc floret; c, section of tubular floret (after Gill and Vear, 1958).

larger and more conspicuous than the hermaphrodite disc florets. The single inferior ovary of each floret contains one ovule and, when this is fertilized, it ripens to form an achene. There may be between 1,000–2,000 florets/flower head when there is only one flower head per plant, but there are many fewer florets per head on plants that produce several heads each.

The disc florets on a sunflower head open from the periphery inwards, two to four circles each day; early in the day that a floret opens, its staminal filaments elongate rapidly and the anther tube, of five united anthers, appears above the top of the corolla; soon afterwards the anthers dehisce and pollen is shed in to the anther tube (Fig. 125). This is followed by elongation of the style and some

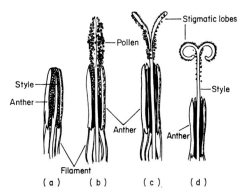

Fig. 125. *Helianthus annuus*, sunflower. Stages in growth of style and stigmas (after Groom, 1906).

contraction of the staminal filaments, which together result in pollen being pushed out of the upper end of the anther tube, but towards the end of the afternoon the tip of the stigma appears above the anther tube, and next morning, the stigmatic lobes have separated and curled back exposing their formerly hidden receptive inner surfaces for pollination. Thus each floret has two stages, the first male and the second female. For most of its flowering time, each head has unopened florets in the centre, surrounded by circles of florets in the female stage, and finally withered florets extending to the periphery of the head.

Each head flowers for about 6–10 days, although for a longer period in unusually wet cool conditions, when unfertilized flowers may remain at the female stage for many days. A crop flowers for about 3–5 weeks, depending on local conditions.

When the stigmas of a floret are not pollinated their lobes curve further downwards so their receptive surfaces touch pollen adhering to the outer surface of the style of their own floret. Thus it is mechanically possible for self-pollination within the floret to occur in the absence of pollination between florets, but proof of the effectiveness of such self-pollination is lacking, although it is sometimes assumed to occur.

Several workers have enclosed sunflower heads in paper or muslin bags and compared their set with that of exposed heads (e.g. Putt, 1940; Rudnev, 1941a; Overseas Food Corporation, 1950; Furgala, 1954a; Luttso, 1956; Kushnir, 1958). The bagged heads usually set about 20–30% seed compared with 70–90% in exposed heads, so it seemed likely that considerable self-pollination could occur. However, in these experiments it is difficult to distinguish between any adverse effects of the bags themselves on seed production and lack of pollination. Furthermore the bags may have rubbed across the tops of the heads and so transferred pollen from one floret to another.

In an attempt to circumvent these difficulties, Free and Simpson (1964) pushed a short length of bamboo into the centre of each of their experimental sunflower heads (variety "Pole Star") so that it held the muslin cover away from the florets; whereas insects were excluded from half the muslin-covered heads, the other half were visited by bumblebees put inside the bags every 2 days. The heads bagged without insects had only 1% of their florets produce seed compared with 35% for the heads bagged with bees and 63% for the heads not enclosed.

It appeared therefore that little fertilization occurs without insects or other external agencies and, that if the reflexed stigmatic lobes do pick up pollen this is usually ineffective, perhaps because the stigmatic surfaces have ceased to be receptive or the pollen has become non-viable. This is supported by Radaeva's (1954) discovery that the ability of a floret to set seed decreases with the length of time it has been open; thus, when he hand-pollinated newly opened florets, florets that had been open 3–4 days and florets that had been open 2 weeks, he found that 87, 69 and 21% of them set seed respectively.

However, Alex (1957a) concluded that the variety "Advance" was self-fertile to some extent. He compared the amount of seed produced in plots of eighteen plants each that were enclosed in large cages with or without honeybee colonies. The plots caged without bees yielded the equivalent of 349 kg/ha compared to 675 kg for plots with bees and 1,044 kg for open plots. Although there is less likelihood of pollen being artificially transferred from flower to flower in large cages than when heads are individually bagged, cageing is less effective at guarding the heads from insects, and perhaps some of the pollination that occurred was caused by small insects inside the cages.

However, possibly varietal differences in self-fertility within the floret occur and, whereas the florets of "Pole Star" cannot be fertilized with their own pollen, those of "Advance" can; this needs investigating.

Pinthus (1959) working in Israel, obtained a low average seed set (varying 0·5–5·0%) of bagged heads from June to September; presumably the vegetable parchment bags he used were less likely to rub against florets than muslin ones. However, the set was much greater (50–59%) in November and October; the reasons for this are obscure although he suggests that the increased self-fertility may be connected with the low temperatures in these months. Perhaps ventilation was inadequate in the parchment bags during the hot summer months. Putt (1940) and Furgala (1954a) found that heads enclosed in muslin bags consistently had a much greater set than heads in paper bags; the latter became pale green, misshapen and pulpy.

Free and Simpson (1964) enclosed half their sunflower heads singly in muslin bags and the other half in pairs. Heads enclosed singly with bees had 24% set, but those in pairs with bees had 45%, indicating that although sunflowers are not self-sterile, greater seed production is likely to occur following cross-pollination between florets of different heads than between florets of the same head. The still greater set (63%) on control heads not enclosed could have resulted from the detrimental effect of the bags themselves on flower and seed development, or because the control heads were visited by more bees from a large apiary nearby; moreover, whereas bagged heads were only cross-pollinated with pollen from one other head, each control head would have received pollen from several other heads, and the compatibility of the pollen may have differed with its source.

Probably cross-pollination is normally favoured. Putt (1940) reported that three strains of sunflowers showed 75, 50 and 78% crossing respectively when grown close together. Furgala (1954a) obtained 53% hybrids in similar conditions and found that the seed produced from cross-pollination contained more oil. Hurt (1946) pointed out that as cross-fertilization readily occurs, a crop grown to produce seed for sowing should not be within 1·6 km of a different variety.

Observations and experiments on field scale confirm the value of bees in pollinating sunflowers and demonstrate that a shortage of bees may limit sunflower seed production.

The Overseas Food Corporation (1950) reported that at Kangwa, Tanganyika, honeybees were in adequate numbers in sunflower fields close to bush, but they were much fewer in fields 3·2 km from bush, and pollination decreased with increasing distance from the bush.

Furgala (1954b) put colonies along the east edge of two sunflower fields in Manitoba at the rate of 2½ colonies/ha; he found that these two fields produced nearly twice as much seed per hectare as four fields isolated by at least 4·8 km from honeybee colonies; furthermore, whereas the seed yields in the two former fields decreased from east to west with increasing distance from the honeybee colonies, no such decrease occurred in the fields without honeybee colonies (Fig. 126). As a result of 3 years' experiments on seventeen farms in Rumania, Cîrnu and Sănduleac (1965) reported that fields with about

1 colony/ha produced 21–27% more seed than fields without honeybee colonies.

Russian workers have also determined the seed production in large sunflower fields at different distances from honeybee colonies. Rudnev (1941a) showed that with increase in distance, from 500 to 1,250 m, of 100 m² plots from honeybee colonies, the number of bees counted daily in each plot fell from 100 to 61, and the weight of seeds produced from 6,000 to 3,700 g.

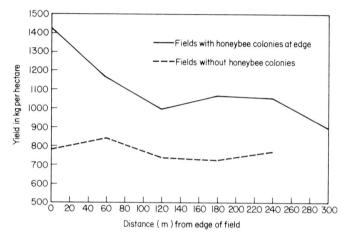

Fig. 126. Yield of *Helianthus annuus*, sunflower, seed per hectare at different distances from edge of fields with and without honeybee colonies (after Furgala, 1954).

Glukhov (1955) found the percentage of florets producing seeds decreased from 92% at 400 m from colonies to 69% at 1,600 m away. Kurennoi (1957) found that at 400 m and 2,000 m from honeybee colonies 89% and 79% of flowers set seed respectively; he also counted more bees foraging, more bees per flower and a greater seed harvest at the nearer site. However, Lecomte (1962b) was unable to find a correlation between the seed set and the number of foragers present in seven fields near Angerville, France, but the lowest set recorded was 92% so pollination was more than adequate in all the fields he sampled.

Unfortunately, the possibility that factors other than the number of bees present may have caused the differential seed set were not effectively eliminated in these observations and more extensive well-planned work is needed. It would be particularly interesting to extend the observations on the seed yield of fields located near to and distant from honeybee colonies; such information should not be too difficult to obtain.

Hand pollination using a glove of hare fur may give a good seed set when done daily, but because of the time involved this is economically less effective than pollination by honeybees (Kushnir, 1958, 1960; Cumakov, 1959).

The amount of nectar produced varies with the different varieties and some may produce three times the amount of others (e.g. Luttso, 1956; Sănduleac, 1962). Burminstrov (1965) found that the average amount of sugar produced per flower of different varieties ranged from 0·4–0·6 mg and increased with the length of the flowering period of the variety concerned. Baculinschi (1957) and Bitkolov (1961) obtained averages of between 0·11 and 0·25 mg per flower per day, and calculated that one hectare of sunflowers should yield between 21 and 75 kg honey on different years. Lesik (1953) and Montgomery (1958) reported that the sugar content of sunflower nectar varied from 49–51% and 48–57% respectively.

It seems to be generally agreed in N. America and Russia that sunflowers are a valuable source of nectar in late summer, and colonies taken to sunflower crops obtain large honey yields; a gain of 47 kg from sunflowers in 15 days in late August and early September has been reported in Manitoba (Mitchener, 1950) and in some parts of Russia sunflowers provide the main honey crop (Burminstrov, 1965).

It is not surprising that several observers have reported that honeybees visit sunflowers mainly for nectar (e.g. Rozov, 1933; Radoev, 1954; Bitkolov, 1961; Lecomte, 1962b) although sunflower pollen is sometimes also collected in abundance; thus Bitkolov (1961) found that 96% of the pollen collected by a colony came from sunflowers although many other pollens were available in the vicinity.

A honeybee visiting a sunflower floret for nectar pushes its tongue and head between the petals and anther tube to reach the nectar at the base of the corolla, and in doing so becomes dusted with pollen, often heavily. About half these bees pack the pollen into their corbiculae but others discard it, usually while hovering in the air (Synge, 1947; Free, 1964a). In addition to these nectar-gatherers that collect pollen incidentally, a few bees deliberately scrape pollen from the anthers, occasionally visiting the florets for nectar as well (Free, 1964a).

Bees with pollen loads, which they are collecting incidentally or deliberately, are most numerous early in the day, probably in association with anther dehiscence; there is a secondary peak in the number of pollen collectors in late afternoon, which is probably associated with the stigma pushing the remaining pollen out of the anther tube. When pollen is in short supply, pollen-gatherers are not deterred from foraging but continue to collect nectar only.

Bees that are deliberately collecting pollen keep to florets in the male stage; nectar-gatherers also mostly visit male stage florets but often stand on florets in the female stage and so may pollinate them (Free, 1964a).

Rudnev (1941b) found a direct correlation between the amount of seed produced per crop and the average number of bee visits each floret received; thus, he obtained an average harvest of 53, 76, 133 and 210 kg seed/million florets from plantations with an average of 1·0, 1·4, 3·4 and 6·1 bee visits/florets respectively.

An average of only two visits per floret should ensure that most florets are visited at least once and Rudnev's results are only understandable when it is realized that the florets actually visited are not those being pollinated, but the longer a bee spends on a flower head, the more florets it walks on and pollinates.

The behaviour of bees on sunflowers represents an isolated example of nectar-gatherers being better pollinators than bees gathering pollen only. In fact, bees that scrabble for pollen may be disadvantageous as they remove pollen with which nectar-gatherers might become dusted. It would be interesting to know whether nectar-gathering bees that collect pollen loads keep their bodies more free of pollen than those that discard it; if so, they would theoretically be less efficient pollinators. Nectar-gatherers that discard pollen may perhaps cause pollination indirectly with the pollen they discard while hovering over a flower head. This behaviour could help account for the small amount of sunflower pollen which Putt (1940) trapped on vaseline-coated slides he had suspended in a sunflower plantation.

Perhaps the reason few bees collect pollen deliberately is that it is relatively unattractive; this would also help to explain why some nectar-gatherers discard any pollen they collect incidentally. The tendency to discard pollen may well vary with the type and amount of the competing flora in the vicinity, and probably on a large sunflower crop, with little other forage available more bees collect pollen loads. However, the habit of discarding pollen is not exclusive to bees working sunflowers, and probably merely characterizes the bees that are least inclined to collect pollen. This receives support from the finding that bees which deliberately scrabbled for pollen were much more likely to collect pollen loads incidentally while visiting florets for nectar than were bees never seen to scrabble (Free, 1964a).

Nectar-gatherers with pollen loads visited more florets per head than bees collecting nectar only (29 : 21), presumably because they were more satisfied; this is also reflected by the fact that they were seen more frequently on the crop (Free, 1964a).

The reason why bees of both groups visit such a small proportion of the florets available per head is not apparent, but this behaviour must favour cross-pollination, especially as the pollen present on a bee's body when it first visits a flower head must soon become diluted with the head's own pollen. The Overseas Food Corporation (1950) found that less nectar was secreted, and more fertilization occurred, in dry than in wet weather; probably this was because when nectar is scarce bees become dissatisfied with the crop, and fly more frequently from head to head and so cross-pollinate more.

Probably bees visit fewer florets on small than on large heads; if so, a greater proportion of the florets visited on small heads are cross-pollinated. This could explain why Radaeva (1954) found that the percentage of florets that produced seed decreased with increase in diameter of the flower head. If this finding is substantiated, selecting plants with smaller but more numerous heads would increase seed yield.

In addition to those honey bees that visit the florets themselves about 10% visit the extrafloral nectaries located in the edges of the bracts beneath the flower heads and in the basal edges of the laminae of the top leaves of the stem. Extrafloral nectaries are especially attractive when flowering is beginning. Free (1964a) found that, although flower-visiting bees readily changed from collecting pollen only to collecting nectar and pollen, or to collecting nectar only, all bees visiting extrafloral nectaries kept constant to them on any one trip and were never dusted with pollen, and 15 of 19 marked bees did so throughout the time they visited the crop, although they readily alternated between leaf and bract nectaries.

No doubt many locally abundant insects (e.g. Furgala, 1954b; Sivori, 1941) help to pollinate sunflowers but there have been few observations on them. *Halictus* spp. are more inclined to collect pollen than honeybees (Radoev, 1954) but *Bombus* spp. are less likely to do so (Free, 1964a); like honeybees they mostly visit florets in the staminal stage but work faster. However, honeybees are the most numerous pollinating insects recorded (e.g. Radoev, 1954). In Tanganyika, honeybees from wild colonies were found in adequate numbers close to bush but 3·2 km from the bush they were far fewer (Overseas Food Corporation, 1950).

Any calculation of the numbers of colonies required to pollinate a hectare of sunflowers must be very approximate only. Kurennoi (1957) found that nectar-gatherers spend between 3·6 and 6·2 sec/floret, and Free (1964a) that they spent an average of 3·2 sec/floret. If it is assumed that on average a bee visits sunflowers for a 5 h period a day, and 4 h of this time is spent actually foraging, and that bees spend 4 sec/flower visit, one bee would visit 1,200 florets/day. Further, if it is accepted that for every six florets visited one is pollinated (Rudnev, 1941a, see above) a bee should pollinate 200 florets/day. Rudnev (1941a) calculated there were 8,670,000 florets/ha. Consequently 43,350 bee-days will be needed to pollinate all the florets per hectare. This is equivalent to 2,064 bees/day per hectare in a 3-week flowering period. The proportion of a colony's foragers that visit a sunflower crop depends upon the competing species, but it would seem likely from the above calculation that one colony could provide sufficient sunflower foragers to pollinate 1 ha of sunflowers.

At least 1 colony/ha is recommended in Russia (Rozov, 1933; Kurennoi, 1957; Kloavev and Ulanichev, 1961) and 2/ha in Rumania (Cîrnu, 1960). Colonies should be moved to a crop directly flowering begins as more than 80% of the heads open during the first 3 days of flowering (Radaeva, 1954).

Carthamus tinctorius L.

The inflorescence of *C. tinctorius*, safflower, consists of a broad receptacle on which there are numerous yellow and orange florets surrounded by bracts. The top of the long corolla tube is divided into five pointed segments. Usually, the anthers dehisce early in the day, and soon afterwards the style elongates and when the stigma appears above the top of the anther tube it is covered with pollen, but sometimes the stigma emerges from the anther tube before pollen is liberated (Howard *et al.*, 1916). Boch (1961) found that the average sugar concentration of the nectar was only 13–17% between 06.00 and 08.00 h, but it was between 24 and 29% during the remainder of the day. Rubis *et al.* (1966) reported an average sugar concentration of 19%.

In India, Howard *et al.* (1916a) observed that many bees visit safflowers early in the day, and in particular visit the newly opened florets for pollen. Boch (1961), in Canada found that honeybees began foraging at about 07.00 h and were most numerous between 09.00 and 11.00 h. During this 2 h period the amount of nectar and pollen in the flowers decreased rapidly, and by 12.00 h nearly all had been collected; thereafter, only nectar and pollen currently being presented was available for collection. In association with this the number of honeybees observed decreased rapidly after 12.00 h. In contrast, the number of syrphids on the crop, although relatively few, steadily increased until 17.00 h; the reason for this was not apparent. Rubis *et al.* (1966) and Levin and Butler (1966) obtained similar results in Arizona, although bees were abundant on the crop about an hour earlier in the day; they also found that pollen collection ceased earlier in the day than nectar-gathering. Boch (1961) and Eckert (1962) found that most bees had pollen loads, while Levin and Butler (1966) found there were more honeybees collecting nectar only than collecting pollen. It is not clear whether pollen-gatherers also collect nectar, or whether nectar- and pollen-gatherers are equally effective in pollinating the flowers.

In Ontario, 90% of the insects visiting safflowers were honeybees, 3% native bees, mostly bumblebees, and 7% insects other than bees, mostly syrphids (Boch, 1961). In different fields in Arizona about 85–90% of the bee visitors were honeybees and the remainder were solitary bees; bees belonging to the genera *Halictus* and *Lasioglossum* were most common during the early part of flowering but bees belonging to *Melissodes*, *Agapostemon* and *Megachile* were as abundant later (Levin and Butler, 1966; Butler *et al.*, 1966).

Howard *et al.* (1916a) found that twenty-five plants enclosed in muslin or net bags produced an average of 35·9 seeds/head compared to 37·4 seeds/head of flowers not bagged, and concluded that insect visits were not necessary for pollination. Boch (1961) caged three plots of safflowers to exclude bees and found that they produced only about half as much seed as plots not caged. Eckert (1962) found that a self-fertile variety of safflower set as much seed when caged to exclude insects as in the open (66 : 64% set), but a variety that was somewhat self-sterile set less when caged than in the open (41 : 59% set). Rubis *et al.* (1966) compared the set in cages with bees, in cages without bees and in open plots; they found that the presence of bees made no difference to the set and crop produced by a normal strain of safflower with abundant pollen. But, when they also used a strain in which anther dehiscence is delayed several hours, so that the stigma projects from the anther tube without the likelihood of being self-pollinated, the presence of honeybees more than doubled the seed yield and the number of seeds per head, although the weight per seed was less. Later experiments (Levin *et al.*, 1967) showed that halictid bees (*Halictus ligatus*, *H. tripartitus* and *Lasioglossum pectoraloides*) and a wasp (*Polistes exclamans*) were effective pollinators.

Therefore it seems that, on the evidence available, there would be little or no increase in seed yield when honeybee colonies are taken to crops of strains that are both self-pollinating and self-fertile, but when they are either not self-pollinating or not self-fertile honeybees will probably increase yield by cross-pollinating the flowers. Rubis *et al.* (1966) demonstrated that honeybees were able to increase substantially the yield of a self-fertile but not self-pollinating strain when little pollen was available; perhaps honeybees would have an even greater effect on strains that are self-sterile with plenty of pollen available, or on strains whose stigmas are not receptive until several hours or more after the anthers have dehisced. Because the amount of natural cross-pollination of safflower plants varies greatly (from 0 to 100%; Claassen, 1950), there are likely to be large differences in the value of honeybees in pollinating different commercial crops. Howard *et al.* (1916a) discovered that extensive crossing occurs in some parts of India but not in others. Kadam and Patankar (1942) found that the amount of contamination of seed crops depends upon the distance from other varieties, and when the varieties were separated by 30 or more metres, and other crops grown between them, contamination was reduced to negligible amounts.

Guizotia abyssinica Cass

The flower head of *Guizotia abyssinica*, niger, has 40–60 tubular hermaphrodite florets surrounded by a marginal row of ligulate florets. The flowering period of each head is from 7–8 days. The florets open and liberate pollen early in the morning, the style emerges about midday and the stigma lobes separate and curl backward toward evening. Howard *et al.* (1919) found that cross-pollination was common; they pointed out that the stigma lobes rarely curled back sufficiently to touch the pollen on their own style and they suggested that this absence of automatic self-pollination explained why isolated heads did not readily set seed. Their results were supported by those of Bhamburke (1958). He caged one plot (1.2×1.2 m) of niger with an *Apis cerana* colony and another without bees. He labelled 40 heads in each and found the average number of seeds produced per head in the cage with bees was 40 compared to only 15 in the cage without bees, indicating the possible value of insect pollination.

Chrysanthemum cinerariaefolium (Trev) Bocc

Chandler (1956) found that the pyrethrins content of fertile achenes of *cinerariaefolium*, pyrethrum, was 1.05% compared to 0.71% of barren achenes. This was because most of the pyrethrins occur in the seed embryo and only a little in the husk, and in a sterile achene only the pyrethrins in the husk are present. Hence pollination would seem to be important in increasing the yield of pyrethrins. According to Kroll (1961), it is generally found that a very great percentage of the achenes are barren, probably because there are insufficient pollinating insects. He found that most of the insects visiting a crop belonged to the Coleoptera and Diptera and their numbers fluctuated greatly; honeybees were not numerous and collected pollen at certain times only. Smith, F. G. (1958) reported that a preliminary experiment in Tanganyika showed the yield of pyrethrins was greater in cages with bees than in cages without bees, and that the yield of uncaged plots was intermediate. However, Kroll also tested the effect of bagging and caging plants with and without honeybees on the percentage of pyrethrins produced, but found the bees were reluctant to work the flowers, and he was unable to show that their presence had any definite effect on seed production. More experiments are needed, if necessary using other insects that will forage on the flowers.

Parthenium argentatum A. Gray

Most strains of *P. argentatum*, guayale, that have been studied are self-incompatible and need cross-pollination. Pollen is released when the disc florets open toward midday. Although the pollen is heavy and sticky, Kalashnikov (1931) found it was carried up to 40 m by the wind; Gardner (1946) found the distance depended on the wind strength and time of day, and in favourable conditions plenty of pollen was carried 274 m from the edge of a guayale field in California, a little was carried 937 m but none was carried 1,097 m. As soon as the wind ceased the pollen fell to the ground and none was airborne on still mornings. Windborne pollen germinated normally, and pollen one day old was as effective as fresh pollen; the viability had diminished greatly in 3-day-old pollen but some even 14-days-old was still viable.

Gardner also put sticky slides inside cloth bags of the type used to isolate flower heads and found that whereas only a few pollen grains penetrated muslin many penetrated coarser cheesecloth bags. By using genetically marked strains of guayale he confirmed that cheesecloth bags were not effective in isolating flowers from undesirable pollen. These experiments could, with advantage, be repeated with other plant species.

Gardner (1947) found that ladybird beetles (mostly *Hippodamia convergens*), cucumber beetles (*Diabrotica soror*), small flies (*Hylemya* spp.), and bugs (*Lygus hesperus*) were very numerous on guayale crops. They all had some guayale pollen on their bodies but the amount varied greatly from about 50 grains on *Hylemya* spp. to about 500 on *L. hesperus*. He tried to test the effectiveness of the various species by confining them with *P. argentatum* plants in bags or small cages. *H. convergens* collected from guayale flowers and enclosed in bags with plants at the rate of 100/plant gave 10·8% fertile achenes compared to only 0·6% by *H. convergens* collected elsewhere. Pollination in bags was not as effective as in small cages and *H. convergens* collected from guayale flowers and enclosed in cages gave the same set (27%) as in the open field.

Fruit flies of the species *Drosophila hydei* gave only 7% set inside cages but *D. melanogaster*, which moved more readily from one flower to another, gave 29% set. However, despite the success Gardner obtained with *H. convergens* and *D. melanogaster* he concluded that the use of insects for making experimental cross-pollinations must wait until better methods have been developed.

Lactuca sativa L.

Each flower head of *L. sativa*, lettuce, has 15 to 25 florets which open simultaneously. On warm bright days the head opens early in the morning and remains open for about 30 min only, never to reopen, but on cool days it may remain open for about 2 hours so there is more opportunity for cross-pollination. Fertilization is accomplished 3–6 h after pollination.

Pollen is shed prior to anthesis; when the pistil elongates the brush hairs on the side of the pistil sweep the pollen grains upward out of the pollen sacs of the dehisced anthers and when the stigmas appear they are already covered with pollen. Although Knuth (1906) stated that the stigmatic lobes make a complete revolution backward and touch the pollen on the brush hairs of the pistil, Jones (1927) reported this occurred only occasionally. He found that 58 of 70 florets bagged to keep away insects during flowering had no pollen grains on their inner stigmatic surfaces and the other 12 had from 1 to 7 grains each, although the edges and backs of every stigma were covered with pollen. All the 70 other florets that were visited by insects had between 4 and 51 pollen grains on their inner stigmatic lobes. However, if pollen on different parts of the stigma was equally effective in fertilization insect pollination would have no advantage over auto-pollination.

In the western U.S.A., the bees *Agapostemon texanus* and *Halictus* spp. frequently visit lettuce flowers from which they mostly collect pollen (Jones, 1927; Thompson, 1933). In Britain, Watts (1958) reported that various species of hoverflies and a few butterflies visited the flowers but bumblebees and honeybees were rarely seen to do so; he tried to use hoverflies to pollinate bagged flowers but without success. Jones and Rosa (1928) observed that, during the flowering of two lettuce plants, 52% of the flower heads of one and 73% of the other were visited by insects, some several times. Hence opportunities for cross-pollination must frequently occur but little has been done to find how extensive this is.

Thompson (1933) planted two varieties with green and red leaves respectively in alternate positions in a row and found that an average of 2·0% crossing occurred in one year and 3·0% in another (range of 1·3–6·2% for individual plants). No more information was available until Thompson *et al.* (1958) found that an average of 0·8% cross-pollination occurred between two varieties in adjacent rows, the maximum amount being 2·9%. When two varieties were in adjacent blocks the percentage of cross-pollination rapidly diminished as the distance separating varieties increased, and none occurred 6·7 m from where the two blocks met. More cross-pollination occurred in the direction of the prevailing wind than against it, but it was not known whether this was directly due to the wind or whether the wind acted indirectly by affecting the insect pollinators. The latter seems more likely, because lettuce pollen is not easily dislodged.

Watts (1958) reported the results of an experiment in which plants of a green variety were each surrounded by eight plants of a pigmented variety. The average amounts that the green variety plants were cross-pollinated on four successive occasions during the season were 0·1, 0·7, 3·2 and 11·5% respectively. He suggested that the increase in cross-pollination shown by the plants that flowered later was probably due to an increase in the hoverfly population. If further experiments prove that such a relationship exists between insect abundance and cross-pollination, the well-established practice of some Californian growers of enclosing plants selected for propagation in cloth bags during flowering (Jones, 1927) will be amply justified.

Chapter 27

Vacciniaceae

Blueberry

Various species of *Vaccinium* known as blueberries in North America, and as bilberry, blaeberry or whortleberry in Britain, grow wild in many areas.

The lowbush blueberry is grown commercially in forest clearings and on barren farm land in Eastern Canada and North Eastern U.S.A. It is not easily propagated but, after the forest has been cleared, the lowbush blueberry appears naturally. It requires periodic burning to prevent reforestation and burning the blueberry fields every second or third year is the recognized method of pruning. There are usually fewer competing bushes and weeds in old pastures and fields, and growing blueberries in them is more productive. Lowbush blueberry species include: *Vaccinium myrtilloides* (*V. canadense*), *V. angustifolium* (*V. pennsylvanicium*), *V. lamarikii*, *V. brittonii*, *V. vacillans* and *V. boreale*.

The highbush, or cultivated blueberry, grows best on permanently wet, acid soil and is planted in various parts of the U.S.A. and Canada. It has been subject to field trials in England (Roach and Perrin, 1959; Perrin and Duggan, 1965; Roach, 1967) and much of Europe. Highbush species include: *V. australe*, *V. atrococeum* and *V. corymbosum*.

Other blueberry species include *V. pallidium*, dryland blueberry, *V. ovatum*, Western evergreen blueberry, *V. membranaceum*, mountain blueberry, and *V. ashei* (rabbiteye blueberry).

The corolla of the blueberry flower is bell-shaped and divided into four or five lobes (Fig. 127). The style is closely encircled by ten shorter stamens; the ovary is inferior and there are several ovules in each locule. According to Coville (1910) and Robbins (1931) only the sticky apex of the stigma is receptive, while its sides have a dry surface to which pollen does not adhere, so that although the flower hangs down, there is little chance that it will be pollinated with its own pollen. However, Merrill (1936) claimed that the stigma is receptive over its entire surface and not merely at its apex.

Nectar is secreted at the base of the style and an insect has to push its tongue between the filaments of the anthers to reach it. Shaw *et al.* (1954) found that the nectar contained an average of 21% sugar; nectar of *V. myrtilloides* and *V. angustifolium* contains equal molar concentrations of glucose and fructose, with small amounts of sucrose and maltose. While collecting nectar, an insect may jar the anthers and dislodge pollen which falls on to its body, and brush against the receptive part of the stigma, pollinating it with pollen collected from a flower previously visited.

The size of flowers differs in different species, and this may affect the ability of some insects to collect nectar; for example, the corolla tube length of *V. boreale* is only about 3 mm compared to 5–7 mm of *V. angustifolium*.

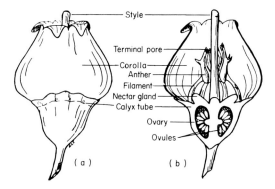

Fig. 127. Flower of *Vaccinium* spp.: a, external view; b, section (after Robbins, 1931, with permission of McGraw-Hill Book Company).

Early attempts to propagate selected highbush blueberry plants by self-pollination failed because either the flowers did not set, or the fruit soon withered and dropped (Coville, 1910, 1921, 1937; Robbins, 1931). This emphasized the need to use cuttings from several different bushes when starting a plantation.

Phipps (1930) put a cage made of copper wire-screen over some bushes in the field and found that, although many flowers set fruit, none matured, whereas about 1,000 berries were produced in an equivalent area outside the cage. Beckwith (1930) did a similar experiment in which he covered bushes of two varieties with one cage, but also had a similar caged plot containing a colony of honeybees; berries matured only in the cage with bees.

These experiments indicated that wind plays little part in blueberry pollination, but less disputable evidence of lack of wind pollination was provided by Merrill (1936). He emasculated all the flowers on one branch of each of several highbush blueberry bushes and found that, although they were exposed to the wind, none set fruit, presumably because insects were not attracted to them. Unfortunately, he does not record whether he hand-pollinated some of these emasculated flowers as a control measure, although in a different experiment he did so and obtained a good set.

Merrill disagreed with previous findings (e.g. Coville, 1937) that highbush blueberries were self-unfruitful. He found that when he kept small branches bagged during flowering the amount of fruit that matured was small (2–45% for different varieties in 1932, and 0–11% in 1933) but when he self-pollinated similarly treated flowers by hand, a large proportion of fruit matured (82–91% in 1932, 20–85% in 1933, 36–79% in 1934, and 82–100% in 1935). He obtained similar results when he hand pollinated flowers in a glasshouse, and when he hand pollinated plants, all belonging to the same variety, that were growing 3·2 km from any other blueberries.

Merrill found that numerous crosses between different varieties gave a slightly lower set than by self-pollination in 1932, but a substantially greater set in 1933 and 1934. However, he pointed out that the berries produced by self-pollination grew as large as those produced by cross-pollination, and no relationship was apparent between the size of a berry and the number of seeds it contained. He concluded that self-pollination gives satisfactory commercial sets in the varieties he investigated.

Merrill and Johnston (1940) produced further evidence showing that the percentage set from self-pollination of bagged flowers was not much different from that of flowers left exposed, and was sometimes even greater. They further claimed that berries produced from self-pollination were no smaller than ones produced from cross-pollinated flowers, but gave no data on this.

A controversy on the possible benefits of cross-pollination continued for many years. Thus, in a series of experiments, Shaw and Bailey (1937a), Bailey (1938) and Shaw *et al.* (1939), enclosed two varieties per cage with bee colonies and found that plenty of fruit was produced on exposed branches in the cages, but on branches covered with muslin little or no fruit was produced. When varieties were caged singly and self-pollinated by hand the result tended to vary with the variety concerned, but all varieties had a greater set outside the cages where cross-pollination was also possible. They supposed that Merrill had obtained different results because he thinned his flowers before each experiment in which flowers were bagged, and the fewer the flowers the greater the percentage set. However, this would not explain why Merrill obtained a good set on plants growing in the open but isolated from other varieties.

The results of White and Clarke (1939) and Morrow (1943) were in some ways between these two extremes. White and Clark (1939) obtained a poor set from the self-pollination of bagged flowers of certain varieties but, on average, the results were almost as good from self- as from cross-pollination;

Schaub and Baver (1942), Morrow (1943) and Meader and Darrow (1947) made tests in glasshouses in which they compared self- and cross-pollination of different varieties, and on average obtained a slightly greater set from cross-pollination than from self-pollination; Meader and Darrow (1947) found that cross-pollination was relatively more effective than self-pollination at low temperatures.

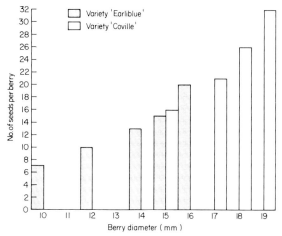

Fig. 128. Correlation between the number of seeds per fruit and fruit size of blueberry (after Filmer and Marucci, 1963).

Extensive experiments were made by Dorr and Martin (1966). They caged: (a) single "Jersey" bushes with bouquets of either "Rubel", "Coville" or "Bluecrop" and honeybee colonies (three replicates of each treatment); (b) single "Jersey" bushes with honeybee colonies but no bouquets (three replicates); (c) single "Jersey" bushes without honeybee colonies or bouquets (four replicates). Bushes caged without bees had only an eighth of the yield of bushes caged with bees (Brewer et al. (1969) have recently obtained similar results). The mean yield of bushes caged with bees was greater when bouquets were present but not significantly so; although as Dorr and Martin point out their experiment might not have been adequate to demonstrate this, any such difference would probably be a comparatively small one.

The most important finding of all these workers was that berries resulting from cross-pollination were consistently larger (9–21% increase) than those resulting from self-pollination and ripened 4–8 days earlier; furthermore, the number of seeds per berry increased with its size. This latter conclusion was substantiated by Filmer and Marucci (1963) and Marucci (1967a) (Fig. 128). However, Eaton (1967) found that most of the variability in the berry weight of cross-pollinated fruits was caused by factors other than differences in the number of seeds present. Perrin and Duggan (1965) reported that R. R. Williams grew blueberry bushes under glass in England and found it was largely self-fruitful, but self-pollination resulted in more seeds per fruit. Brewer *et al.* (1969) found that bushes whose flowers were pollinated by bees gave larger berries that matured earlier than bushes isolated from bees.

Morrow (1943) pointed out that growers wanting early-maturing, large-sized fruits should provide for cross-pollination in their plantations; whereas growers interested in late-maturing berries might find it advantageous to plant one variety only per plantation, provided that any decrease in set and berry size did not offset the advantages gained from late maturation. Moreover, planting large blocks of a single variety facilitates cultivation and harvesting.

It is difficult to reconcile these different results especially as some of the varieties used by the different workers were the same. The results may have been influenced partly by differences in clonal material, partly by environmental conditions and partly by differences in technique; thus in his later tests Merrill (Michigan) used "glassine" bags, Bailey (Massachusetts) used cloth bags, White and Clark (New Jersey) used manilla paper bags and Morrow (North Carolina) used unbagged flowers in a glasshouse.

Hence it seems that in our present state of knowledge it would be wise to test the pollination requirements of varieties, and hence the need to interplant different varieties, in each main highbush blueberry area. Unless there is information to the contrary, at least two varieties of blueberry should be present per plantation.

Results from studying the pollination requirements of lowbush blueberries appear to show a similar wide variation. Because of the difficulty of establishing lowbush blueberry plants in new fields, commercial plantations are developed in areas where blueberries are already established, by destroying competing plants. Thus, many genetically different clones are present in a commercial field, and often more than one species is present, so there is no lack of sources for cross-pollination, and it is not so important, as for highbush blueberries, to determine any need for cross-pollination.

Lee (1958) made a series of experiments in which he enclosed two clones of lowbush blueberries in each of two adjacent cages, one cage with bees and the other without. In the cages without bees no fruit was ever produced, but in the cages with bees the fruit set varied greatly (from 36–98%); if it is assumed that pollination was adequate in each cage with bees, the differences must have resulted from physiological differences between the clones. He noticed a similar variation between uncaged clones scattered throughout the 8 ha field.

Hall and Aalders (1961) confirmed these differences in fertility, and concluded that there are moderately self-fertile clones, slightly self-fertile clones and self-sterile clones. They also made detailed studies of the relationship between berry weight and seed content. On occasions when all the berries were harvested together, the weight of a berry increased by an average of 11 mg for every additional seed it contained but when the berries were harvested as they matured the average weight increase per additional seed was only 5 mg. The weight of a seed itself is only 0·3 mg, so presumably a hormone produced by the seed must have stimulated the remaining increase in weight of the fruit.

Karmo (1958) reported that in Nova Scotia the presence of honeybee colonies in lowbush blueberry fields increased yields, but Wood (1961) was only able partially to confirm these results in New Brunswick. In each of 3 years, Wood put honeybee colonies beside the fields of one group but not those of another; there were 5–10 fields/group, and $2\frac{1}{2}$–$7\frac{1}{2}$ colonies/ha were used in 1957 and 1958, and $2\frac{1}{2}$ colonies/ha in 1959. Whereas the presence of honeybees made no difference to set in 1957 and 1958, it doubled the crop in 1959 (from 24–53%). The only obvious difference between 1959 and the other 2 years was the duration of flowering which was 24 and 16 days in 1957 and 1958, but only 9 days in 1959. Perhaps the number of wild pollinating insects was not sufficient to pollinate the crop adequately in 1959 and, as with other commercially grown fruit and seed crops, the presence of plenty of honeybees is especially beneficial when conditions are not ideal for pollination. However, Wood rightly concluded that taking honeybees to blueberry crops does not necessarily increase the crop.

Experiments by Hall and Aalders (1961) in Nova Scotia have done much to explain why increasing the number of honeybees visiting a lowbush blueberry crop is only sometimes beneficial. They noted that, although *V. myrtilloides* (diploid) and *V. angustifolium* (tetraploid) occur together in many fields, no hybridization between the two species occurs. They found that when *V. angustifolium* is pollinated with *V. myrtilloides* pollen its ovules develop normally until the endosperm is partially developed, but the ovules then abort and the berries drop off; 12 days after pollination, all the fruits produced from the cross-pollinated flowers had dropped. Even when *V. angustifolium* was pollinated with pollen from another *V. angustifolium* clone, the additional presence of *V. myrtilloides* pollen significantly reduced fruit set, berry size, and number of seeds, and significantly increased the time taken for the fruit to mature. The flowering periods of the two species almost completely coincide and pollinating insects do not discriminate between them (see also Wood, 1965), so where they grow together, the *V. myrtilloides* pollen not only dilutes the *V. angustifolium* pollen and decreases the chance of fruit set, but fertilization by *V. myrtilloides*, which is subsequently abortive, prevents effective pollination by *V. angustifolium*. Pollen of both species grows equally fast in the style, so when they are present in equal abundance they are equally likely to effect fertilization.

In fields, such as used by Wood (1961), which are developed from forest, *V. angustifolium* and *V. myrtilloides* occur in about equal proportions so, as Hall and Aalders (1961) point out, even the maximum amount of pollination would not produce a set greater than about 50%. Where sufficient wild pollinators are present to achieve this, the addition of honeybee colonies makes no difference, and any correlation between nectar productivity and fruit set of different clones would be unlikely (see Wood and Wood, 1963). But *V. angustifolium* is predominant in fields developed from abandoned farm lands, as used by Karmo (1958), so sets well in excess of 50% would be possible and there is a greater likelihood that the presence of honeybees would be beneficial.

Hall and Aalders (1961) concluded that if plantations could be restricted to two or more good pollen yielding clones of the same species, which were planted in alternate rows, the problem of obtaining sufficient pollination of lowbush blueberries would largely disappear.

Little is known about the pollination requirements of other types of blueberry. Meader and Darrow (1944) compared the effects of self-pollinating and cross-pollinating by hand ten varieties of *Vaccinium ashei*, rabbiteye blueberry. They found that most varieties, including the most important ones, were either partially or completely self-sterile, but one variety was completely self-fertile. Berries produced from cross-pollination matured more quickly and were 18–178% larger, depending on the variety concerned, than berries produced from self-pollination.

Because blueberries are grown near bogs, woodland, deserted farm land or other derelict country, the population of wild pollinators is often relatively large and sufficient to adequately pollinate small fields although not big ones (e.g. Lee, 1958; Filmer and Marucci, 1963; Wood *et al.*, 1967). However, in areas where blueberry plantations are concentrated, there may be too few wild bees, and inadequate pollination has been demonstrated by caging bushes with honeybees and in such areas comparing the set with uncaged bushes in the same field (Filmer and Marucci, 1963; Marucci, 1967a). Dorr and Martin (1966) found that on seven plantations of highbush blueberry, the berry size and number of seeds per berry was correlated with the number of honeybees present.

Honeybees and bumblebees are usually by far the most abundant pollinators present, occurring in about equal numbers, but solitary bees (including *Colletes*, *Halictus*, *Andrena* and *Nomada*) and flies (including *Bombylius*, *Calliphora*, *Eristalis*) are also common (e.g. Phipps, 1930; Shaw and Bailey, 1937b; Beckman and Tannenbaum, 1939; Shaw *et al.*, 1939). Dorr and Martin (1966) found that bumblebees (nine species) were sometimes more numerous than honeybees in areas where suitable nesting and hibernating sites were available. They noted that the carpenter bee *Xylocopa virginica*, which was sometimes quite common, bit holes at the bases of the corolla tubes to obtain nectar, and that other bees used these holes and so failed to pollinate the flowers.

There is little information on the behaviour of bees when visiting the flowers. Bumblebees visit 10–20 flowers/min but honeybees and solitary bees only 5 (Beckman and Tannenbaum, 1939; Shaw *et al.*, 1939; Wood, 1965). The long tongue of the bumblebee reaches the base of a blueberry flower, but it has been suggested that the honeybee with its shorter tongue cannot reach the nectaries of some varieties. Merrill (1936) observed that honeybees select large flowers with wide mouths, through which they can insert their heads and so reach the nectaries with their tongues, but he supposed that they cannot work varieties with relatively long narrow flowers (e.g. "Pioneer" and "Cabot"). Many of the visits honeybees make to blueberry flowers do not result in pollination. Thus a bee can obtain nectar from the wide-mouthed flowers of some varieties (e.g. "Earliblue") without touching the stigmas, and flowers that are too old to be fertilized are usually preferred to younger flowers that have a lower nectar content (Marucci, 1967a). Although it appears therefore that, in general, honeybees are less valuable pollinators than bumblebees, more evidence is needed.

Only 4% of the honeybees observed by Shaw *et al.* (1954) collected pollen. Karmo (1958) reported that pollen is readily gathered by bumblebees which shake the flowers to release it but honeybees either do not, or cannot, do this and so the pollen is largely unobtainable to them. Therefore, it would seem advisable to use honeybee colonies with adequate pollen stores, or without excessive pollen requirements, so the bees are not diverted to pollen-yielding crops. Karmo (1958) reported that colonies taken to blueberry fields, in New Brunswick may store up to 18 kg honey during flowering time, but larger yields have been reported from many parts of the U.S.A.

It is agreed that honeybees prefer some blueberry varieties to others, and Marucci (1967a) found there was a relationship between the attractiveness of a variety to honeybees and the amount of berries produced. The more attractive varieties had more abundant but not more concentrated nectar. The differences in attractiveness were not as pronounced in the evening as at midday, presumably because the nectar supply had become depleted. Filmer and Marucci (1963) reported that bumblebees visit all varieties indiscriminately, whereas Dorr and Martin (1966) found that bumblebees also prefer certain varieties. In the latter circumstances, increased pollination will have a greater effect on the varieties that were previously less favoured (Brewer *et al.* 1969). The reluctance of honeybees to work certain varieties can be overcome by increasing the population of bees present so that greater competition causes them to become less fastidious in their choice of forage. It will probably also help if the unattractive varieties are arranged at intervals among the attractive ones, although this will increase management and harvesting problems.

It is important to have sufficient pollinators present to visit all the flowers soon after they have opened as the capacity of a flower to set fruit rapidly decreases as it gets older (Fig. 129; Wood and Wood, 1963).

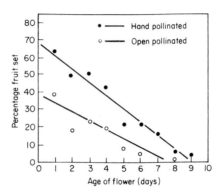

Fig. 129. Relationship between age of *Vaccinium* spp., lowbush blueberry, flower and fruit set (after Wood and Wood, 1963).

Because of the apparent inability of honeybees to pollinate some varieties, the large populations of wild pollinators sometimes present, and the factors limiting pollination in mixed fields of *V. angustifolium* and *V. myrtilloides*, the need to import honeybee colonies will vary more from one blueberry field to another than for most other crops. Recommendations of $2\frac{1}{2}$ colonies/ha (e.g. Lathrop, 1954) should not be regarded as generally applicable. Kinsman (1957) suggested that if there is less than one pollinating insect per square metre of lowbush blueberries during favourable weather—"sunny and calm, and the temperature above 16°C (60°F)"—there will probably be insufficient pollination, and this may be remedied by importing honeybee colonies. He calculated that one strong honeybee colony should be able to produce just enough foragers to pollinate 0·8 ha even when no other pollinators are present. However, for varieties that are relatively unattractive 5 colonies/ha has been recommended (Marucci, 1967a).

Cranberry

The "large" or "American" cranberry (*Vaccinium macrocarpum* L.) is native to North America and has been grown there commercially for about 150 years. It is also grown a little in England and Holland. A closely related species of cranberry (*V. oxycoccus* L.) (Fig. 130) grows wild in Europe but has such small berries that it is not worth cultivating.

The cranberry vine bears flowers which hang downwards, both on its upright branches and runners. When a flower opens, each of the four petals curls back on itself exposing a ring of eight stamens enclosing the style. A ring of small nectaries occur at the base of the flower just inside the stamens.

According to Roberts and Struckmeyer (1942), the stigma of a flower that has just opened is the same length as the stamens and, if the flower is jarred, large quantities of pollen are liberated close to the stigma which is receptive at this stage. They reported that when honeybees visited the flowers they did not usually touch the stigmas but they jarred the flowers sufficiently to cause self-pollination. However, they thought that wind pollination was of prime

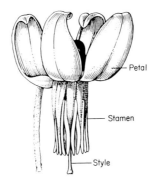

Fig. 130. Flower of *Vaccinium oxycoccus*, cranberry (after Ross-Craig, 1963).

importance, although not over great distances, as the pollen grains adhered in groups of four and emasculated flowers rarely set fruit. They reported that plants left undisturbed in muslin cages set few fruits, but when they agitated the plants with a stick good sets were obtained. Their ideas received considerable publicity and in an attempt to increase pollination by jarring the vines it became a practice to drag long ropes through cranberry plantations.

However, the views of Roberts and Struckmeyer (1942) are no longer generally accepted (see Cross, 1966; Marucci, 1967b). It is now thought that, when a flower opens, the stigma is dry and unreceptive and well hidden within the ring of stamens; if a bee visits a flower in this stage to collect nectar, pollen is shaken onto its body. Two days later the flower will probably have shed all its pollen but the stigma will by then project beyond the anthers and be sticky and receptive, and so can be pollinated by a bee that has previously visited a younger flower.

Recent experiments have also demonstrated that any wind pollination of cranberries is negligible. Filmer (1949) exposed a series of agar coated glass plates in a cranberry plantation during flowering and found that whereas plates at ground level had 20–50 grains/cm^2 of cranberry pollen those at 15,

30 and 45 cm above ground had only an occasional pollen grain, showing that the heavy grains were not transported much, if at all by wind. He pointed out that his results consistently showed a decrease in pollination from the edge to the centre of a cranberry bog and he supposed that this would not be so if wind pollination were of major importance; however, it seems that the concentration of insect visits toward the edges of plantations could produce such a pollination gradient even if wind were the more important factor. Filmer also caged six plots, and at noon on each day during flowering, he jarred the flowers in one half of each cage with a bamboo cane, but this treatment increased the mean set from 5·8 to only 7·1 %. In this experiment none of the pollen had been removed by insects; when some or all of the pollen has already been removed by insects before the stigmas become receptive any effect of wind would be even less pronounced.

Substantial evidence has now accumulated showing the need for insect pollination. Hutson (1925) found that in an area of cranberry bog caged without bees only 8% of the flowers set fruit compared to 56% in an equal area caged with bees; Farrar and Bain (1947) found that, in 2 years, plots caged without bees gave only 108 and 32 berries/m^2 compared to 1840 and 721 in a cage with bees, and 1,335 and 484 in uncaged plots. Filmer and Dochlert (1959) found only 161 berries/m^2 on plots caged to exclude insects but 969–1,636 berries/m^2 on adjacent uncaged plots; Marucci (1967b) reported an experiment in which plots caged without bees, caged with bees and not caged gave 53, 1,067 and 1,130 berries/m^2. Sharma (1961) found that the average fruit set on nine bagged branches was only 5% compared to 26% on open branches on the same three bushes. Although the cage walls undoubtedly reduced wind speed to some extent, the low set in the cages without pollinating insects is further evidence of lack of wind pollination.

Marucci and Filmer (1964) noticed that in areas, where several varieties grew together, an exceptionally large yield sometimes occurred and in preliminary trials they caged three varieties, each growing in a separate location, with honeybee colonies. In each cage they also included a section, about 30 cm^2, of cranberry turf of a different variety to provide for cross-pollination. Although they found that the set was greater in all of the cages than in areas situated just outside, and that the number of seeds per berry was greater in two of them, this could merely be because there were more bees per plot inside the cages than in the open rather than the result of cross-pollination; further tests are needed. A positive correlation has been found for several varieties between the diameter of a berry and its seed content (Eaton, 1966; Marucci, 1967b) and cross-pollination has resulted in larger berries and more seed than self-pollination (Marucci and Filmer, 1964).

However, even under the best conditions, it is rarely possible to exceed a 50% set, presumably because the bearing capacity of the plant has been reached; it has generally been found that uprights that had a smaller number of flowers than usual also had a greater percentage set (Marucci and Filmer, 1957). Bergman (1954) suggested that a major reason why a large percentage of flowers often fail to set is not due to lack of pollination but because the flower buds are injured by spring frosts, the developing ovules being the parts most susceptible to damage. Under such conditions the presence of even large numbers of pollinators is, of course, ineffectual. Frost also often injures the developing nectaries to various extents. Bergman found that when the injury was slight the amount of nectar secreted was only slightly diminished and there was little effect on the number of insect visitors but, when the injury was severe, many fewer insects were attracted to the flowers.

As cranberries are grown in natural bogland surrounded by waste land, it is not surprising that they are visited by many wild insects, especially bumblebees in North America (Hutson, 1925; Franklin, 1940; Johansen and Hutt, 1963) and honeybees, *Apis cerana*, in India (Sharma, 1961). However, in large bogs the native insect population may be inadequate to set maximum crops and this deficit may be overcome by importing honeybees. An average of $2\frac{1}{2}$ colonies/ha is used in Wisconsin (Shoemaker, 1955), and Cross (1966) found it difficult to envisage a situation when one colony for every 0·8 ha would not be worthwhile.

The need for importing honeybees has increased in recent years. In New Jersey, Hutson (1925) found there was an average of 12 syrphids, 7 leaf-cutting bees and 1,107 bumblebees (mostly *B. impatiens*) per hectare. On the assumption that a bumblebee visited 12 flowers/min for 12 h a day he calculated that there were enough bumblebees to provide more than adequate pollination and concluded that honeybees were not needed in New Jersey cranberry bogs. More recently Filmer and Doehlert (1959), also of New Jersey, supposed that although wild bees can be depended upon to pollinate cranberry in bogs of small or moderate size and in large narrow bogs, all parts of which are within easy access of uncultivated land where wild bees nest, there are not sufficient to pollinate compact bog areas of 50 acres or more. They suggested that when there is fewer than one bumblebee per 8 m² during good foraging weather honeybee colonies should be imported, at the rate of one or two colonies per hectare depending on the wild bee population. By 1964, Marucci and Filmer counted a mean of only 2·5 bumblebees but 20·0 honeybees per 5 min period on cranberry flowers and reported that many New Jersey growers had noticed a decrease in the bumblebee population.

In India, the sugar content of cranberry nectar varies from 16–51% (Singh, 1954). Shaw *et al.* (1956) found that the mean sugar concentration of nectar in the honeystomach of bees working three cranberry varieties was 46, 50 and 55% respectively. Filmer and Doehlert (1959) reported that about 50% of the pollen collected by honeybee colonies close to cranberry bogs came from cranberry flowers, and Shimanuki *et al.* (1967) found that the mean percentage of cranberry pollen collected by six colonies varied from 12–99%. However, few of the bees collected by Shaw *et al.* (1956) had pollen loads, and Marucci (1967b) found that cranberry flowers are poor producers of nectar and pollen and honeybees were not eager to visit them especially in dry conditions and when flowering had only recently begun. Fortunately the long flowering period, which may exceed 4 weeks, ensures that a sufficient percentage of the flowers are pollinated even when the early flowers are neglected and the weather is poor during much of flowering; provided the bees visited the plantations for a week during the peak of flowering sufficient pollination occurred (Marucci, 1967b).

Chapter 28

Solanaceae

The family Solanaceae includes many valuable crop plants including *Lycopersicon esculentum*, tomato; *Capsicum annuum* and *C. frutescens*, sweet pepper and chillies; *Nicotiana tabacum*, tobacco; *Solanum melongena*, egg-plant; and *Solanum tuberosum*, potato.

Lycopersicon esculentum Mill

The inflorescence of *L. esculentum*, wild tomato, contains four to twelve pendent flowers, although the inflorescences of the more important commercial varieties have four or five only. The calyx and corolla have five to

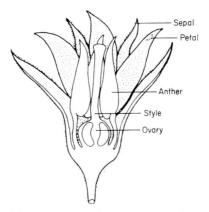

Fig. 131. Flower of *Lycopersicon esculentum*, tomato (after Purseglove, 1968).

ten, but mostly six, pointed lobes and those of the yellow corolla are reflexed (Fig. 131). There are usually six stamens, attached to the base of the corolla tube; the short filaments are surmounted by long bright yellow anthers which are partly united into a cone surrounding the pistil; the top of each anther is prolonged into a sterile beak. The ovary has five to nine chambers.

The flowers are self-fertile. The anthers dehisce longitudinally towards the flower centre one or two days after the flower opens, so that when the style is short, as in most cultivated varieties, self-pollination occurs. Probably the pendent position of the flowers facilitates self-pollination in varieties whose stigmas project beyond the cone of anthers. In some varieties the style does not elongate and grow through the anther tube until the time of anther dehiscence or afterwards, so the chances of self-pollination and self-fertilization are increased. However, the stigma becomes receptive about the time the flower opens, so when the style elongates before the anthers dehisce, there is the possibility of cross-pollination occurring, especially when the stigma projects beyond the anthers (see Robbins, 1931; Pashchenko, 1940; Cobley, 1956). Hence differences in the morphology of the flowers of different varieties can have important influences on pollination. There is some evidence that the length of the style is influenced by environmental factors, including day length, temperature and humidity (Burk, 1930; Smith, 1932). The optimum temperature for pollen tube growth is $30°C$ ($86°F$) but even at this temperature growth is slow and the minimum time from pollination to fertilization is 50 h (Smith, 1935; Smith and Cochran, 1935).

Early experiments in N. America by Fletcher and Gregg (1907) with six different varieties of tomato showed that cross- and self-pollination gave similar sets but, despite the apparently effective mechanism for self-pollination many varieties of tomatoes grown in glasshouses do not set well unless artificially pollinated, either by hand or by shaking the plant itself. Thus, Jones (1916) found that flowers bagged in the bud stage and left undisturbed usually failed to set fruit, whereas jarring the plant when the anthers had dehisced usually resulted in fertilization; White (1918) found that the differences in set between hand-pollinated plants and those left undisturbed in glasshouses was much greater for some varieties than others. In New South Wales, it has been found that varieties whose pistils are shorter than their stamens are easier to set in glasshouses than varieties whose pistils are longer than their stamens (Wenholz, 1933). Fletcher and Gregg (1907) found that the size of the fruit produced depended within limits on the amount of pollen transferred to the stigma. When only one to five pollen grains were applied, small solid fruits with only one or two seeds, if any, developed, and when a small amount of pollen was applied to one side only of the stigma, small lop-sided fruits were produced. They concluded that the small irregular tomatoes grown under glass are caused by insufficient pollination and recommended more care in hand pollination to remedy this.

Verkerk (1957) demonstrated that for several varieties the size and weight of the fruit increased with the number of seeds present, although the more seeds per fruit, the smaller the ratio of weight to seeds. The more often the flowers were pollinated, up to a total of four times, the greater the number of fruits set and the number of seeds per fruit. He also pollinated male sterile flowers with mixtures of pollen and lycopodium powder containing 0·1, 1, 10, and 100% pollen. The results (Table 30) showed a positive correlation between the concentration of pollen applied and the number of seeds and weight of fruit produced, but the increase was small relative to that of the concentration of pollen used. Thus a thousand-fold increase in the number of pollen grains used gave a twenty-fold increase in seeds per fruit, but only a two-fold increase in fruit set. The greater the concentration of the pollen used, the faster the fruit developed, and hence the earlier and more profitable the crop.

TABLE 30

% pollen in pollen/ lycopodium mixture	0·1	1	10	100
Mean No. seeds/fruit	7	12	72	153
Mean wt (g)/fruit	63	72	108	136

Auxins are sometimes sprayed onto clusters when fruit set is poor. They have the same effect on fruit growth as pollination, except no seed is produced (Verkerk, 1957). Electric devices also are now commonly used in glasshouses in Europe and N. America for vibrating flower clusters or flowers (Cottrell-Dormer, 1945; Kerr and Kribs, 1955) and they have been found more effective than other methods of artificial pollination especially early in the season. Thus Verkerk (1957) found that the fruits grown from vibrated flowers ripened earlier, had more seeds and were larger than those from untreated flowers, all factors indicative of more successful pollination. Numerous tests also showed that plants whose flowers were vibrated also had a larger total yield. These results were confirmed by Christensen (1960) who found that using vibrators increased the yield of the first and second clusters in particular.

Because of the large amount of labour involved in pollinating glasshouse tomatoes, Neiswander (1954) in Ohio tried to compare the effect of bees and electric vibrators in producing fruit. Unfortunately, the design of most of his experiments was inadequate and did not allow decisive results to be obtained, but in all six experiments, the use of a vibrator and bees was superior to the use of bees alone. In his final experiment, one compartment of a glasshouse contained a colony of honeybees and another was without bees; a vibrator was used on half of the plants in each compartment. The results are given in Table 31.

More experiments of this type are needed before honeybees can be confidently recommended for the pollination of tomatoes in glasshouses. Most varieties grown in glasshouses in Britain, and some of those in N. America, are parthenocarpic so the problem of pollination does not arise.

TABLE 31

Treatment of plants		No. fruits/ plant	Wt (g)/ fruit	Wt (kg) fruit/plant
Compartment A.	Bees only	27·1	159	4·3
	Bees and vibrator	29·4	170	5·0
Compartment B.	Vibrator only	26·1	145	3·8
	Neither vibrator nor bees	28·5	88	2·5

The effect of wind in jarring the flowers must contribute to the self-pollination of plants grown in the open, but the extent to which field crops are cross-pollinated depends on the variety concerned and on the abundance of insect pollinators (Fig. 132). Jones (1916) found between 2 and 4% cross-pollination occurred in the field, and Lesley (1924) found there was about 5% cross-pollination of a long styled variety compared to 0·6% of a short styled variety. Rick (1949) measured the set of male sterile plants under widely different environmental conditions in California, and found that it varied from 2–47% of the set of fully fertile plants in the same plots. The set was greatest where the plants were near undisturbed land which provided suitable nesting sites for wild bees. However, these observations do not indicate the amount of cross-fertilization likely to occur with male fertile plants, as the pollen transferred to male fertile plants will have to compete with the much more abundant pollen provided by the recipient flower's own anthers.

The amount of cross-pollination found by Richardson and Alvarez (1957) in fields in Mexico ranged from 3–12%, and they supposed these differences were caused by the noticeable differences in the insect population in the different fields, but they did not provide data on this. The principal pollinator was the halictid bee, *Augochloropis ignata*, which nests in moist banks, and they supposed that the availability of such sites could account for the different populations in the fields.

The tomato is a native of the Peru–Equador area where frequent hybridization occurs between plants of the cultivated tomatoes, and between the cultivated and the wild or "currant" tomato, *Lycopersicon pimpinellifolium* Jusl., and other *Lycopersicon* species. Numerous halictid bees visit the tomato flowers in this area and averages of 15 and 26% cross-pollination have been recorded (Rick, 1958). In contrast to many commercial varieties, the stigma of the wild species protrudes beyond the ring of anthers and no doubt this helps cross-pollination. In fact, commercial varieties have probably been selected for self-pollination and self-fertilization.

Fig. 132. Bumblebee vibrating *Lycopersicon esculentum*, tomato, flower. (Photo: W. P. Nye.)

The amount of cross-pollination decreases rapidly with increasing distance from the pollen source. For example, Currence and Jenkins (1942) found that the maximum amount of cross-pollination was 5·2% and the amount rapidly diminished with distance, and when rows were 1·8 m apart little or no crossing occurred beyond an interval of 12 rows (Fig. 133). Rick (1947) found that the amount of cross-pollination varied on different parts of male sterile plants that were adjacent to fertile plants and the greatest amount, (4%), occurred where fertile and male sterile branches intermingled. Isolation of seed fields is necessary to prevent the limited amount of cross-pollination that might otherwise occur. Hawthorn and Pollard (1954) suggested 15 m should be the minimum distance between varieties and, if possible, a greater distance when stock seed is required.

The amount of cross-pollination and the distance over which it occurs are interesting, not only in connection with the isolation of seed crops, but also with regard to producing seeds of first generation hybrid tomatoes, the use of

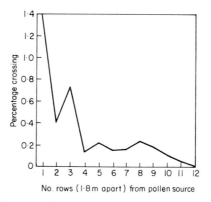

Fig. 133. The amount of cross-pollination of *Lycopersicon esculentum*, tomato, at increasing distances from the pollen source (after Currence and Jenkins, 1942).

which is increasing (Rick, 1947). Schneck (1928) reported that tomato flowers were attractive to bumblebees in New York State; but they are now found to be unattractive to honeybees and wild bees in California, and although suitable male sterile lines have been developed insufficient pollinators are attracted to them, and the hybrid seed is laboriously produced by hand pollination. Bohart and Todd (1961) suggested that either male sterile lines of tomato should be used effectively in their Peruvian homeland, or that the wild bees that readily visit the flowers should be established in California.

Capsicum

There is little information on the pollination of *Capsicum annuum*, sweet pepper, and *C. frutescens* L., bird chilli, and this tends to be contradictory. The corolla, 8–15 mm diameter, is deeply divided into five or six parts, and has five or six stamens inserted near its base; the two-celled ovary is surmounted by a single style which is usually longer than the stamens (Fig. 134).

Cochran (1936) pointed out that the dropping of buds, flowers and partially mature fruits of *C. frutescens* is generally common and can seriously diminish yield. However, he found that 94% of flowers set fruit on plants isolated in a glasshouse. Hand-pollinating some of the flowers failed to increase set, and there was no reduction in the set of flowers that were emasculated and bagged prior to anthesis. Consequently, he concluded that pollination did not seem to be necessary. In contrast, Sampson (1936) reported that seed setting of bagged flowers was very sparse unless they were agitated at short intervals of time to distribute the pollen. He noticed that, although the flowers opened between 08.00 and 10.00 h, the stigma was frequently covered with pollen before the anthers dehisced 30 min to 5 h later, indicating that cross-pollination was frequent; he claimed that only 20% of flowers not bagged bred true. Odland and Porter (1941) grew plots of a variety of *C. frutescens* which has dominant genes for several distinguishing characteristics; they

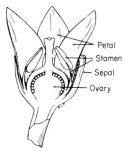

Fig. 134. Median section of *Capsicum annuum*, pepper, flower (after Robbins, 1931, with permission of McGraw-Hill Book Company).

removed a central plant from each plot and replaced it with a plant of a recessive variety, which thus had the greatest possible chance of being cross-pollinated. The six recessive varieties tested in this way received 9·1, 10·1, 11·0, 14·1, 23·0 and 31·8% cross-fertilization respectively. The greater amount of cross-fertilization in some varieties than in others, was probably associated with their tendency to produce insufficient viable pollen. Odland and Porter concluded that enough natural cross-pollination occurs to necessitate isolating seed plots for breeding. This was confirmed by Murthy and Murthy (1962) who reported that 18% of the 1,356 progeny they grew from sixteen plants resulted from cross-pollination and that eleven of the sixteen plants were cross-pollinated to some extent. Purseglove (1968) stated that about 16% cross-pollination occurs.

Odland and Porter (1941) observed that honeybees frequently visited the flowers and thrips were also often present in them. Ants have also been mentioned as possible cross-pollinators. There is no indication of the proportion, if any, of cross-pollination for which wind is responsible.

Nicotiana tabacum L.

Each inflorescence of *N. tabacum*, tobacco, contains up to 150 flowers. The corolla consists of five pink, red, or white, petals joined together into a tube about 5 cm long which expands into five lobes at the top (Fig. 135). The superior ovary has two chambers with numerous ovules. The long slender style is closely surrounded by five stamens which are joined to the base of the corolla. The anthers dehisce longitudinally, pollen falls on to the stigma and self-fertilization usually occurs. Darwin (1876) showed that self-pollination was as effective as cross-pollination in producing seed.

Peens (1958) determined the amount of cross-pollination that occurred either by using plants with a dominant corolla colour, or by using male sterile or emasculated flowers. It seldom exceeded 0·5% between plants only 0·9 m

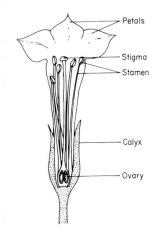

Fig. 135. Median section of *Nicotiana tabacum*, tobacco, flower (after Poehlman, 1959).

apart and reached only 0·3% between plants 7·3 m apart. However, Poehlmam (1959) and Purseglove (1968) reported that up to 4% crossing can occur as a result of visits by nectar-seeking hummingbirds, bees and other insects. Although not all cross-pollination is effective, as cross-incompatibility sometimes occurs, in general the pollen tube grows faster following cross-pollination than following self-pollination. Even when cross-pollination occurs two hours after self-pollination it accounts for 27% of the progeny, and 10% when it occurs as much as 10 h later. Therefore, to ensure pure seed it is necessary to bag the inflorescences or isolate the plants.

Nicotiana rustica L.

Howard *et al.* (1910) classified the varieties of *N. rustica*, yellow-flowered tobacco, nicotine tobacco, into three categories according to whether the stamens were about the same length as the style, much longer than the style, or much shorter than the style. They found that plants of the first category could readily be self- and cross-pollinated. But cross-pollination of plants in the second category was rare and self-pollination and self-fertilization of bagged flowers readily occurred. Self-pollination of plants of the last category was very difficult and very few flowers set seed naturally, although they did so when self-pollinated by hand.

Solanum tuberosum L.

Because *S. tuberosum*, potato, is propagated vegetatively, the flower is only important to plant breeders, and many commercial varieties either rarely flower, or have distorted flowers, or have flowers that are soon shed, sometimes in the bud stage.

The perfect potato flower has a tubular calyx and corolla which are lobed at the top (Fig. 136). The five stamens alternate with the corolla lobes and

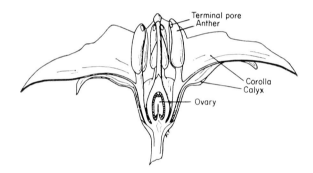

Fig. 136. Median section of *Solanum tuberosum*, potato, flower (after Robbins, 1931, with permission of McGraw-Hill Book Company).

bear large fleshy anthers which project to form a cone. The anthers dehisce by terminal pores and jarring the flower causes pollen to fall out. There is a two-chambered ovary, with numerous ovules, surmounted by a single style and stigma. In varieties which have the stigma the same length as the stamens self-pollination readily occurs; in other varities in which the stigma projects beyond the anthers, it has been supposed that self-pollination is facilitated either by the stigma curving downward to bring it into line with the falling pollen, or by the flower drooping so that pollen falls onto the stigma, or perhaps by the corolla folding and transferring pollen which has fallen onto it to the stigma (Knuth, 1906; Stevenson and Clark, 1937).

Most cross-pollination is by wind. The potato flower secretes no nectar, and because of the method of dehiscence, little pollen is available, so it is not surprising that insect visits are few. However, Müller (1882) observed bumblebees visiting potato flowers and collecting pollen, and Stevenson and Clark (1937) reported that in some localities bumblebees and honeybees are often seen visiting them, and they thought that insects may cause much more cross-pollination than is commonly supposed.

Solanum melongena L.

The floral structure of *S. melongena*, egg-plant, is similar to that of the potato and the cone-like formation of the anthers favours self-pollination. Kakizaki (1924) (quoted by Jones and Rosa, 1928) found that plants grown in a cage without insects produced no fruits; from this limited evidence it seems that insect visits may be needed to transfer the pollen of some varieties; if so, this is probably because the stigma projects beyond the anthers. Kakizaki also grew plants with dominant purple fruits among others with recessive white fruits and, by determining the progeny of 63 white-fruited plants the following year, he found that an average of 7% (range 0·2–47%) cross-pollination by the purple-fruited plants had occurred. Presumably the total amount of cross-pollination, including that from the white-fruited plants as well, was about twice this.

Chapter 29

Chenopodiaceae

Economically important species are present in the genera *Beta* L., *Spinacia* L. and *Tetragonia* L.

Beta vulgaris L.

The species *B. vulgaris* includes several distinct cultivated types including sugar beet, garden beet, spinach beet and mangold. The small greenish flowers occur in clusters of two or three. Each has five narrow, incurved sepals, five stamens inserted at the base of the calyx lobes, an inferior ovary of three fused carpels containing a single ovule, and a short style with three stigmatic lobes (Fig. 137). The flowers open in the morning and, if the day

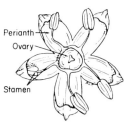

Fig. 137. Flower of *Beta vulgaris*, beet (after Robbins, 1931, with permission of McGraw-Hill Book Company).

is warm and sunny, the anthers dehisce before midday. In most varieties the stigmatic lobes begin to open gradually in the afternoon and are not fully exposed until the following day or even 2 days later. The anthers may have shrivelled completely before the stigmatic lobes open. Hence there is usually pronounced protandry but, very rarely, anther dehiscence and opening of the stigmatic lobes occur simultaneously (Artschwager, 1927). The stigmas are reported to be receptive for 17 days at the beginning of the flowering season and 24 days at its end.

Not only does the behaviour of the flower usually favour cross-pollination, but self-incompatibility seems to be the general rule (Shaw, 1916). Following self-pollination the pollen grains germinate, even when the stigmas have not spread open, but the pollen tube growth rate is slow compared to that of foreign pollen and is soon greatly diminished or stops completely, usually failing to reach the ovules. Even when fertilization occurs the newly formed zygote soon degenerates. However, isolated plants sometimes set a limited amount of seed after self-fertilization and some highly fertile strains, which have pollen tubes that grow fast and the ability to undergo self-fertilization without degeneration of the embryo, have been selected (Archimowitsch, 1949; Savitsky, 1950).

Soon after the anthers dehisce the pollen becomes dry and powdery and is dispersed by wind and insects. It seems to be generally supposed that wind pollination is the most important; beet pollen has been collected 4,500 m from a crop and at a height of 5,000 m above it, although most airborne pollen occurs up to 750 m above ground level (Meier and Artschwager, 1938).

Archimowitsch (1949) planted 300–600 genetically marked beet plants in the centre of a field occupied by plants of other crops and grew small groups of recessive beet plants at different distances and directions from the centre. The percentage of cross-pollination of plants in the small groups varied with their direction from the centre of the field and, in accordance with the direction and strength of the wind during flowering, but the average percentage of hybrids at various distances from the centre was: 0–80 m, 7·7%; 80–200 m, 1·2%; more than 200 m, 0·3%. Archimowitsch (1949) also tested the amount contamination was reduced by growing relatively tall plants (e.g. hemp and sorghum) between blocks of beet. When the plant screens were 6, 10 and 12 m wide the contamination between adjacent blocks was 17·1, 5·4 and 0·7%; hence a 12 m wide plant screen between blocks is equivalent to 200 m of open space. However, when the pollen source is more extensive and consists of a whole field of plants instead of a few hundred, contamination is likely to be much greater. Poole (1937) reported that cross-pollination can occur over a distance of several kilometres and, because all forms of *B. vulgaris* inter-breed freely, a system of zoning the different strains for seed production has been developed in the U.S.A. as the most practical solution.

Stewart (1946) grew isolated, recessive plants among a crop which consisted mostly of genetically marked plants, and enclosed part of each recessive plant in a cage whose walls allowed access to small but not to large insects. In each of two years the sets on the exposed and caged flowering branches of the recessive plants were similar. He concluded that his experiment indicated that wind alone is sufficient to effect the necessary transfer of pollen in a crop. Unfortunately, however, he did not record whether the exposed parts of the recessive plants were in fact visited by large insects and, as he pointed out, the role of thrips, and other small insects that could enter the cages, was not evaluated.

Indeed, other workers, and especially Shaw (1914), have stressed the importance of thrips in cross-pollinating sugar beet flowers. Shaw discovered that, although he emasculated and protected single flowers with paper bags, they still became fertilized, and he supposed that thrips were the only pollen-bearing insects that were small enough to gain entrance of the flower through gaps that occurred where the mouth of the paper bag was tied round the stem. When the cloth used to enclose beet plants was not closely woven, wind and thrips carried pollen between its meshes (Shaw, 1916). He collected thrips from flowers by exposing them to chloroform vapour and discovered that the most abundant species in Utah were *Heliothrips fasciatus*, *Frankliniella fusca*, *Frankliniella tritici* and *Thrips tabaci*. Thrips were also abundant in beet fields in Idaho, Indiana and Michigan. In some localities, in some seasons, they were especially numerous, and there were sometimes as many as five or six per flower. Both the larva and adult stages fed on the nectar and pollen and had averages of 40 and 140 pollen grains on their bodies. Pollen was transferred from one insect to another when they brushed against each other. Because thrips walked over all the parts of the flowers, including the stigmas, Shaw thought they probably contributed toward pollination and he made experiments to test this. When he freed flowers of thrips, emasculated and bagged them, none set seed, but when he introduced thrips from other flowers about 20% did so, which was similar to the percentage set of flowers not bagged. He obtained similar results in two other trials. However, in these experiments he actually transferred thrips from one flower to another, whereas the amount they cross-pollinate in natural conditions will depend on how often they move between flowers. He observed that this does in fact frequently happen and that the thrips carry pollen grains with them; it was very noticeable that thrips migrated away from blocks of beet that had ceased to flower onto other beet plants that were just beginning to do so.

Unfortunately, the value of thrips as pollinators is diminished by their tendency sometimes to injure the floral organs, particularly when they occur in such large numbers that the available nectar and pollen is insufficient to support them. Hence there is some doubt as to whether, on balance, they are beneficial, although it would certainly be interesting to discover whether there is any correlation between the abundance of thrips and seed production.

Treherne (1923) reported that in Canada syrphids were the most abundant visitors to sugar beet and mangold flowers, but honeybees, solitary bees and various Hemiptera were also important and coccinellids were sometimes locally numerous. Archimowitsch (1949) classified insects that were possible pollinators of beet flowers in the Ukraine into the following groups: (a) those that suck sap, e.g. thrips and *Aphis fabae*; (b) those that are attracted to plants by the presence of *Aphis fabae*, e.g. ladybird beetles (*Coccinella septempunctata*) and their larvae, and ants; (c) those that eat pollen, e.g. beetles of the genera *Zonabris*, *Leptura* and *Cerocoma*; (d) those that collect nectar only, e.g. the wasp fly *Sphaerophoria scripta*; (e) those that collect both nectar and pollen, e.g. the honeybee, *Andrena* spp. and *Halictus* spp. Popov (1952) observed that solitary bees belonging to the Halictidae, Megachilidae, and Anthophoridae were the most abundant visitors to beet flowers. Archimowitsch (1949) pointed out that, although honeybees are reluctant to visit beet, they will do so in large numbers when no other sources of nectar and pollen are available to them. Mikitenko (1959) also found this, and obtained data suggesting that bee visits may increase yield. However, although it seems possible that insect pollination supplements wind pollination of beet flowers the evidence is very inconclusive.

Spinacia oleracea L.

The flowers of *S. oleracea*, spinach, are unisexual; male and female flowers are usually on separate plants but occasional monoecious plants do occur. The female flower has a two to four lobed perianth, four to five short styles and an ovary with a single ovule. The male flower has four to five perianth segments and four to five stamens which produce abundant pollen. Because the flowers of a single plant open over several days and a stamen does not shed all its pollen on one day, the pollen is available over a long period. Cross-pollination, probably by wind, can occur over considerable distances and it is recommended that varieties being grown for seed should be separated by 1·6 km or more (see Hawthorn and Pollard, 1954). Any insects that contribute to cross-pollination presumably collect nectar as pollen-gatherers would probably tend to confine their visits to male plants.

Tetragonia expansa Murr

T. expansa, New Zealand Spinach, has small greenish-yellow perfect flowers, with four perianth segments, numerous stamens and an ovary with several ovules. I can find no information about its pollination.

Chapter 30

Polygonaceae

Fagopyrum esculentum Moench

Most of the recent work on the pollination of *F. esculentum*, buckwheat, has been done in the U.S.S.R. The flowers, which occur in racemes, each have five petaloid perianth segments, and eight stamens which occur in an inner whorl of three closely surrounding the style and an outer whorl of five. There are eight small nectaries which alternate with the filaments of the stamens. The inner whorl of stamens dehisces outwards, and the outer whorl dehisces inwards, so an insect probing for nectar becomes dusted with pollen on both sides of its body. The superior ovary is of three united carpels and is surmounted by three styles.

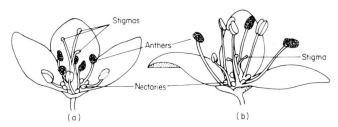

Fig. 138. Flower of *Fagopyrum esculentum*, buckwheat. a, long-styled form; b, short-styled form (after Knuth, 1909).

Flowers occur in two forms (Fig. 138); one has long stamens and short styles which are about level with the middle of the filaments of the stamens, and the other has short stamens but long styles which project 2–3 mm above the anthers; an insect visiting one type of flower accumulates pollen on the parts of its body which is about level with the stigmas of the other type. Pollen grains of the short-styled flowers are larger than those of the long-styled flowers (for details see Davydova, 1954) and Rozov and Skrebtsova (1958) reported that the predominant pollens on the thoraces and abdomens of honeybees foraging on buckwheat were of the long and short pistil types respectively. They observed that a bee usually touched the stigma of a long pistil flower with its abdomen but the stigma of a short pistil flower with its head and thorax.

Very rarely, infertile flowers with styles and stamens of equal length are produced, but usually an individual plant has only one type of flower. Such heterostyly tends to encourage cross-pollination, and individual pollen loads of honeybees contain both types of pollen showing that they visit plants of both types during a single foraging trip (Davydova, 1954). Furthermore, it has been shown that so-called "legitimate pollination" (i.e. the transfer of pollen to a flower of the opposite type) results in fertilization within 48–60 h after pollination, but that when "illegitimate pollination" (i.e. the transfer of pollen to a flower of the same type) has occurred pollen tube growth appears to be inhibited and only a few flowers become fertilized, even after 96 h, although self-fertilization is sometimes possible (see Darwin, 1876; Garber and Quisenberry, 1927; Hector, 1938). It has also been reported that pollen from the inner whorl of stamens is more effective in achieving fertilization than pollen from the outer whorl (Pausheva, 1961).

Buckwheat flowers are visited by numerous insects but honeybees are usually abundant; Kopel'kievskii (1953) found that honeybees constituted 63–72% of the insects visiting buckwheat crops, and that they occurred most abundantly between 09.00 and 12.00 h daily. A honeybee visited an average of 14 flowers/min, and worked buckwheat for 4–5 h/day, during which time it made an average of about 5 trips.

The amount of nectar produced, and its concentration varies with the variety of buckwheat and the year concerned; the best of ten varieties that Demianowicz and Ruszkowska (1959) tested gave 292 kg/ha of nectar on one year but only 90 on another. When buckwheat was more extensively grown in the U.S.A. it was an important source of honey (e.g. Phillips and Demuth, 1922; Pellet, 1923). It is claimed that providing an adequate supply of moisture, lime, phosphorus and nitrogen in the soil increases the nectar sugar concentration by 20–50% and as a result, increases the seed crop (Kopel'kievskii, 1955, 1964; Skrebtsova, 1957b). Fertile soils with adequate moisture are said to help maintain nectar secretion at low temperatures.

Bees do not visit buckwheat during its first 2–3 days of flowering but they are most abundant soon afterwards when the peak period of flowering occurs, and the nectar concentration is greatest; thereafter, the nectar concentration gradually diminishes (Leshchev, 1952; Solov'ev, 1960). Butler (1945a) and Shaw (1953) found that buckwheat nectar had averages of 35 and 45% sugar concentration. Kopel'kievskii (1955, 1960) found that 70% of the total sugar nectar was secreted during the first half of flowering. Therefore, in order to obtain maximum honey crops it is necessary to take honeybee colonies to the buckwheat fields soon after flowering begins.

Solov'ev (1960) recommended sowing a small percentage of an early flowering variety, or sowing some of the field earlier than the rest, to attract bees to the crop initially. This may explain why Mel'nichenko (1962) found that crops produced by a sequence of sowings had a longer flowering period and were better pollinated than usual although the increased ratio of flowers to bees was also probably important.

Unfortunately, the amount of seed, if any, produced by auto-pollination does not seem to have been determined under field conditions, and cage experiments are needed to determine the yields when honeybee colonies are put in some cages but not others. Elagin (1953) found that, in the Ukraine, there was a correlation between honeybee abundance and seed production; thus on crops with 1 and 5 colonies/ha, 58 and 80% respectively of the flowers set seed. He also discovered that the seed yield of crops decreased from 850 to 575 kg/ha as their distance from honeybee colonies increased from a few to 1,500 m. Very similar results were obtained by Kopel'kievskii (1960) who reported seed yields of 740 kg/ha near an apiary but 420 kg/ha on a crop 2 km away; he also demonstrated that the number of bees working a crop decreased with increasing distance from their colonies. Materikina (1956) moved colonies from beside a buckwheat field to 1·5 km distance and found this caused the seed yield to decrease from 1,600 to 640 kg/ha; he suggested using 2 colonies/ha for pollination but, when no colonies are available, the customary practice (i.e. dragging a rope with sacks attached to it across the field) should be resorted to. Kashtovskii (1958) discovered that when plenty of bees were present such mechanical pollination did not increase the seed yield and Mel'nichenko (1962) stated that it actually reduced seed yields when plenty of insects were present. Mel'nichenko also reported that 4–5 colonies/ha are necessary to obtain the greatest yields.

Naghski (1951) pointed out that *Fagopyrum tataricum*, tartary buckwheat, and its tetraploid variety (*F. tetratataricum*), which had been planted extensively in the U.S.A. to obtain the drug, rutin, are self-fertile and unattractive to bees. He thought that the flowers produce deither no nectar or nectar containing little sugar. Durnig each of 4 years he grew *F. tataricum*, *F. tetratataricum*, *F. emarginatum*, and two varieties of *F. esculentum*, in adjacent plots but no bees visited the first two types, few visited *F. emarginatum* and many visited *F. esculentum*.

Rheum rhaponticum L.

The inflorescence of *R. rhaponticum*, rhubarb, contains many small flowers each with a six partite calyx, but no corolla, and outer and inner whorls of six and three stamens (Fig. 139). The three styles are not receptive until after the anthers have shed their pollen, so self-pollination within the same flower is impossible although it is thought that flowers may receive pollen from ones higher in an inflorescence. However, protandry, together with possible self-sterility, facilitates cross-pollination. Insects, especially Diptera, visit

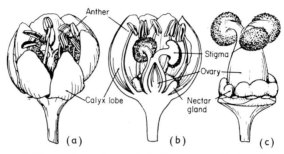

Fig. 139. Flower of *Rheum rhaponticum*, rhubarb: a, external view; b, median section; c, with calyx removed (after Robbins, 1931, with permission of McGraw-Hill Book Company).

the flowers but the relative amount of pollination done by insects, wind and gravity, and the relative amount of self- and cross-pollination does not seem to be known (see Darwin, 1878; Robbins, 1931; Hawthorn and Pollard, 1954).

Chapter 31

Lauraceae

Persea americana Mill

P. americana, the avocado, produces numerous small green flowers in dense panicles. There are no petals but the six sepals are in two consecutive whorls (Fig. 140). The nine stamens are in three whorls of three stamens each; the

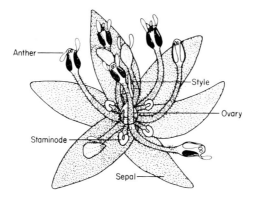

Fig. 140. Flower of *Persea americana*, avocado (after Purseglove, 1968).

inner three stamens are larger than the others and each has two nectaries at its base. Each stamen has four anthers which dehisce by means of small hinged flaps. An inner whorl of staminodes immediately surrounds the style. The style itself is short with a simple stigma; the ovary has a single chamber.

An individual avocado flower goes through two stages. When it opens in stage I the pistil is receptive, and pollination and fertilization can occur. The flower closes after stage I and opens again in stage II when the anthers dehisce. Avocado varieties fit into two general types according to the time of day their flowers are in the different stages. The flowers of type A varieties are in stage I in the morning of the first day and stage II in the afternoon of the following day, so that the flower's opening cycle lasts about 36 h. Type B varieties are in stage I in the afternoon of the first day and stage II in the morning of the following day so the flower's opening cycle lasts about 20 h only (see Bringhurst, 1951). In natural conditions, a group of trees would contain types A and B varieties so that, in the morning, type A variety trees are pollinated with pollen from type B variety trees, and in the afternoon type B variety trees receive pollen from type A varieties.

However, some self pollination is possible. Stout (1933) recorded the daily behaviour of flowers of twenty-five varieties, and found there was usually a short period when stages I and II overlapped so some of the flowers might have become self-pollinated then. But, he thought it probable that the pistils were not fully receptive throughout stage I and he noted that pollen not collected by insects during stage II dried into little clumps that fell to the ground. Traub et al. (1941) confirmed that some varieties showed a sufficient overlap in stages I and II to allow self-fertilization, and they also found that other varieties produced a small percentage of flowers which completed their male and female stages of development during a single opening. Even when self-pollination is possible insects are needed to transfer the pollen and when trees are caged singly, or when types A and B are caged together without insect pollination, few fruits are produced (Lesley and Bringhurst, 1951; Peterson, 1955). Furthermore, there is evidence that some varieties are self-incompatible as Stout (1933) and Lesley and Bringhurst (1951) found, respectively, that only 2 of 8 and 4 of 8 varieties set fruit when self-pollinated.

Unfavourable cool weather delays flower opening; when the opening of a type A flower is delayed, stage I occurs in the afternoon when insects are still active so some pollination can usually occur, but, when type B opening is delayed stage I opening may occur so late in the day that insects are no longer present, or it may be omitted altogether; in both circumstances fruit set is diminished. This is especially likely to occur in orchards on the coast where they are subject to lower temperatures than inland (Peterson, 1956).

During prolonged unfavourable weather, the development of flowers is so retarded that synchronous and alternating opening of the flowers is scarcely in evidence (Stout, 1933). Experiments in which branches are bagged to determine the amount of self-pollination that can occur, probably alter the flower opening behaviour so drastically that they are best ignored. Similar, but less pronounced effects are probably produced when an entire tree is caged.

When a tree of merit is propagated by grafting all the trees produced will have the same flower opening sequence so, in commercial plantations, it is necessary to choose varieties that are not only cross-fertile but also have reciprocal floral behaviour. Planting types A and B trees alternately has been suggested (Stout, 1933; Peterson, 1955) and Lecomte (1961) thought that a tree should not be more than 15 m from one of a complementary flowering group.

It seems to be generally agreed that the heavy sticky pollen is unlikely to be carried by wind and that bees are the principal pollinating agents (e.g. Stout, 1933; Cobley, 1956; Lecomte, 1961). The arrangement of varieties in an orchard and the number of colonies provided, should be influenced by the consideration that insects only pollinate when they go from trees whose flowers are shedding pollen to trees whose flowers are in the pistillate stage, i.e. from type B to type A in the morning and from type A to type B in the afternoon. Stout (1933) stated that some bees collected pollen only, while others collected nectar only, and that only nectar-gatherers go to flowers in the male and female stages, pollen-gatherers restricting their visits to the times when pollen is shed in flowers of the male stage. If these statements are confirmed, it would seem that nectar-gatherers are the more valuable pollinators, and that pollen-gatherers might even be detrimental to pollination as they remove pollen which nectar-gatherers might otherwise transfer. However, it would be surprising if some bees that were primarily nectar-gatherers did not also collect some pollen when it was available. Moreover, bees gathering pollen only probably visit a few flowers on female stage trees before getting discouraged and deserting them; this might not detract from their value as pollinators because probably only the first few female flowers they visit would become pollinated anyway, and thereafter most of the pollen would have disappeared from their bodies. In fact, the segregation of male and female stages might well ensure that bees visit more trees per trip than if each tree simultaneously contained flowers in both male and female stages.

It is also possible, of course, that some of the pollen collected by honeybees and other insects during stage II is retained on their bodies and pollinates the same variety when next its flowers are in stage I. If so, this could help explain why a reasonable set is sometimes obtained on trees caged singly with honeybee colonies (e.g. Peterson, 1955). Pollen is more likely to retain its viability on the insect's body from stage II to stage I of type B, which occur on the same day, than of type A which occur on separate days.

Because the flower behaviour can restrict opportunities for cross-pollination, and because other crops, particularly citrus, are sometimes preferred by bees, Stout (1933) suggested there should be more than $2\frac{1}{2}$ colonies/ha of avocado plantation, while Lecomte (1961) recommended 2 colonies/ha.

Cinnamomum zeylanicum Breyn

The small flowers of *C. zeylanicum*, cinnamon, consist of a bell-shaped calyx tube of six pointed sections, nine stamens, three staminodes and a single-celled superior ovary. It is reported to be pollinated by insects, especially flies (Purseglove, 1968).

Chapter 32

Euphorbiaceae

Aleurites

Commercial species of tung trees include *Aleurites montana* (Lour) Wils, which occurs in the tropics, and *A. fordii* Hemsl which is native to China but is also grown in south-eastern U.S.A. Male and female flowers occur on the same tree; racemes are either composed of male flowers only, female flowers only or both; when both occur in the same raceme the upper flowers are female and the lower ones are male. Flowers are bell-shaped and have a two- or three-lobed calyx and five petals. The male flower, which is smaller than the female, has eight to twenty stamens in three whorls. The female flower has a three- to four-celled ovary with one ovule in each; it is surmounted by a divided style. Flowering of *A. fordii* extends for a month or more. There is no evidence that self-incompatibility occurs.

Brown and Fisher (1941) and Angelo *et al.* (1942) have studied the pollination of *A. fordii* in U.S.A. Brown and Fisher bagged 90 flowers and self-pollinated 10 of them by hand on each day from 0–9 days after opening; the numbers that matured fruit were 10, 9, 8, 9, 8, 7, 8, 8, 6 and 4 respectively. Fruit from flowers pollinated within 5 days of opening contained an average of 3·4 seeds, but the average numbers produced from flowers pollinated when 6, 7, 8 and 9 days old were 2·7, 3·0, 1·4 and 1·0 respectively. Angelo *et al.* (1942) found that the percentage sets using pollen collected 1, 2, 3 and 4 days after anthesis were 100, 93, 38 and 54% respectively. Hence, pollination between male and female flowers on the day they open is likely to be the most successful. Only 4% of bagged pistillate flowers set fruit compared to about 80% of flowers left exposed. The latter contained an average of about 3·5 seeds per fruit.

Angelo *et al.* found that microscope slides, covered with petroleum jelly, put 4·6–12·2 m from the nearest tung tree often collected groups of tung pollen, but only two pollen grains were caught on four slides at 18·3–30·5 m from the trees. As a result, they concluded that wind pollination alone was unlikely to be responsible for the high proportion of flowers that are usually pollinated in a tung orchard. They also enclosed two trees in muslin cages and put a honeybee colony with one of them; this tree had a good set, but the only flowers that set on the tree caged without bees were those that were hand-pollinated, even though the experimenters shook all the branches to distribute as much pollen as possible. The insects usually responsible for pollination are not known. Possibly small insects, such as aphids and thrips, play a part as 52% of pistillate flowers set fruit when enclosed in wire screens with 6 meshes/cm, although wind pollination may also have been responsible for the set that occurred in these circumstances. More experiments to determine the relative importance of wind and insects in pollination, and the insects concerned, are needed.

The amount of self- and cross-pollination that usually occurs in a tung orchard is not known with certainty, although Potter (1959) suggested that tung is probably largely self-pollinated, and that only 5–10% cross-pollination occurs.

Hevea brasiliensis Muell-Arg

A *H. brasiliensis*, para rubber, tree produces numerous flowers in inflorescences; the female flowers are situated at the end of the conical axis and the ends of main side branches, while the more numerous male flowers occur

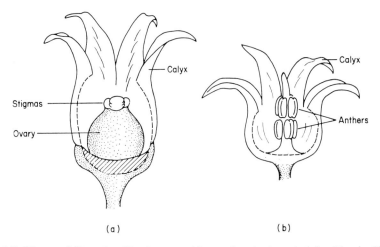

Fig. 141. Flower of *Hevea brasiliensis*, para rubber: a, female; b, male (after Morris, 1929).

on the lesser shoots. The flowers each have a bell-shaped calyx of five yellow lobes but no corolla. The male flower, which is 5 mm long, has two rings of five anthers each, and the female flower, which is 8 mm long, has three stigmas and a three-celled ovary (Fig. 141). Some of the male flowers of a tree open first but drop after about a day; the female flowers then open and remain open for 3–5 days; finally the remainder of the male flowers open; the whole flowering period of a tree lasts about 2 weeks.

Normally, only a very small percentage of female flowers set fruit, and about half of these fall after a few weeks. Early work (reviewed by Morris, 1929) showed that 0·5–1·7% of flowers set when self-pollinated compared to 5·6–10·9% when cross-pollinated. Rao (1961) reported that, in 3 years, averages of 0·3, 1·6 and 0·5% of flowers respectively set fruit in a plantation in Malaya, whereas hand pollination gave 8·4 and 46% mature fruit during the main and late flowering periods respectively. Morris (1929) found that no flowers of thirty bagged inflorescences set fruit, and Muzik (1948) confirmed that auto-pollination of bagged inflorescences does not occur and no flowers set fruit even when the clones concerned are not entirely self-sterile. These, and similar findings, indicate that para rubber needs pollination and that cross-pollination is beneficial for some strains at least.

Para rubber pollen is sticky and not transported by wind. Rao (1961) failed to collect any pollen with a Hirst spore trap that was located for 3 days within 15·2 m of a heavily flowering tree. Morris (1929) suggested that insect pollination occurs, but although he saw bees, flies, moths and beetles visit male flowers for pollen, he saw only two small beetles and one small fly enter female flowers and these did not touch the stigmas. Muzik (1948) also thought that insect pollination must occur but failed to see any insects visiting the flowers during daytime and investigations to find whether pollination occurred at night revealed only a few ants and these carried no pollen.

The mystery of the insects responsible for pollinating para rubber in Puerto Rico was solved by Warmke (1951). He observed that small brown insect hairs were present on the stigmatic surfaces of most open flowers. Thus, on one occasion he examined 128 stigmas and found 101 had hairs and 98 had pollen grains adhering to them. On 88 of the 98 stigmas with pollen grains, hairs were also present. In an attempt to find the insect to which the hairs belonged cardboard cylinders, spread with adhesive compound, were wrapped round the branches just below the inflorescences. These cards captured many insects but of these, the midges of the genera *Dasyhelea*, *Atrichopogan* and *Forcipomyia* had body hairs that were the most similar to those on the stigmas. The adhesive compound also trapped these midges when it was applied to the entrances of female flowers. When midges were brushed against the stigmas experimentally they were found to shed some of their hairs; indeed, it was found later that sometimes the midges themselves stuck to the stigmas. Warmke (1952) obtained evidence that midges were also important pollinators of para rubber in Brazil but, whereas *Dasyhelea* spp. were the most abundant midges in Puerto Rico, *Atrichopogan* spp. were the most abundant in Brazil, although midges of the genera *Dasyhelea*, *Stilobezzia* and

Culicoides were also present. Sometimes the midges were very numerous and were especially active for about an hour after sunrise and at sunset, but few were seen during the rest of the day, or during wind or rain. Their main interest seemed to be the long hairs present inside the calyx and when they had alighted near the mouth of a flower they walked into it with their backs toward the anthers or stigmas. All individuals carried numerous pollen grains, particularly on their antennae, thoraces and wings. When they entered a female flower, through the narrow opening between the stigma lobes and the calyx, they deposited pollen on the former. Thrips also occurred in the flowers; there were few of them in Brazil and many in Puerto Rico; however, they carried little pollen on their bodies.

In Malaya, Rao (1961) collected insects of thirty-six species belonging to nineteen families of the orders Diptera, Hymenoptera and Thysanoptera from para rubber flowers. However, only midges, thrips, scatopsid flies, and various parasitic Hymenoptera which visited both male and female flowers were possible pollinators; the other insects merely rested on the unopened flowers. He believed that, although thrips were abundant, they were incapable of sufficient active flight to be valuable as cross-pollinators, and that the parasitic Hymenoptera were insufficiently hairy to carry much pollen. However, he supposed that the minute size, long hairs and sustained flight of midges made them especially suitable as pollinators and concluded that they were probably the most important pollinators of para rubber in Malaya as well as in Puerto Rico and Brazil.

Ricinus communis L.

The flower of *R. communis*, castor, has three to five united sepals and no petals. The male flower has numerous branched stamens, and the female flower has a superior three-celled ovary terminated by three styles. Both sexes occur in the same inflorescence. The pollination requirements and the pollinating agents are unknown. Purseglove (1968) thought that although castor is mostly wind pollinated, insects may play some part. However, Alex (1957a) reported that, although honeybees visit staminate flowers of castor and obtain large amounts of pollen, tests showed they are of no value as pollinators.

Chapter 33

Moraceae

Ficus carica L.

Ficus carica, the wild fig, or caprifig, and the small wasps, *Blastophaga psenes*, (Hymenoptera, Agaonidae) provide an example of complete adaptation between a flower and its pollinator. (See Condit, 1920, 1947; Condit and Enderud, 1956.)

The unisexual flowers of the caprifig occur on the inner lining of the hollow cup-shaped receptacle of the fruit (Fig. 142). A narrow tunnel, known

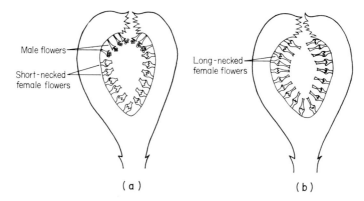

Fig. 142. a, Diagramatic median section of *Ficus carica*, caprifig, showing the male flowers near the ostiole and below them the short-necked female flowers, where the galls are formed; b, diagramatic median section of edible fig which contains only long-necked flowers (after Knoll, 1926).

as the ostiole, is located at the apex of the fruit and leads into the interior of the receptacle; it is lined with scales which spread apart as the fruit matures and leave the tunnel open.

The male flower (Fig. 143) consists of a four-lobed perianth, and one to three united filaments bearing a total of two to six anthers; the number of stamens and the amount of pollen produced is different in different varieties.

When male flowers are present in a fruit, they are usually located near the ostiole, although in some varieties they are found scattered indiscriminately among the pistillate flowers.

Each pistillate flower (Fig. 143) consists of a single ovary, a short style and a stigma. *Blastophaga psenes* develops in the ovary of the pistillate flowers which, as a result, are sometimes known as "gall" flowers. However, this is a misnomer and their short style and shape is characteristic of the caprifig

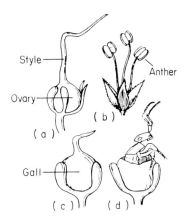

Fig. 143. Flowers of *Ficus carica*, fig. a, female flower; b, male flower; c, female flower with "gall"; d, *Blastophaga psenes*, fig wasp, escaping from gall (after Robbins, 1931, with permission of McGraw-Hill Book Company).

pistillate flower whether it is pollinated or not, and contrasts with the long styled pistillate flowers of the "Smyrna" and "Common" figs. Rudimentary pistils rarely occur in staminate flowers, and rudimentary stamens rarely occur in pistillate flowers. Self-pollination is impossible because the pistillate flowers are receptive several weeks before the staminate flowers in the same receptacle present their pollen; this behaviour is effectively co-ordinated with the development of *B. psenes*.

The caprifig gives three crops of fruit each year. The figs of the spring crop (Profichi) appear in March and bear both pistillate and staminate flowers, about seven or eight pistillate to one staminate. The figs of the summer crop (Mammoni) and winter crop (Mamme) have few, if any, staminate flowers, but numerous pistillate ones. *B. psenes* overwinters in the larval stage in the pistillate flowers of the Mamme fruits. In early spring the larvae pupate and the adults emerge at the beginning of April. The males emerge first by biting through the ovary wall and immediately seek and mate with females while they are still inside the flowers' ovaries. The males are wingless, nearly blind and have attenuated abdomens (Fig. 144); they die soon after mating without ever leaving the fig. About ten times as many females as males are produced so each male probably mates several times. After being impregnated the winged females emerge from the ovaries and leave the figs through the ostioles. When the temperature is relatively great, the first females leave the figs at about 07.00 h and all may have left by 12.00 h but, under relatively cool conditions, they may not begin leaving until 09.00 h and continue to do so until 14.00 h. After leaving, the females usually crawl over the leaves and fruits and may make short flights from branch to branch although in windy conditions they may be carried for several kilometres.

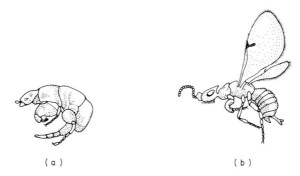

Fig. 144. *Blastophaga psenes*, figwasp. a, wingless male; b, winged female (after Wigglesworth 1964).

The overwintered females emerge at a time when the Profichi crop is about a quarter grown. In attempting to force an entry into these small figs, the females' wings are usually broken off; once inside, they begin to deposit their eggs down the short styles and into the ovaries of the pistillate flowers. Oviposition, which lasts about 55 sec, injures the cells in the stylar canal and these produce a toxin which inhibits pollen tube growth. A single female may enter more than one receptacle and continue laying until her 300–400 eggs are depleted. The adult wasp does not eat and so she is unable to replenish her supply of eggs; females may remain alive overnight in the shelter of a receptacle but when exposed in the orchard they live only a few hours.

The adults emerge about two months after the eggs have been laid; this coincides with anther dehiscence and, when the females leave the fig, they become covered with pollen from the staminate flowers near the ostiole. Although the wasps attempt to clean the pollen from their bodies, they carry ample with them to figs of the Mammoni crop which have just begun to develop. Usually many wasps enter each Mammoni fig and as a result most of the pistillate flowers are probably pollinated, but seed is prevented from developing in those in which eggs are laid and fertile seeds develop from only those flowers that are not inhabited by the wasp larvae. Even so, more than 1,000 fertile seeds per fig are commonly found and a single female wasp pollinates an average of 850 flowers. Fertilization stimulates development of the surrounding endosperm, but this also occurs parthenogenetically in flowers whose ovaries contain wasp larvae.

The adult wasps that emerge from Mammoni figs enter other small figs and oviposit but, because these figs contain few staminate flowers, they are unlikely to effect pollination. Some varieties of caprifigs bear fruits in all stages of development from August to early winter, and the winter or Mamme figs can be regarded as Mammoni figs that remain on the tree throughout the winter. All fruits that continue development during the winter normally contain wasp larvae. Thus, the life cycles of *Ficus carica* and *Blastophaga psenes* are closely interwoven.

The caprifig is the wild or native fig of south-western Asia and south-eastern Europe and, although the fruits of most varieties of it are inedible, both because of the numerous insects inside them and because of the numerous male flowers near the ostioles, the three main edible classes of fig (i.e. Symrna, Common and San Pedro figs) have been derived from it by selection and vegetative propagation. Fully developed staminate flowers are found only in caprifigs and in intermediate types, and in order to avoid crop failure of Smyrna figs, it has been the practice of native growers for many centuries to suspend caprifigs in the Smyrna trees each spring; this procedure is known as caprification and if it is not done the figs fail to set and are shed. In California, orchards of Smyrna fig were planted from imported cuttings in the last 20 years of the nineteenth century but crops were obtained only where caprifigs containing *B. psenes* were also present.

Smyrna figs require the stimuli of pollination and fertilization to set fruit, and this cannot be accomplished unless the fig wasp transfers pollen to them from the caprifigs. When the pollen covered wasps emerge from the caprifigs they enter Smyrna figs in search of a place to lay their eggs but, because the styles of the Smyrna fig are much larger than those of the caprifig, they are unable to oviposit successfully and, after wandering among the flowers, become exhausted and die. However, while inside the Smyrna figs they pollinate the flowers.

Experience has shown that Smyrna fig growers should make arrangements for caprification from the time the Smyrna figs are 5·5 mm in diameter, and should continue to do this at intervals of 4 days, for about 3 weeks. The caprifigs should be used when their anthers are dehiscing and the male wasps are emerging; at this stage their colour changes from green to greenish-yellow and the ostioles have started to open. It is common practice to pick the caprifigs after the wasps have finished emerging for the day and to store them in a cool place overnight. Early next day they are strung on raffia or string, or put in wire or wooden baskets, and suspended in the Smyrna trees. Some varieties of caprifig produce more pollen than others and so fewer fruits are required for caprification but, in general, the number used depends on the size of the trees (Fig. 145). Increased caprification can result in a bigger crop, but it also tends to increase disease and insect infestation of the trees.

Fig. 145. Illustration of number of caprifigs and containers necessary for caprification of "Lob Injir" trees of different diameters on the basis of four distributions of caprifigs per season. Thus a 3 m tree would have one basket with two figs in it on four occasions (after Condit, 1920).

Most growers plant three to five caprifig trees to every hundred Smyrna trees; they are commonly planted to windward of the Smyrna trees so that fig wasps from caprifig fruits that have not been picked stand some chance of being blown on to the Smyrna trees. Attempts to dispense with hand distribution of caprifigs by interplanting caprifig among Smyrna trees have not been entirely successful, irregular results being obtained, some Smyrna trees being over-caprified and others under-caprified. Furthermore, these two classes of fig have different water requirements. Similarly, it has been found impossible to regulate the rate of caprification when caprifig branches have been grafted onto Smyrna trees. Plant breeders find hand pollination is easy to accomplish either by introducing strange pollen on a needle or on a fine glass rod, or by blowing it through a fine glass tube into the receptacle. Enclosing a caprifig in a bag with the branch to be pollinated is also effective, but likely to cause an excessive set.

The Common fig, which is the class to which most varieties grown in Europe and America belong, also has pistillate flowers with long styles, but these do not require pollination to set fruit. However, most varieties of Common fig give larger fruits when caprified, and it is generally thought that the presence of fertile seeds improves the flavour. Unfortunately, it also results in more spoilt fruit.

The San Pedro class of fig sets fruit without caprification when producing the first crop of the year, but caprification is necessary for flowers of the second crop to set fruit.

Therefore, to avoid being shed, and to complete their development, figs of the caprifig class need to contain *Blastophaga psenes*, and figs of the Smyrna class, and the second crop of the San Pedro class, need to be pollinated by *B. psenes*. Figs of the Common fig class and the first crop of the San Pedro class, undergo parthenocarpic development apparently without extrinsic stimulation.

Chapter 34

Rosaceae: *Prunus* and *Pyrus*

Flowers of the genus *Prunus* have five petals, numerous stamens, a single style and an ovary with a single carpel which contains two ovules (Fig. 146).

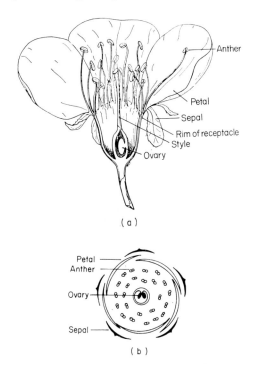

Fig. 146. Flower of *Prunus cerasus*, sour cherry: a, median section; b, floral diagram (after Robbins, 1931, with permission of McGraw-Hill Book Company).

Those of the genus *Pyrus* also have five petals and numerous stamens, but the ovary consists of five carpels each containing two ovules (Fig. 147); the five styles are either separate or united at their bases.

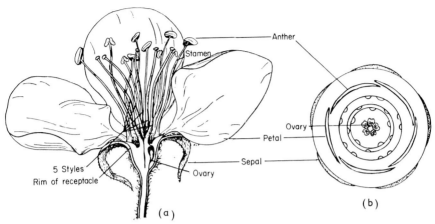

Fig. 147. Flower of *Pyrus malus*, apple: a, median section; b, floral diagram (after Robbins, 1931, with permission of McGraw-Hill Book Company).

Pollination requirements

Most species and varieties need pollination. Flowers that are not fertilized soon drop, or the newly-formed fruits soon do so. When flowers of some varieties are inadequately fertilized they develop into misshapen fruits (Fig. 148). Immature fruits with relatively few seeds are more inclined to be shed later than ones with many seeds (e.g. Murneek, 1937; Roberts, 1945; Horticultural Education Association, 1961).

"Self-fruitful" varieties or clones, produce fruit when pollinated with their own pollen; "self-unfruitful" varieties only produce fruit when pollinated with pollen from another variety. Many varieties are only partially self-fruitful, and produce more and better quality fruit when cross-pollinated than when self-pollinated (e.g. Hutson, 1926; Gould, 1939; Gardner *et al.*, 1952). Certain varieties vary in self-fruitfulness in different years and in different locations. Most varieties of apple (*Pyrus malus* L.), pear (*Pyrus communis* L.), plum (e.g. *Prunus domestica* L., *P. insititia* L., *P. salicina* L., *P. americana* L.), sweet cherry (*Prunus avium* L.) and almond (*Prunus amygdalus* Bartsch L.) are self-unfruitful. Peach and nectarine (*Prunus persica* L.), apricot (*Prunus armeniaca* L.), and sour cherry (*Prunus cerasus* L.) are largely self-fruitful. Gardner *et al.* (1952) recommended that provision should be made for cross-pollination when planting an orchard, unless it is definitely known to be unnecessary for the variety and local conditions concerned.

When fruit is produced from the cross-pollination of one self-unfruitful or partially self-fruitful variety by another, the two varieties are called "cross-fruitful". Numerous experiments have been made to determine the inter-fruitfulness of different varieties and the results are summarized in general

382 INSECT POLLINATION OF CROPS

textbooks and advisory bulletins (e.g. Snyder, 1946; Griggs, 1953; Ministry of Agriculture, Fisheries and Food, 1958a and 1961; Shoemaker and Teskey, 1959; Horticultural Education Association, 1961; Harris, 1962). A "pollinizer" variety chosen to pollinate a main variety must have a flowering period

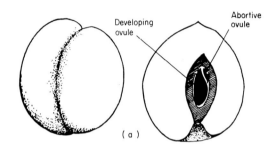

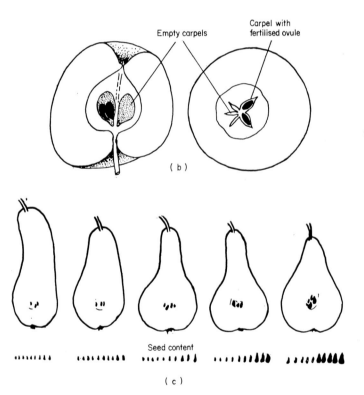

Fig. 148. Effect of inadequate fertilization on fruit shape: a, *Prunus persica*, peach (after Tukey, 1936); b, *Pyrus malus*, apple (after Horticultural Education Association, 1961); c, *Pyrus communis*, pear (after Schander, 1955).

coinciding or substantially overlapping with that of the main variety. It is also usually recommended that one is selected which produces ample, viable pollen, flowers annually and bears fruit of commercial value.

It seems that the stigmas of all species and varieties of *Prunus* and *Pyrus* are receptive as soon as the flowers open. Snyder (1942) found that when temperatures were high the anthers began to dehisce before the flowers opened, but in cool conditions the flowers were sometimes well open before dehiscence occurred. Seaton and Kremer (1939) determined that in fairly still air, the air temperature increased approximately 0·6°C (1°F) with every 3 m above ground level, and the flowers at the top of a tree presented pollen and were visited by insects before those on the lower branches. The optimum temperture for pollen tube growth and for fertilization occurs between 18 and 27°C (64 and 81°F) and growth is retarded or nearly stopped at temperatures below 16°C (61°F). Dry, desiccating winds are thought to be associated with poor fruit set in California; but rain, unless continuous during flowering, has no direct adverse effect (Griggs, 1958).

Early pollination, soon after the flower opens, seems desirable for the species and varieties that have been studied. Hartman and Howlett (1954) discovered that the ovules of the apple "Delicious" are often late in developing and showed signs of early degeneration, so that adequate and early pollination is vital to secure a good set; delays of 0, 24, 48 and 72 h between anthesis and pollination produced sets of 61, 27, 11 and 3% respectively. Indications were obtained that degeneration of the ovules tended to occur later in the terminal flowers than in the lateral flowers of clusters; this might be one reason why the set of terminal flowers is often superior. Eaton (1959, 1962) reported rapid degeneration of the ovules following anthesis in some sweet cherry varieties. At high temperatures degeneration was more rapid. The longer cross-pollination was delayed the smaller the proportion of embryo sacs that could be fertilized. In the variety "Schmidt", which is often unproductive, degeneration of all but a few ovules was well advanced at anthesis. Griggs and Iwakiri (1964) found that almond flowers were most receptive to pollination on the day immediately after they opened; set was significantly reduced when pollination was delayed until the third day after opening, and practically no set occurred when pollination did not occur until five or more days after opening. Williams (1966) reported that the effective pollination period may vary with different varieties of apple and pear, and with different years.

Murneek (1937) pointed out that application of nitrogenous fertilizers often increased set, but he did not know why this effect was obtained. Recent work (Hill-Cottingham and Williams, 1967) indicates that application of additional nitrogen in the later summer or autumn can delay degeneration of the ovules and so give more time for pollen tube growth to be completed before this happens.

These studies on the rapidity of embryo sac degeneration and the effect of low temperatures on the growth rate of pollen tubes emphasize the desirability of attempting to obtain early pollination. Because unfertilized flowers may be visited by insects long after their period of receptivity is past, the percentage of useful visits bees make to flowers is less than is at first apparent.

Wind pollination

It has long been supposed that wind sometimes effects self-pollination of *Prunus* and *Pyrus*, or even cross-pollination when the branches of adjacent trees are closely interwoven (Sprenger, 1916; MacDaniels and Heinicke, 1929; Auchter and Knapp, 1937). Waugh and Blackhouse (cited by Fletcher, 1916) reported that practically no plum or pear pollen, which is characteristically moist and sticky, was distributed by wind, but that apple pollen, which is drier, was distributed by wind to some extent. Several workers have suspended glass slides smeared with glycerine at various distances from fruit trees and have measured the amount of fruit pollen deposited on them. Lewis and Vincent (1909) working on apple, Chittenden (1914) on apple and pear, Howlett (1927) on apple, and Kobel (1942) on cherry found that slides near trees collected little or no pollen. Hooper (1931) stated that apple, plum and cherry pollens were not even carried by wind the short distance between one tree and another. In contrast, Hockey and Harrison (1930) and Brittain (1933) found that apple pollen was deposited on slides placed up to 61 m from apple trees, Stephen (1958) found pear pollen on slides up to 1·2 km from pear trees and Smith and Williams (1967) trapped pollen on rotorod samples 48 m from apple trees; in all observations the amount decreased with increasing distance from the trees. Stephen, however, supposed that pollen must be forcibly brushed onto a stigma before it adheres sufficiently to germinate, and that wind-borne pollen is therefore ineffective. The other workers mentioned above concluded that wind pollination of fruit is not important.

The methods most of the above workers used to collect air-borne pollen were not very efficient, whereas Burchill (1963) using a Hirst volumetric spore trap found much apple pollen in the air of orchards containing "Cox's Orange Pippin" and "Bramley's Seedling". His findings encouraged Free (1964b) to reinvestigate the relative importance of wind and insect pollination. Small apple trees of varieties which need cross-pollination were enclosed in nylon cages either: (a) with bouquets of a pollinizer variety suspended in their branches and a small honeybee colony; (b) with bouquets only; (c) with bees only; (d) with neither. The four trees with bouquets and bees had 38% of their flowers set fruit, but ten of the other trees set no fruit and the very little set on the other two trees soon dropped off. None of the trees caged with bouquets but no bees set fruit although the cage walls only slowed down the wind speed by about 30%. More recently, Smith and Williams (1967) found that trees enclosed in cages with a small mesh that prevented insects from entering had no flowers set fruit although some airflow through the cage walls still occurred. In contrast, Wertheim (1968) reported substantial fruitset on branches or trees caged to exclude insects although less than on uncaged controls. Future experiments of this type should establish whether airborne pollen does in fact enter the cages from outside.

Additional evidence that wind pollination is not important has been provided by experiments in which petals and stamens have been removed from apple flowers leaving the stigmas intact. Sax (1922) found that of several hundred such flowers only one set fruit, and Free (1964b) found that only 0·3% of about 2,000 emasculated flowers set fruit compared to 7·3% of the control flowers which were not emasculated. Because emasculation might be harmful to set (see Howlett, 1926), Free also hand-pollinated the emasculated flowers on two trees and obtained a set of 13·6%. The well-exposed stigmas of the emasculated flowers were at least as likely to receive airborne pollen as the stigmas of flowers not emasculated, but because they lacked their petals they were probably less attractive and visited less by insects than ordinary flowers (see later). Hence these experiments confirmed that apple flowers are normally insect pollinated, but produced no evidence of wind pollination. Attempts were also made to "wind" pollinate pear flowers by blowing air from a spraying machine through a row of "Conference" (pollinizer) trees onto an adjoining block of "Comice" but this failed to increase the set compared to that on control trees.

Insect pollination

The major role of insects, especially honeybees, in pollinating *Prunus* and *Pyrus* has long been recognized (Waite, 1895, 1898). Early work demonstrated apparent correlations between the number of honeybees present and the fruit crop subsequently produced (e.g. Sprenger, 1916; Gates, 1917; Ewert, 1921). More recently, Mommers (1952) and Weiss (1957) found positive correlations over several years, between the number of colonies moved annually into a cherry-growing area and the fruit crop obtained. Stephen (1958) obtained a positive correlation between the numbers of bees and amount of fruit set in five pear orchards, and Free (1962c) found that plum trees near to honeybee colonies were visited by more bees and had a greater set than trees further away.

Much circumstantial evidence has also accumulated showing that increased crops are obtained by moving honeybee colonies to orchards previously having relatively few pollinating insects (Auchter, 1924; Hootman and Cale, 1930; Loewel, 1943; Menke, 1950; Clark, 1959).

Information on the pollinating ability of honeybees has also been obtained by numerous experiments in which individual trees of self-unfruitful varieties with appropriate pollinizers were enclosed in insect proof cages, honeybee colonies being put in some cages but not others. In many experiments, flowers on one branch of each tree were hand pollinated. In general, the percentage of fruit set on a tree enclosed with a honeybee colony was much greater than that on a tree from which pollinating insects were excluded, and of about the same order as that obtained by hand pollination (e.g. Hendrickson, 1916; Auchter, 1924; Hutson, 1926; MacDaniels and Heinicke, 1929; Howlett, 1927; Hootman and Cale, 1930; Brittain, 1933; Tsygankov, 1953).

Hootman (1930) stated that yields of self-fertile sour cherry orchards may be increased by taking honeybee colonies to them, and Hooper (1939) reported that certain varieties of pear, morello cherry and plum can mature fruits without insect visits because their stamens touch their stigmas, but there is little information on the extent to which self-fertile or partially self-fertile varieties depend upon insect visitors to transfer the pollen, and how much this occurs without outside agents. Experiments in which branches or clusters of flowers are enclosed in cloth or muslin bags (e.g. Sharma, 1961) do not give the answer because the bags themselves could help transfer the pollen. Bulatovic and Konstantinovic (1960) used bags of parchment, which are less likely to transfer pollen than muslin bags, to enclose flowers of self-fertile peach varieties. The average percentages of fruit set of bagged and control flowers during 2 years are given in Table 32.

Table 32

	"Alexander"	"May-flower"	"Morteltini"	"Red bird"	"Vadel"
Flowers not bagged	36·8	29·2	41·1	28·3	38·8
Bagged flowers	34·4	17·2	28·1	6·8	11·5

The amount of auto-pollination that occurred appears to have differed with the variety, although differences in self-fertility may also have been responsible. Additional experiments in which some of the bagged flowers are self-pollinated by hand are needed. However, ideally entire trees of self-fertile varieties should be caged and honeybee colonies put in some cages but not others. The results might well depend on the relative positions of the anthers and stigmas of the varieties concerned. A preliminary experiment of this kind was made by Laere (1957) on two self-fertile sweet cherry trees; the one caged with bees had about 67% more flowers set fruit than the one caged without bees.

Lötter (1960) pointed out that the set of the self-fertile apple "Rare Beauty" and pear "Bon Chretien", was very dependent on weather conditions during flowering, and that in California there was a positive correlation between the number of hours during flowering when temperatures exceeded 16°C (60°F) and the size of crop produced. This correlation could be attributed to temperature influencing self-compatibility directly, or influencing pollination by regulating the number of insect pollinators present on the trees. "James Grieve" apple sets well when self-pollinated (D. W. Way, personal communication) and Free (1966a) found that 61% of 161 flowers that received single visits from honeybees set fruit compared to 10·3% of 155 flowers that were bagged and not visited by bees; presumably some of the set of the latter flowers resulted from the bags themselves transferring the pollen. Hence, taken as a whole, the small amount of evidence available indicates that many self-compatible varieties benefit from insect pollination, but more experiments are needed.

Most surveys have shown that honeybees form a high percentage of the insects visiting *Prunus* and *Pyrus* flowers (e.g. Menke, 1952c, 75%; Smith, 1952, 60%; Tsygankov, 1953, 85%; Roberts, 1956, 99%; Wafa and Ibrahim, 1957b, 88–90%; Dyce, 1958, 95%; Free, 1966a, 87%). But many other insects also visit the flowers. Notable amongst these are bumble bees (Bombus spp.) and solitary bees of the families Andrenidae and Megachilidae (e.g. Hooper, 1931; Wilson, 1929; Brittain, 1933; Menke, 1951). Various Diptera (e.g. Syrphidae, Calliphorinae, Bibionidae, and Muscidae) have also been found in abundance on *Prunus* and *Pyrus* trees (Hutson, 1925; Atwood, 1933; Brittain, 1933; Vansell, 1942; Webster *et al.*, 1949; Brown, 1951; Bohart, 1952; Lewis and Smith, 1969). Hutson (1925) captured Chironomidae and Muscidae in considerable numbers from the trees, but saw none on the flowers themselves. Some observers (e.g. Brittain, 1933; Vansell, 1942; Palmer-Jones and Clinch, 1968) believed that neither the structure nor the behaviour of Syrphidae or Calliphoridae favoured pollination. However, Atwood (1933) pointed out that although many flies visiting apple flowers alight on the petals, merely eat a little pollen and do not touch the stigmas, others, especially the Syrphidae and Bombyliidae, walk over the anthers and stigmas and transport pollen from flower to flower. Bohart (1952) supposed flies were important in some localities, notably for pollinating pears, and Brown (1951) recorded that Calliphoridae appeared to be very effective pollinators of plum. In Japan, Noro and Yago (1934) found that Muscidae visited the flowers mainly between 08.00 and 09.00 h and between 14.00 and

15.00 h whereas Anthomyyidae and bees were found mainly between 09.00 and 14.00 h. Massee (1936) supposed that various small beetles might contribute to pollination. It is difficult to get precise information on the relative abundance of different insect species in *Prunus* and *Pyrus* orchards as the characters that separate closely related species cannot be observed without first capturing them, and some are much easier to catch than others. Furthermore, the relative abundance of the various species must be linked with their tendency to transfer pollen before their presence can be evaluated.

Wild pollinating insects are often scarce when *Prunus* and *Pyrus* trees are in flower, fewer being avilable for early flowering species (e.g. cherries, peaches, plums) than for later flowering ones (e.g. apples, pears). However, Wilson (1929) showed that, in the absence of honeybees, there were enough wild pollinators to give an excellent fruit set in an orchard at Wisley (Surrey), and Howlett (1934) noted that in certain parts of Ohio there were sufficient wild bees to pollinate apples adequately. Brittain (1933) concluded that solitary bees were the most important pollinators of apple in the Annapolis Valley, Nova Scotia, because in 4 years out of 5 the wild bee population was adequate to produce a satisfactory fruit set in most orchards. However, the numbers of wild bees and flies may fluctuate greatly from year to year (Atwood, 1933; Dunham, 1939d; Menke, 1951; Mommers, 1952; Karmo, 1958).

It is commonly believed that because of the intensive cultivation of land, and of other related factors, the numbers of bumblebees, solitary bees and other wild pollinators, have generally declined in recent years (see pages 2 and 3). This supposition is supported by reports that they are more numerous in orchards near uncultivated land than near cultivated land (Murneek, 1937; Vansell, 1942; Menke, 1952c; Webster et al., 1949), and that small orchards near woods or swamps, where insects overwinter in large numbers, usually have a satisfactory fruit set in the absence of honeybees (Hootman and Cale, 1930). Hutson (1926) counted only four bumblebees in an apple orchard surrounded by cultivated land, but thirty-one in an apple orchard in an area otherwise uncultivated. Luce and Morris (1928) stated that orchards nearest the foothills or open country had the heaviest set of fruit, and Karmo (1958) found that apple orchards on mountain slopes, where there were ample nesting sites for wild bees, had over twice as many bumblebees and solitary bees as orchards on the valley floor where nesting sites were scarce. However, Menke (1951) found that even orchards adjacent to waste-land had few wild bees, and when wild bees did occur they tended to confine their activities to the peripheries of the orchards.

Wild bees, especially bumblebees, are probably more valuable than their numbers imply. Compared with honeybees, bumblebees work faster, in more inclement weather and probably for longer hours (Brian, 1954; Free and

Butler, 1959). In parts of eastern Canada and New England the weather during the period of apple flowering is often unfavourable for honeybee flight, and bumblebees and a few other wild insects, which are active at lower temperatures, are probably more satisfactory pollinators (Bohart, 1952). However, visits to flowers at temperatures lower than the minimum for anthesis may be of little value. Because of their larger size, bumblebees are more efficient pollinators of certain fruit varieties than honeybees (see below); solitary bees have the advantage of carrying pollen on larger areas of their bodies and carrying it in a drier condition, so it is more likely to be rubbed off onto the stigmas of the flowers they visit. From time to time it has been suggested that the breeding of various Diptera for *Prunus* and *Pyrus* pollination should be enouraged, but this has not yet been shown to be practical. The only known satisfactory method of increasing the number of pollinating insects in an orchard is by introducing honeybee colonies.

Pollen

Prunus and *Pyrus* trees often provide much of the pollen collected by honeybee colonies in spring (e.g. Synge, 1947; Louveaux, 1954; Free, 1959). Todd and Bretherick (1942) found that the percentage of protein in peach and almond pollens was well above the mean value for twenty-six flower species examined; they pointed out, however, that the protein content of pollen does not necessarily indicate its nutritive value. Maurizio (1951) showed that pollen from certain flowers (including *Pyrus* spp.) is particularly valuable to honeybees, stimulating hypopharyngeal gland and fat-body development, and resulting in greater length of life.

Nearly all *Pyrus* and *Prunus* varieties produce pollen, but there are exceptions, and Dickson (1942) listed varieties of apple, pear and peach which produce little or no pollen. Different species and varieties vary greatly in the amount of pollen they produce. Percival (1955) found the amount produced per flower to be: apple, 1·7 mg; pear, 1·2 mg; wild cherry, 0·3 mg. Some varieties of almond (Tufts, 1919) and apple (Webster, 1947; Webster *et al.*, 1949) produce more pollen than others, and Larsen and Tung (1950) found that triploid apple varieties produced 2·6–8·5 times as much pollen per anther as diploid varieties, although it had a lower germination. Rozov (1957) found that a bee carried up to 5 million pollen grains on its body, the number varying for different varieties of the same species.

Percival (1955) found that the temperature range for free anthesis was about the same for cherry and pear [5–14°C (41–57°F)] but higher for apple (10–19°C (50–66°F)). The duration of anther dehiscence in single flowers was 1–2 days for cherry, 1–5 days for apple, and 2–7 days for pear. The different species released their pollen almost at the same periods of the day (apple and cherry, 08.00–17.00 h; pear 07.00–18.00 h). Similar periods were recorded by Parker (1926) for cherry, apple, pear and plum. Percival found that the peak of presentation of cherry pollen was from 08.00–12.00 h, whereas the peak of pollen presentation of apple and pear was from 12.00–16.00 h. With minor fluctuations, there were positive correlations between the periods of presentation of apple and cherry pollens and the periods during which they were collected by honeybees, but the peak of pear pollen collection occurred after the peak of pear pollen presentation. Apple flowers exposed to bee visits in either the morning or afternoon set similar amounts of fruit (Menke, 1952c).

Vansell (1942) counted honeybees on a pear tree from which they were gathering pollen but no nectar. He found that visits began at 08.00 h, increased to a peak before midday and then rapidly decreased. This early decline in pollen collection contrasts with the results of Parker (1926) and Percival (1955), and may have been because the bees forsook pear for more attractive crops. Johansen (1956), using pollen traps, found that most apple pollen was collected between 09.00 and 14.00 h. Probably the duration and peak of pollen presentation of a species vary under different conditions and in different places.

Nectar

The nectar of *Prunus* and *Pyrus* flowers is valuable in helping honeybee colonies to grow in spring although fruit trees are only occasionally major sources of honey (e.g. Pellett, 1923; Oertel, 1939). Nectar secretion occurs only above a threshold temperature which varies with the species concerned; Behlen (1911) showed that wild cherry only secretes nectar at temperatures of 8°C (46°F) or above.

Ewert (1940) found apple and cherry nectar to be most concentrated in old flowers about to wither. In most *Prunus* and *Pyrus* flowers, as in other flowers with relatively exposed nectaries (Park, 1929), nectar concentration fluctuates widely in accordance with relative humidity throughout the day (Vansell, 1934, 1942—pears and plums; Karmo and Vickery, 1954—apples). Vansell (1934) observed that in some flowers with an open structure, notably apricot, nectar was diluted by dew and rain. Vansell and Griggs (1952) found that evaporation increased in wind; a dry wind produced a marked increase in apricot nectar concentration and bee visitation.

Fruit trees must have proper nutrition for a good crop to be produced. The supply of soil nutrients may influence the growth of fruit directly, or indirectly by affecting nectar production, and thus the number of bees visiting and pollinating the flowers. Plass (1952) found that nectar secretion by two apple trees, that had received no mineral fertilizer for several years, was less than half that by two trees that had received fertilizers. Apple trees have a high potash requirement, and Ryle (1954) found that extra potash significantly increased nectar secretion by apple flowers, but that extra nitrogen or phosphate did not.

Differences in environmental conditions probably account for differences in the amount of nectar secreted by flowers of the same variety of apple tree in different areas (Beutler, 1953), and by flowers in different parts of the same tree (Ryle, 1954).

The sugar concentration of nectar also varies with the species. Ranges of average percentages of sugar recorded are: apple, 25–55%; apricot, 5–25%; peach, 20–38%; pear, 2–37%; plum, 10–40%; sour cherry, 15–40%; sweet cherry, 21–60% (see Vansell, 1934, 1942; Butler, 1945a; Beutler, 1949; Singh, 1954; Rymashevskii, 1956; Sharma, 1958).

Information given by Beutler (1949) shows no relationship between the amount of nectar produced by different fruit species and its concentration, but Mommers (1966) discovered that with increase in the quantity of nectar produced by six different apple varieties the total amount of sugar also increased but its concentration decreased. There were large differences in the composition and abundance of nectar in the six varieties he tested but, in general, the number of honeybees that visited them was directly correlated with the amount of nectar they produced.

The amount of nectar produced also varies with the species. Sazykin (1955) found the following amounts secreted per flower per day: apple, 3·26–7·09 mg; cherry, 0·81–2·30 mg; plum, 0·96–1·74 mg; pear, 0·84–0·85 mg. Rymashevskii (1957), however, found the average amounts of nectar produced per flower per day to be: apricot, 5·1 mm^3; cherry, 3·6 mm^3; apple, 1·3 mm^3—more from cherry than from apple. Studies on apples (Beutler and Schöntag, 1940; Butler, 1945a), cherries (Beutler and Schöntag, 1940), pears (Vansell, 1942, 1952; Webster, 1946) and plums (Vansell, 1942, 1952; Brown, 1951; Percival, 1965) show that at the same time and in the same locality different varieties of the same species may differ in the amount and concentration of their nectar.

Foraging behaviour of bees on flowers

Darwin (1876) suggested that, although relatively few of the numerous flowers on a large apple tree set fruit, their profusion increases the attractiveness of the tree to pollinating insects, and Brittain (1933) found that orchards of trees with many flowers attracted more bees than orchards of trees with few flowers. Lewis and Vincent (1909) removed the petals from all the 1,500 flowers of an apple tree; only eight bees visited the tree, and only five flowers set fruit. Auchter (1924) removed the petals from 250 apple flowers;

he found that no bees visited them, and that none set fruit, although control flowers on the same branches were visited by bees and set fruit. However, Wilson (1926, 1929) found that apple flowers denuded of petals were still visited by honeybees although in reduced numbers, but not by bumblebees. Numerous honeybees sometimes visited apricot and nectarine flowers from which Vansell (1942) had removed the petals and stamens. Free (1960a) found that honeybees continued to visit plum and sweet cherry flowers from which the petals had fallen.

Parker (1926) showed that honeybees visiting apple, pear, plum and cherry flowers collected either pollen only, or nectar only, or both. He described how a bee collecting pollen only scrabbles over the anthers, pulling them towards its body and frequently biting them. However, Vansell (1942) observed that honeybees collecting nectar from cherry and peach pushed through the stamens and pistil to reach the nectary, and became covered with pollen in the process, and Stephen (1958) observed that bees collecting nectar from pears had abundant pollen on their heads and thoraces.

The behaviour of bees when visiting flowers determines their efficiency as pollinators. Although all bees that deliberately scrabble for pollen touch the stamens and stigmas and so may pollinate the flowers (Fig. 149), whether or not nectar-gatherers pollinate depends upon where they stand on the flowers. Thus, Free (1960a) observed that when a nectar-gatherer stood on the anthers and pushed its tongue and the front part of its body toward the nectaries it touched the stigmas and stamens and so could pollinate (Fig. 150) but, when it stood on the petals and pushed between the stamens to reach the nectaries, it did not touch the stigmas and so did not pollinate (Fig. 151) although, when visiting flowers with spreading stamens (e.g. plum, pear, peach and apricot), its body became dusted with pollen. About half of the nectar-gatherers collected pollen as well as nectar on the same flower visit, scrabbling over the anthers in the same way as bees collecting pollen only. Although during a single trip most bees were consistent in the type of flower visit they made, a few bees visited some flowers for pollen only and others for nectar.

In contrast to the spreading stamens of the flowers of other fruit species, those of apple are relatively upright, and bees that obtained nectar by pushing their tongues between the filaments from the side of the flower, often fail to touch the anthers as well as stigmas (Brittain, 1933; Roberts, 1945; Free, 1958d, 1960a). Roberts found that the degree of spread of the anthers largely determined the proportions of honeybees which worked the flowers from the top or side. Free found that the proportion of nectar-gatherers that approached the nectaries from the sides depended on the structure of the stamens of the variety concerned. When they were short and comparatively flexible, bees preferred to approach the nectaries from the top of the stamens and so may have pollinated. There may be a correlation between the degree of spread of stamens and their flexibility, but more probably flexible stamens become spread out by the activities of

Fig. 149. Honeybee scrabbling for pollen over anthers of *Pyrus malus*, apple.

Fig. 150. Honeybee entering top of *Pyrus malus*, apple, flower to collect nectar.

visiting bees. Fewer of the nectar-gatherers seen standing on the petals than on the anthers had pollen in their corbiculae, probably reflecting the tendency of bees to stand either on the petals or on the anthers during successive flower visits. Preston (1949) found that the tall thick stamens of "Bramley" apple flowers prevented honeybees from reaching the nectaries; this has been confirmed by Free (1960a) and Free and Spencer-Booth (1964a) who found that a lower percentage than usual of bees visiting "Bramley" collected nectar only.

Fig. 151. Honeybee approaching nectaries of *Pyrus malus*, apple, flower from the side. (Photo: M. V. Smith).

Because honeybees collecting pollen necessarily contact the anthers and stigmas, and usually work faster than nectar-gatherers (see below), they are generally regarded as more efficient pollinators. Free (1966a) discovered that "James Grieve" and "Cox's Orange Pippin" flowers that received single visits from bees that scrabbled for pollen set averages of 17·3 and 13·5% more flowers respectively than those visited by bees that did not scrabble. However, the relationship between the length of the pistil and stamens is constant within a variety (Forshey, 1953), and even pollen-gatherers may not pollinate certain varieties, whose stigmas are too far above or below the stamens (Crane, 1945; Roberts, 1945; Brown, 1951; Löken, 1958). Where it is above, bumble bees may be effective because their large bodies bridge the gap. Brittain (1933) pointed out that because female solitary bees are mainly pollen-gatherers, they are proportionally more valuable as pollinators than honeybees. Andrenids become covered with pollen as a result of inserting their heads into the middle of the staminal column of a flower and curling their abdomens round the anthers (Menke, 1952c).

The proportions of honeybees collecting nectar or pollen, or both, from fruit blossoms depends on the relative availability of nectar and pollen at the time, and on the food requirements of their colonies, so it is not surprising that the records of different authors vary considerably (e.g. Parker, 1926; Filmer, 1932; Roberts, 1945; Overley and O'Neill, 1946; Free, 1958d). Parker (1926) observed that on warm days, when the relative humidity decreased rapidly, the available pollen was soon collected; Johansen (1956) found that the proportion of honeybees gathering apple pollen was greatest at the beginning of the day. Gagnard (1954) recorded that there were more nectar-gatherers than pollen-gatherers on almonds early and late in the day. Brown (1951) found that practically all the honeybees that visited plum flowers were, at any one time, either collecting nectar only or pollen only. Brittain (1933), working on apple, and Free (1960a) on apple, apricot, peach, pear, plum and sweet cherry, found that the ratio of nectar-gatherers to pollen-gatherers varied greatly on different days and at different times on the same day, and that bees often collected pollen when nectar was either unavailable or unattractive. Free and Spencer-Booth (1964a) attempted to discover whether changes in the ratio of nectar- to pollen-gatherers reflects a change in the behaviour of the bees, or a change in the population present. They recorded the behaviour of marked foragers in an orchard of dwarf apple trees, and found that although most tended to be constant to the type of forage they were collecting, bees that were originally scrabbling for pollen did not desert the crop when pollen became scarce, but instead changed to collecting nectar only. Hence, it seems that changes in the proportions of nectar- to pollen-gatherers in orchards is mostly associated with changes in the behaviour of individual bees. Probably the transition between scrabbling for pollen and collecting nectar is easier to make on apple flowers than on flowers of many other species.

The rate at which bees visit fruit flowers depends on the amounts of nectar and pollen present, and these vary with the type of flower and the stage of its development, with climatic conditions, and with the number of foraging insects present. The average number of flowers bees have been seen to visit per minute is given below in Table 33.

Hutson (1926) found the rate of visitation to apple flowers to differ with different varieties and from year to year, and Wilson (1926, 1929) and Free (1960a) found that adverse weather increased the time spent per flower.

Some wild bees work faster than honeybees. Sax (1922) found that bumblebees visited 20 apple flowers/min, and Wilson (1929) found that *Bombus lucorum* and *B. terrestris* working on apple, cherry, pear and plum flowers visited 15·6 flowers/min, and that *B. pratorum* worked slightly faster. Menke (1951) reported that anthophorid bees visited 15 apple flowers/min, and that one bumblebee visited 18·7 flowers/min. Löken (1958) noted that bumblebees visiting apple flowers worked twice as fast as honeybees, but solitary bees only half as fast.

Table 33

	Nectar-gatherers	Pollen-gatherers	Un-classified	Author
Apple	—	—	6·0	Sax (1922)
	—	—	6·0–9·0	Hutson (1926)
	—	—	7·1	Menke (1951)
	8·1	15·8	8·8	Rymashevskii (1956)
	6·7	10·9	—	Free (1960a)
	7·2	7·1	—	Free and Spencer-Booth (1964a)
Apricot	—	—	8·2	Rymashevskii (1956)
	5·6	7·9	—	Free (1960a)
Peach	2·7	4·0	—	Free (1960a)
Pear	—	—	16·0	Hutson (1926)
	9·1	10·2	—	Stephen (1958)
	7·7	6·6	—	Free (1960a)
Plum	3·8	10·7	—	Free (1960a)
Cherry	—	—	7·4	Rymashevskii (1956)
	6·0	6·7	—	Free (1960a)

There is little information on the number of fruit flowers honeybees visit per trip. McCulloch (1914) observed 2 bees which visited 53 and 61 apple flowers respectively, and Webster (1947) suggested that about 100 apple flowers supply a nectar load. Vansell (1942) observed one honeybee that visited 84 pear flowers for a nectar load and another that visited the same number of flowers for a pollen load. Free (1960a) watched 2 nectar-gatherers and 2 pollen-gatherers for what were probably complete trips; the pollen-gatherers visited 89 apricot and 38 pear flowers and the nectar-gatherers visited 76 pear and 82 sweet cherry flowers. There is little relevant information on the duration of foraging trips or the number made per day. Dyce (1929), Brittain (1933) and Zander (1936) agreed that a low estimate of the number of apple flowers visited per honeybee per day would be about 720, and Karmo and Vickery (1954) stated that in good weather honeybees make 7 trips/day when visiting apple flowers. Menke (1952c) suggested that a bumblebee visits about 240 apple flowers per trip.

Number and strength of honeybee colonies required for pollination

Honeybees may travel 3 km or more to visit fruit flowers (Wilson, 1929). Gagnard (1954) found that an almond orchard was visited by equal numbers of bees from a colony situated in the orchard and from one 200 m away, but by fewer bees from a hive 500 m away (see also page 69).

Estimates of the number of colonies necessary to pollinate a given area of orchard are based on experiences and assumptions of fruit growers and beekeepers, rather than on experimental results, and at best can only be approximate. Tufts (1919) recommended $2\frac{1}{2}$ colonies/ha of mature orchard and his advice has frequently been repeated. The Horticultural Education Association (1961) calculated that to obtain an economic yield of apples 5% of the flowers, or approximately 136,000 flowers/ha must set and mature. From the little evidence available (see above) it seems that a bee visits about 700 flowers/day so that 194 bee days are needed to visit the flowers in 1 ha of orchard. However, only few of the visits bees make to flowers needing cross-pollination are beneficial and it would be surprising, even in a well-planned orchard, if more than an average of 1 in 6 flowers visited set fruit as a result (see Free, 1966; Kurennoi, 1967; Petkov and Panov, 1967). Therefore, at least 1,164 (194 × 6) bee days are needed to pollinate a hectare. Because many visits will be to flowers that are already pollinated, or that are incapable of being pollinated, probably twice as many (i.e. 2,328) bee days are necessary. At the time of fruit blossom in England and Scotland, colonies have an average of about 10,000 bees each (Jeffree, 1955; Free and Racey, 1968a). In good weather a third of the bees (3,300) might forage and perhaps about a third of the foragers (1,100) might visit fruit (Free, 1958d, 1959). Hence, during a single day of good foraging weather $2\frac{1}{2}$ colonies should be able to pollinate just over 1 ha. However, for much of the flowering time of fruit trees the weather is often unsuitable for anther dehiscence or foraging so the recommendation of $2\frac{1}{2}$ colonies/ha is probably not excessive.

The number of colonies needed will depend partly on their size, and recommendations on the minimum size of colony for orchard pollination are usually made in terms of number of brood combs present, e.g. 4 combs (Grout, 1950; Menke, 1951; Webster et al., 1949); 4 or 5 combs (Griggs, 1953); 5 combs (Snyder, 1946; Coggshall, 1951; Ministry of Agriculture, 1958b); 4–6 combs (Karmo and Vickery, 1954); 6 combs (Dunham, 1939d); 7–8 combs (Dirks, 1946).

The relative merits, for orchard pollination, of overwintered colonies and colonies of package bees of various sizes have often been discussed (e.g. Hutson, 1928; Farrar, 1931b; Filmer, 1932; Brittain, 1933; Woodrow, 1934; Vansell, 1942; Reese, 1951; Karmo and Vickery, 1954). It is difficult to reach any general conclusions from the experiments that have been made because of variation in the size of the packages and colonies used. However, overwintered colonies have brood, whereas package colonies at first have none, and the amount of brood probably influences the percentage of foragers which collect pollen. Furthermore, whereas package colonies decrease in population for a few weeks after their installation in hives, the population of overwintered colonies increases. The time that elapses between the installation of packages of bees and the flowering of the fruit trees in the area must, therefore, also be considered.

The bee population required in an orchard depends on many factors, which vary in different orchards and different places. The age of the orchard is important. Howlett (1927) and Murneek (1930) suggested that 1 colony/1–2 ha is sufficient to pollinate orchards of young trees. Dickson (1942) suggested that one strong colony is adequate to pollinate 3–4 ha until the orchard is 10–12 years old.

The bee population in surrounding orchards and countryside must also be considered (Brittain, 1933, 1935). Orchards surrounded by areas without bees require the introduction of more colonies than those in areas where many colonies are kept. Probably, large orchards usually have fewer wild pollinators per tree than small ones. It is generally believed that orchards exposed to strong or cold winds have fewer wild pollinators than sheltered ones (see pages 81 and 400).

Because fewer bee visits are necessary to self-fertile than self-unfertile varieties to give the same percentage of fruit set (Free, 1966a), fewer colonies will be needed in orchards of self-fertile varieties.

The species of fruit tree is particularly important, because the set required for a commercial crop differs for the different species. Thus a greater bee population is recommended for cherry orchards, which need a high fruit set, than for other orchards (Schuster, 1925; Mommers, 1951). The structure of "Delicious" apple flowers does not facilitate pollination (Roberts, 1945), and more than $2\frac{1}{2}$ colonies/ha may be needed to pollinate this variety (Kelty, 1948). Certain varieties may be visited comparatively little by bees if more attractive flowers are nearby (see below), and consequently a higher bee population than normal will be required.

Finally, there is the effect of weather. Hutson (1926) and Brittain (1933) reviewed observations on the effects on bee flight of wind, temperature, sunlight, humidity and precipitation. Some authors (e.g. Philp and Vansell, 1944; Webster *et al.*, 1949; Grout, 1950) supposed that $2\frac{1}{2}$ colonies/ha may result in over-pollination under ideal weather conditions, particularly with some varieties, but produce under-pollination in unfavourable weather. Brittain (1933) concluded from his experiments, which were rather unsatisfactory because of inclement weather, that in favourable conditions the set obtained on an apple tree after 1 h exposure to bee visits was sometimes sufficient to produce a commercial crop, and that the crop after 5 h exposure was sometimes great enough to require thinning. Townsend *et al.* (1958) found indications that 5 colonies/ha, provided with dispensers (see below), might pollinate an orchard sufficiently during only 2 h of favourable weather. Thus, the same colonies could possibly be used in more than one orchard in the same season.

However, until methods are available for estimating accurately the amount of pollination that has already taken place, and satisfactory arrangements exist between beekeepers and fruit growers to move colonies at short notice, the risk of over-pollination is clearly preferable to under-pollination, and sufficient bees should be used to ensure adequate pollination under unfavourable conditions (i.e. Schuster, 1925; Hooper, 1929; Brittain, 1933; Mommers, 1952; Nevkryta, 1957).

Competition between flowers for bee visits

If individual bees keep to only one variety of fruit tree cross-pollination is not facilitated. Honeybees prefer some varieties to others (e.g. Hutson, 1926; MacDaniels, 1931; Brittain, 1933; Shaw and Turner, 1942; Karmo, 1958), so when cross-fruitful varieties are selected for planting in the same orchard, their relative attractiveness should be considered, as should that of varieties in surrounding orchards.

The preference of bees for particular varieties seems to be related to differences in nectar secretion (see above), and correlations have been found between the numbers of honeybees visiting different varieties and the amount and/or concentration of nectar present (e.g. Butler, 1945a, apples; Crane, 1945 and Brown, 1951, plums). Inaccessibility of nectar, due to floral structure, also influences attractiveness (Preston, 1949; Brown, 1951; Löken, 1958). Brown (1951) found that the variety of plum, "Utility", which attracted most nectar-gatherers, also attracted most pollen-gatherers; he supposed that this might generally be true. However, some varieties may usually be visited for pollen only (e.g. Vansell, 1942, "Bartlett" pear), or they may be particularly attractive to pollen-gatherers (e.g. Brittain, 1933, "Golden Russet" apple). It is likely that abundance of pollen (see above) makes a variety more attractive to pollen-gatherers. Different varieties may be attractive during different periods of the day, presumably because of variations in nectar and pollen availability (Overley and O'Neill, 1946). Differences in flower colour and scent, and in the number of flowers per tree, are probably minor factors contributing towards varietal preferences.

Filmer (1941) found that when an orchard of different varieties had a low honeybee population, some varieties were visited by many more bees than others, but that when the population was high, bees were distributed almost uniformly on the various varieties, presumably because of increased competition. He concluded that uniformity of bee distribution is a reliable criterion for judging whether there are sufficient pollinators in such an orchard.

Different species of fruit trees also differ in their attractiveness. Wilson (1926) saw bees fly from pear to plum flowers and from apple to cherry flowers, and it has frequently been observed that honeybees prefer apple to pear flowers (e.g. Vansell, 1942; Hambleton, 1944; Webster, 1947; Webster et al., 1949; Grout, 1950), and this has been attributed to the normally low sugar concentration of pear nectar. Vansell (1952) found that bees visited sweet but not sour cherries in the same orchard, their mean nectar concentrations being 55% and 20% respectively. Glowska (1958) observed, during three flowering seasons, the relative attractiveness of three varieties each of plum, sour cherry, apple, sweet cherry and pear and found that the total amount of nectar secreted was greatest for plum and decreased in the above order. On almost every day of flowering the pear flowers secreted less nectar than the other species; fewer bees visited pears and most of those that did so collected pollen. Singh (1954) found that honeybees concentrated on apple and sweet cherry to the exclusion of apricots and pears. Roberts (1956) noted that honeybees generally prefer the pollens of other species to that of plum, and that the pollen of different plum varieties differs in its attractiveness.

Honeybees may neglect fruit in favour of other crops. Philp and Vansell (1944) found that filaree growing in orchards was highly attractive to honeybees until the flowers closed, which often happened as early as 10.00 h on sunny days. Vansell (1942) once saw honeybees desert plum for *Arctostaphylos manzanita* at 10.00 h but return to plum in mid-afternoon when the *A. manzanita* nectar was almost exhausted. When a *Brassica alba* cover crop is grown in pear orchards, the bees prefer it to the pear because the concentration of its nectar is several times as great (Vansell, 1942, 1952; Hambleton, 1944). Stephen (1958) found that pear orchards with plenty of *B. alba* and *Stellaria* spp. growing in them had poor crops, and he observed that honeybees only visited pear flowers abundantly when the ground flora was overpopulated. The sugar concentrations of *B. alba* and *Stellaria* spp. were 48–64% and 51–58% respectively, and that of pear 7–34%, usually in the lower part of this range. *B. alba* can also be a serious competitor to plum (Vansell, 1934, 1952).

Butler (1945a) described competition between apple and *Taraxacum officinale*, and between pear and *Crataegus monogyna*, in which the fruit flowers had a lower nectar concentration and had fewer visits than the other species. He thought it probable that nectar concentration principally determines the species visited, and nectar abundance, the size of the bee population. Roberts (1956) found that bees preferred nectar of *Leptospermum scoparium*, *Sophora microphylla*, *Hakea saligna*, and various brassicas to plum nectar, and *Ulex europaeus* and brassica pollen to plum pollen.

It has long been supposed that *Taraxacum officinale*, dandelion, is a major competitor to fruit (e.g. Brittain, 1933; Filmer, 1941; Coggshall, 1951; Karmo and Vickery, 1954) but Kremer (1950) suggested that this might not be so. He found that on cold days bees preferred dandelion to fruit, but pointed out that as the fruit flowers had no nectar or pollen available they would not have been visited anyway. On days with rapidly rising temperature the dandelion closed at midday and the bees deserted it, whereas the amount of fruit pollen collected was greatest at 15.00 h. He supposed that these two species competed only on days warm enough for them both to present nectar and pollen, but not so warm that the dandelion flowers closed. He therefore, regarded the presence of dandelion as inconsequential to fruit growers. However, this does not seem to be true in England, although the peak periods of pollen presentation of dandelion and apple are from 10.00–11.00 h and from 12.00–16.00 h respectively (Percival, 1955). Thus, Free (1968a) found that dandelion pollen formed a large proportion of the pollen trapped from colonies in orchards and furthermore, because most bees visiting dandelion are primarily nectar-gatherers that only collect pollen incidentally when it is available, dandelion is probably a more serious competitor to fruit than is apparent from information on pollen collection only. Some bees continued to collect nectar from dandelion long after it had finished yielding pollen for the day. By using marked bees, it was found most bees remained constant to dandelion or fruit. Even on warm days when the dandelion closed soon after midday, very few of the pollen-gatherers that had been visiting them moved to fruit flowers. Hence, the dandelion even when it closed early, prevented many bees from visiting the fruit. *Andrena haemorrhoa*, which was the most abundant species of solitary bee foraging on the dandelion, also collected apple pollen, so dandelion also competes with the fruit for visits from this species.

Filmer (1941) found that by increasing the total bee population in a district, the percentage of it that visited an apple orchard was increased, presumably because of overpopulation on competing crops. Others (e.g. Coggshall, 1951; Clark, 1959; Raphael and Cunningham, 1960; Free, 1968a) have suggested that growers should eliminate dandelion and other competing crops from their orchards by mowing or using selective herbicides, and growing only those cover crops that flower after the fruit has been pollinated (Vansell, 1942). Such management will also lessen the risk of poisoning bees from any insecticides that may have become deposited on the floor of the orchard. Because most dandelion pollen is presented in the morning and most apple pollen is presented in the afternoon, it seemed probable that if hives moved into an orchard, were not opened until the dandelion had ceased yielding for the day, fewer bees would become conditioned to dandelion and more to apple. An experiment in which one group of colonies moved into an orchard were allowed to forage at 09.00 h and another at 13.00 h demonstrated that this was so (Free and Nuttall, 1968b). However, the gain in the pollinating efficiency of the colonies must be balanced against any possible damage that results from confining them to their hives.

Foraging areas of bees

Information on the foraging areas of individual honeybees in *Prunus* and *Pyrus* orchards, and an understanding of the factors affecting their size, are important when trying to solve many pollination problems, because bees that keep to one fruit tree during a trip are valueless as cross-pollinators.

Minderhoud (1931) found that during a single trip the foraging area of a honeybee was not greater than 100 m^2, and supposed that large fruit trees, standing well apart from others, might not be sufficiently cross-pollinated. MacDaniels (1931) observed that individual honeybees tended to keep to a single tree, or to two adjacent trees, and Roberts (1956) said that, when trees are large with plenty of flowers, a bee is likely to keep to one tree per trip. Singh (1950) observed 66 bees and found that 45 of them kept to one tree, 16 to two trees, 2 to three trees, 2 to four trees, and the remaining bee to five trees. But the average time for which he was able to follow the bees was greater in the latter circumstances, so the foraging areas were probably less restricted than his figures imply. Free's (1960a) data indicated that on average bees visited about two standard trees per trip, and when they did move they went to the nearest tree. Singh (1950) observed that bees flew twice as readily between trees 3 m apart as between trees 5 m apart, and Rymashevskii (1956) and Free (1960a) found that, in orchards where trees within rows were closer together than trees in adjacent rows, bees changed trees more frequently within rows than between rows. Kurennoi (1965) noticed that bees frequently passed from one tree to another where branches were intermeshed, but were reluctant to do so in orchards where the crowns of the trees remained discrete. Free and Spencer-Booth, (1964a) found that in an apple orchard with dwarf trees arranged in continuous "hedges", a bee tended to keep to about 3 m of one row per trip and rarely moved to another row; when bees did move to another row it was nearly always adjacent to the one they had been working.

The size of a bee's foraging area fluctuates considerably and is partly determined by the amount of nectar and pollen available; when supplies become short, bees become restless and search more widely for forage. Singh (1950) supposed that a honeybee's foraging area can extend to several fruit trees when forage is scanty, and Free (1960a) found that, during unfavourable weather, bees visited fewer flowers per tree than in good weather. Both authors noted that wind, and disturbance by other insects, could cause extension of foraging areas. Although windbreaks may increase the population of insects on nearby trees (Lewis and Smith, 1969) the insects concerned may move from tree to tree less often, and so give no increase in fruit set.

Conditions that result in large individual foraging areas, and consequently most cross-pollination, are therefore not necessarily those associated with easy foraging; when they occur, presumably relatively few bees are attracted to the crop. Menke (1951) saw honeybees fly from one tree to another when their population in the orchard was not large. It seems likely that enough bees usually wander from tree to tree in an orchard to effect adequate cross-pollination, and that the foraging areas of individual bees are largest at the beginning and end of the flowering period, and possibly at the beginning and end of each day. Probably, in relatively cold springs, when the flowering season is prolonged and forage is sparse, the bees range over larger areas than usual. If the foraging areas of individual honeybees could be increased, more cross-pollination would result. Butler (1943, 1945b), Butler and Simpson (1953) and Ribbands (1953) discussed whether increasing the number of bees present, and thus the competition between them, would result in larger or smaller foraging areas; although the former seems more probable, it has not yet been demonstrated, and Free (1966a) found no evidence that a change in the size of the foraging population affected the size of foraging areas of the individual bees.

Betts (1931) suggested that pollen carried by a honeybee to a flower on the outskirts of its foraging area might become transferred to another bee with an overlapping foraging area, and eventually be used to pollinate a flower outside the foraging area of the first bee. This hypothesis is supported by experiments of Townsend *et al.* (1958). These workers marked bees leaving their hives with a fluorescent powder which subsequently became distributed on the flowers; they found that wild bees became marked with the fluorescent powder, showing that it was further distributed after being deposited on the flowers. However, the fluorescent powder may have been more readily distributed than pollen, and it is uncertain to what extent pollen placed on flowers by hand is subsequently distributed by bees (see below). Mittler (1962) tried dusting one or more trees with fluorescent powder to find the extent to which it was distributed by bees, but he found this was difficult to do without contaminating adjacent trees. Furthermore, he reported that the presence of the fluorescent powder on the flowers influenced the bees' behaviour. Likewise his technique of producing radioactive pollen by introducing radioactive phosphate solution into the trunks of fruit trees, does not seem to have given useful results. He obtained most success by examining the proportion of diploid and triploid pollen grains in the pollen loads of bees foraging in an apple orchard where the diploid and triploid varieties occurred in separate rows; all but a small proportion of bees captured were carrying only pollen of the type characteristic of the particular trees on which they were foraging.

So far this discussion has been limited to the size of the foraging area of an individual honeybee during a single trip. Over consecutive trips its foraging area may be greater. Mommers (1948) marked bees on the centre tree of a row (trees 2 m apart) in an apple orchard and later found them widely distributed along the row, although they remained most abundant near its centre. In contrast, the marked bees moved comparatively little between rows (trees 4 m apart), but his results may have been influenced by the fact that no two consecutive rows were of the same variety.

Free and Spencer-Booth (1964a) recorded the locations in a dwarf apple orchard at which marked honeybees were foraging during consecutive days and they calculated that the average foraging area of an individual bee increased from 339 m^2 during 2 consecutive days foraging to 1,016 m^2 during 8 consecutive days. Furthermore, the bees did not appear to discriminate between the two main varieties ("James Grieve" and "Cox's Orange Pippin") comprising the orchard, and so would have readily transferred pollen from one to the other, but discriminated against "Bramley".

In contrast, Free (1966a) found that, in more favourable weather conditions, bees foraging for several consecutive days in an orchard of standard apple trees tended to have much smaller overall foraging areas; many bees were seen throughout the observation period only on or near the same trees they were first seen to visit. This orchard contained five varieties, arranged in separate rows, and the proportion of honeybees and solitary bees visiting each variety depended on the number of its flowers and its stage of flowering. When the early flowering varieties began to lose their petals, the proportion of bees visiting them decreased, and they moved to other varieties, preference being given to those nearby. In fact, it seems that the attractiveness of a late flowering variety may be enhanced by its proximity to a variety the bees have been visiting. Bees originally visiting a variety that retained or increased its attractiveness continued to visit it consistently, and sometimes even preferred it to a more attractive adjacent variety. Hence, bees can distinguish between some varieties and they do not select a tree solely on the basis of the amount of food it is producing. Because bees visiting an attractive variety remain consistent to it, they can only pollinate its flowers if it is self-fertile, and are unlikely to cross-pollinate different varieties. Although the more attractive varieties receive pollen from the less attractive ones, the opposite is much less likely to happen. The interchange of bees between different varieties is probably most frequent when the varieties concerned are equally attractive, have concurrent flowering periods, and the bees do not appear to differentiate between them. However, when two varieties have different amounts of bloom and overlapping flowering times, bees will interchange between them most when the early flowering variety has more blossom, so that at first bees move to it from the other variety, but movement in the opposite direction occurs a few days later.

When a forager leaves its hive it still carries several thousand pollen grains on its body. To test whether this pollen is still viable, Free and Durrant (1966b) directed foragers that were leaving their hives into cages containing "Cox's" apple trees and found that the set on these trees was similar to that of trees not caged and much greater than that of trees caged without bees. Hence, a bee may distribute viable pollen more widely than it would during a single trip and can cross-pollinate if it visits different varieties on consecutive trips, even though during a single trip it keeps to one variety. However, even when bees remain constant to a variety over several trips it is still possible for them to cross-pollinate. Stadhouders (1949) and Karmo (1960 and 1961b) supposed that, when in their hives, foragers may take up on their bodies pollen of species or varieties they have not visited, and Free (1966a) reported that bees captured on flowers or leaving their hives nearly always carried some pollen of species other than the one they were currently working. Perhaps the pollen became brushed onto them from the combs or from other adult bees inside the hive. Although pollen on a bee's body loses its viability after about 48 h (Latimer, 1936) at least some of that carried on the bees' bodies may be viable and could result in cross-pollination between varieties while the carrying bee remains constant to a single variety.

It is generally agreed that bumblebees, and particularly solitary bees, work less methodically than honeybees and fly more readily from tree to tree (e.g. Wilson, 1929; Brittain, 1933; Brittain and Newton, 1933; Brown, 1951; Menke, 1951), and prefer the upper parts of trees (Atwood, 1933; Menke, 1951). If so, they are more efficient cross-pollinators than honeybees, but evidence in support of these contentions is scarce (see pages 262 and 284).

Arrangement of pollinizers

A few decades ago, numerous varieties were often grown in the same orchard and cross-pollination was no problem. Nowadays, concentration on a few commercial varieties means that suitable pollinizer varieties have to be especially planted. The limited foraging areas of individual bees should largely determine the planting arrangement of main variety trees and pollinizers. Recommendations on the number of rows of one self-unfruitful variety which can be planted together without preventing adequate cross-pollination vary from two to six. Although it is more convenient to harvest the fruit if each row contains one variety only, it is also commonly recommended that every third tree in every third row (one tree in nine) should be a pollinizer so that every tree is adjacent to a pollinizer (e.g. Tufts, 1919; Schuster, 1925; Philp and Vansell, 1944; Gardner *et al.*, 1952; Griggs, 1953; Dickson and Smith, 1958).

Stephen (1958) found that the cross-pollination was greater, the greater the intermixture of varieties throughout an orchard, and several authors (e.g. Tukey, 1924; Schuster, 1925; MacDaniels, 1931; Roberts, 1947; Stephen, 1958; Townsend *et al.*, 1958) reported that the percentage of fruit can decrease as the number of rows between the main crop trees and a row of pollinizers increases, and that this effect may be quite marked even for trees only one row away from pollinizers, especially when unfavourable weather occurs during flowering. However, to prove that this depends on nearness to pollinizers and not on other possible factors, such as soil conditions and nearness to honeybee colonies, it is necessary to show that the fruit set increases again as another row of pollinizers is approached or that the set decreases with increasing distance on either side of the pollinizers.

Free (1962c) and Free and Spencer-Booth (1964b) demonstrated the effect of increased distance from pollinizer varieties on decreasing the initial and final fruit sets obtained in plum, apple, pear and sweet cherry orchards. For

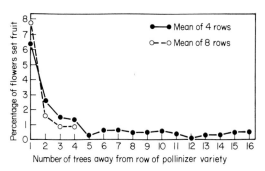

Fig. 152. Effect of nearness of a main variety *Prunus domestica*, plum tree to a pollinizer tree on the percentage of its flowers that set fruit (after Free, 1962c).

example, in the plum orchard (variety "Wydale") which had two rows of a pollinizer variety ("Victoria") bisecting it, about 7% of the flowers on trees adjacent to the pollinizer set fruit, but only 1% or less did so on trees four rows or more away (Fig. 152). Williams and Smith (1967) extended these observations in pear and apple orchards and confirmed that such gradients in fruit set sometimes occur. Data collected by Roach (1965) is important in illustrating how the distribution of pollinizer varieties can influence the crop produced. During a period of six years he recorded the crops on individual Cox's apple trees in a block of six rows which were bordered on either side by rows of pollinizer trees. The annual average number of trays of apples collected from trees that were: (a) adjacent to the pollinizer rows; (b) in rows one removed from the pollinizers; (c) in rows two removed from the pollinizers, was 313, 277 and 262 respectively.

It has also been noted (e.g. MacDaniels and Heinicke, 1929; Brittain, 1933; Roberts, 1945; Singh, 1953; Congdon and Woodhead, 1959) that the flowers on sides of trees adjacent to pollinizers usually set more fruit than those on the opposite sides. Brown (1951) found that only 2·8% of the flowers in the central part of a wall-trained plum tree set fruit, whereas flowers of the two outer parts, each adjacent to a pollinizer, had sets of 6% and 9%. Free (1962c) found that plum trees adjacent to pollinizers had greater sets on their sides facing pollinizers than on their far sides (10·8 : 4·3%). Some of the plum trees had bouquets of a pollinizer variety against their east and west sides

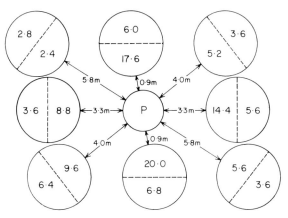

Fig. 153. Average percentages of "Cox" apple (*Pyrus malus*) flowers that set fruit, with "Worcester" pollinizers (P) planted as every third tree in every third row (after Free, 1962c).

but not their north and south sides; the sets on the east and west parts of these trees were more than three times greater than those on their north and south sides, and on nearby trees without bouquets. In one of the apple orchards studied, the main variety was "Cox" with "Worcester" as every third tree in every third row so that eight "Cox" trees surrounded each "Worcester"; the sides of "Cox" trees nearest the "Worcesters" had a greater fruit set (10·4 : 4·8%), more seeds per fruit (4·7 : 3·3) and more carpels with seeds per fruit (3·2 : 2·5) than on their opposite sides. Furthermore the set on a particular tree depended on its nearness to a pollinizer (Fig. 153). Williams and Smith (1967) reported that the sides of "Comice" pear trees facing a pollinizer variety had a greater set than the distant sides.

These findings seem to reflect the limited foraging areas (i.e. two adjacent trees per trip) of individual bees, and indicate that when a bee moves from a pollinizer to a main variety tree it probably pollinates only the first few flowers it visits; presumably the compatible pollen it is carrying soon becomes unavailable because it is either packed into the pollen baskets or greatly diluted with main variety pollen.

To ensure even pollination throughout an orchard, trees should be planted so that every tree needing pollination is surrounded by pollinizers; this can be achieved if the pollinizer and main variety trees are planted alternately, their relative positions being changed in each successive row as in Plan "A" below. The arrangement of trees in Plan "B" would facilitate harvesting,

```
    Plan A              Plan B              Plan C

   .  P  .  P  .     .  P  .  P  .  P     P     P     P     P
   P  .  P  .  P     .  P  .  P  .  P     .     .     .     .
   .  P  .  P  .     .  P  .  P  .  P     .     .     .     .
   P  .  P  .  P     .  P  .  P  .  P     P     P     P     P
   .  P  .  P  .     .  P  .  P  .  P     .     .     .     .
   P  .  P  .  P     .  P  .  P  .  P     .     .     .     .
   .  P  .  P  .     .  P  .  P  .  P     P     P     P     P
   P  .  P  .  P     .  P  .  P  .  P     .     .     .     .
   .  P     P  .     .  P  .  P  .  P     .     .     .     .
```

but the set may not be as great on sides of self-compatible trees facing their own variety, as on sides facing pollinizer trees, particularly when a greater distance separates rows than separates trees within a row. Furthermore, the presence of varieties in discrete rows undoubtedly helps bees to orientate themselves and remain constant to varieties. In order to encourage bees to make mistakes and hence to cross-pollinate, trees of different varieties should occur within the same row. When the pollinizer fruit is of lower commercial value than main variety fruit, the grower needs to choose between more main variety trees each producing less fruit and fewer main variety trees each producing more fruit. The fewer the pollinizer trees, the fewer the foraging trips on which bees will pollinate the flowers. A useful compromise would be to plant a pollinizer as every third tree in every row and to stagger the rows as in Plan C; in this arrangement every main variety tree is adjacent to three pollinizers.

Additional studies with dwarf apple trees have shown that rows adjacent to a pollinizer have greater initial and final sets than rows separated from pollinizers, and in association with this the pollinating efficiency per bee visit is much greater to flowers in the former rows (Free, 1966a). In an orchard of dwarf pyramid "Comice" pear trees every tenth tree had a graft of "Conference" pear and these trees set more fruit than the rest, thus again demonstrating the local effect of the pollinizer. These, and other studies, indicated that every row in orchards of cordon or dwarf trees should have pollinizer trees interspersed at frequent intervals among those of the main variety (Free and Spencer-Booth, 1964a).

In many of the orchards studied by Free (1962c), Free and Spencer-Booth (1964b) and Williams and Smith (1967) the average set was much less than that necessary to achieve the maximum possible crop. In a recent survey (Williams, 1969) showed that hand pollination increased the fruit set in 2 of 10 "Comice" pear orchards and 12 of 21 "Cox's" apple orchards. Insufficient set would probably result primarily from insufficient flower visits by pollinating insects, or from poor orchard planning with too few pollinizer trees, or from both. Williams (1969) concluded that a high density of honeybees, and a main variety/pollinizer tree ratio of 1 : 1 or 2 : 1 was necessary to provide adequate pollination in the orchards concerned. B. D. Moreton (personal communication) found that in Kent, England, a fruit set of 10% or more occurred only in those apple orchards that were well provided with honeybee colonies and had a ratio of one or more pollinizer trees to every five main variety trees.

Temporary aids to pollination

Cross-pollination is not always facilitated in existing orchards, for the following reasons: (a) the only variety present is self-unfruitful; (b) more than one self-unfruitful variety is present but they are not cross-fruitful; (c) the number of trees of the pollinizing variety is inadequate; (d) the arrangement of trees of the pollinizing variety is unsatisfactory. These faults can be overcome by planting new pollinizers. Alternatively, scions from pollinizers can be topgrafted onto existing trees. There is greater success, and less difficulty in harvesting, when whole trees are grafted over, rather than one or two branches per tree. The latter method is, however, likely to lead to more cross-pollination (e.g. Tufts, 1919; Schuster, 1925; Snyder, 1946; Griggs, 1953; Corner et al., 1964) although its effect may be localized (Free and Spencer-Booth, 1964b). If grafts are used their fruit should be distinguishable from the main variety fruit so no confusion arises during picking. Alternatively, the fruits can be stripped off the grafts; this also helps to ensure blossom the following season.

Temporary measures to ensure cross-pollination are necessary for some years while the new trees or grafts are growing, or when unusual weather conditions destroy the synchronous flowering of the varieties concerned (e.g. Griggs, 1958), although applications of growth retarding substances can delay flowering for several days and so help to synchronize the flowering of different varieties (see Batjer *et al.*, 1964; Griggs *et al.*, 1965). Flowering branches of cross-fruitful varieties can be placed in containers of water throughout the orchard (e.g. Tufts, 1919; Schuster, 1925; MacDaniels, 1931; Lötter, 1960). They should be cut when the "king" blossom of the main variety trees open, and replaced every 2 or 3 days. When these "bouquets" are on the ground, bees tend not to visit them freely (Brittain, 1933), and it is often recommended that a bouquet be hung in every tree (e.g. Snyder, 1946; Griggs, 1953; Dickson and Smith, 1958). MacDaniels (1930a) found that three apple trees with bouquets produced an average of four times as much fruit per tree as trees without bouquets. However, bouquets may only affect the fruit set of flowers within a metre or so of them (MacDaniels, 1930a; Free, 1962c).

Kremer and Hootman (1930) and Overley and Bullock (1947) stated that bees work best on the warmest and most protected parts of trees, and Free (1960a) counted more bees on the sunny than the shady side of large trees. The suggestion of Snyder (1946) that bouquets be hung on the south side of trees is, therefore, probably important.

However, considerable labour is involved in securing bouquets and in attempting to keep them fresh, and the provision of enough for a large orchard involves considerable cutting of the trees from which they are obtained (Brittain, 1933).

The hand application of pollen to flowers is sometimes used as a temporary measure to ensure pollination. MacDaniels and Heinicke (1929) and Snyder (1946) suggested that it might be of value in unfavourable weather even in well planned orchards, and Gardner *et al.* (1952) stated that it has become a standard practice in some parts of North America, particularly in "Delicious" apple orchards. Hand pollination of a limited number of flowers is a valuable method of determining whether a low fruit set has resulted from insufficient natural pollination (see MacDaniels, 1930b; Williams, 1969).

Methods by which pollen can be collected from flowers, stored and applied, have been described (e.g. Schuster, 1925; MacDaniels, 1930b; Overley and Bullock, 1947; Griggs *et al.*, 1953). It is usually collected by rubbing clusters of newly opened flowers over wire screen (about 5 meshes/cm) to separate the anthers. Because anthers are more difficult to remove from wilting flowers than from turgid ones a wide-mouth jar, with a piece of wire screen substituted for the cap, is sometimes attached to the belt of each collector so the anthers may be removed directly the flowers are picked. After collection the anthers and pollen are put in shallow trays at 21–27°C (70–81°F) for 36–48 h to induce the anthers to dehisce. The pollen can then be stored in small cardboard containers at 0°C (32°F) (Snyder, 1942). Before application, the pollen is often mixed with a suitable carrier, lycopodium spores being most commonly used, although Bullock (1948) found that powdered skimmed milk and egg albumen were as effective. Recently, comparisons have been made on the storage and germinating ability of hand- and bee-collected pollen, the latter being more readily procurable in quantity (Kremer, 1949; Griggs and Vansell, 1949; Singh and Boynton, 1949; Griggs et al., 1950; Vansell and Griggs, 1952; Johansen, 1956). Bee-collected pollen is inferior to hand-collected pollen in both respects, although some methods of partially overcoming this disadvantage have been found (Griggs et al., 1953).

It is usually recommended that, when hand pollination is undertaken, the pollen should be applied to one flower in every fourth or fifth cluster, about an hour being necessary to pollinate a mature tree (Overley and Bullock, 1947; Vansell and Griggs, 1952; Griggs, 1953), so the process is time consuming. According to Snyder (1942), a mature tree of 12·2 m spread can be hand pollinated in about 45 min using 140–280 g of pollen/ha. Bullock (1948) stated that 90–100% of the flowers that are hand pollinated by an experienced person set fruit, and when blossom is abundant hand pollination of one king blossom in every five clusters is sufficient; he and Karmo and Vickery (1960) found that it took 40–50 man hours and about 57 g of pollen to hand pollinate 0·4 ha of orchard. Oberle and Moore (1957) found the set on trees in blocks of single varieties was similar whether bees and bouquets or hand pollination were employed, but because of the additional labour involved in hand pollination they advocated the former.

Hand pollination might come into more general use if, as suggested by Karmo and Vickery (1954), the application of pollen to only five or ten flowers per tree is sufficient, the pollen being adequately spread over the remaining flowers by bees, but there is conflicting evidence about this. Phillips (cited by Singh, 1950) found that after pollen had been applied by hand to a single tree, bees sometimes distributed it to neighbouring trees, the fruit set decreasing with increase in distance from the hand-pollinated tree. However, Overley and Overholser (1938) found that, although the percentage of fruit set on hand pollinated branches increased with the number of flowers that had been hand pollinated, the fruit set of adjacent branches did not, so probably little of the hand applied pollen was transferred to other parts of the tree, although honeybee colonies were in the orchard. Snyder (1946) thought it risky to depend on insects for distributing artificially applied pollen; he considered that if time or labour is scarce, the pollen should be applied only to flowers on the north side of trees, or on the side facing the prevailing wind. Overley and Bullock (1947) and Lötter (1960) recommended that particular attention should be given to the tops and shady sides of trees because bees are more likely to forage on their sunny sides (Free, 1960a) and the set there is likely to be greater (Free and Spencer-Booth, 1964b). Johansen and Degmen (1957) found that honeybees failed to transfer pollen effectively from hand-pollinated flowers to the remaining flowers of the tree. Perhaps hand applied pollen deteriorates very rapidly and the extent to which it is effectively spread depends upon the number of insects avaliable at the time. There seems to be no advantage in hand-pollinating with a mixture of pollen in sugar syrup as recommended by Sadamori *et al.* (1958).

Various labour-saving methods of applying pollen have been tried. Brittain (1933) placed branches of pollinizers against apple trees and blew currents of air through them on to the trees. High fruit sets sometimes resulted, but he doubted the economic value of this method. Overley and Bullock (1947) applied pollen to trees with a small bellows-type hand duster, and obtained a heavy fruit set. They concluded that the method was very wasteful of pollen and gave excessive pollination on some branches but insufficient on others. However, Lötter (1960) reported more favourable results from applying pollen with bellows and he found that about 500 pollen grains/cm^2 were distributed 45 cm from the mouth of the bellows; 25 trees ($2 \cdot 7 \times 3 \cdot 7$ m) could be pollinated per hour. Unfortunately no experiments seem to have been done to compare the set of clusters, to which pollen has been applied with bellows, with that of control clusters. Bullock (1948) and Blasberg (1951) sprayed suspensions of pollen in water onto fruit flowers but obtained unsatisfactory results because, when applied in water, the pollen did not adhere to the stigmas and was injured after 15–30 min.

Others have tried applying pollen from aeroplanes, or from so-called pollen-dispensing "bombs" which blow pollen onto the trees; in these methods the pollen is usually mixed with a diluent. Bullock and Snyder (1946) selected paired branches on each of several apple and cherry trees and covered one of each pair with muslin while pollen was applied from aircraft; however, there was no difference between the percentage set on exposed and protected branches although the pollen itself was uniformly dispersed and germinated well. Thus, this experiment gave additional evidence that any pollen transported by wind is probably ineffective in pollinating the flowers. Pollen-dispensing "bombs" did not influence fruit set and the viability of the pollen from bombs was low. Overley and Bullock (1947) obtained suggestive results from aeroplane application with one apple variety, but negative results with another, and the use of "bombs" decreased fruit set compared with controls. Snyder (1947) and Bullock and Overley (1949) obtained negative results by both methods. In contrast, Traynor (1966) claimed that good sets were obtained from dusting a mixture of pollen and lycopodium powder on trees, either from the ground or from a helicopter, and that the best results were obtained with species that have large open flowers to catch the pollen. He pointed out that, when the pollen has been deposited on the petals it may be transferred to the stigmas by bees, so it seems possible that the different results obtained might partly have been due to differences in the abundance of bees present. However, the transference of dusted pollen by bees to the stigmas needs to be demonstrated, either by using dyed pollen, or by examining the bees for lycopodium powder.

Recently, several workers have tested "pollen dispensers" or "pollen inserts". These are devices fitted to hive entrances, and are so constructed that outgoing foragers are forced to walk through pollen of the required compatible variety and carry some with them to the trees they visit (Fig. 154).

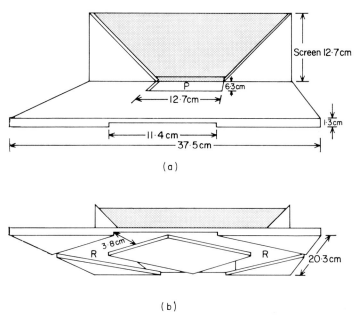

Fig. 154. A pollen dispenser: a, top view from rear, showing pollen tray (P); b, underneath view from rear showing passage-ways for returning bees (R) (after Townsend *et al.*, 1958).

Burrel and King (1932) developed a pollen dispenser in which they placed undiluted pollen. They enclosed two self-unfruitful apple trees individually in insect-proof cages, each cage containing a colony of bees; one colony had a pollen dispenser fitted to its hive entrance, but the other had not. The tree enclosed with the former colony gave six times the fruit yield of the other. King and Burrel (1933) described a modified pollen dispenser designed to force incoming and outgoing bees to use separate paths. They did similar experiments to that mentioned above, but on a larger scale, and obtained very suggestive results.

Overley and O'Neill (1946) tested two types of dispenser, and found that both greatly decreased the rate of exit and entry of bees, although this could have been because the bees drifted to a nearby colony without a pollen dispenser. Furthermore, most of the pollen from the dispensers was stored in the combs, and outgoing foragers carried very little of it. They concluded that the dispensers tested were of questionable value. Kremer (1948) also devised a pollen dispenser and suggested using *Lycopodium* spores as a diluent for the pollen, but gave no information about its value. Telford *et al.* (cited by Webster *et al.*, 1949) found that the Kremer dispenser greatly reduced the number of bees leaving the hive, although they carried some pollen with them. These authors also investigated the modified Harwood-Antles dispenser (Antles, 1953) and found that very few or none of the bees leaving it carried pollen. Griggs *et al.* (1952) tested Kremer's dispenser on cherry and almond trees enclosed in cages, and obtained very poor fruit sets compared with those obtained by hand-pollination or natural cross-pollination. They concluded that this was because some bees collected the pollen from the dispenser and carried it back into the hive, and others fanned the pollen away so quickly that bees could pass through the dispenser without touching the pollen.

Karmo and Vickery (1954) developed a dispenser which they claimed was very beneficial in orchard pollination, and recommended dilution of the pollen with *Lycopodium* spores. They, and Karmo (1958), observed that outgoing foragers sometimes gathered small amounts of pollen from the dispenser and re-entered their hive with it, but claimed that this stimulated the colony to forage. Lötter (1960) stated, without giving evidence, that bees were stimulated into seeking the specific pollen in the dispensers, and that the dispenser pollen was especially useful when there was competition from other crops.

Johansen and Degman (1957) found that the King-Burrel and Karmo-Vickery dispensers segregated incoming and outgoing bees better than the Harwood-Antles one. When they put pollen mixed with *Lycopodium* spores in the dispensers, the bees became very agitated and cleaned themselves thoroughly; when outgoing bees were collected about 7 m from their hives there was no pollen on their bodies. When pure pollen was provided, the bees behaved normally, and outgoing bees carried pollen on their body hairs and in their corbiculae. In a subsequent experiment they put pollen, stained with a fluorescent dye, into a dispenser; they found that it was distributed throughout the orchard. No fluorescence was found in the cells of the colonies, whichever of the above three types of dispenser was used, thus indicating that, contrary to previous experience (e.g. Overley and O'Neill, 1946; Webster *et al.*, 1949; Griggs, 1953; Karmo, 1958), the bees had not stored the pollen. Johansen and Degman (1957) and Johansen (1960b) also found that when colonies in hives fitted with dispensers were placed in cages containing fruit trees, the fruit set obtained equalled that of adjacent trees not in cages, and was greater than that of a control tree caged with a colony whose hive had no dispenser. Similar tests were done by Karmo (1960); two apple trees caged with colonies of bees gave 15% set and 26% malformed fruit whereas two apple trees caged with colonies of bees and pollen dispensers gave 25% set and 12% malformed fruit.

It is difficult to obtain the right conditions for a properly controlled experiment to test pollen dispensers on a field scale. Townsend et al. (1958), who modified the Karmo-Vickery dispenser to facilitate more rapid entry of foragers, used a pear orchard which was inadequately provided with pollinizers and had never produced a commercial crop of fruit (Fig. 155). During 2 years

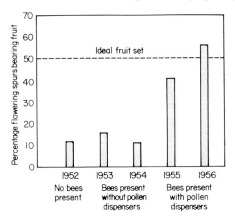

Fig. 155. Fruit set in a "Bartlett" pear (*Pyrus communis*) orchard on years when pollen dispensers were, and were not, used (after Townsend et al., 1958).

(1953 and 1954) colonies of bees without dispensers were put in the orchard during flowering but this failed to increase the fruit set. During the final 2 years (1955 and 1956) of the experiment, when honeybee colonies with pollen dispensers were used, large increases in the quantity and uniformity of fruit were obtained. The set of a similar orchard located a mile away, and belonging to the same grower, remained at 8–14% throughout the 4 years of the experiment. Johansen (1960b) reported that a Californian grower obtained excellent crops during 2 years in which he used bee dispensers in an almond orchard that had never before produced a commercial crop. Lötter (1960) claimed that using colonies with pollen dispensers for 2 years in a one variety apple orchard increased the yield by 50% compared to that of the 4 previous years when honeybee colonies without dispensers had been used. Unfortunately, he published no details. Jaycox and Owen (1965) also did an experiment, similar to that of Townsend et al. (1958), in an apple orchard whose varieties were not suitable for cross-pollinating each other and which, they reported, had consistently poor fruit yields. In 1963, they put colonies without dispensers in the orchard but failed to get increased set; in 1964 they used colonies with dispensers and stated that one variety gave a 68% increase and two others a 113% increase over the average of the previous 6 years.

Griggs and Iwakiri (1960) overcame the difficulty of simultaneous controls by using the following method. On each of several experimental trees they chose branches which during flowering were either: (a) kept bagged continuously; (b) kept bagged except for a few minutes when their flowers were hand pollinated; (c) kept bagged except for a 2–5 day period when dispensers were used in the orchard; (d) bagged only after the dispensers has been used; (e) bagged only until use of the dispensers ceased; (f) never bagged. The experiments were done in almond, cherry and apple orchards and the dispensers were always used when the weather was favourable and during peak flowering periods. However, the set on branches exposed during the time the dispensers were operative was about the same as that on other branches exposed for similar periods at other times during flowering. Moreover, hand pollination showed that the pollen was viable, and that a high proportion of the flowers were capable of producing fruit. Griggs and Iwakiri pointed out that a bee is much more likely to get pollen on its body when forcibly brushing itself against dehisced anthers, than when walking through a pollen dispenser. They noted that, although bees soon clear a path through the pollen in a dispenser, the first bees to walk through it after it had been charged with pollen did carry some to the flowers they visited. They suggested that this could account for the positive results obtained by Johansen and Degman (1957) when one colony was confined with a tree and given excess pollen. However, when a colony is caged, few bees actually work the tree in the cage and it is an open question as to whether more bees worked the trees in Johansen's and Degman's cages than normally visit trees in the open.

Thus, there is still some disagreement as to the value of pollen dispensers which will probably only be resolved by more years of experiment and experience. It is possible that in some circumstances the reduction in foraging caused by the dispensers might more than offset any gain caused by increased pollination per bee. Probably, therefore, pollen dispensers should be regarded, at present, as only temporary aids to pollination. It is doubtful whether the use of honeybee colonies with pollen dispensers in an orchard without pollinizers would give more cross-pollination than the use of honeybee colonies alone in an orchard adequately provided with pollinizers. Bees of colonies with dispensers may possibly distribute compatible pollen only on the first few flowers they visit on each trip. However, further experiments are needed to determine this point, and also to find out whether the use of pollen dispensers results in increased pollination in orchards that are well planned for natural pollination. If so, pollen dispensers will no doubt become used much more extensively. Perhaps they will prove useful for crops other than tree fruits, and particularly for the production of hybrid seed.

Chapter 35

Rosaceae: *Fragaria*

Fragaria × *ananassa* Duchesne

The cultivated strawberry of today, *F.* × *ananassa*, is derived from two N. American species. Originally the male and female flowers were on different plants but the North American Indians selected a variety with hermaphrodite flowers (Darrow, 1937). Most modern varieties are self-fertile and have hermaphrodite flowers (Fig. 156a), although a few varieties have either: (a) pistillate flowers only; (b) flowers with few stamens; (c) stamens that fail to produce pollen so they are practically self-sterile (e.g. Hughes, 1951; Shoemaker, 1955; Horticultural Education Association, 1961; Eaton and Smith, 1962; Hyams, 1962).

The strawberry is an aggregate fruit (Fig. 156b); the conical receptacle of a strawberry flower contains numerous pistils each with one carpel from which the true fruits or achenes of the strawberry are formed. Achenes containing fertilized ovules release a hormone that stimulates growth of the receptacle; when an achene does not contain a fertilized seed it remains small and the receptacle in its area fails to grow (see Robbins, 1931; Nitsch, 1952); when a group of such achenes occur together the fruit is noticeably deformed.

Although, as suggested by Knuth (1906) and Darrow (1927), some automatic pollination of self-fertile strawberries probably occurs by pollen falling or being blown from the anthers onto the stigmas, the stigmas are receptive long before the anthers dehisce so cross-pollination by insects is favoured. However, the pollination of strawberries by insects has only recently been investigated.

Skrebtsova (1957c) bagged flowers in the bud stage, and each day during flowering exposed them for a time to bee visits. She found that the percentage of flowers that set fruit increased with the number of bee visits per flower up to fifteen to twenty visits. More visits had little or no effect on set, but until about sixty visits had been made the mean weight of the berries continued to increase. The weight of berry produced also depended on the stage of development of the flower when it was visited, and visits to flowers at times other than when the maximum number of their stigmas were receptive resulted in fruits of less than maximum size being produced; the same principle applied when flowers were hand pollinated.

I cannot find any record of crop yields being increased by taking honeybee

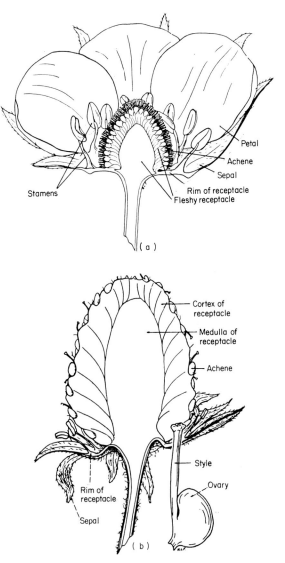

Fig. 156. *Fragaria* × *ananassa*, strawberry: a, median section of flower; b, median section of fruit (after Robbins, 1931, with permission of McGraw-Hill Book Company).

colonies to strawberry plantations, although Kronenberg *et al.* (1959) stated that there was often insufficient pollination in the centre of a strawberry plantation and that weather unsuitable for bees increased the percentage of malformed fruit.

In preliminary experiments, Hughes (1961, 1962) caged plants of several varieties of strawberry to exclude insects and she found that this decreased yield and produced malformed fruits some of which were quite unsaleable. Petkov (1965) reported that only 31–39% of flowers isolated from insects set fruit compared to 55–60% of exposed flowers that did so; furthermore, 60–65% of the fruit from the isolated flowers was malformed.

Mommers (1961) enclosed strawberry plants growing in a glasshouse in two cages and put a honeybee colony in one cage. In four tests, the cages with and without bees produced means of 86·9 kg and 74·0 kg of well formed fruit, and 3·3 kg and 4·2 kg of malformed fruit, but the differences were not statistically significant. However, the differences between the first pickings, which are the most profitable, were somewhat larger.

In a more extensive test, Free (1968f) covered twelve plots (3×3 m) containing plants of the variety "Favourite", with cages half of which contained small colonies of honeybees. Plants in the plots caged with bees had a slightly but significantly greater set than plants in the plots from which insects were excluded (66 : 56%) and also had slightly but significantly larger berries (8·3 : 6·7 g), but a much smaller proportion of the berries were malformed (20·7 : 48·6%); malformed and well formed berries had similar average weights. There was a progressive decrease in the size of berry produced with each successive harvest. This is a common experience of growers, and Valleau's (1918) observation that the earlier flowers contain a greater number of flower parts, including many more ovules and hence greater seed potential than those produced later, probably accounts for this; obviously every effort must be made to pollinate the earliest flowers of each cluster.

Valleau (1918) found that the primary flowers of a strawberry plant also had the greatest percentage set and produced the greatest percentage of well-formed fruits, and that both the amount of fruit set and well-formed fruits that matured progressively decreased in the later flowers of an inflorescence. He supposed that this was because the fertility of the ovules decreased in the later opening flowers of an inflorescence, and he did not think that lack of pollination was responsible for the lack of set or the production of malformed fruits. It now seems more likely that fruit production in his experiments was limited by nutritional factors rather than by lack of fertility of the ovules or pollination, especially as Free (1968f) found that the percentage set and the percentage of malformed berries showed no reduction as flowering progressed.

In two "pickings" Free (1968f) found that the plots not caged had significantly fewer malformed strawberries than in plots caged with bees. This may have been because the plants in the open were visited more freely by pollinating insects, but the cages themselves may have had an adverse effect on plant growth (page 121). Moreover, the plants in the open plots could have been cross-pollinated with pollen from other varieties growing close by, and the cross-pollination of some varieties may either increase yield as suggested by Šaškina (1950) or be essential for varieties producing little or no pollen. Vinson (1935) gave examples of both types of variety and suggested that in commercial plantations the varieties to be cross-pollinated should be in blocks of eight to ten rows wide, although in private gardens where keeping the varieties separate when picking is not a major problem, the varieties should be in alternate rows. Shoemaker (1931) found that the yield of a variety with imperfect anthers decreased rapidly with increase in distance from a perfect variety, and Shoemaker (1955) advocated a row of pollinizers to every three to five rows of an imperfect variety which needed cross-pollination, but pointed out that the closer the two types are planted the fewer the malformed berries.

Honeybees are the chief pollinators of strawberry flowers (Petkov, 1965) and the plantings suggested above are in conformance with our knowledge of the foraging areas of honeybees on strawberries. For example, Free (1968b) observed honeybees foraging on "Cambridge Favourite" strawberry plants (planted 0·3 m apart in rows and 1 m between rows), and found that, on average, a bee visited only about 12 % of the flowers open on each plant before going to another. His data indicate that nectar-gatherers and pollen-gatherers visited about ten and fifteen plants, and five and seven rows respectively per trip. Thus it seems that with a 5 : 1 planting of pollinizer and main variety plants a bee would carry pollen from a pollinizer to a main variety row an average of at least once per trip, but the greater the frequency of pollinizer rows the more the cross-pollination. Skrebtsova (1957c) found that, although bees readily move from one variety to another, they preferred some varieties to others, and she was able to correlate this preference with the amount of pollen the different varieties produced, and the amount of pollen on the bodies of the bees working them.

Shaw et al. (1954) reported that on 3 consecutive days strawberry nectar contained 28, 36 and 26% sugar. Petkov (1965) found that in different years an average of 0·6–0·8 mg of nectar of 26–30% sugar concentration was secreted per flower. The nectar is secreted by a narrow ring of fleshy tissue in the receptacle of a strawberry flower between the stamens and stigmas, and Free (1968b) observed that on nearly every flower visit nectar-gatherers touched both stigmas and anthers. Although bees sometimes alighted on the petals of a flower and approached the nectary from the side, they nearly always proceeded to walk over the stigmas. Some bees that were collecting nectar also had pollen loads and so were collecting pollen incidentally.

However, some bees also collected pollen deliberately; such bees either walked round the ring of anthers and scrabbled for pollen while doing so, or stood on the central stigmas and pivoted their heads and fore-parts of their thoraces over the ring of anthers. Some bees scrabbling for pollen also collected nectar. It has been found that bees spend about 7 sec (Petkov, 1965) or 10 sec (Free, 1968b) per flower visit although those collecting pollen worked slightly faster than those collecting nectar only.

Free anthesis of strawberry does not occur below 14°C (57°F) (Percival, 1955) and honeybee activity on strawberries increased with temperature and as conditions in general became more favourable for foraging; on the days when many honeybees were counted their numbers reached a peak in the early afternoon and the numbers of pollen-gatherers and the amounts of pollen collected were greatest near midday. Whereas bumblebees are abundant on raspberry and more numerous than honeybees on blackcurrant, very few bumblebees visit strawberry and relatively cold weather readily deters honeybees from doing so (Free, 1968b). Probably when it is cool during flowering such pollination as occurs is mainly done by insects other than bees, and especially by various Diptera (Hooper, 1932).

Although it is probably wise for a strawberry grower to import honeybee colonies into his plantation, especially when it is large, because honeybee visits tend to be limited to good weather he might not derive much economic benefit from doing so on some years. However, honeybee colonies could profitably be employed to ensure strawberries grown in glasshouses are pollinated and so avoid the need for tedious hand pollination (e.g. Hyams, 1962; Bewley, 1963); if tests show that blowflies are as efficient pollinators in glasshouses, they might be more convenient and cheaper to use.

Chapter 36

Rosaceae: *Rubus*

Because of their similarity the raspberry (*Rubus idaeus* L. of Europe, *R. strigosus* Michx and *R. occidentalis* L. of the U.S.A.), blackberry (*R. fruticosus* L.) and dewberry (*R. caesius* L.) will be considered together.

The flowers resemble those of strawberry but each carpel contains two ovules instead of only one (Fig. 157a). After fertilization, however, the receptacle remains small, and the carpels enlarge and conceal it, thus forming a compound fruit (Fig. 157b).

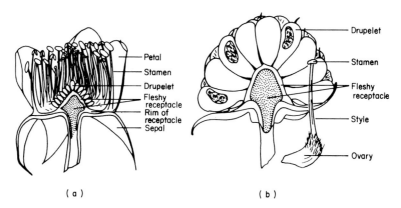

Fig. 157. *Rubus strigosus*, American red raspberry: a, median section of flower; b, median section of fruit (after Robbins, 1931, with permission of McGraw-Hill Book Company).

Numerous species and varieties of blackberry and raspberry, together with their hybrids, occur in the wild. Some wild species are self-sterile, and others have separate male and female plants and so need cross-pollination, but self-fertile hermaphrodite varieties have been selected for cultivation (e.g. Grubb, 1935; Darrow, 1937 and Shoemaker, 1955).

Crane and Lawrence (1931) found that cultivated raspberry varieties set fruit abundantly when selfed or crossed. Hardy (1931) hand pollinated four varieties of raspberry with pollen from their own and other varieties. For one variety ("Latham") the set was similar for each treatment, but the other three varieties ("Cuthbert", "King" and "Lloyd George") had a greater set from self-pollination than from cross-pollination (mean 83 : 68%). The mean weight of fruit produced from self-pollination was also greater (2·3 : 1·9 g). However, Robbins (1931) reported that better yields of dewberry are obtained when facilities for cross-pollination by another variety are provided, but he gave no evidence.

The outermost anthers are the first to dehisce and at the same time the stigmas become receptive; only the dehisced anthers of the innermost stamens sometimes touch the outermost stigmas, and may pollinate them if this has not already been accomplished, so an external pollinating agent is needed.

Johnston (1929) caged four plants of "Cuthbert" and four plants of "Cumberland" varieties of raspberry to exclude insects. Compared to similar plants outside the cages, fewer of the flowers set fruit (27 : 66% for "Cuthbert" and 90 : 100% for "Cumberland") and more of the berries were imperfect (43 : 3% for "Cuthbert" and 22 : 2% for "Cumberland"). He concluded that to obtain increased pollination and quality of fruit, honeybee colonies should be present in raspberry plantations during flowering. Wellington (1930) also reported that when bees were excluded from two varieties of raspberry a smaller crop containing a greater proportion of inferior berries was produced but gave no evidence. Couston (1963) caged two 5 m long plots of "Malling Jewel" raspberry to exclude pollinating insects; the caged plots flowered for a longer period and more abundantly than two control plots which were visited by numerous bees from nearby colonies. As a consequence, an average of more berries were harvested per caged than uncaged plot (3,588 : 2,423) but they weighed less (4·3 : 5·8 kg) and had fewer seeds per berry (32 : 89). Shanks (1969) made similar experiments with varieties "Puyallup" and "Sumner". He obtained 28 and 12% fewer berries per caged than uncaged plot of the two varieties respectively, and the weight per berry per uncaged plot was 10 and 27% less. Shaking some of the caged plants or blowing a constant stream of air through them made no difference to the crop produced. Plants of "Sumner" variety that were caged had only 39 drupelets per berry compared to 66 per berry on uncaged plots. Obviously these experiments on raspberry would well repay repeating more extensively with honeybee colonies in some caged plots.

Abundant nectar is secreted by a ring of tissue located on the receptacle between the stamens and stigmas. The stamens diverge widely so even short-tongued insects can reach the nectar, and many species of Hymenoptera, Diptera and Coleoptera have been recorded visiting the flowers (e.g. Hooper, 1929). Because the nectaries of blackberry and raspberry are quite exposed, the volume of nectar they contain is probably greatly influenced by changes in the relative humidity of the atmosphere.

Percival (1946) found that blackberry flowers began secreting nectar as soon as their petals started to unfold and finished when the petals had fallen and the filaments of the stamens were beginning to curl inwards over the carpels, although about three-quarters of the nectar was secreted in the first half of the 6-day flowering period. A mean of 14·5 mg of sugar was secreted per flower, but the amount for individual flowers varied from 3·7–19·5 mg; part of the reason for this large variation was that flowers borne on robust shoots secrete more nectar than those borne on thinner shoots. A single raspberry flower has been found to secrete an average of 8·5 m^3 of nectar (Rymashevskii, 1957) and 1·41 mg sugar (Sănduleac and Baculinschi, 1959) per day, and 13 mg nectar during its whole flowering period (Haragsimová-Neprašová, 1960).

Fig. 158. Honeybee foraging on *Rubus idaeus*, raspberry.

The average sugar concentration of blackberry nectar has been reported as 28–31% and raspberry as 21–30% (e.g. Butler, 1945a; Shaw *et al.*, 1954; Sazykin, 1952; Sănduleac and Baculinschi, 1959). Sazykin (1952) found that nectar secretion by raspberry differed with the variety, soil and climatic conditions, but averaged 4–6 mg/flower per day; he calculated that the nectar from 1 ha of raspberry plantation would produce 208–268 kg of sugar.

Because raspberry crops are so attractive to insects, adequate pollination may be achieved without importing honeybee colonies, although an increased yield and better quality fruit are sometimes obtained when this is done (Shoemaker, 1955).

A honeybee foraging on a raspberry flower (Fig. 158) follows the ring of nectar tissue and, as it does so, one side of its body touches the stigmas. All nectar-gatherers become dusted with pollen, but very few bees deliberately scrabble for it or collect pollen only (Shaw *et al.*, 1954; Free, 1968b).

Raspberry pollen is collected throughout the day, and the proportion of foragers collecting pollen does not vary much at different times of the day, but tends to be greater in the afternoon than in the morning (Free, 1968b). However, because free anthesis of raspberry occurs at relatively low temperatures (12°C (54°F); Percival, 1955) the percentage of raspberry pollen collected by foragers tends to be greater at the beginning and end of the day's foraging rather than during the peak of the foraging period (Free, 1968b).

Whereas some bees pack the pollen they have collected into their pollen baskets, others clean it from their bodies and discard it. Free (1968b) marked some honeybees collecting pollen and others collecting nectar only, and recorded their behaviour during the next 2 days. An average of 70% of the bees kept constant to the type of forage they were collecting when marked. More bumblebees than honeybees had pollen loads, so perhaps they are less likely to discard pollen; this needs investigating.

Honeybees spent 9 sec/flower when collecting pollen and 8 sec when collecting nectar only; they visited about 50 flowers/trip each of which was about 7 min duration. Bumblebees worked slightly faster, and although they were fewer than honeybees their numbers fluctuated less. Petkov (1963b) observed that honeybees spent 11 sec/flower on one raspberry variety, and 13 sec/flower on another; the latter variety tended to have more nectar and was visited for slightly longer each day.

Chapter 37

Liliaceae

The family Liliaceae includes several cultivated members of the genus *Allium* (i.e. *A. cepa* L., common onion or shallot; *A. ampeloprasum* L., leek; *A. chinese* L., rakkyo; *A. sativum* L., garlic; *A. fistulosum* L., Welsh onion; *A. schoenoprasum* L., chive and *Asparagus officinalis* L., asparagus).

Allium cepa L.

The inflorescence of *A. cepa*, onion, is a terminal umbel borne on a stalk 1–2 m tall. It contains 50–200 individual flowers arranged in smaller inflorescences (cymes) of 5–10 flowers each. Within each small cyme the flowers open in a definite sequence with a delay between each. A plant may flower for 30 days or more.

A single flower (Fig. 159) has three inner and three outer perianth lobes and six stamens in two whorls of three each. The anthers of the inner whorl of stamens dehisce at irregular intervals before those of the outer whorl and the process is usually completed in 24–36 h and before the stigma becomes receptive. Most pollen is presented between 09.00 and 17.00 h. Nectaries occur at the bases of the three inner stamens, and drops of nectar accumulate between their broad filaments and the ovary wall. The superior three-celled ovary has two ovules in each cell. The style, which has a three-lobed stigma, is approximately 1 mm long when the flower opens and does not reach its maximum length of about 5 mm until 1 or 2 days after the pollen has been shed. The stigma then becomes receptive.

The delay in stigma receptivity decreases the likelihood of individual flowers being self-pollinated, but because an umbel contains flowers at different stages of development and an insect usually visits many of the flowers of an umbel before leaving it (Jones, 1937), self-pollination between flowers of the same umbel frequently occurs; most plants are highly self-compatible (Jones, 1923). It seems that wind and gravity play minor roles in pollination. Jones and Emsweller (1933) reported that 157 seeds were produced per bagged head compared to 712 seeds per head not bagged; bagged heads that were staked to minimize wind action produced only 54 seeds per head. Wind may also influence the direction of flight of pollen-carrying insects and thus the direction in which the pollen is transferred (see Agati, 1952). Plant breeders accomplish selfing by enclosing one or more umbels in paper or muslin bags, and tap the bags every day to help distribute the pollen, usually in the afternoon when the pollen is dry. It is common practice to put blowflies into the bags to help transfer the pollen (page 127).

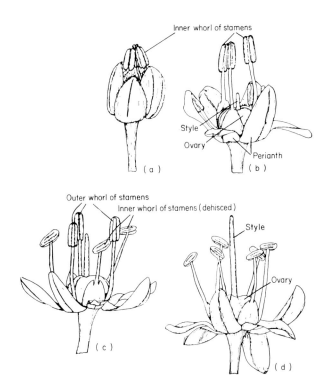

Fig. 159. Anthesis and anther dehiscence of *Allium cepa*, onion: a, perianth expanding and inner whorl of stamens elongating; b, inner whorl of stamens about to dehisce; c, inner whorl has dehisced and outer whorl is elongating; d, both whorls have dehisced (after Jones and Rosa, 1928).

Under natural conditions, onions are commonly cross-pollinated by insects, especially honeybees (Treherne, 1923) and solitary bees and Diptera (Lederhouse *et al.*, 1968). Jones (1923) thought that the numerous thrips present in flowers might aid pollination. In Russia and Italy, it has been found that honeybees are the most frequent visitors to onion crops, but other insects, especially Hymenoptera and Diptera, are also present (Trofimec, 1940; Agati, 1952).

In New York State Lederhouse *et al.* (1968) found that wild bees and Diptera comprised 52% and 34% respectively of insects visiting onion flowers. They considered that the nectar sugar concentration, of between 30 and 50%, was sufficiently great to ensure that onion flowers competed favourably with other flower species for insect visits.

According to Peto (1950) excellent yields of hybrid seed have been obtained by alternating four rows of male sterile and four rows of male fertile lines. Lederhouse *et al.* (1968) suggested that two female rows should alternate with one male row; they pointed out that leaving gaps in the male rows might encourage bees to move from male to female plants; the value of such a procedure should be investigated.

Cross-pollination between varieties is also common. Darwin (1876) grew 4 varieties of onions near each other and raised 46 progeny from the seed, 31 of which were hybrids. Jones and Rosa (1928) recommended that when varieties are grown for seed they should be planted at least 0·8 km apart, but Jones and Mann (1963) concluded that complete isolation is practically impossible as insects carry pollen from one field to another over great distances, and they suggested that in seed-growing areas different localities should be designated for certain varieties only, with localities at least 4·8 km apart.

The floral mechanisms of other cultivated *Allium* species appear to be similar to that of *A. cepa* and the limited information available also indicates they are pollinated by bees. Hybridization between *A. cepa* and *A. fistulosum* can occur, the latter species being the female parent the more frequently. The other species do not hybridize between themselves or with *A. cepa* and *A. fistulosum*.

Asparagus officinalis L.

As a rule, the staminate and pistillate flowers of *A. officinalis*, asparagus, (Fig. 160) are borne on different plants, although hermaphrodite flowers sometimes do occur. The staminate flowers are slightly larger than the pistillate and open first. The flower has six perianth segments, six stamens which are rudimentary in the pistillate flower, and a style with three stigmas which are rudimentary in the staminate flower. The nectaries are at the base of the perianth segments.

Because the flowers are dioecious cross-pollination and cross-fertilization occurs. Although male and female plants are equally abundant in natural populations only one male is needed for every five or six female plants when producing seed crops (see Hawthorn and Pollard, 1954). However, too few male plants may be detrimental to seed production as Huyskes (1959) found that in an asparagus crop in Holland, a male plant usually pollinated only those female plants within a 1·5 m radius of it.

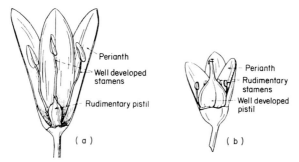

Fig. 160. Flower of *Asparagus officinalis*, asparagus: a, median section male flower; b, median section female flower (after Robbins, 1931 with permission of McGraw-Hill Book Company).

Jones and Rosa (1928) thought that wind pollination of asparagus was negligible and that bees were the main insect pollinators. Eckert (1956) caged one female and two male asparagus plants together to exclude insects, and obtained only 6 g of seed compared to 775 g from a similar female plant outside the cage. He obtained indications that taking honeybee colonies to asparagus crops increased their seed yield but more experiments are needed to prove this. He suggested that $2\frac{1}{2}$–5 colonies of honeybees should be provided per hectare of asparagus.

Chapter 38

Other Families

I have included in this chapter crops belonging to families that have not already been discussed. Most of them are only of minor importance and there is little information about their pollination requirements. I have listed them systematically.

Papaveraceae

Papaver somniferum L.

The flower of *P. somniferum*, medicinal, opium or oil poppy, consists of 2 sepals which fall when the flower opens, 4 petals, numerous stamens, a single-chambered ovary with many ovules, and a large multi-lobed stigma (Fig. 161).

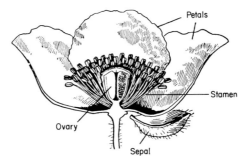

Fig. 161. Median section of *Papaver somniferum*, oil or opium poppy, flower (after Groom, 1906).

Before anthesis the stigma is thickly covered with pollen (Knuth, 1906). There are no nectaries but like other species of poppy, *P. somniferum* is vigorously worked by bees and other insects for pollen.

An experiment by Darwin (1876) indicates that automatic self-pollination and self-fertilization of this species occurs; 30 capsules each from covered and uncovered plants gave 16·5 and 15·6 g seed respectively. However, according to Howard *et al.* (1910), although some self-pollination occurs before the flowers open, cross-pollination by insects also occurs, and when flowers are bagged, they set seed less than normal. Further experiments are needed to determine any possible benefit of self- and cross-pollination by insects.

Theaceae

Camellia sinensis (L.) O. Kuntze

The flowers of *C. sinensis*, tea, which are 2·5–4 cm diameter, occur either singly or in groups of up to four. The calyx and corolla are each comprised of five to seven parts (Fig. 162); the petals are united with each other at their

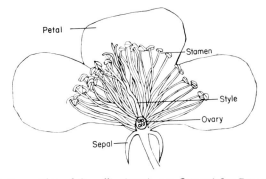

Fig. 162. Median section of *Camellia sinensis*, tea, flower (after Purseglove, 1968).

bases and also to the filaments of the outer group of numerous stamens. The superior ovary has three or four carpels, each containing four to six ovules; the styles which are equal in number to the carpels are united at their bases. The flowers open in the afternoon and remain open for 2 days.

Whereas it is generally agreed that tea is largely, if not completely, self-sterile and needs cross-pollination by insects to produce seed (e.g. Cobley, 1956; Eden, 1965; Purseglove, 1965), I can find no information as to which insects are responsible.

Bombacaceae

Ceiba pentandra Gaertn

Flowers of *C. pentandra*, kapok or silk-cotton tree, occur in dense clusters, and each is on a pedicel 8 cm long. The flower has a five-lobed calyx, and five fleshy petals which form a short broad tube and spread out to form a showy flower 5 or 6 cm in diameter. Nectar is secreted at the base of the flower so copiously that it tends to run out of the corolla at anthesis. The sexual parts consist of five stamens, which produce sticky pollen, a five-celled ovary with many ovules, and a single five-lobed stigma, which is either placed among the anthers or bent so that it projects obliquely from the flower. In the latter circumstance cross-pollination is favoured but, as isolated trees fruit and seed freely, self-pollination and self-fertilization commonly occur. A flower opens rapidly about half an hour after dark, remains open all night and persists until late afternoon of the next day when the corolla and stamens fall.

Baker and Harris (1959) observed that in Ghana the flowers were visited by the fruit bats *Epomophorus gambianus*, *Nanonycteris veldkampii* and *Eidolon helvum*, and at least forty bats occupied a tree in full flower. The bats crawled over the branches to visit the flowers from which they lapped the nectar and in so doing knocked off some of the anthers and accumulated much pollen on their fur. Baker and Harris suggested that bat pollination of kapok might also be important in the New World tropics. They also saw a large hawkmoth and several small moths visiting the flowers, but supposed these insects made only a minor contribution to pollination. Although, soon after dawn the flowers were visited by numerous bees (see also Pynaert, 1953) they thought that most of the flowers would have been pollinated by then. However, no experiments to compare the value of bat and insect visits have been made. To set fruit, 20–120 ovules need fertilizing per flower and the size of pod and the amount of kapok increases with the number of seeds present (see Purseglove, 1968).

Tiliaceae

This family includes *Corchorus capsularis* L. and *C. olitorius* L. which are the source of commercial jute.

Corchorus capsularis L.

The flowers of *C. capsularis*, white jute or round-podded jute, are small and inconspicuous with five sepals, five petals, ten or more stamens, a short style, and a superior five-celled ovary with numerous ovules. Howard *et al.* (1919) reported that in India the flowers open between 08.00 and 10.00 h and remain open about 5 h. Anther dehiscence occurred in the bud about an hour before the flower opened; automatic self-pollination and fertilization seemed the general rule and isolated flowers set abundant seed. Only a small amount of cross-pollination occurred; although bees visited the flowers they did not do so in large numbers. Kundu *et al.* (1959) found that up to 10% cross-pollination may occur although usually only 2 or 3%.

Corchorus olitorius L.

According to Howard *et al.* (1919) the small flowers of *C. olitorius* (Tossa jute, Jew's mallow, long-podded jute) open between 06.00 and 09.00 h and remain open for 3 h only. Just before the flower opens the filaments of the stamens elongate so the anthers surround and surmount the stigma. Anther dehiscence occurs in the bud or as the flower is opening. Automatic self-pollination and self-fertilization occur. Whereas Howard *et al.* found that few bees visited the flowers and cross-pollination was rare, Kundu *et al.* (1959) found that an average of 12% cross-pollination occurred.

Malpighiaceae

Malpighia glabra L.

Yamane and Nakasome (1961) found that in Hawaii *M. glabra*, Barbados West Indian cherry, normally has a low set which can be increased by hand pollination. However, little of its pollen was carried by wind and introduction of honeybee colonies into orchards failed to increase the set. Hence its pollinating mechanism is still unknown.

Sapindaceae

Litchi chinensis Sonn.

The flower of *L. chinensis*, lychee, has a cup-shaped calyx with four or five divisions but no corolla. There are three types of flower (Mustard *et al.*, 1953). The male flower has five to eight stamens, with functional anthers arising from a fleshy disc which secretes nectar but there is no style. One type of hermaphrodite flower has functional anthers but the style remains small and the stigma lobes never separate; pollen from this type of flower is, on average, more viable than from the male flowers. The other type of hermaphrodite flower has a style which grows to full size and the stigma opens to two or three lobes, but the anthers never open.

In India, Khan (1929) observed that many insects of several different species visited lychee flowers, and recently Pandey and Yadav (1970) have reported that honeybees *Apis dorsata*, *A. cerana*, *A. florea* and *Melipona* spp. were predominantly present in two Indian orchards, but hoverflies, black ants, *Musca* spp. and *Vespa* spp. were also regular visitors. The honeybees were most abundant between 06.30 and 12.00 h when the peak of anther dehiscence also occurred; after a decline in numbers during the heat of midday, foraging increased again at about 15.00 h but the bees were not as abundant as in the morning. In Florida honeybees, the screw-worm fly, *Callitroga macellaria*, and the soldier beetle, *Chauliognathus marginatus*, were most abundant (Butcher, 1957). Honeybees worked the flowers in the morning but deserted them in the afternoon when no nectar was present. Hence it does not seem that lychee flowers are a particularly attractive source of pollen in Florida; the absence of wild bees on the flowers tends to support this assumption.

Despite the abundance of insect visitors, fruit set in general is low. Khan (1929) reported that only 1·6% of female flowers set fruit in India. Although Mustard *et al.* (1953) found that initial set was high in Florida, a large loss, probably due to embryo abortion and lack of fertilization, occurs during the first 2–4 weeks after setting, and an average of only about 5% of female flowers produce mature fruit. Butcher (1956) found that a lychee tree caged to exclude insects produced only one fruit compared to an average of 99 fruits produced by four uncaged adjoining trees, which could be visited by a nearby honeybee colony. Pandey and Yadav (1970) found that averages of between 2·7 and 4·9% flowers set fruit on unbagged inflorescences compared to only 0·03–0·1% set on bagged inflorescences.

Caricaceae

Carica papaya L.

C. papaya, pawpaw, is the only economically important member of the Caricaceae and is cultivated throughout the tropics. Sometimes, male and female flowers are on separate trees, sometimes they are on the same tree, and sometimes there are hermaphrodite flowers. Sometimes, also, a tree may

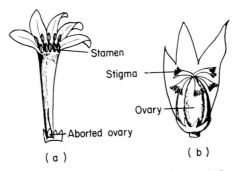

Fig. 163. Flower of *Carica papaya*, papaw: a, median section male flower; b, median section female flower (after Nicholls and Holland, 1940).

change from producing one type of flower to producing another, and in particular may change from producing male to producing hermaphrodite flowers. The male flower (Fig. 163a) has ten stamens, the female flower (Fig. 163b) a single pistil and five stigmas, and the hermaphrodite flower has five stamens and a pistil, but many variations and gradations occur between these three types (see Storey, 1958). Very little is known about the pollination of *C. papaya* flowers, although it has been suggested that they are pollinated by wind, by small insects such as thrips, and by nocturnal moths. Allan (1963) demonstrated that, in South Africa, pawpaw is pollinated by insects, especially honeybees. It has been reported that female trees 730 m from the nearest male tree were pollinated and set fruit, and in a plantation one male per ten female trees is regarded as sufficient (Greenway and Wallace, 1953; Purseglove 1968).

Apocynaceae

Carissa grandiflora A. DC.

The pollination of *C. grandiflora*, Natal plum or African carissa, in California has been studied by Schroeder (1951). Varieties either have short-styled flowers with functional anthers or long-styled flowers with sterile

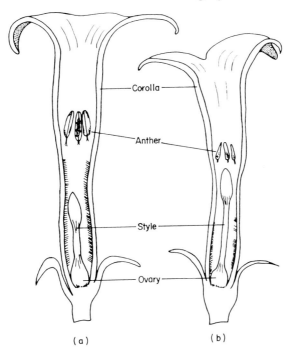

Fig. 164. Flower of *Carissa grandiflora*, Natal plum: a, median section of flower with short style and functional anthers; b, median section of flower with long style and sterile anthers (after Schroeder, 1951).

anthers (Fig. 164). The latter type of variety tended to give poor fruit yields, particularly when planted in isolation, and artificial cross-pollination gave increased yields. However, few insects were observed to visit the flowers.

Honeybees occasionally collected nectar from the nectary at the base of the pistil, but usually only after the corolla tube had fallen off, and very few entered the flower to do so. Many of the flowers contained thrips but their value as pollinators has not been ascertained.

Convolvulaceae

Ipomoea batatas L.

I. batatas, sweet potato, is the only economically important species in this family. Each of the isolated flowers (Fig. 165) has a five-lobed calyx, a funnel shaped corolla, five stamens attached to the base of the corolla, a single style with a two-lobed stigma and a two-celled ovary. The flowers open during the latter part of the night and usually remain open during the morning only. Thereafter they close and wilt, but they may remain open all day in cool weather. The stigma is receptive in the evening before opening and the anthers dehisce at about midnight.

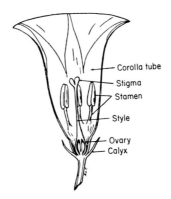

Fig. 165. Median section of *Ipomoea batatas*, sweet potato, flower (after Robbins, 1931, with permission of McGraw-Hill Book Company).

The stamens are of different lengths and the two largest are either level with or taller than, the stigma. Thompson (1925) supposed that this facilitates self-pollination although he noted that a large proportion of flowers failed to set seed and the great variation in progeny that were produced when different varieties were grown together indicated that cross-pollination occurred. It has since been found that all cultivars are self-sterile and some show various degrees of cross-incompatibility; fertilization only occurs from cross-pollination between compatible cultivars (see Purseglove, 1958).

Thompson (1925) saw several Hymenoptera, including *Melissodes rufodentata*, *M. foxi*, and the honeybee visiting the flowers. There is apparently no other information on the agents responsible for cross-pollination.

Pedaliaceae

Sesamum indicum L.

The important crop of this family is *S. indicum*. In India Howard *et al.* (1919) reported that the flowers open between 03.00 and 04.00 h and are shed about 12 h later. The flower consists of a tubular corolla, four stamens, and a superior ovary of two united carpels containing many ovules, surmounted by a single style. A disc of nectariferous tissue surrounds the ovary, and there is an extrafloral nectary on either side of the base of the flower stalk.

The anthers dehisce and the stigma is receptive at about the time the flower opens. Insects do not visit the flowers until about 06.00 h by which time self-pollination has occurred and the stigmas are usually covered with pollen. Howard *et al.* found that covered flowers readily set seed so self-fertilization probably generally occurs. However, they noticed that many flowers had aborted anthers and so would need to be cross-pollinated and they suggested that bee visits accomplished this.

Langham (1944) reported that under field conditions 4·6% cross-pollination occurred, and attributed this to visits by insects, mostly honeybees.

Piperaceae

Piper nigrum L.

The inflorescence of *P. nigrum*, pepper, consists of a spike, up to 25 cm long of 50 to 100 flowers borne on the axils of ovate fleshy bracts. Hermaphrodite flowers predominate in most cultivated varieties, although varieties with unisexual flowers do occur. The hermaphrodite flower has no perianth, but has two to four short fleshy stamens, a single-celled globose ovary with a single ovule, and a 3 to 5 branched stigma (Fig. 166). The flowers are protogynous and the anthers do not dehisce until several days after the stigmas are first exserted.

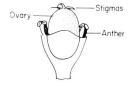

Fig. 166. Flower of *Piper nigrum*, pepper (after Hutchinson, 1959).

Pollination has been studied by Martin and Gregory (1962) in Puerto Rico. They found that the stigma remains receptive for 10 days but peak receptivity occurs 3–5 days after exsertion. A spike flowers for 7–8 days; the flowers at the base of the spike open first and opening proceeds in sequence toward the tip. As the spike is pendent, self-pollination from anthers of the older flowers to the stigmas of the younger flowers probably occurs by gravity, possibly aided by wind and rain. Martin and Gregory tried to test the supposition (see Anandan, 1924, Menon, 1949) that wind and rain are the main pollinating agents, and that the rain drops break up aggregations of pollen grains and distribute them. They found that the abundance and viability of pollen differed with the variety concerned, but even within a variety the set was erratic and differed greatly on different inflorescences of the same plant; within a spike fruit set was haphazard and not related to the sequence of anthesis. These findings indicated that set was dependent on some irregular process and mainly occurred within a spike. Bagging tended to damage the stigmas and give a lower set than usual, but bagged flowers of hermaphrodite varieties set fruit showing that they were self-fertile and that self-pollination could occur without the intervention of wind or rain. They also recorded very high sets during the dry months, indicating that rain is not necessary for pollination to occur. Furthermore, heavy rains washed most of the pollen from the spikes onto the ground. Six spikes protected by polyethylene covers from heavy rain, but not from dew or condensation, had 69% of their flowers set compared to 55% on unsheltered spikes. However, drops of dew and light rain on the spikes contained many pollen grains and so could have helped distribute the pollen.

A clone with unisexual flowers had a poor set compared to clones with bisexual flowers (15 : 86%) even though plants were located only 1·2–15·2 m from adequate pollen sources. No insects were seen to visit the female flowers, so such pollination that did occur was probably due solely to wind; clearly this was not very efficient and indicates that wind is responsible for little cross-pollination.

Many small insects were found on hermaphrodite spikes, the most common being a springtail (Collembola) that carried pollen on its antennae and legs. However, their importance as pollinators has yet to be determined. Possibly their greater abundance in favourable conditions may help compensate for any decreased pollination by wind or rain.

Myristicaceae

Myristica fragrans Houtt

The only economically important species of this family is *M. fragrans*, nutmeg, mace. The small bell-shaped flowers have no petals and the calyx consists of three sepals which are fused to form a tube narrowing at the top (Fig. 167). The female flowers have a single-celled ovary and two short stigmas and the male flowers have eight to twelve anthers, whose fused filaments form a central column. Both male and female flowers have nectaries at the base of the calyx tube. Male and female flowers are usually on separate trees, although occasionally trees with both occur. Occasional hermaphrodite flowers also occur; male trees sometimes tend to produce hermaphrodite flowers as they age.

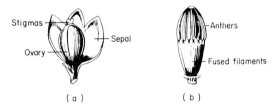

Fig. 167. *Myristica fragrans*, nutmeg: a, female flower; b, anther column (after Nicholls and Holland, 1940).

Pollination is said to be accomplished by small insects and by wind. In plantations one male tree is provided for every eight to ten female trees, and it has been suggested that it is especially important for male trees to be on the windward side of the plantation (Nicholls and Holland, 1940).

Orchidaceae

Vanilla

Vanilla planifolia And. is the principal source of vanilla, but *V. pompona*, Schiede, and *V. tahitensis*, Moore, are also cultivated commercially. The greenish-yellow flowers occur in racemes of twenty or more; one to three flowers open per raceme per day and each flower lasts for one day only.

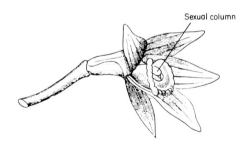

Fig. 168. Flower of *Vanilla* spp. (after Nicholls and Holland, 1940).

The flower (Fig. 168) has three sepals and three petals, one of which is modified and enlarged to form the trumpet-shaped lip; in the centre of the flower is the column, or gynondrium, which supports the sexual organs.

The anthers which occur at the top of the column produce pollen grains which adhere together to form masses called pollinia. The stigma is lower on the column and a beak-like projection, the rostellum, separates it from the pollinia so natural self-pollination is impossible. The three-celled ovary has numerous ovules. Cross-pollination produces fertile seed but artificial self-pollination produces sterile seed.

In Central and South America, where *V. planifolia* is indigenous, the flowers are visited by bees of the genus *Melipona* and by hummingbirds, which collect nectar secreted at the base of the lip, and while doing so transfer the pollinia to the stigmas of other flowers. However, in other parts of the world where natural cross-pollination does not occur, seed is produced by hand pollination (Fig. 169). In this process a pointed stick is used to detach the hood which

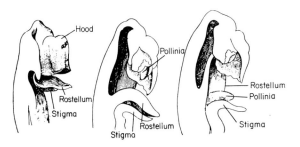

Fig. 169. Diagrams of sexual column of *Vanilla* spp. showing stages in artificial fertilization (after Nicholls and Holland, 1940).

covers the anthers and pollinia, and to slip the rostellum out of the way under the anthers; there is then no obstacle between the pollinia and stigma and they are gently pressed together. Artificial pollination is done during the morning the flowers open and to ensure maximum seed production the plantation must be hand pollinated every day during the flowering season. A trained worker can pollinate an average of 1,000–2,000 flowers/day (see Stanford, 1937; Nicholls and Holland, 1940; Correll, 1953). Perhaps the importation of natural pollinators of *V. planifolia* into other areas where it is grown commercially would be worthwhile, but even in Mexico only a small percentage of seed is set naturally, so hand pollination is practised to increase the crop.

Bromeliaceae

Ananas sativus Schult. f.

The flower of *A. sativus*, pineapple, has three sepals and three petals which are free from each other but form a long narrow tube enclosing the six stamens and the style with three stigmatic lobes. The ovule is of three united carpels with ten to fifteen ovules each. Flowers open early in the morning and begin to wither in the late afternoon; there are about 150 flowers per inflorescence and, as only five to ten open per day, an inflorescence is in flower for about 3 weeks.

The narrow mouth of the corolla only allows small insects to enter the flower. Bees are too large to reach the nectar from the mouth of the flower, but sometimes learn to obtain it by pushing their tongues between the petals at the flower base. In Central and South America, various species of humming birds visit the flowers.

All pineapple varieties are self-incompatible, but set seed when cross-pollinated. However, the pineapple is parthenocarpic and, except for breeding purposes, seed production is generally undesirable and the presence of seeds lowers the economic value of the fruit. Varieties which regularly produce seed in Central and South America are seedless in Hawaii, where no natural pollinators exist. In an attempt to maintain this situation the importation of humming birds into Hawaii is forbidden (Collins, 1960).

Musaceae

Musa spp. L.

Pollination is essential for fruit set of the wild seeded banana, but the edible bananas produce fruit parthenocarpically; their ovules shrivel at an early stage of growth and they do not normally produce viable pollen (see Simmonds, 1957); therefore, although they are visited by honeybees and hummingbirds (Mahadevan and Chandy, 1959; Chapman, 1964) to collect their copious nectar, this does not influence fruit production.

Palmae

Cocos nucifera L.

C. nucifera, coconut, is monoecious and male and female flowers are born on the same spadix. Nearly all of the hundreds, or thousands, of flowers on a spadix are male but there are usually a few female flowers at its base, although sometimes only one occurs, the number depending upon climatic and environmental conditions.

The male and female flowers each have three sepals and three petals (Fig. 170). The male flower has an abortive pistil whose apex is divided into three parts, each with a nectar gland, and six stamens which shed large amounts of powdery pollen before, or simultaneously with, the splitting of the perianth lobes, and by the time the flowers are fully open most of the pollen has been discharged. The female flower has a short style with three stigmas and three ovules, only one of which is fertile. Each type of flower has nectaries. The coconut flower develops very slowly; about a year passes from the time a female flower is differentiated to the opening of the spathe, and another year before the nuts are ripe. The following description is based on reviews by Copeland (1931), Menon and Pandalai (1958) and Whitehead (1965).

Usually only one spadix per plant flowers at a time. In the tall varieties the male flowers of a spadix start opening about a month earlier than the female, the uppermost flowers opening first. Each male flower opens for one day only. As the female flowers usually do not open until 3–6 days after the

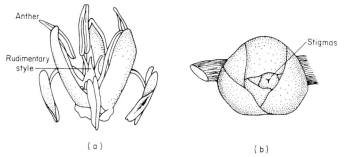

Fig. 170. Flower of *Cocos nucifera*, coconut: a, male; b, female (after Tammes, 1937).

end of the male phase, no self-pollination can occur within a spadix. However, pollination from the male flowers of the next spadix to flower is sometimes possible, and in some climatic conditions several inflorescences are likely to flower within a short time. Normally, however, the succession of inflorescences is such that cross-pollination from another tree is necessary.

The female flowers of dwarf varieties start flowering about a week after the male ones so, for much of the flowering period of a spadix, both types of flower are open and self-pollination generally occurs. Because pollen remains viable for 2–9 days it is possible to save pollen from the male flowers to hand pollinate the female ones should adverse weather delay opening of the female flowers; under the right conditions pollen can be stored for several months (Whitehead, 1963).

Although the flowers need pollination to produce nuts, there seems to be little information on the amount of set which normally occurs, or on the agents responsible. The relative importance of wind and insects is a matter of controversy. Menon and Pandalai (1958), Mahadevan and Chandy (1959) and Chapman (1964) reported that ants, bees and other insects visit the flowers for nectar and Whitehead (1965) observed 103 bees collect nectar from a single female flower during 30 min and after each visit the nectar was rapidly replaced. Hence, it seems likely that insect pollination is important, although its value has yet to be demonstrated.

Phoenix dactylifera L.

The flowers of *P. dactylifera*, date palm, are dioecious and only very rarely are hermaphrodite flowers present or both sexes present on the same tree. The clusters of flowers are ensheathed in spathes at the axils of the leaves; when the leaves mature the spathes split open. The female flower consists of a spherical ovary of three carpels surmounted by three hooked stigmas enclosed within six perianth segments. The male flower consists of six stamens and a rudimentary ovary also enclosed in six perianth segments, the outer three of which are much diminished. Artificial pollination of the date palm is traditionally practised in Arabic-speaking countries; sprigs of male flowers are secured to each female inflorescence as soon as it has burst from its spathe, sometimes after first shaking the male sprig over the female flowers. Popenoe (1913) reported that, in America pollen was carried 0·4 km or more, possibly by insects; but information on the agents responsible for pollen transfer seems to be lacking. Nicholls and Holland (1940) thought that at least one male tree should be planted to every hundred female trees.

References

Adlerz, W. C. (1966). Honey bee visit numbers and watermelon pollination. *J. econ. Ent.* **59,** 28–30.
Afzal, M. and Khan, H. (1950). Natural crossing in cotton in western Punjab. *Agron. J.* **42,** 14–19; 89–93; 202–205; 236–238.
Agati, G. (1952). Indagini ed osservazioni sulla biologia fiorale della cipolla. *Riv. Ortoflorofruttic. ital.* **36,** 67–77.
Akamine, E. K. and Girolami, G. (1957). Problems in fruit set in yellow passion fruit. *Hawaii Fm Sci.* **5,** 3–5.
Akamine, E. K. and Girolami, G. (1959). Pollination and fruit set in the yellow passion fruit. *Tech. Bull. Hawaii agric. Exp. Stn.* **39.**
Åkerberg, E. (1952). Factors in obtaining high yields of legume seed. *Proc. Int. Grassld Congr.* **6,** 827–834.
Åkerberg, E. (1953). Blombiologien hos rödklövern och pollineringen. *Medd. Sverig. Fröodl Förb.* **2** 16–33.
Åkerberg, E. and Hahlin, M. (1953). Undersokningar rörande tillgangen på pollinerande insekter i rödklöverfröodlingar under aren 1949–1952. *Medd. Sverig. Fröodl Förb.* **1** 122–130.
Åkerberg, E. and Lesins, K. (1949). Insects pollinating alfalfa in Central Sweden. *K. LantbrHögsk. Annlr* **16,** 630–643.
Åkerberg, E. and Stapel, C. (1966). A survey of pollination and seed growing of red clover in Europe. *Bee Wld* (Suppl.), **47,** 15–42.
Åkerberg, E. and Umaerus, M. (1960). Pollination and seed setting in red clover, lucerne and rape under Northern conditions. *XVII Int. Beekeep. Congr.* **2,** 185–201.
Akhter, A. R. (1932). Studies in Indian Brassicae. I. Sterility and selective pollen tube growth. *Ind. J. agric. Sci.* **2,** 280–292.
Alex, A. H. (1957a). Pollination of some oilseed crops by honeybees. *Progr. Rep. Texas agric. Exp. Stn.* No. 1960.
Alex, A. H. (1957b). Honeybees aid pollination of cucumbers and cantaloupes. *Glean. Bee Cult.* **85,** 398–400.
Alex, A. H., Thomas, F. L. and Warne, B. (1950). Importance of bees in vetch seed production. *Prog. Rep. Texas agric. exp. Stn.* No. 1306.
Allard, H. A. (1910). Preliminary observations concerning natural crossing in cotton. *Am. Cott. Breeders. Mag.* **1,** 247–261.
Allan, P. (1963). Pollination of pawpaws. *Fmg S. Afr.* No. 81, 1–3.
Allen, M. Delia (1955). Observations on honeybees attending their queen. *Brit. J. anim. Behav.* **3,** 66–69.

Allen, M. Delia (1959). Respiration rates of worker honeybees of different ages and at different temperatures. *J. exp. Biol.* **36,** 92–101.
Allen, M. Delia (1965). The role of the queen and males in the social organization of insect communities. *Symp. zool. Soc. Lond.* No. 14, 133–157.
Allen, M. Delia and Jeffree, E. P. (1956). Influence of stored pollen and of colony size on the brood rearing of honeybees. *Ann. appl. Biol.* **44,** 649–656.
Alpatov, V. V. (1948). Bee races and red clover pollination. *Bee Wld* **29,** 61–63.
Amaral, E. (1952). Ensáio sôbre a influência de *Apis mellifera* L. na polinização do cafeeiro (Nota Prévia). *Bolm Esc. agŕc. 'Luis Queiroz* **9.**
Amaral, E. (1960). Ação dos insetos na polinização do cafeeiro Caturra. *Rev. Agric., Piracicaba* **35,** 139–147.
Amos, J. M. (1943). A measurement of the value of bees in the pollination of lima beans. *Am. Bee J.* **83,** 240–241.
Amos, J. M. (1951). The effect of honeybees on the pollination of crimson clover. *Amr. Bee. J.* **91,** 331–333.
Anandan, M. (1924). Observations on the habits of the pepper vine with special reference to the reproductive phase. *Yb. Dep. Agric. Madras* 49–69.
Anderson, E. J. (1958). Pollination of crown vetch. *Glean. Bee Cult.* **86,** 281–283, 318.
Anderson, E. J. (1959). Pollination of crown vetch. *Glean. Bee Cult.* **87,** 590–593.
Anderson, E. J. and Wood, M. (1944). Honeybees and red clover pollination. *Am. Bee J.* **84,** 156–157.
Angelo, E., Brown, R. T. and Ammen, H. J. (1942). Pollination studies with tung trees. *Proc. Am. Soc. hort. Sci.* **41,** 176–180.
Ankinovič, G. B. and Ljubimov, A. A. (1954). Bees in glasshouses and frames. *Sad Ogorod* (5), 24–27.
Antles, L. C. (1953). New methods of orchard pollination. *Am. Bee J.* **93,** 102–103.
Antsiferova, T. A. (1957). Pollination of lucerne by bees. *Pchelovodstvo, Mosk.* **34,** 28–32.
Archimowitsch, A. (1949). Control of Pollination in Sugar-Beet. *Bot. Rev.* **15,** 613–628.
Arizona University. (1961). Melon pollination, bees and insecticides. *Folder Univ. Ariz. agric. Exp. Sta.* **90.**
Armstrong, J. M. and Jamieson, C. A. (1940). Cross-pollination of red clover by honeybees. *Sci. Agric.* **20,** 574–484.
Armstrong, J. M. and White, W. J. (1935). Factors influencing seed-setting in alfalfa. *J. agric. Sci.* **25,** 161–179.
Armstrong, W. D. (1936). New varieties and pollination of Muscadine grapes. *Proc. Am. Soc. hort. Sci.* **35,** 450–452.
Artschwager, E. (1927). Development of Flowers and seed in the sugar beet. *J. agric. Res.* **34,** 1–25.
Atkinson, W. T. and Constable, E. F. (1937). A honeybee technique in seed production of selected cruciferous plants. *Agric. Gaz. N.S.W. Misc. Publ.* **48,** 554–555, 592.
Atwood, C. E. (1933). Studies on the *Apoidea* of Western Nova Scotia with special reference to visitors to apple bloom. *Can. J. Res.* **9,** 443–457.
Atwood, S. S. (1943). Natural crossing of white clover by bees. *J. Am. Soc. Agron.* **35,** 862–870.
Auchter, E. C. (1924). The importance of proper pollination in fruit yields. *N.J. St. hort. Soc. News* 133–142.
Auchter, E. C. and Knapp, H. B. (1937). "Orchard and Small Fruit Culture." John Wiley, New York.

Avetisyan, G. (1958). Bees—the cotton farmers' helpers. *Indian Bee J.* **20**, 28–29.
Baculinschi, H. (1957). Productivitatea de nectar a culturilor de floareasoarelui în zona de Stepă. *Apicultura, Bucuresti* **30**, 9–10.
Bader, K. L. and Anderson, S. R. (1962). Effect of pollen and nectar collecting honeybees on the seed yield of birdsfoot trefoil, *Lotus corniculatus* L. *Crop Sci.* **2**, 148–149.
Bailey, J. S. (1938). Pollination of the cultivated blueberry. *Proc. Am. Soc. hort. Sci.* **35**, 71–72.
Bailey, M. E., Fieger, E. A. and Oertel, E. (1954). Paper chromatographic analyses of some southern nectars. *Glean. Bee Cult.* **82**, 401–403, 472–474.
Baker, H. G. (1961). The adaptation of flowering plants to nocturnal and crepuscular pollinators. *Q. Rev. Biol.* **36**, 64–73.
Baker, H. G. (1963). Evolutionary mechanisms in pollination biology. *Science, N.Y.* **139**, 877–883.
Baker, H. G. and Harris, B. J. (1959). Bat pollination of the silk-cotton tree (*Ceiba pentandra* (L.) Gaerth. (*Sensu lato*), in Ghana. *Jl. W. Afr. Sci. Ass.* **5**, 1–9.
Baker, H. G. and Hurd, P. D. (1968). Intrafloral Ecology. *A. Rev. Ent.* **13**, 385–414.
Baldini, E. and Pisani, P. (1961). Ricerche sulla biologia florale e di fruttificazione del ribes nero. *Riv. Ortoflorofruttic. ital.* **45**, 6, 619–639.
Balfour-Browne, F. (1925). "Concerning the Habits of Insects." Cambridge University Press, London.
Balls, W. L. (1912). "The Cotton plant in Egypt." MacMillan, London.
Balls, W. L. (1929). The Natural Crossing of Cotton Flowers in Egypt. *Bull. Minist. Agric. Egypt* **89**.
Banks, C. J. (1951). Syrphidae as pests of cucumbers. *Entomologist's mon. Mag.* **86**, 239–240.
Barbier, E. (1964). Pollinisation et fructification du Clémentinier. *Annls Abeille* **7**, 63–80.
Barrons, K. C. (1939). Natural crossing in beans at different degrees of isolation. *Proc. Am. Soc. hort. Sci.* **36**, 637–640.
Barskii, Ya. S. (1956). Training bees to pollinate grape vines. *Sad Ogorod* (4), 64.
Bateman, A. J. (1947a). Contamination in seed crops. III. Relation with isolation distance. *Heredity, Lond.* **1**, 303–336.
Bateman, A. J. (1947b). Contamination in seed crops, I. Insect pollination. *J. Genet.* **48**, 257–275.
Bateman, A. J. (1951). The taxonomic discrimination of bees. *Heredity, Lond.* **5**, 271–278.
Batjer, L. P., Williams, M. W. and Martin, G. C. (1964). Effects of N-dimethyl amino succinamic acid (B-nine) on vegetative and fruit characteristics of apples, pears and sweet cherries. *Proc. Am. Soc. hort. Sci.* **85**, 11–16.
Battaglini, M. B. (1968). Importanza delle api nella fruttificazioni di "*Cucurbita pepo*" L. *Apic ital.* **35**, 9–12.
Bawolski, S. (1961). Wplyw terminu koszenia pierwszego pokosu na plon nasion koniczyny czerwonej. *Biul. Instytut Hodowli I Aklimatyzacji Roslin* **4**, 36–39.
Beach, S. A. (1898). Self-fertility of the grape. *Bull. N.Y. St. agric. Exp. Stn.* **157**.
Beach, S. A. (1899). Fertilizing self-sterile grapes. *Bull. N.Y. St. agric. Exp. Stn.* **169**.
Beard, D. F., Dunham, W. E. and Reese, C. A. (1948). Honeybees increase clover and seed production. *Ext. Bull. Ohio agric. Exp. Stn.* No. 258.
Beckham, C. M. and Girardeau, J. H. (1954). A progress report on a study of honeybees as pollinators of crimson clover. *Mimeogr. Univ. Ga agric. Exp. Stns* **70**.

Beckman, W. and Tannenbaum, L. (1939). Insects pollinating cultivated blueberries. *Am. Bee J.* **79**, 436–437.
Beckwith, C. S. (1930). Cranberry and blueberry investigations. *Rep. New Jers. St. agric. Exp. Stn.* No. 43, 170–174.
Beghtel, F. E. (1925). The embrogeny of *Pastinaca sativa*. *Am. J. Bot.* **12**, 327–337.
Beling, Ingeborg, (1929). Über das Zeitgadächtnis der Bienen. *Z. vergl. Physiol.* **9**, 259–338.
Behlen, F. (1911). Die Honigabsonderung der Pflanzen und ihre Ursachen. Eine Literaturstudie. *Leipzig. Bienenztg* **26**, 163–167, 179–183.
Belozerova, E. I. (1960). Bees increase the seed crop from winter rape. *Pchelovodstvo, Mosk.* **37**, 38–40.
Benoit, P., Gillard, A. and Brande, J. (1948). Bijdrage tot de studie van de zaadzetting bij rode klaver (*Trifolium pratense* L.) in verband met de zaadproductie. *Meded. LandbHoogesch. Opzoek Stns Gent* **13**, 297–346.
Bergman, H. F. (1954). Flowering and fruiting characteristics of the Cranberry in New Jersey. *Proc. Am. Cranberry Grow. Ass.* Feb., 17–27.
Berkel, N. van. (1960). Zaadkoppen bij komkommers. *Jversl. Proefstn Groente-en Fruitteelt Glas Naaldwijk* (1959), 122–123.
Berkel, N. van and Vriend, S. (1957). Zaadkoppen bij komkommers. *Jversl. Proefst. Groent-en Fruitteelt Glas Naaldwijk* 1956, 117–120.
Betts, A. D. (1920). The constancy of the pollen-collecting bee. *Bee Wld* **2**, 10–11.
Betts, A. D. (1931). Research notes. *Bee Wld* **12**, 58.
Betts, A. D. (1935). The constancy of the pollen-collecting bee. *Bee Wld* **16**, 111–113.
Beutler, Ruth (1930). Biologisch-chemische Untersuchungen am Nektar von Immenblumen. *Z. vergl. Physiol.* **12**, 72–176.
Beutler, Ruth (1949). Ergiebigkeit der Trachtquellen. *Imkerfreund* **4**, 207–208.
Beutler, Ruth (1953). Nectar. *Bee Wld* **34**, 106–116, 128–136, 156–162.
Beutler, Ruth and Schöntag, A. (1940). Ueber die Nektarabscheidung einiger Nutzpflanzen. *Z. vergl. Physiol.* **28**, 254–285.
Bewley, W. F. (1963). *Commercial Glasshouse Crops*. Country Life Books, London.
Bhambure, C. S. (1957). Importance of honeybees in the pollination of *Luffa aegyptiaca*. *Indian Bee J.* **19**, 144–151.
Bhambure, C. S. (1958a). Further studies on the importance of honeybees in pollination of *Cucurbitaceae*. *Indian Bee J.* **20**, 10–12.
Bhambure, C. S. (1958b). Effect of honey bee activity on niger seed production. *Indian Bee J.* **20**, 189–191, 195.
Bieberdorf, G. A. (1949). Some observations on pollination of alfalfa hay. *Proc. Okla. Acad. Sci.* **30**, 49–51.
Bieberdorf, G. A. (1954). Honey bees and vetch pollination. *Proc. Okla. Acad. Sci.* **33**, 36–37.
Bigger, M. (1960). *Selenothrips rubrocinctus* (Giard) and the floral biology of cashew in Tanganyika. *E. Afr. agric. J.* **25**, 229–234.
Billes, D. J. (1941). Pollination of *Theobroma cacao* L. in Trinidad, B.W.I. *Trop. Agric., Trin.* **18**, 151–156.
Bingefors, S. (1958). Svalöfs Ulva tetraploid rödklöver. Erfarenheter från försök och odling i Mellansverige. *Sver. Utsädesför. Tidskr.* **68**, 7–32.
Bingefors, S. and Eskilsson, L. (1962). Pollination problems in tetraploid red clover in central Sweden. *Pflanzenz.* **48**, 205–214.
Bingefors, S., Eskilsson, L. and Fridén, F. (1960). Insektförekomst och frösättning i fröodlingar i Mälar-Hjälmarområdet år 1959. *Svensk Frötidn.* **29**, 11–16, 17–21.
Bird, J. N. (1944). Seed setting in red clover. *J. Am. Soc. Agron.* **36**, 346–357.

Bitkolov, R. Sh. (1961). Sunflower and bees. *Pschelovodstvo, Mosk.* **38**, 20–21.

Blackwall, F. L. C. (1969). Effects of weather, irrigation, and pod-removal on the setting of pods and the marketable yield of runner beans (*Phaseolus multiflorus*). *J. hort. Sci.* **44**, 371–384.

Blackwall, F. L. C. (1964). Adequate pollination is a must for early yields. *Grower,* **61**, 1272–1273.

Blagoveshchenskaya, N. N. (1955). Forcing honeybees to pollinate lucerne. *Pchelovodstvo, Mosk.* **32**, 51–53.

Blake, G. H. (1958). The influence of honey bees on the production of crimson clover seed. *J. econ. Ent.* **51**, 523–527.

Blasberg, C. H. (1951). A preliminary report on spraying pollen to apple trees in commercial orchards. *Proc. Am. Soc. hort. Sci.* **58**, 23–25.

Blinov, N. M. (1960). Training bees for red clover. *Pchelovodstvo, Mosk.* **37**, 35–36.

Boch, R. (1956). Die Tänze der Bienen bei nahen und fernen Trachtquellen. *Z. vergl. Physiol.* **38**, 136–167.

Boch, R. (1961). Honeybee activity on safflower (*Carthamus tinctorius* L.). *Can. J. Pl. Sci.* **41**, 559–562.

Boch, R. and Shearer, D. A. (1962). Identification of geraniol as the active component in the Nassanoff pheromone of the honey bee. *Nature, Lond.* **194**, 704–706.

Boch, R. and Shearer, D. A. (1964). Identification of nerolic and geranic acids in the Nassanoff pheromone of the honey bee. *Nature, Lond.* **202**, 320–321.

Bogoyavlenskii, S. G. (1953). Bees and lucerne. *Pchelovodstvo, Mosk.* **30**, 36–41.

Bogoyavlenskii, S. G. (1955). Bees and sainfoin. *Pchelovodstvo, Mosk.* **32**, 10–14.

Bogoyavlenskii, S. G. and Kovarskaya, A. L. (1956). Determination of nectar productivity in plants. *Pchelovodstvo, Mosk.* **33**, 41–43.

Bohart, G. E. (1951). Alfalfa seed growers of Utah should protect their wild bees. *Utah Farm Home Sci.* **12**, 32, 33, 37.

Bohart, G. E. (1952). Pollination by native insects. *Yb. Agric. U.S. Dep. Agric.* 107–121.

Bohart, G. E. (1955). Notes on the habits of *Osmia* (*Nothosmia*) *seclusa* Sandhouse. *Proc. ent. Soc. Wash.* **57**, 235–236.

Bohart, G. E. (1957). Pollination of alfalfa and red clover. *A. Rev. Ent.* **2**, 355–380.

Bohart, G. E. (1958a). Transfer and Establishment of the Alkali Bee. Report, 16th Alfalfa Improvement Conference, Ithaca, N.Y., pp. 94–98.

Bohart, G. E. (1958b). Alfalfa pollinators with special reference to species other than honey bees. *Proc. 10th Int. Congr. Ent., Montreal* (1956), **4**, 929–937.

Bohart, G. E. (1960). Insect Pollination of Forage Legumes. *Bee Wld* **41**, 51–64; 85–97.

Bohart, G. E. (1962). How to manage the alfalfa leaf-cutting bee (*Megachile rotundata* Fabr.) for alfalfa pollination. *Circ. Utah agric. Exp. Stn.* 144.

Bohart, G. E. (1966). Studies on pollination of diploid and tetraploid red clover varieties in Utah. *Bee Wld* **47** (Suppl.), 43–50.

Bohart, G. E. (1967). Management of wild bees. *Agric. Handb. U.S. Dep. Agric.* No. 335, 109–118.

Bohart, G. E. and Cross, E. A. (1955). Time relationships in the nest construction and life cycle of the alkali bee. *A. ent. Soc. Am.* **48**, 403–406.

Bohart, G. E. and Knowlton, G. F. (1968). Alkali Bees. *Ext. Leafl. Utah St. Univ. agric. exp. Stn.* **78**.

Bohart, G. E. and Pedersen, M. W. (1963). The alfalfa leaf-cutting bee, *Megachile rotundata* F., for pollination of alfalfa in cages. *Crop Sci.* **3**, 183–184.

Bohart, G. E. and Todd, F. E. (1961). Pollination of seed crops by insects. *Yb. Agric. U.S. Dep. Agric.* 245.

Bohart, G. E., Nye, W. P., and Levin, M. D. (1955). Growing alfalfa for seed. III. Pollination. *Circ. Utah agric. Exp. Stn.* **135**, 42–59.

Bohart, G. E. and Nye, W. P. (1960). Insect pollinators of carrots in Utah. *Bull. Utah agric. Exp. Stn.* 419.

Bohart, G. E., Moradeshaghi, M. J. and Rust, R. W. (1967). Competition between honey bees and wild bees on alfalfa. *XXI Int. Beekeep. Congr. prelim. Sci. Mtg. Summ. Paper* **32**, 66–67.

Bohn, G. W. and Davis, G. N. (1964). Insect pollination is necessary for the production of muskmelons (*Cucumis melo* v. *reticulatus*). *J. apic. Res.* **3**, 61–63.

Bohn, G. W. and Mann, L. K. (1960). Nectarless, a yield-reducing mutant character in the muskmelon. *Proc. Am. Soc. hort. Sci.* **76**, 455–459.

Bolhuis, G. G. (1951). Natuurlijke Bastaardering bij de Aardnoot (*Arachis hypogaea*). *Landbouwk. Tijdschr.*, '*S-Grav.* **63**, 447–455.

Bolton, J. L. (1948). A study of combining ability in alfalfa in relation to certain methods of selection. *Scient. Agric.* **42**, 236–238.

Bolton, J. L. (1962). "Alfalfa: Botany, Cultivation and Utilization." Leonard Hill, London.

Bond, D. A. and Fyfe, J. L. (1962). Breeding field beans. *Rep. Pl. Breed. Inst.* 1960–61, 4–26.

Bond, D. A. and Hawkins, R. P. (1967). Behaviour of bees visiting male-sterile field beans (*Vicia faba*). *J. agric. Sci., Camb.* **68**, 243–247.

Bond, D. A., Fyfe, J. L. and Toynbee-Clarke, G. (1966). Yields of mixtures of single-cross hybrids with one of the parental inbred lines. *J. agric. Sci., Camb.* **67**, 235–237.

Bonnier, G. (1906). Sur la division du travail chez les abeilles. *C. R. hebd. Séanc. Acad. Sci., Paris* **143**, 941–946.

Bonnier, G. (1909). Les sensations des abeilles et des fourmis. *Apiculteur* **53**, 172–176, 204–209, 247–251, 285–286, 325–328.

Booth, N. O. (1902). A study of grape pollen. *Bull. N.Y. St. agric. Exp. Stn.* **224**.

Borthwick, H. A. and Emsweller, S. L. (1933). Carrot breeding experiments. *Proc. Am. Soc. hort. Sci.* **30**, 531–533.

Bradner, N. R., Frakes, R. V. and Stephen, W. P. (1965). Effects of bee species and isolation distances on possible varietal contamination in alfalfa. *Agron. J.* **57**, 247–248.

Braines, L. and Istomina-Tsvetkova, K. P. (1956). Utilizing reflected ultraviolet light for directing the flight of bees. *Pchelovodstvo, Mosk.* **33**, 48–53.

Brandenburg, W. (1961). Broad beans: causes of poor yields sought. *N.Z. Jl Agric.* **102**, 277, 279, 280.

Braun, E., MacVicar, R. M., Gibson, D. A., Pankiw, P and Guppy, J. (1953). Studies in red clover seed production. *Can. J. agric. Sci.* **33**, 437–447.

Brewer, J. W., Dobson, R. C. and Nelson, J. W. (1969). Effects of increased pollinator levels on production of the highbush blueberry, *Vaccinium corymbosum*. *J. econ. Ent.* **62**, 815–818.

Brian, A. D. (1951). The pollen collected by bumblebees. *J. Anim. Ecol.* **20**, 191–194.

Brian, A. D. (1954) The foraging of bumble bees. *Bee Wld* **35**, 61–67, 81–91.

Brian, A. D. (1957). Differences in the flowers visited by four species of bumblebees and their causes. *J. Anim. Ecol.* **26**, 69–96.

Bringhurst, R. S. (1951). Influence of glasshouse conditions on flower behaviour of Hass and Anaheim avocados. *Yb. Calif. Avocado Soc.* 164–168.

Brittain, W. H. (1933). Apple pollination studies in the Annapolis Valley, N.S., Canada 1928–1932. *Bull. Dep. Agric. Can. New Ser.* No. 162.

Brittain, W. H. (1935). Studies in bee activity during apple bloom. *J. econ. Ent.* **28**, 553–559.

Brittain, W. H. and Newton, Dorothy, E. (1933). A study in the relative constancy of hive bees and wild bees in pollen gathering. *Can. J. Res.* **9**, 334–349.

Brittain, W. H. and Newton, Dorothy E. (1934). Further observations on the pollen constancy of bees. *Can. J. Res.* **10**, 255–263.

Brown, A. G. (1951). Factors affecting fruit production in plums. *Fruit Yb.* 1950, 12–18.

Brown, H. B. (1927). Vicinism or Natural Crossing in Cotton. *Bull. Mo. agric. Exp. Stn.* **13**.

Brown, H. B. and Ware, J. O. (1958). Cotton (3rd edition). McGraw-Hill, New York.

Brown, R. T. and Fisher, E. G. (1941). Period of stigma receptivity in flowers of the tung tree. *Proc. Am. Soc. hort. Sci.* **39**, 164–166.

Buffet, M. (1960). Contribution a l'etude de l'allogamie du cotonnier. *Coton. Fibr. trop.* **15**, 391–396.

Bulatovic, S. and Konstantinovic, B. (1960). The role of bees in the pollination of the more important kinds of fruit in Serbia. *Proc. 1st Int. Symp. Pollination*, Copenhagen, pp. 167–172.

Bullock, R. M. (1948). Is commercial artificial pollination practical? *Proc. Am. pomol. Soc.* **62**, 39–49.

Bullock, R. M. and Overley, F. L. (1949). Handling and application of pollen to fruit trees. *Proc. Am. Soc. hort. Sci.* **54**, 125–132.

Bullock, R. M. and Snyder, J. C. (1946). Some methods of tree fruit pollination. *Proc. Wash. St. hort. Ass.*, 215–226.

Bunting, E. S. (1967). Break Crops: Oilseed Rape. *Agriculture, Lond.* **74**, 465–470.

Burchill, R. T. (1963). Air-borne Pollen in Apple Orchards. *Rep. E. Malling Res. Stn.* (1962), 109–111.

Burk, E. F. (1930). The role of Pistil length in the development of Forcing Tomatoes. *Proc. Am. Soc. hort. Sci.* **26**, 239–240.

Burkart, A. (1947). Adelantos recientes en la técnica de mejoramiento genético de la alfalfa. *An. Acad. Cienc. exact fis. nat. B. Aires* **12**, 39–57.

Burminstrov, A. N. (1965). The melliferous value of some sunflower varieties. Apimondia. XX Int. Beekeep. Jubilee Congress, Bucharest, 1965.

Burrel, A. B. and King, G. E. (1932). A device to facilitate pollen distribution by bees. *Proc. Am. Soc. hort. Sci.* **28**, 85–86.

Butcher, F. G. (1956). Bees pollinate lychee blooms. *Proc. Fla. Lychee Gr. Ass.* **3**, 59–60.

Butcher, F. G. (1957). Pollinating insects on lychee blossoms. *Proc. Fla. St. hort. Soc.* **70**, 326–328.

Butler, C. G. (1941). A study of the frequency with which honeybees visit red clover (*Trifolium pratense*), together with an examination of the environmental conditions. *A. appl. Biol.* **28**, 125–134.

Butler, C. G. (1943). Work on bee repellents. Management of colonies for pollination. *A. appl. Biol.* **30**, 195–196.

Butler, C. G. (1945a). The influence of various physical and biological factors of the environment on honeybee activity. An examination of the relationship between activity and nectar concentration and abundance. *J. exp. Biol.* **21**, 5–12.

Butler, C. G. (1945b). The behaviour of bees when foraging. *J. R. Soc. Arts* **93**, 501–511.

Butler, C. G. (1951). The importance of perfume in the discovery of food by the worker honeybee (*Apis mellifera* L.). *Proc. R. Soc. B.* **138**, 403–413.

Butler, C. G. (1954). The method and importance of the recognition by a colony of honeybees (*A. mellifera*) of the presence of its queen. *Trans. R. ent. Soc. Lond.* **105**, 11–29.

Butler, C. G. (1957). The control of ovary development in worker honeybees (*Apis mellifera*). *Experientia* **13**, 256–257.

Butler, C. G. (1964). Pheromones in sexual processes in insects "Insect reproduction". Symp. No. 2. *R. ent. Soc. Lond.*, 66–67.

Butler, C. G. (1967). Insect pheromones. *Biol. Rev.* **42**, 42–87.

Butler, C. G. (1969). Some pheromones controlling honeybee behaviour. *Proc. VI Congr. IUSSI, Bern*, 19–32.

Butler, C. G. and Calam, D. H. (1969). Pheromones of the honeybee—the secretion of the Nassanoff gland of the worker. *J. Insect Physiol.* **15**, 237–244.

Butler, C. G. and Haigh, J. C. (1956). A note on the use of honeybees as pollinating agents in cages. *J. hort. Sci.* **31**, 295–297.

Butler, C. G. and Simpson, J. (1953). Bees as pollinators of fruit and seed crops *Rep. Rothamst. exp. Stn.* (1953), 167–175.

Butler, C. G. and Simpson, J. (1967). Pheromones of the queen honeybees (*Apis mellifera* L.) which enable her workers to follow her when swarming. *Proc. R. ent. Soc. Lond.* (A) **42**, 149–154.

Butler, C. G., Free, J. B. and Simpson, J. (1956). Some problems of red clover pollination. *A. appl. Biol.* **44**, 664–669.

Butler, C. G., Jeffree, E. P. and Kalmus, H. (1943). The behaviour of a population of honeybees on an artificial and on a natural crop. *J. exp. Biol.* **20**, 65–73.

Butler, G. D. Jr., Todd, F. E., McGregor, S. E. and Werner, F. G. (1960). *Melissodes* bees in Arizona cotton fields. *Tech. Bull. Ariz. agric. Exp. Stn.* No. 139.

Butler, G. D., Tuttle, D. M. and Todd, F. E. (1956). Honeybees increase alfalfa seed yields. *Progve Agric. Ariz.* **8**, 3.

Butler, G. D. Jr., Werner, F. G. and Levin, M. D. (1966). Native bees associated with safflower in south central Arizona. *J. Kans. ent. Soc.* **39**, 434–436.

Buzzard, C. N. (1936a). De l'organisation du travail chez les abeilles. *Bull. Soc. Apic. Alpes-Marit.*, **15**, 65–70.

Buzzard, C. N. (1963b). Bee organisation. *Bee Wld.* **17**, 133–135.

Cale, G. H. (1968). Pollen-gathering relationship to honey collection and egg-laying in honey bees. *Am. Bee J.* **108**, 8–9.

Callow, R. J., Johnston, N. C. and Simpson, J. (1959). 10-hydroxy-Δ^2-decenoic acid in the honeybee (*Apis mellifera*). *Experientia* **15**, 421–425.

Canadian Department of Agriculture (1961). Rape. *Res. Rep. Saskatoon Res. Stn.* 1957–1960, 20.

Canadian Department of Agriculture (1963). Rape. *Res. Rep. Saskatoon Res. Stn.* 1961–1962, 13–14.

Canadian Department of Agriculture (1961). Effects of Honey Bees on Cucumber Production. *Charlottetown Exp. Farm, P.E.I. Res. Rep.* 1958–61, 17.

Carlson, J. W., Evans, R. J., Pedersen, M. W. and Stoker, G. L. (1950). Growing alfalfa for seed in Utah. *Circ. Utah. agric. Exp. Stn.* 125.

Carmin, J. (1959). First year with peanuts—1955. *Bull. Independent Biol. Labs, Palest.* **13**.

Carvalho, A. and Krug, C. A. (1949). Agentes de polinização da flor do cafeéiro (*Coffea arabica* L.). *Bragantia* **9**, 11–24.

Cejtlin, M. G. (1956). A method for obtaining a more complete fertilization of flowers and better development of berries in bi-sexual varieties of vine. *Nauč. Tr. Uzb. s.-h. In-t.* **9**, 197–215.
Chambers, V. H. (1946). An examination of the pollen loads of *Andrena:* the species that visit fruit trees. *J. Anim. Ecol.* **15**, 9–21.
Chandler, S. E. (1956). Botanical aspects of pyrethrum III. The natural history of the secretary organs; the pyrethrins content of the fertile achenes. *Pyrethrum Post* **4**, 10–15.
Chapman, G. P. (1964). Pollination and the yields of tropical crops: an appraisal. *Euphytica* **13**, 187–197.
Chapman, G. P. (1965). A new development in the agronomy of pimento. *Caribb. Q.* **2**, 2–12.
Chapman, G. P. and Glasgow, S. K. (1961). Incipient Dioecy in Pimento. *Nature, Lond.* **192**, 1205–1206.
Chauvin, R., Darchen, P. and Pain, Janine (1961). Sur l'existence d'une hormone de construction chez les abeilles. *C.R. hebd. Séanc. Acad. Sci., Paris* **253**, 1135–1136.
Chittenden, F. J. (1914). Pollination in orchards. *A. appl. Biol.*, **1**, 37–42.
Christensen, S. A. (1960). Afprøvning af tekniste hjaelpenùdler. 2. "Vibrator og tågesprøjte ved bestøvning af tomater". ProduktUdvalg. Gartneri Fruglavl, Copenhagen.
Christopher, E. P. (1958). "Introductory Horticulture." McGraw-Hill, New York.
Cîrnu, I. (1960). Rezultatele polenizării culturilor de floarea-soarelui cu ajutorul albinelor. *Apicultura, Bucuresti* **33**, 18–20.
Cîrnu, I. and Sănduleac, E. (1965). Eficienţa economică a polenizării-soarelui cu ajutorul albinelor. *Lucr. ştiinţ. Stat. cent. Serţ. Apic.* **5**, 37–51.
Claassen, C. E. (1950). Natural and controlled crossing in safflower, *Carthamus tinctorius* L. *Agron. J.* **42**, 381–384.
Clark, K. K. (1959). Honeybees in the Orchard. *Agriculture, Lond.* **65**, 572–575.
Clark, O. I. (1926). Pollination of feijoas. *Yrb. Calif. Avocado Ass.* 1925–1926, 94–95.
Clements, F. E. and Long, F. L. (1923). Experimental pollination: an outline of the ecology of flowers and insects. Carnegie, Washington.
Cobley, L. S. (1956). "An Introduction to the Botany of Tropical Crops." Longmans, London.
Cochran, H. L. (1936). Some factors influencing growth and fruit setting in the pepper (*Capsicum frutescens* L.). *Mem. Cornell agric. Exp. Stn.* **190**.
Coe, H. S. and Martin, J. R. (1920). Sweet clover seed. *Bull. U.S. Dep. Agric.* No. 844.
Coggshall, W. L. (1951). Some pollination suggestions. *Leafl. U.S. Dep. Agric. Ext. Serv.*
Coit, J. E. (1915). "Citrus Fruits." Macmillan, New York.
Colby, A. S. (1926). Notes on self-fertility of some gooseberry varieties. *Proc. Am. Soc. hort. Sci.* **23**, 138–140.
Collins, J. L. (1960). "The Pineapple." London, Leonard Hill.
Condit, I. J. (1920). Caprifigs and Caprification. *Bull. Calif. agric. Exp. Stn.* **319**, 341–377.
Condit, I. J. (1947). "The fig". Chronica Bot., Waltham, Mass., U.S.A.
Condit, I. J. and Enderud, J. (1956). A bibliography of the fig. *Hilgardia* **25**, 1–663.
Congdon, N. B. and Woodhead, C. E. (1959). Erratic cropping of winter Nelis pears in Hawke's Bay. *N.Z. Jl Agric.* **98**, 575–576.

Cook, O. F. (1909). Suppressed and intensified characters in cotton hybrids. *Bull. U.S. Bur. Pl. Ind.* No. 147.
Copaitici, M. (1955). Sulfina. *Apicultura, Bucuresti* **28**, 12–15.
Cope, F. W. (1939). Studies in the mechanism of self-incompatibility in Cacao I. *Rep. Cacao Res.* **8**, 20–21.
Cope, F. W. (1940a). Studies in the mechanism of self-incompatibility in Cacao. II. *Rep. Cacao Res.* **9**, 19–23.
Cope, F. W. (1940b). Agents of pollination in cacao. *Rep. Cacao Res.* **9**, 13–19.
Cope, F. W. (1958). Incompatibility in *Theobroma cacao*. *Nature, Lond.* **181**, 279.
Cope, F. W. (1962). The effects of incompatibility and compatibility on genotype proportions of populations of *Theobroma cacao* L. *Heredity, Lond.* **17**, 183–196.
Copeland, E. B. (1931). "The Coconut" (3rd edit.). Macmillan, London.
Corner, J., Lapins, K. O. and Arrand, J. C. (1964). Orchard and honeybee management in planned tree-fruit pollination. *Apiary Circ., Victoria* No. 14, 18 pages.
Correll, D. S. (1953). Vanilla, its botany, history, culture and economic importance. *Econ. Bot.* **7**, 291–358.
Cottrell-Dormer, W. (1945). An electrical pollinator for tomatoes. *Qd. J. agric. Sci.* **2**, 157–169.
Courtois, G. and Lecomte, J. (1958). Sur un procédé de marquage des abeilles butineuses au moyen d'un radioisotope. *C.R. hebd. Séanc. Acad. Sci., Paris* **247**, 147–149.
Couston, R. (1963). The influence of insect pollination on raspberries. *Scott. Beekeep., Glasg.* **40**, 196–197.
Coville, F. V. (1910). Experiments in blueberry culture. *Bull. U.S. Bur. Pl. Ind.* **193**.
Coville, F. V. (1921). Directions for blueberry culture. *Bull. U.S. Dep. Agric.* **974**.
Coville, F. V. (1937). Improving the wild blueberry. *Yb. Agric. U.S. Dep. Agric.* 559–574.
Cox, I. E. (1957). Flowering and pollination of passion fruit. *Agric. Gaz. N.S.W.* **68**, 573–576.
Crane, M. B. (1945). What are the best plums to grow? *John Innes Bull.* **1**, 44–53.
Crane, M. B. and Lawrence, W. J. C. (1931). Inheritance of Sex, Colour and Hairiness in the Raspberry, *Rubus Idaeus* L. *J. Genet.* **24**, 243–255.
Crane, M. B. and Mather, K. (1943). The natural cross-pollination of crop plants with particular reference to the radish. *A. appl. Biol.* **30**, 301–308.
Cross, C. E. (1966). Cranberry flowers and pollination. *Publ. Mass. agric. Exp. Stn.* no. 435, 27–29.
Cruchet, P. (1953). Distances à observer dans les cultures de haricots porte-graines. *Rev. hort. suisse* **56**, 46–47.
Crum, C. P. (1941). Bees on clover, value of bees as pollinators. *Am. Bee J.* **81**, 270–272.
Cumakov, V. (1955). Usmerenie včiel na d'atelinu lúčnu pomocov aromatických olejov. *Za soc. Zeměd.* **5**, 747–754.
Cumakov, V. (1959). Vyznam medonosých včiel pri pestovaní slnečnice na semeno. *Naša veda* **6**, 20–23.
Currence, T. M. and Jenkins, J. M. (1942). Natural crossing in tomatoes as related to distance and direction. *Proc. Am. Soc. hort. Sci.* **41**, 273–276.
Cuypers, J. (1968). Using honeybees for pollinating crops under glass. *Bee Wld.* **49**, 72–76.
Dade, H. A. (1962). Anatomy and dissection of the honeybee. Bee Research Association, London.

D'Aguilar, I., Della Giustina, W. and Lecomte, J. (1967). Problèmes de pollinisation des cultures sous abri. *Bull. tech. Inf. Ingrs Servs agric.* No. 217, 185–188.

Damisch, W. (1963). Über den Einfluss biologischer Faktoren auf den Samenansatz bei Rotklee. *Biol. Zbl.* **82**, 303–341.

Dann, B. (1930). Ueber die Befruchtungsverhältnisse der Bastardluzerne (*Medicago media*), anderer *Medicago*-Arten und Steinklee (*Melilotus*). *Z. Zucht. A* **15**, 366–418.

Darrow, G. M. (1927). Sterility and fertility in the strawberry. *J. agric. Res.* **34**, 393–411.

Darrow, G. M. (1937). Strawberry improvement. *Yb. U.S. Dep. Agric.* 445–495.

Darwin, C. (1858). On the agency of bees in the fertilisation of Papilio-naceous Flowers and on the crossing of kidney beans. *Ann. Mag. nat. Hist.* **2**, 459–465.

Darwin, C. (1876). "The Effects of Cross and Self Fertilisation in the Vegetable Kingdom." Murray, London.

Davey, V. M. (1959). Cultivated Brassiceae: Information available to the breeder. *Rep. Scott. Soc. Res. Pl. Breed.* 23–62.

Davis, J. H. (1952). Soyabeans for honey production. *Am. Bee J.* **92**, 18–19.

Davydova, N. S. (1954). Analysis of honeybee pollen loads from buckwheat. *Uchen. Zap. kishinev. Univ.* **13**, 167–173.

Demianowicz, Zofia (1965). Comparative Investigations on nectar secretion, Honey and Seed Yield of 5 Polish Varieties of Rape (*Brassica napus* var. Oleifera Metzger). Apimondia, XX Int. Beekeeping Jubilee Congress, Bucharest, Aug. 26–31 (1965).

Demianowicz, Zofia and Jabloński, B. (1966). Nektarowanie i wydajność miodowa 4 gatunków roślin o drobnych kwiatach. *Pszczel. Zesz. nauk.* **10**, 87–94.

Demianowicz, Zofia and Ruszkowska, B. (1959). Gryka jako roślina pozytkowa. *Pszczel. Zesz. nauk.* **3**, 11–24.

Dennis, B. A. and Haas, H. (1967). Pollination and seed-setting in Diploid and Tetraploid Red Clover (*Trifolium pratense* L.) under Danish conditions. II. Studies of floret morphology in relation to the working speed of honey and bumble bees (*Hymenoptera Apoidea*). *Årsskr. k. Vet.-Landbohøjsk.* 118–133.

Dessart, P. (1961). Contribution à l'étude des Ceratopogonidae (Diptera) Les Forcipomyia pollinisateurs du cacaoyer. *Bull. agric. Congo belg.* **52**, 525–540.

Dickson, G. H. (1942). Pollination in relation to orchard planning. *Bull. Ont. Dep. Agric.* No. 424.

Dickson, G. H. and Smith, M. V. (1958). Fruit pollination. *Circ. Ont. Dep. Agric.* No. 172.

Dillman, A. C. (1938). Natural Crossing in Flax. *J. Am. Soc. Agron.* **30**, 279–286.

Dillman, A. C. (1953). Classification of Flax Varieties, 1946. *Tech. Bull. U.S. Dep. Agric.* **1064**.

Dirks, C. O. (1946). Keeping bees in Maine. *Maine ext. Bull.* No. 346.

Dorr, J. and Martin, E. C. (1966). Pollination studies on the highbush blueberry, *Vaccinium corymbasum* L. *Q. Bull. Mich. St. Univ. agric. Exp. Stn.* **48**, 437–448.

Doull, K. M. (1961). Studies in the efficiency of pollination of Lucerne in South Australia. *Aust. J. agric. Res.* **12**, 593–599.

Doull, K. M. (1966). The relative attractiveness to pollen-collecting honeybees of some different pollens. *J. apic. Res.* **5**, 9–13.

Doull, K. M. and Purdie, J. D. (1960). Lucerne for seed. *Bull. Dep. Agric. S. Aust.* No. 463.

Drake, C. J. (1949). Influence of insects on alfalfa seed production in Iowa. *J. econ. Ent.* **41**, 742–750.

Drayner, Jean M. (1959). Self- and cross-fertility in field beans (*Vicia Faba* Linn.). *J. agric. Sci. Camb.* **53**, 387–402.

Dunavan, D. (1953). Insect pollination of ladino clover in South Carolina. *Am. Bee J.* **93**, 468–469.

Dunham, W. E. (1931). A colony of bees exposed to high external temperatures. *J. econ. Ent.* **24**, 606–611.

Dunham, W. E. (1939a). Collecting red clover pollen by honeybees. *J. econ. Ent.* **32**, 668.

Dunham, W. E. (1939b). Insect pollination of red clover in Western Ohio. *Glean. Bee Cult.* **67**, 486–488, 525.

Dunham, W. E. (1939c). The importance of honeybees in alsike seed production. *Glean. Bee Cult.* **67**, 356–358, 394.

Dunham, W. E. (1939d). Insect pollination of orchard fruits. *Glean. Bee Cult.* **67**, 281–285.

Dunham, W. E. (1943). Honey-bees increase clover seed production 15 times. *Am. Bee J.* **83**, 310–311.

Dunham, W. E. (1957). Pollination of clover fields. *Glean. Bee Cult.* **85**, 218–219.

Dunne, T. C. (1943). Pollen-containing sprays for the cross-pollination of Ohanez grapes. *J. Agric. W. Aust.* **19**, 210–213.

Dunning, J. W. (1886). The importation of Humble bees into New Zealand. *Trans. R. ent. Soc. Lond.* **6**, 32–34.

Durham, G. B. (1928). Pollen carriers on summer squash, *Cucurbita pepo*. *J. econ. Ent.* **21**, 436.

Dwyer, R. E. (1931). Seed setting in lucerne. Some observations on the controlling factors. *Agric. Gaz. N.S.W.* **42**, 703–708.

Dwyer, R. E. and Allman, S. L. (1932). Further observations on pollination and seed setting in lucerne. *Agric. Gaz. N.S.W.* **43**, 141–146.

Dyce, E. J. (1929). Bees help the fruit grower. *Can. Hort. Home Mag.* **52**, 143–144.

Dyce, E. J. (1958). Honeybees and the pollination problem in New York State. *Glean. Bee Cult.* **86**, 140–143.

Eaton, F. M. (1957). Selective gametocide opens way to hybrid cotton. *Science, N.Y.* **126**, 1174–1175.

Eaton, G. W. (1959). A study of the megagametophyte in *Prunus avium* and its relation to fruit setting. *Can. J. Pl. Sci.* **39**, 466–476.

Eaton, G. W. (1962). Further studies on sweet cherry embryo sacs in relation to fruit setting. *Rep. hort. Exp. Stn Prod. Lab. Vineland.*

Eaton, G. W. (1966). The effect of frost upon seed number and berry size in the cranberry. *Can. Pl. Sci.* **46**, 87–88.

Eaton, G. W. (1967). The relationship between seed number and berry weight in open-pollinated highbush blueberries. *Hort. Sci.* **2**, 14–15.

Eaton, G. W. and Smith, M. V. (1962). Fruit Pollination. *Publ. Dep. Agric. Ont.* **172**.

Eckert, J. E. (1933). The flight range of the honeybee. *J. agric. Res.* **47**, 257–285.

Eckert, J. E. (1942). The pollen required by a colony of honeybees. *J. econ. Ent.* **35**, 309–311.

Eckert, J. E. (1956). Honeybees increase asparagus seed. *Am. Bee J.* **96**, 153–154.

Eckert, J. E. (1962). The relation of honey bees to safflower. *Am. Bee J.* **102**, 349–350.

Eden, T. (1965). "Tea." Longmans, Green, London.

Edgecombe, S. W. (1946). Honeybees as pollinators in the production of hybrid cucumber seed. *Am. Bee J.* **86**, 147.

Edwards, A. J. (1961). Pollination of field beans. Honeybees as tripping agents. M.A.A.F., N.A.A.S., East Midland Region. Report on field experiments and observation studies 1961. Pp. 79–80.

Einset, O. (1930). Open pollination versus hand pollination of pollen-sterile grapes. *Tech. Bull. N.Y. St. agric. Exp. Stn.* **162**.

Elagin, I. (1953). (Influence of pollination by bees on the yield from buckwheat.) *Pchelovodstvo, Mosk.* **30**, 31–33.

Elders, A. T. (1926). Some pollination and cytological studies of sweet clover. *Scient. Agric.* **6**, 360–365.

Entwistle, Helen (1956). Cacao Pollination. Proc. Cacao Breed. Conf., W. Afr. Cacao Res. Inst., Tafo, pp. 19–21.

Ermakova, I. A. (1959). Nectar productivity of white mustard. *Pchelovodstvo, Mosk.* **36**, 29–31.

Ewert, R. (1921). Die Einfluss der Bienenzucht auf Befruchtung und Ertrag der Obstpflanzungen. *Arch. Bienenk.* **3**, 83–93.

Ewert, R. (1929). Die Befruchtung der Cruciferenblüte durch die Biene. *Arch. Bienenk.* **10**, 310–312.

Ewert, R. (1940). Das Honigen unserer Obstgewächse. Leipziger Bienenzeitung, Leipzig.

Fabre, J. H. (1915). "Bramble-bees and others, translated by A. T. de Mattos." Hodder and Stoughton, London.

Faegri, K. and Pijl, L. van der. (1966). "The Principles of Pollination Ecology." Pergamon Press, Oxford.

Fahn, A. (1949). Studies in the ecology of nectar secretion. *Palest. J. Bot. Jerusalem Ser.* **4**, 207–224.

Farrar, C. L. (1931a). A measure of some factors affecting the development of the honeybee colony. Unpublished doctoral thesis, Massachusetts State College, Amherst.

Farrar, C. L. (1931b). The evaluation of bees for pollination. *J. econ. Ent.* **24**, 622–627.

Farrar, C. L. (1932). The influence of the colony's strength on brood-rearing. *Rep. Beekeep. Ass. Prov. Ont.* (1930 and 1931), 126–130.

Farrar, C. L. (1936). Influence of pollen reserves on the surviving population of overwintered colonies. *Am. Bee J.* **76**, 452–454.

Farrar, C. L. (1937). The influence of colony population on honey production. *J. agric. Res.* **54**, 945–954.

Farrar, C. L. (1963). Large cage design for insect and plant research. *U.S. Dep. Agric., Agric. Res. Serv. Ars.* 33–77.

Farrar, C. L. and Bain, H. F. (1947). Honeybees as pollinators of the cranberry. *Cranberries.* **11**.

Faulkner, G. J. (1962). Blowflies as pollinators of brassica crops. *Comml. Grow.* (3457), 807–809.

Fechner, E. (1927). Untersuchungen über die Einwirkung eines Rückganges der Bienenzucht auf den Samenertrag einiger landwirtschaftlichen Kulturpflanzen. *Arch. Bienenk.* **8**, 1–72.

Ferwerda, F. P. (1948). Coffee breeding in Java. *Econ. Bot.* **2**, 258–272.

Filmer, R. S. (1932). Brood and colony size as factors in activity of pollination units. *J. econ. Ent.* **25**, 336–343.

Filmer, R. S. (1941). Honeybee population and floral competition in New Jersey orchards. *J. econ. Ent.* **34**, 198–199.

Filmer, R. S. (1949). Cranberry pollination studies. Am. Cranberry Growers' Assoc., Proc. 8th Ann. Convention, pp. 14–20.

Filmer, R. S. and Doehlert, C. A. (1952). Use of honeybees in cranberry bogs. *Bull. N.J. agric. Exp. Stn.* No. 764.

Filmer, R. S. and Doehlert, C. A. (1959). Use of honeybees in cranberry bogs. *Circ. N.J. agric. Exp. Stn.* No. 588, 4.

Filmer, R. S. and Marucci, P. E. (1963). The importance of honeybees in blueberry pollination. Proc. 31st Annual Blueberry Open House, New Jersey Agric. Exp. Stn., pp. 14 21.

Finkner, M. D. (1954). Random activity of pollen vectors in isolated plots of upland cotton. *Agron. J.* **46**, 68–70.

Fischer, R. L. (1953). Native pollinators of alfalfa *Medicago sativa* L. in Northern Minnesota. *Minn. Beekpr.* **6**, 8–9.

Fischer, R. L. (1954). Honeybees aid production of alsike clover seed. *Minn. Home Fm Sci.* **11**, 7–9.

Fletcher, S. W. (1916). Pollination. *In* "Standard Cyclopedia of Horticulture" (ed. Bailey, L. H.), pp. 2734–2737. Macmillan, New York.

Fletcher, S. W. and Gregg, O. I. (1907). Pollination of forced tomatoes. *Spec. Bull. Mich. agric. Exp. Stn.* **39**, 2–10, 294–301.

Folsam, J. W. (1922). Pollination of red clover by *Tetralonia* and *Melissodes*. *Ann. ent. Soc. Am.* **15**, 181–184.

Fomina, K. Ya. (1961). Effect of shelter belts on nectar production and seed yield of sainfoin and sunflower. *Dokl. TSKhA* **62**, 531–536.

Fontanilla-Barroga, S. (1961). Insects associated with cacao pollination: a progress report. *Coffee cacao J.* **4**, 208, 219.

Foot Guimarães, R. and Kerr, W. E. (1959). Auto-fecundação em *Eucalyptus alba*, Reinw. *Bol. Co. paulista Estradas Ferro* **11**.

Forshey, C. G. (1953). Blossom structure and its relation to fruit set in Delicious and several other commercially important apple varieties. *Proc. Am. Soc. hort. Sci.* **62**, 154–158.

Forster, I. W. and Hadfield, W. V. (1958). Effectiveness of honey bees and bumble bees in the pollination of Montgomery red clover. *N.Z. Jl. agric. Res.* **1**, 607–619.

Foster, R. E. and Levin, M. D. (1967). F_1 Hybrid Muskmelons. II. Bee activity in seed fields. *J. Ariz. Acad. Sci.* **4**, 222–225.

Françon, J. (1938). "L'esprit des Abeilles." Gallimard, Paris.

Françon, J. (1939). "The Mind of the Bees." Methuen, London.

Franklin, H. J. (1940). Cranberry growing in Massachusetts. *Bull. Mass. agric. Exp. Stn.* **371**.

Franklin, W. W. (1951). Insects affecting alfalfa seed production in Kansas. *Tech. Bull. Kans. agric. Exp. Stn.*, No. 70.

Fresnaye, J. (1963). Les erreurs d'orientation des abeilles (dérive) dans le rucher moderne. *Annls Abeille* **6**, 185–200.

Free, J. B. (1955). The collection of food by bumble bees. *Insects soc.* **2**, 303–311.

Free, J. B. (1956). A study of the stimuli which release the food begging and offering responses of worker honeybees. *Brit. J. Anim. Behav.* **4**, 94–101.

Free, J. B. (1957). The transmission of food between worker honeybees. *Br. J. Anim. Behav.* **5**, 41–47.

Free, J. B. (1958a). The collection of food by bumblebees. *Anim. Behav.* **3**, 147–153.

Free, J. B. (1958b). The ability of worker honeybees (*Apis mellifera*) to learn a change in the location of their hives. *Anim. Behav.* **6**, 219–223.

Free, J. B. (1958c). The drifting of honey bees. *J. agric. Sci., Camb.* **51**, 294–306.

Free, J. B. (1958d). Attempts to condition bees to visit selected crops. *Bee Wld* **39**, 221–230.

Free, J. B. (1959). The effect of moving colonies of honeybees to new sites on their subsequent foraging behaviour. *J. agric. Sci., Camb.* **53**, 1–9.

Free, J. B. (1960a). The behaviour of honeybees visiting the flowers of fruit trees. *J. Anim. Ecol.* **29**, 385–395.

Free, J. B. (1960b). The distribution of bees in a honeybee (*Apis mellifera* L.) colony. *Proc. R. ent. Soc. Lond.* (A) **35**, 141–144.

Free, J. B. (1961). Hypopharyngeal gland development and division of labour in honeybee (*Apis mellifera* L.) colonies. *Proc. R. ent. Soc. Lond.* (A) **36**, 5–8.

Free, J. B. (1962a). The attractiveness of geraniol to foraging honeybees. *J. apic. Res.* **1**, 52–54.

Free, J. B. (1962b). The behaviour of honeybees visiting field beans (*Vicia faba*). *J. Anim. Ecol.* **31**, 497–502.

Free, J. B. (1962c). The effect of distance from pollinizer varieties on the fruit set on trees in plum and apple orchards. *J. hort. Sci.* **37**, 262–271.

Free, J. B. (1963). The flower constancy of honeybees. *J. Anim. Ecol.* **32**, 119–131.

Free, J. B. (1964a). The behaviour of honeybees on sunflowers (*Helianthus annus* L.). *J. appl. Ecol.*, **1**, 19–27.

Free, J. B. (1964b). Comparison of the importance of insect and wind pollination of apple trees. *Nature, Lond.* **201**, 726–727.

Free, J. B. (1965a). The ability of bumblebees and honeybees to pollinate red clover. *J. appl. Ecol.* **2**, 289–294.

Free, J. B. (1965b). Attempts to increase pollination by spraying crops with sugar syrup. *J. apic. Res.* **4**, 61–64.

Free, J. B. (1965c). The behaviour of honeybee foragers when their colonies are fed with sugar syrup. *J. apic. Res.* **4**, 85–88.

Free, J. B. (1965d). The effect on pollen collecting of feeding honeybee colonies with sugar syrup. *J. agric. Sci., Camb.* **64**, 167–168.

Free, J. B. (1966a). The pollinating efficiency of honeybee visits to apple flowers. *J. hort. Sci.* **41**, 91–94.

Free, J. B. (1966b). The pollination requirements of broad beans and field beans (*Vicia faba*). *J. agric. Sci., Camb.* **66**, 395–397.

Free, J. B. (1966c). The pollination of the beans *Phaseolus multiflorus* and *Phaseolus vulgaris* by honeybees. *J. apic. Res.* **5**, 87–91.

Free, J. B. (1967a). The production of drone comb by honeybee colonies. *J. apic. Res.* **6**, 29–36.

Free, J. B. (1967b). Factors determining the collection of pollen by honeybee foragers. *Anim. Behav.* **15**, 134–144.

Free, J. B. (1968a). Dandelion as a competitor to fruit trees for bee visits. *J. appl. Ecol.* **5**, 169–178.

Free, J. B. (1968b). The foraging behaviour of honeybees (*Apis mellifera*) and bumblebees (*Bombus* spp.) on blackcurrant (*Ribes nigrum*), raspberry (*Rubus idaeus*) and strawberry (*Fragaria* × *Ananassa*) flowers. *J. appl. Ecol.* **5**, 157–168.

Free, J. B. (1968c). The conditions under which foraging honeybees expose their Nasanov glands. *J. apic. Res.* **7**, 139–145.

Free, J. B. (1968d). The behaviour of bees visiting runner beans (*Phaseolus multiflorus*). *J. appl. Ecol.* **5**, 631–638.

Free, J. B. (1968e). The pollination of black currants. *J. hort. Sci.* **43**, 69–73.

Free, J. B. (1968f). The pollination of strawberries by honeybees. *J. hort. Sci.* **43**, 107–111.

Free, J. B. (1969). The influence of the odour of a honeybee colony's food stores on the behaviour of its foragers. *Nature, Lond.* **222**, 778.

Free, J. B. (1970a). The effect of flower shape and nectar guides on the behaviour of foraging honeybees. *Behaviour*. In press.

Free, J. B. (1970b). The flower constancy of bumblebees. *J. Anim. Ecol.* **39**, 395–402.

Free, J. B. and Butler, C. G. (1955). An analysis of the factors involved in the formation of a cluster of honeybees (*Apis mellifera*). *Behaviour* **7**, 304–316.

Free, J. B. and Butler, C. G. (1959). "Bumblebees." Collins, London.

Free, J. B. and Durrant, A. J. (1966a). The dilution and evaporation of the honeystomach contents of honeybees at different temperatures. *J. apic. Res.* **5**, 3–8.

Free, J. B. and Durrant, A. J. (1966b). The transport of pollen by honeybees from one foraging trip to the next. *J. hort. Sci.* **41**, 87–89.

Free, J. B. and Nuttall, P. M. (1968a). The pollination of oilseed rape (*Brassica napus*) and the behaviour of bees on the crop. *J. agric. Sci., Camb.* **71**, 91–94.

Free, J. B. and Nuttall, P. M. (1968b). Effect of the time of day at which honeybee colonies are first allowed flight in a new location on their choice of flower species. *Nature, Lond.* **218**, 982.

Free, J. B. and Preece, D. A. (1969). The effect of the size of a honeybee colony on its foraging activity. *Insectes Soc.* **16**, 73–78.

Free, J. B. and Racey, P. A. (1966). The pollination of *Freesia refracta* in glasshouses. *J. apic. Res.* **5**, 177–182.

Free, J. B. and Racey, P. A. (1968a). The effect of the size of honeybee colonies on food consumption, brood rearing and the longevity of bees during winter. *Ent. exp. appl.* **11**, 241–249.

Free, J. B. and Racey, P. A. (1968b). The pollination of runner beans (*Phaseolus multiflorus* Willd.) in a glasshouse. *J. apic. Res.* **7**, 67–69.

Free, J. B. and Simpson, J. (1964). The pollination requirements of sunflowers (*Helianthus annuus* L.). *Emp. J. exp. Agric.* **32**, 340–342.

Free, J. B. and Smith, M. V. (1961). The foraging behaviour of honeybees from colonies moved into a pear orchard in full flower. *Bee Wld* **42**, 11–12.

Free, J. B. and Spencer-Booth, Yvette (1959). The longevity of worker honey bees (*Apis mellifera*). *Proc. R. ent. Soc. Lond.* (A) **34**, 141–150.

Free, J. B. and Spencer-Booth Yvette (1960). Chill-coma and cold death temperatures of *Apis mellifera*. *Ent. exp. appl.* **3**, 222–230.

Free, J. B. and Spencer-Booth, Yvette (1961). The effect of feeding sugar syrup to honeybee colonies. *J. agric. Sci., Camb.* **57**, 147–151.

Free, J. B. and Spencer-Booth, Yvette (1962). The upper lethal temperatures of honeybees. *Ent. exp. appl.* **5**, 249–254.

Free, J. B. and Spencer-Booth, Yvette (1963a). The foraging areas of honey-bee colonies in fruit orchards. *J. hort. Sci.* **38**, 129–137.

Free, J. B. and Spencer-Booth, Yvette (1963b). The pollination of mustard by honeybees. *J. apic. Res.* **2**, 69–70.

Free, J. B. and Spencer-Booth, Yvette (1964a). The foraging behaviour of honeybees in an orchard of dwarf apple trees. *J. hort. Sci.* **39**, 78–83.

Free, J. B. and Spencer-Booth, Yvette (1964b). The effect of distance from pollinizer varieties on the fruit set of apple, pear and sweet cherry trees. *J. hort. Sci.* **39**, 54–60.

Free, J. B. and Williams, Ingrid H. (1970a). The exposure of Nasanov glands by water-collecting honeybees. *Behaviour*. In press.

Free, J. B. and Williams, Ingrid H. (1970b). Preliminary investigations on the occupation of artificial nests by *Osmia rufa* L. *J. appl. Ecol.* **7**, 559–566.

Free, J. B., Free, Nancy W. and Jay, S. C. (1960). The effect on foraging behaviour of moving honey bee colonies to crops before or after flowering has begun. *J. econ. Ent.* **53,** 564, 566.

Free, J. B., Needham, P. H., Racey, P. A. and Stevenson, J. H. (1967). The effect on honeybee mortality of applying insecticides as sprays or granules to flowering field beans. *J. Sci. Fd. Agric.* **18,** 133–138.

Frey-Wyssling, A., Zimmermann, M. and Maurizio, Anna (1954). Uber den enzymatischen Zuckerumbau in Nektarien. *Experienta,* **10,** 490–492.

Frick, K. E., Potter, H. and Weaver, H. (1960). Development and maintenance of alkali bee nesting sites. *Stn. Circ. Wash. agric. Exp. Stn.* **366.**

Fridén, F., Eskilsson, L. and Bingefors, S. (1962). Bumble bees and red clover pollination in central Sweden. *Medd. Sverig. Fröodlförb.,* No. 7, 17–26. Örebro.

Frisch, K. von (1914). Der Farbensinn und Formensinn der Biene. *Zool. Jb. Abt.* **35,** 1–188.

Frisch, K. von (1919). Ueber den Geruchsinn der Bienen und seine Blütenbiologische Bedeutung. *Zool. Jb., Abt. 3,* **37,** 1–238.

Frisch, K. von (1923). Ueber die "Sprache" der Bienen. *Zool. Jb., Abt. 3,* **40,** 1–186.

Frisch, K. von (1934). Über den Geschmackssinn der Biene. Ein Beitrag zur Vergleichenden, Physiologie des Geschmacks. *Z. vergl. Physiol.* **21,** 1–156.

Frisch, K. von (1946). Die Tänze der Bienen, *Öst. Zool. Z.* **1,** 1–48.

Frisch, K. von (1947). "Duftgelenkte Bienen im Dienste der Landwirtschaft und Imkerei". Springer, Vienna.

Frisch, K. von (1954). "The Dancing Bees". Methuen, London.

Frisch, K. von (1955). Die Sinne der Bienen im Dienst ihrer sozialen Gemeinschaft. *Nova Acta Leopoldina* **17,** 472–482.

Frisch, K. von (1965). "Tanzsprache und Orientierung der Bienen." Springer-Verlag, Berlin.

Frisch, K. von (1967). "The Dance Language and Orientation of Bees." Oxford University Press, London.

Frisch, K. von (1968). The role of dances in recruiting bees to familiar sites. *Anim. Behav.* **16,** 531–535.

Frisch, K. von and Rösch, G. A. (1926). Neue Versuche über die Bedeutung von Duftorgan und Pollenduft für die Verständigung im Bienenvolk. *Z. vergl. Physiol.* **4,** 1–21.

Frison, T. H. (1926). Experiments in attracting queen bumblebees to artificial domiciles. *J. econ. Ent.* **19,** 149–155.

Frison, T. H. (1927). Experiments in rearing colonies of bumblebees (*Bremidae*) in artificial nests. *Biol. Bull. Wood's Hole* **52,** 51–67.

Fronk, W. D. and Painter, L. I. (1960). Some characteristics of alkali bee nesting sites. *J. econ. Ent.* **53,** 424–425.

Fronk, W. D. and Slater, J. A. (1956). Insect fauna of cucurbit flowers. *J. Kans. ent. Soc.* **29,** 141–145.

Frost, S. W. (1943). An observation box for solitary bees and wasps. *J. econ. Ent.* **36,** 803–804.

Fryxell, P. (1956). Effect of varietal mass on percentage of outcrossing in Gossypium hirsutum in New Mexico. *J. Hered.* **47,** 299–301.

Fujita, K. (1957). The effect of honey bees on the fruiting and appearance of Satsuma oranges. *Bull. Kanagawa agric. Exp. Stn. hort. Branch* **5.**

Fujita, M. (1939). (Influence of honeybees on the fructification of rape.) *Bull. imp. Zootech. Exp. Stn. Chiba-Shi* **34,** 1.

Furgala, B. (1954a). The effect of the honey bee *Apis mellifera* (L.) on the seed set, yield and hybridisation of the cultivated sunflower, *Helianthus annus* L. Ph.D. Thesis, Manitoba University.

Furgala, B. (1954b). Honey bees increase seed yields of cultivated sunflowers. *Glean. Bee Cult.* **82,** 532–534.

Furgala, B., Gochnauer, T. A. and Holdaway, F. G. (1958). Constituent sugars of some northern legume nectars. *Bee Wld* **39,** 203–205.

Furgala, B., Tucker, K. W. and Holdaway, F. G. (1960). Pollens in the proboscis fossae of honeybees foraging certain legumes. *Bee Wld* **41,** 210–213.

Fye, R. E. and Medler, J. T. (1954). Field domiciles for bumblebees. *J. econ. Ent.* **47,** 672–676.

Fyfe, J. L. and Bailey, N. T. J. (1951). Plant breeding studies in leguminous forage crops. I. Natural crossing in winter beans. *J. agric. Sci.* **41,** 371–378.

Gaag, H. C. van der (1955). Het kweken en het gebruiken van vleesoliegen bij het verdelingswerk. *Zaadbelangen* **9,** 233–236.

Gagnard, J. M. (1954). Recherches sur les caractères systématiques et sur les phénomènes de stérilité chez les variétés d'amandiers cultivées en Algérie. *Annls Inst. agric. Algér.* **8,** 1–163.

Garber, R. J. and Quisenberry, K. S. (1927). Self-fertilisation in buckwheat. *J. agric. Res.* **34,** 185–190.

Gardner, E. J. (1946). Wind-pollination in guayule, *Parthenium argentatum* Gray. *J. Am. Soc. Agron.* **38,** 264–272.

Gardner, E. J. (1947). Insect-pollination in guayule, *Parthenium argentatum* Gray. *J. Am. Soc. Agron.* **39,** 224–233.

Gardner, V. R., Bradford, F. C. and Hooker, H. D. (1952). "The Fundamentals of Fruit Production." McGraw-Hill, New York.

Gary, N. E. (1962). Chemical mating attractants in the queen honey bee. *Science, N.Y.* **136,** 773–774.

Gates, B. N. (1917). Honey bees in relation to horticulture. *Trans. Mass. hort. Soc.* **1,** 71–88.

Gill, N. T. and Vear, K. C. (1958). "Agricultural Botany." Gerald Duckworth, London.

Giltay, E. (1904). Ueber die Bedeutung der Krone bei den Blüten und über das Farbenunterscheidungsvermögen der Insekten. *Jb. wiss. Bot.* **40,** 368–402.

Girardeau, J. H. (1958). The mutual value of crimson clover and honey bees for seed and honey production in South Georgia. *Mimeogr. Ser. Ga. agric. Exp. Stn. N.S.* **63.**

Girardeau, J. H. and Leuck, D. B. (1967). Effect of mechanical and bee tripping on yield of the peanut. *J. econ. Ent.* **60,** 1454–1455.

Gladwin, F. E. (1937). Pollination with particular reference to the grape. *Am. Fruit Grow. Mag.* **57** (2), 9, 24–25, 35; (3), 16, 35.

Glendinning, D. R. (1958). Natural pollination. *Rep. W. Afr. Cocoa Res. Inst.* (1957–1958), 50–51.

Glowska, Z. (1958). Porównanie intensywności nektarowania i oblatywania przez pszczoły pięciu gatunków drzew owocowych. *Pszczel. Zesz. Nauk.* **2,** 121–148.

Glukhov, M. M. (1955). "Bee Plants". State Publishing House of agricultural Literature, Moscow.

Glushkov, N. M. (1958). Problems of beekeeping in the U.S.S.R. in relation to pollination. *Bee Wld* **39,** 81–92.

Glushkov, N. M. and Skrebtsov, M. F. (1960). Increasing the cotton crop by saturation bee pollination. *Pchelovodstvo, Mosk.* **37,** 29–30.

Glushkov, N. M. and Skrebtsov, M. F. (1965). Efficacy of using Honey Bees in Cotton Pollination. XX International Beekeeping Congress (II/14).

Goetze, G. (1930). Chemische und biologische Prüfung von Bienentrachtpflanzen. Bienenweide (A. Koch). Leipzig Bienenztg, Leipzig.

Goetze, G. (1948). Versuche zur Ausnutzung des Rotklees durch die Honigbienen. *Beitr. Agrarwiss.* **2,** 1–16.

Goff, C. G. (1937). Importance of Bees in the production of watermelons. *Fla Ent.* **20,** 30–31.

Gooderham, C. B. (1948). Honeybees increase clover seed production. *Am. Bee J.* **88,** 304.

Gooderham, C. B. (1950). Overwintered colonies versus package bees for orchard pollination. *Progr. Rep. Dom. Apiarist Can. Dep. Agric.* (1937–1948), 7–8.

Goplen, B. P. (1960). Honeybees and seed production of sweet clover. *Forage Notes* **6,** 31–32.

Goplen B. P. and Cooke, D. A. (1965). Isolation distance study in sweet clover. *Forage Notes* **15,** 15–16.

Goplen, B. P. and Pankiw, P. (1961). Note on a greenhouse technique for crossing legumes by honey bees. *Can. J. Pl. Sci.* **41,** 679–682.

Gordienko, V. A. (1960a). Production of fertile hybrids of soya with the help of bees. *Trud. mold. Akad. Nauk.* **2,** 31–39.

Gordienko, V. V. (1960b). Sexual hybrids of soya beans obtained by directed bee pollination. In "Pollination of agricultural plants by bees", Vol. 3, (ed. A. N. Mel'nichenko and others), pp. 400–407. Moldavian Agricultural Institute, Moldavia S.S.R., U.S.S.R.

Gould, H. P. (1939). Why fruit trees fail to bear. *Leafl. U.S. Dep. Agric.* No. 172.

Grant, V. (1949). Pollination systems as isolating mechanisms in angiosperms. *Evolution,* **3,** 82–97.

Grant, V. (1950). The flower constancy of bees. *Bot. Rev.* **16,** 379–398.

Green, H. B. (1956). Some factors affecting pollination of white dutch clover. *J. econ. Ent.* **49,** 685–688.

Green, H. B. (1957). White clover pollination with low honeybee population. *J. econ. Ent.* **50,** 318–320.

Greenway, P. J. and Wallace, M. M. (1953). The Papaw, its botany, cultivation, diseases and chemistry. *Pamphl. Dep. Agric. Tanganyika* **52.**

Griffin, H. H. (1901). The Cantaloup. *Bull. Colo. agric. Exp. Stn.* **62.**

Griggs, W. H. (1953). Pollination requirements of fruits and nuts. *Circ. Calif. agric. Exp. Stn.* No. 424.

Griggs, W. H. (1958). Little known facts about pollination. *Am. Fruit Grow. Mag.* **13,** 68–69.

Griggs, W. H. and Iwakiri, B. T. (1960). Orchard tests of beehive pollen dispensers for cross-pollination of almonds, sweet cherries and apples. *Proc. Am. Soc. hort. Sci.* **75,** 114–128.

Griggs, W. H. and Iwakiri, B. T. (1964). Timing is critical for effective cross/pollination of almond flowers. *Calif. Agric.* **18,** 6–7.

Griggs, W. H. and Vansell, G. H. (1949). The use of bee-collected pollen in artificial pollination of deciduous fruits. *Proc. Am. Soc. hort. Sci.* **54,** 118–124.

Griggs, W. H., Vansell, G. H. and Reinhardt, J. F. (1950). The germinating ability of quick-frozen, bee-collected apple pollen stored in a dry ice container. *J. econ. Ent.* **43,** 549.

Griggs, W. H., Vansell, G. H. and Iwakiri, B. T. (1952). The use of beehive pollen dispensers in the pollination of almonds and sweet cherries. *Proc. Am. Soc. hort. Sci.* **60,** 146–150.

Griggs, W. H., Vansell, G. H. and Iwakiri, B. T. (1953). Pollen storage—high viability of pollen obtained after storage in home freezer. *Calif. Agric.* **7,** 12.

Griggs, W. H., Iwakiri, B. T. and Bethell, R. S. (1965). B-nine fall sprays delay bloom and increase fruit set on Bartlett pears. *Calif. Agric.* **19,** 8–11.

Groom, P. (1906). "Elementary Botany." George Bell, London.

Groot, A. P. de and Voogd, S. (1954). On the ovary development in queenless worker bees (*Apis mellifica* L.). *Experientia* **10,** 384–385.

Grout, R. A. (1949). "The Hive and the Honey Bee." Dadant, Hamilton, Illinois.

Grout, R. A. (1950). "Planned Pollination—an Agricultural Practice." Dadant, Hamilton, Illinois.

Grout, R. A. (1955). Honey Bees make Hybrid Cotton possible. *Am. Bee J.* **95,** 10–11.

Grubb, N. H. (1935). Raspberry Breeding at East Malling—1922–34. *J. Pomol.* **13,** 108–134.

Gubin, A. F. (1936). Bestäubung und Erhöhung der Samenernte bei Rotklee *Trifolium pratense* L. mit Hilfe der Bienen. *Arch. Bienenk.* **17,** 209–264.

Gubin, A. F. (1945a). Cross pollination of fibre flax. *Bee Wld* **26,** 30–31.

Gubin, A. F. (1945b). Bee training for pollination of cucumbers. *Bee Wld* **26,** 34–35.

Gubin, A. F. and Smaragdova, N. P. (1940). The directing action of aromatic feeding and the role of the scent-producing organ of *Apis mellifera* L. in searching a source of honey-flow by bees. *Zool. Zh.* **19,** 790–800.

Gubin, A. F. and Verdieva, M. G. (1956). Pollination of cotton and a new method of training bees. *Pchelovodstvo, Mosk.* **33,** 52–54.

Haarer, A. E. (1956). "Modern Coffee Production." Leonard Hill, London.

Haas, H. (1966). Preliminary investigations on pollination, seed setting and seed yield on 2n and 4n red clover throughout the complete flowering period. Proc. 2nd Int. Symp. Pollination. *Bee Wld* **47** (Suppl.), 71–82.

Hadfield, J. W. and Calder, R. A. (1936). Lucerne (*Medicago sativa*). Investigations relative to pollination and seed production in New Zealand. *N.Z. Jl Sci. Technol.* **17,** 577–594.

Hagberg, A. (1952). Undersökning över graden av korsberniktning hos sötlupin. *Sver. Utsädesför. Tidskr.* **62,** 301–310.

Haig, J. C. (1953). Plant breeding report. *Rep. Natn. Veg. Res. Stn.* **4,** 10–13.

Haigh, J. C. (1956). Plant breeding report. *Rep. Natn. Veg. Res. Stn.* **7,** 11–15.

Hale, C. R. and Jones, L. T. (1956). The pollination of Ohanez grapes. *J. Agric. W. Aust.* **5,** 565–567.

Hall, C. J. J. van (1932). "Cacao". (2nd Edit.) Macmillan, London.

Hall, C. J. J. van (1938). Coffee selection in the Netherlands Indies. *Bull. colon. Inst. Amst.* **2,** 135–145.

Hall, I. V. and Aalders, L. E. (1961). Cytotaxonomy of lowbush blueberries in eastern Canada. *Am. J. Bot.* **48,** 199–201.

Hambleton, J. I. (1925). The effect of weather upon the change in weight of a colony of bees during the honey flow. *Bull. U.S. Dep. Agric.* **1339.**

Hambleton, J. I. (1944). The role of bees in the production of fruits and seeds. *J. econ. Ent.* **34,** 522–525.

Hamilton, R. (1960). Trucking packages from south to north. *Glean. Bee Cult.* **88,** 142–145.

Hammer, O. (1949). Om Konkurrencen Mellem Blomstrende Afgroder. *Ugeskr. Landm.* (10).
Hammer, O. (1950). Biernes bestovningsarbejde og froudbyttets storrelse. *Tidsskr. Frøarl Saetryk* **19**.
Hammer, O. (1952). Om rapsavlen, bierne og frøavlen. *Dansk, Landbr.* **71**, 67–69.
Hammer, O. (1961). Lidt om afstandens betydning for honningproduktionen. *Nord. Bitidskr.* **13**, 20–23.
Hammer, O. (1962). Om honningudbytte og afstand til troekkilden *Tidsskr. Biavl.* **96**, 115–117.
Hammer, O. (1963). Om sommerrapsen som konkurrent ved kløverarternes bestøvning. *Dansk. Frøavl.* (14).
Hammons, R. O., Krombein, K. V. and Leuck, D. B. (1963). Some bees (Apoidea) associated with peanut flowering. *J. econ. Ent.* **56**, 905.
Hammons, R. O. and Leuck, D. B. (1966). Natural cross-pollination of the peanut *Arachis hypogaea* L. in the presence of bees and thrips. *Agron. J.* **58**, 396.
Hansen, P. (1934). Humlebiernes og honningbiernes betydning for rødkløverens bestøvning. *Ugeskr. Landm.* **79**, 232–234.
Hansson, A. (1960). Försök med doftstyrning av bin. *Nord. Bitidskr.* **12**, 25–28.
Haragsim, O., Vesely, V., Sědivý, J., Taimr, L., Dočkal, J. and Balcar, J. (1965). Activity of Honey Bees marked with Radioisotopes and moved to fields of Lucerne (*Medicago sativa*). 20th International Beekeepers Congress (II/4).
Haragsimová-Neprašová, L. (1960). Zjištování nektarodárnosti rostlin. *Věd. Pr. vyzk. Úst. vcelař. CSAZV*, **2**, 63–79.
Hardy, M. B. (1931). Self and cross fertility of red raspberry varieties. *Proc. Amer. Soc. hort. Sci.* **28**, 118–121.
Hare, Q. A. and Vansell, G. H. (1946). Pollen collection by honeybees in the Delta, Utah, alfalfa seed-producing area. *J. Am. Soc. Agron.* **38**, 462–469.
Harland, S. C. (1925). Studies in Cacao. Part I. The method of pollination. *Ann. appl. Biol.* **12**, 403–409.
Harle, A. (1948). Ist der Rapsglanzkafer (*Meligethes aeneus* Fabr.) nur ein Schädling? *Nachr-Bl. dt. Pfl. Schutzdienst, Berl.* **2**, 40–42.
Harris, W. B. (1962). Fruit Tree Pollination. *J. Dep. Agric. S. Aust.* **65**, 257–263.
Hartman, F. O. and Howlett, F. S. (1954). Fruit setting of the Delicious apple. *Bull. Ohio agric. Exp. Stn.* No. 745.
Hartwig, E. E. (1942). Effects of self-pollination in sweet clover. *J. Am. Soc. Agron.* **34**, 376–387.
Haskell, G. (1943). Spatial isolation of seed crops. *Nature, Lond.* **152**, 591–592.
Hasler, A. and Maurizio, Anna (1950). Ueber den Einfluss verschiedener Nährstoffe auf Blütenansatz, Nektarsekretion und Samenertrag von honigenden Pflanzen, speziell von Sommerraps (*Brassica napus* L.). *Schweiz. landw. Mh.* (6), 201–211.
Hassanein, M. H. (1953). Studies on the effect of pollinating insects, especially the honeybee on the seed yield of clover in Egypt. *Bull. Soc. Fouad Im. Ent.* **37**, 337–344.
Hassanein, M. H. (1955). The value of pollinating insects to flax seed production in Egypt. *Ann. agric. Sci. Univ. A'in Shams*, 773–784.
Hassanein, M. H. and El Banby, M. A. (1956). Studies on the ability of the Egyptian honeybee on carrying nectar and pollen. *Ann. agric. Sci. Fac. Agric. A'in Shams* **1**, 23–36.
Hassanein, M. H. and Ibrahim, M. M. (1959). Studies on the importance of insects, especially the honeybee, in pollination of citrus in Egypt. *Agric. Res. Rev., Cairo* **37**, 390–409.

Hasselrot, T. B. (1952). A new method for starting bumble bee colonies. *Agron. J.* **44,** 218–219.

Hasselrot, T. B. (1960). Studies on Swedish bumblebees (genus *Bombus* Latr.): their domestication and biology. *Opusc. ent.* (Suppl.) **17.**

Hawkins, R. P. (1956). A preliminary survey of red clover seed production. *Ann. appl. Biol.* **44,** 657–664.

Hawkins, R. P. (1958). A survey of late-flowering and single cut red clover seed crops. *J. natn. Inst. agric. Bot.* **8,** 450–461.

Hawkins, R. P. (1961). Observations on the pollination of red clover by bees. I. The yield of seed in relation to the number and kinds of pollinators. *Ann. appl. Biol.* **49,** 55–65.

Hawkins, R. P. (1965). Factors affecting the yield of seed produced by different varieties of red clover. *J. agric. Sci.* **65,** 245–253.

Hawkins, R. P. (1966). The pollination of red clover by bees in England and Wales. *Bee Wld* (Suppl.) **47,** 51–57.

Hawkins, R. P. (1968). Honeybees as pollinators in greenhouses. *J. apic. Res.* **49,** 157.

Haws, B. A. and Holdaway, F. G. (1957). Insects and sweetclover seed production. *Minn. Fm. Home Fact Sh. Ent.* **3.**

Hawthorn, L. R., Bohart, G. E. and Toole, E. H. (1956). Carrot seed yield and germination as affected by different levels of insect pollination. *Proc. Am. Soc. hort. Sci.* **67,** 384–389.

Hawthorn, L. R., Bohart, G. E., Toole, E. H., Nye, W. P. and Levin, M. D. (1960). Carrot seed production as affected by insect pollination. *Bull. Utah agric. Exp. Stn.* **422.**

Hawthorn, L. R. and Pollard, L. (1954). "Vegetable and Flower Seed Production." Blakiston, New York.

Haydak, M. H. (1932). Division of labour in the colony. *Wis. Beekeeping* **8,** 36–39.

Haydak, M. H. (1945). Value of pollen substitutes for brood rearing of honeybees. *J. econ. Ent.* **38,** 484–487.

Haydak, M. H. (1957). Changes with age of the appearance of some internal organs of the honeybee. *Bee Wld* **38,** 197–207.

Haydak, M. H. (1958). Pollen—pollen substitutes—bee bread. *Am. Bee J.* **98,** 145–146.

Haydak, M. H. (1963). Age of nurse bees and brood rearing. *J. apic. Res.* **2,** 101–103.

Hector, J. M. (1938). "Introduction to the Botany of Field Crops" Vol. II, Non-Cereals. Central News Agency, Ltd., Johannesburg, S.A.

Heide, F. F. R. (1923). Biologische onderzoekingen bij Landbouwgewassen. I. Biologische Waarnemingen bij *Arachis hypogaea* L. *Meded. alg. Proefstn Landb. Buitenz.* **14,** 5–19.

Heinrichs, D. H. (1967). Seed increase of alfalfa in growth chambers with *Megachile rotundata* F. *Can. J. Pl. Sci.* **47,** 691–694.

Hendrickson, A. H. (1916). The common honey bee as an agent in prune pollination. *Bull. Calif. agric. Exp. Stn.* No. 274, 127–132.

Henry, A. W. and Tu, C. (1928). Natural crossing in Flax. *J. Am. Soc. Agron.* **20,** 1183–1192.

Henslow, G. (1878). Fertilisation of the Scarlet Runner by Humblebees. *Gdnrs' Chron.* **10,** 561.

Heran, H. (1952). Untersuchungen über den Temperatursinn der Honigbiene (*Apis mellifica*) unter besonderer Berücksichtigung det Wahrnehmung strahlender Wärme. *Z. vergl. Physiol.* **34,** 179–206.

Hilkenbäumer, F. and Klämbt, H. D. (1958). Die Berüchsichtigung der Befruchtungsverhältnisse bei Schwargen und Roten Johannisbeeren für den erwerbsmässigen Anbau. *Rhein. Mschr. Gemüse, Obst. Gartens* **46**, 11–12.

Hill-Cottingham, D. G. and Williams, R. R. (1967). Effect of time of application of fertilizer nitrogen on the growth, flower development and fruit set of maiden apple trees, var. Lord Lambourne, and on the distribution of total nitrogen within the trees. *J. hort. Sci.* **42**, 319–338.

Hills, K. L. (1941). Red clover seed production at Moss Vale, N.S.W. *J. Coun. scient. ind. Res. Aust.* **14**, 249–252.

Hirashima, Y. (1959). On the distribution of sexes in the nests of *Osmia excavata* Alfken (Hymenoptera, Megachilidae). *Sci. Bull. Fac. Agric. Kyushu Univ.* **17**, 45–54.

Hitchings, J. M. (1941). Observations on bee behaviour in a cucumber greenhouse. *Rep. St. Apiarist, Iowa* 1940, 76–81.

Hobbs, G. A. (1950). Pollinating species of bees in the irrigated regions of southern Alberta. Report 12th Alfalfa Improvement Conference, Lethbridge, Alberta, pp. 47–49.

Hobbs, G. A. (1956). Ecology of the Leaf-cutter Bee, *Megachile perihirta* Ck11. (Hymenoptera: Megachilidae) in relation to production of alfalfa seed. *Can. Ent.* **88**, 625–631.

Hobbs, G. A. (1957). Alfalfa and red clover as sources of nectar and pollen for honey, bumble, and leaf-cutter bees (Hymenoptera: Apoidea). *Can. Ent.* **89**, 230–235.

Hobbs, G. A. (1958). Factors affecting value of bees (Hymenoptera: Apoidea) as pollinators of alfalfa and red clover. *X Int. Congr. Ent.* **4**, 939–942.

Hobbs, G. A. (1964). Importing and managing the alfalfa leaf-cutter bee. *Publs Dep. Agric. Can.* **1209**.

Hobbs, G. A. (1967a). Obtaining and protecting red-clover pollinating species of *Bombus* (Hymenoptera: Apidae). *Can. Ent.* **99**, 943–951.

Hobbs, G. A. (1967b). Domestication of alfalfa leaf-cutter bees. *Publs Dep. Agric. Can.* **1313**.

Hobbs, G. A. and Lilly, C. E. (1955). Factors affecting efficiency of honeybees as pollinators of alfalfa in Southern Albertas. *Can. J. agric. Sci.* **35**, 422–432.

Hobbs, G. A., Virostek, J. F. and Nummi, W. O. (1960). Establishment of *Bombus* spp. (Hymenoptera: Apidae) in artificial domiciles in southern Alberta. *Can. Ent.* **92**, 868–872.

Hobbs, G. A., Nummi, W. O., and Virostek, J. F. (1961). Food-gathering behaviour of honey, bumble, and leaf-cutter bees (Hymenoptera: Apoidea) in Alberta. *Can. Ent.* **93**, 409–419.

Hobbs, G. A., Nummi, W. O. and Virostek, J. F. (1962). Managing colonies of bumble bees (Hymenoptera: Apidae) for pollination purposes. *Can. Ent.* **94**, 1121–1132.

Hockey, J. F. and Harrison, K. A. (1930). Apple pollen may be wind-borne. *Rep. Iowa St. hort. Soc.* **65**, 248–250.

Hodges, Dorothy (1952). "The Pollen Loads of the Honeybee." Bee Research Association, London.

Holm, S. N. (1960). Experiments on the domestication of bumblebees (*Bombus* Latr.) in particular *B. lapidarius* L. and *B. terrestris* L. *K. VetHøjsk. Aarsskr*, 1–19.

Holm, S. N. (1962). De vigtigste baelgplanters konkurrence om de blomstersøgende bier. *Ugeskr. Landm.* **18**, 267–73.

Holm, S. N. (1964). Bladskaerebien (*Megachile rotundata*). *Ugeskr. Landm.* **45**.
Holm, S. N. (1966a). The utilization and management of bumble bees for red clover and alfalfa seed production. *A. Rev. Ent.* **11**, 155–182.
Holm, S. N. (1966b). Problems of domestication of bumblebees. *Bee Wld* (Suppl.), **47**, 179–186.
Holm, S. N. and Haas, H. (1961). Erfahrungen und Resultate dreijähriger Domestikationversuche mit Hummeln (*Bombus* Latr.). *Albrecht-Thaer-Arch.* **5**, 282–304.
Hooper, C. H. (1919). Notes on insect visitors to fruit blossoms. *J. Pomol.* **1**, 116–124.
Hooper, C. H. (1929). The study of the order of flowering and pollination of fruit blossoms applied to commercial fruit growing. *J. R. Soc. Arts*, **77**, 424–442.
Hooper, C. H. (1931). Insect visitors to fruit blossoms. *J. S.-E. agric. Coll. Wye*, **28**, 211–215.
Hooper, C. H. (1932). The insect visitors of fruit blossoms. *J. R. Soc. Arts.* **81**, 86–105.
Hooper, C. H. (1939). Hive bees in relation to commercial fruit production. *J. S.-E. agric. Coll. Wye.* **44**, 103–108.
Hootman, H. D. (1930). The importance of pollination and the honey bee in fruit yields. *Proc. N.Y. St. hort. Soc.* **75**, 49–58.
Hootman, H. D. and Cale, G. H. (1930). A practical consideration of fruit pollination. *Am. Bee J. Bull.* 579.
Hopkins, C. Y., Jevans, A. W. and Boch, R. (1969). Occurrence of octadeca-*trans*-2, *cis*-9, *cis*-12-trienoic acid in pollen attractive to the honey bee. *Can. J. Biochem.* **47**, 433–436.
Horber, E. (1961). Beitrag zur Domestikation der Hummeln. *Vjschr. naturf. Ges. Zürich*, **106**, 424–447.
Horn, C. W. van and Todd, F. E. (1954). Bees, bouquets and better tangerines. *Progr. Agric. Ariz.* **6**, 11.
Horticultural Education Association (1961). Pollination of fruit crops. *Scient. Hort.* **14**, 126–150; **15**, 82–122.
Howard, A. and Howard, Gabrielle, L. C. (1910). Studies in Indian tobaccos. I. The types of *Nicotiana rustica* L. Yellow flowered tobacco. *Mem. Dep. Agric. India bot. ser.* **3**, 1–59.
Howard, A. and Howard, Gabrielle, L. C. (1915). First report on the improvement of Indigo in Bihar. *Bull. agric. Res. Inst., Pasa*, **51**.
Howard, A., Howard, Gabrielle, L. C. and Abdur Rahman, K. (1910). The economic significance of natural cross-fertilization in India. *Mem. Dep. Agric. India bot. ser.* **3**, 281–330.
Howard, A., Howard, Gabrielle L. C. and Abdur Rahman, K. (1916a). Studies in Indian oil-seeds. I. Safflower and Mustard. *Mem. Dep. Agric. India. bot. ser.* **7**, 237–272.
Howard, A., Howard, Gabrielle L. C. and Abdur Rahman, K. (1916b). Some varieties of Indian gram. (*Cicer arietinum* L.). *Mem. Dep. Agric. India. bot. ser.* **7**, 214–235.
Howard, A., Howard, Gabrielle L. C. and Abdur Rahman, K. (1919). Studies in the Pollination of Indian crops. *Mem. Dep. Agric. India. bot. ser.* **10**, 195–220.
Howlett, F. S. (1926). Methods of procedure in pollination studies. *Proc. Am. Soc. hort. Sci.* **23**, 107–119.
Howlett, F. S. (1927). Apple pollination studies in Ohio. *Bull. Ohio agric. Exp. Stn.* No. 404, 1–84.
Howlett, F. S. (1934). Pollination of the apple in Ohio. *Bull. Ohio agric. Exp. Stn.* No. 167, 65–70.

Hua, H. (1943). (Natural crossing in *Vicia faba.*). *Chin. J. scient. Agric.* **1**, 63–65.
Hudson, R. (1929). Bee investigations. *Rep. N.J. agric. Exp. Stn.* 198–203.
Hudson, R. (1930). Bee investigations. *Rep. N.J. agric. Exp. Stn.* 176–182.
Hughes, Hilary M. (1951). "Fruit cultivation for amateurs." Collingridge, London.
Hughes, Hilary M. (1961). Preliminary studies on the insect pollination of soft fruits. *Expl. Hort.* **6**, 44.
Hughes, Hilary M. (1962). Pollination studies. Progress Report. M.A.A.F., N.A.A.S., *Rep. Efford Exp. Hort. Stn.* 1961.
Hughes, Hilary M. (1963). Black currant. Pollination experiment 1962. Final Report. *Rep. Efford exp. Hort. Stn.* 1962, 13–14.
Hughes, Hilary M. (1966). Investigations on the pollination of blackcurrants var. Baldwin. *Expl. Hort.* No. 14, 13–17.
Hughes, J. H. (1943). The alfalfa plant bug. *Tech. Bull. Minn. agric. Exp. Stn.* 161.
Hume, H. H. (1957). "Citrus Fruits." MacMillan, New York.
Hurd, P. D. (1964). The pollination of pumpkins, gourds and squashes (genus *Cucurbita*). II. Int. Symp. Pollination. *Bee Wld* (Suppl.), **47**, 97–98.
Hurd, P. D. and Linsley, E. G. (1964). The squash and gourd bees—Genera *Peponapis* Robertson and *Xenoglossa* Smith—inhabiting America north of Mexico (*Hymenoptera: Apoidea*). *Hilgardia* **35**, 375–477.
Hurd, P. D. and Linsley, E. G. (1966a). Squash and Gourd bees of the genus *Xenoglossa* (*Hymenoptera: Apoidea*). *Ann. ent. Soc. Am.* **60**, 988–1007.
Hurd, P. D. and Linsley, E. G. (1966b). South American squash and gourd bees of the genus *Peponapis* (*Hymenoptera: Apoidea*). *Ann. ent. Soc. Am.* **60**, 647–661.
Hurt, E. F. (1946). "Sunflower." Faber and Faber, London.
Hutchinson, J. (1959). The families of flowering plants. Clarendon, Oxford.
Hutson, R. (1924). Bees in fruit pollination. *Glean. Bee Cult.* **52**, 290–292.
Hutson, R. (1925). The honey bee as an agent in the pollination of pears, apples and cranberries. *J. econ. Ent.* **18**, 387–391.
Hutson, R. (1926). Relation of the honeybee to fruit pollination in New Jersey. *Bull. N.J. agric. Exp. Stn.* No. 434.
Hutson, R. (1928). Package versus overwintered bees for orchard use as pollinizers. *Am. Bee J.* **68**, 128.
Hutson, R. (1929). The effect of moving bees at orchard blooming time. *J. econ. Ent.* **22**, 522–526.
Huyskes, J. A. (1959). The value of comparative tests of progenies from open-pollinated female asparagus plants. *Euphytica* **8**, 141–144.
Hyams, E. (1962). "Strawberry Growing Complete." Faber and Faber, London.
Ibrahim, S. H. and Selim, H. A. (1962). Studies on pollen collection from *Eucalyptus* spp. and Compositae plants by the honeybee. *Agric. Res. Rev. Cairo* **40**, 116–123.
Istomina-Tsvetkova, K. P. and Skrebtsov, M. F. (1964). A new method of stimulating the pollinating activity of bees. *Trud. nauch.-issled. Inst. Pchelov.*, 205–222.
Ivanoff, S. S. (1947). Natural self-pollination in cantaloupes. *Proc. Am. Soc. hort. Sci.* **50**, 314–316.
Izmailova, A. V. (1934). Pollination of lucerne by insects. *Semenovodstvo* (1), 47–49.
Jablónski, B. (1961). Wyniki badań nad wartością pszczelarską rukwi siewnej. *Pszczel. Zesz. Nauk* **5**, 33–51.
James, E., Massey, J. H. and Corley, W. L. (1960). Effect of a plant arrangement on cross-pollination of muskmelons. *Proc. Am. Soc. hort. Sci.* **75**, 480–484.
Jamieson, C. A. (1950). Pollination studies. *Can. Bee J.* **58**, 20–22.
Jany, E. (1950). Der "Einbruch" von Erdhummeln (*Bombus terrestris* L.) in die Blüten der Feuerbohne (*Phaseolus multiflorus* Willd.). *Z. angew. Ent.* **32**, 172–183.

Jay, S. C. (1963). The development of honeybees in their cells. *J. apic. Res.* **2**, 117–134.

Jay, S. C. (1965). Drifting of honeybees in commercial apiaries. I. Effect of various environmental factors. *J. apic. Res.* **4**, 167–175.

Jay, S. C. (1966). Drifting of honeybees in commercial apiaries. II. Effect of various factors when hives are arranged in rows. *J. apic. Res.* **5**, 103–112.

Jay, S. C. (1969). Drifting of honeybees in commercial apiaries. V. Effect of drifting on honey production. *J. apic. Res.* **8**, 13–17.

Jaycox, E. R. (1970). Honey bee queen pheromones and worker foraging behaviour. *Ann. Ent. Soc. Am.* **63**, 222–228.

Jaycox, E. R. and Owen, F. W. (1965). Honeybees and pollen inserts can improve apple yields. *Am. Bee J.* **105**, 96–97.

Jeffree, E. P. (1955). Observations on the decline and growth of colonies. *J. econ. Ent.* **48**, 723–726.

Jeffree, E. P. and Allen, M. Delia (1957). The annual cycle of pollen storage by honey bees. *J. econ. Ent.* **50**, 211–212.

Jenkins, J. M. (1942). Natural self-pollination in cucumbers. *Proc. Am. Soc. hort. Sci.* **40**, 411–412.

Jenkinson, J. G. and Glynne-Jones, G. D. (1953). Observations on the pollination of oil rape and broccoli. *Bee Wld* **34**, 173–177.

Johannson, T. S. K. (1959). Tracking honey bees in cotton fields with fluorescent pigments. *J. econ. Ent.* **52**, 572–577.

Johansen, B. R. (1968). Isolation distances in lucerne. *K. Vet.-og Landbohøisk Arsskr.* 153–158.

Johansen, C. A. (1956). Artificial pollination of apples with bee-collected pollen. *J. econ. Ent.* **49**, 825–828.

Johansen, C. A. (1960a). Insect Pest Control and Pollination on Red Clover grown for seed in Central Washington. *E.M. Wash. St. Univ. agric. Ext. Serv.* 1985.

Johansen, C. A. (1960b). Pollination of tree fruits in Eastern Washington. *Proc. Wash. St. hort. Ass.* **56**, 17–19.

Johansen, C. A. (1966). Pollination of clovers raised for seed in Washington. *Am. Bee J.* **106**, 298–300.

Johansen, C. A. and Degman, E. (1957). Progress report on hive inserts for apple pollination. *Proc. Wash. St. hort. Ass.* **53**, 77–81.

Johansen, C. A. and Eves, J. (1966). Parasites and nest destroyers of the alfalfa leaf-cutting bee. *Circ. Wash. agric. Exp. Stn.* **469**.

Johansen, C. A. and Hutt, R. (1963). Encouraging the bumble bee pollinator of cranberries. *Agric. ext. Serv. Wash. St. Univ.* No. E.M. 2262.

Johnson, A. P. (1944). Honey from soyabeans. *Am. Bee J.* **84**, 306.

Johnson, W. C. and Nettles, W. C. (1953). Pollination of crimson clover: 1952 demonstration results. *Misc. Publ. S.C. Ext. Serv.* (unnumbered).

Johnston, S. (1929). Insects aid fruit setting of raspberry. *Q. Bull. Mich. agric. Exp. Stn.* **9**, 105–106.

Jones, D. F. (1916). Natural cross-pollination in the tomato. *Science, N.Y.* **43**, 509–510.

Jones, G. A. (1912). The structure and pollination of the Cacao flower. *W. Ind. Bull.* **12**, 347–350.

Jones, H. A. (1923). Pollination and self-fertility in the Onion. *Proc. Am. Soc. hort. Sci.* **20**, 191–197.

Jones, H. A. (1927). Pollination and Life History Studies of Lettuce (*Lactuca sativa* L.). *Hilgardia* **2**, 425–479.

Jones, H. A. (1937). Onion improvement. *Yb. Agric. U.S. Dep. Agric.* 233–250.
Jones, H. A. and Emsweller, S. L. (1933). Methods of breeding onions. *Hilgardia* **7**, 625–642.
Jones, H. A. and Emsweller, S. L. (1934). The use of flies as onion pollinators. *Proc. Am. Soc. hort. Sci.* **31**, 161–164.
Jones, H. A. and Mann, L. K. (1963). "Onions and Allies." Leonard Hill, London.
Jones, H. A. and Rosa, J. T. (1928). "Truck Crop Plants." McGraw-Hill, New York.
Jones, M. D. and Tamargo, M. A. (1954). Agents concerned with natural crossing of kenaf in Cuba. *Agron. J.* **46**, 459–462.
Jones, H. A., Bailey, S. F. and Emsweller, S. L. (1934). Thrips resistance in the onion. *Hilgardia* **8**, 215–232.
Jorgensen, C. O. (1921). Om bestøvnings- og befrugtningsforhold hos nogle graesmarksbaelgplanter med henblik paa deres foraedling. *K. VetHøjsk. Aarsskr.* 218–244.
Julén, G. (1953). Erfarenheter från fröodlings-försök med rödklöverstammer i Götaland. Undersökningar rörande insektspollinerade kulturväxter II. *Medd. Sveng. FröodlFörb.* **2**, 83–101.
Julén, G. (1953b). Speciella problem rörande frösättningen hos tetraploid rödklöver. *Medd. N:o 2 fran Sveriges Fröodlareförb.* 79–82.
Julén, G. (1954). Aspects on the Breeding of Tetraploid Red Clover with Special Reference to the Seed Setting problem. European Grassland Conference, Paris 1954.
Julén, G. (1956). Practical Aspects on Tetraploid Clover. *Proceedings Seventh Int. Grassl. Congr.* **11**, 44.
Kadam, B. S. and Patankar, V. K. (1942). Natural cross-pollination in safflower. *Indian J. Genet. Pl. Breed.* **2**, 69–70.
Kakizaki, Y. (1922). Self-sterility in Chinese cabbage. *J. Hered.* **13**, 374–376.
Kakizaki, Y. (1930). Studies on the genetics and physiology of self- and cross-incompatibility in the common cabbage (*Brassica oleracea* L. var. *capitata* L.). *Jap. J. Bot.* **5**, 133–208.
Kakizaki, Y. and Kasai, T. (1933). Bud pollination in cabbage and radish. Some examples of conspicuous "pseudo-fertility" in normally self-incompatible plants. *J. Hered.* **24**, 359–360.
Kalashnikov, V. M. (1931). A contribution to the methodics of breeding the guayule (*Parthenium argentatum* Gray). *Bull. appl. Bot. Pl.-Breed, Petrograd* **27**, 489–560.
Karlson, P. and Butenant, A. (1959). "Pheromones (Ectohormones) in insects." *A. Rev. Ent.* **4**, 39–58.
Karmo, E. A. (1958). Honey bees as an aid in orchard and blueberry pollination in Nova Scotia. *X Int. Congr. Ent.* 1956, **4**, 955–959.
Karmo, E. A. (1960). Report on pollination studies on the apple set by using honeybees in combination with pollen from different sources. *Rep. Nova Scotia Fruit Grow. Ass.* **97**, 125–128.
Karmo, E. A. (1961a). Increasing the pollination efficiency of the honeybee through colony rotation. *Mimeo. Circ. Nova Scotia Dep. Agric.* No. 102.
Karmo, E. A. (1961b). Report on pollination research in Nova Scotia. *Proc. Can. Ass. Apiculturists*, 21–23.
Karmo, E. A. and Vickery, V. R. (1954). The place of honey bees in orchard pollination. *Mimeogr. Circ. Nova Scotia Dep. Agric. Mktg* No. 67.
Karmo, E. A. and Vickery, V. R. (1957). Bees to the rescue. *Am. Fruit Grow.* (4), 42–45.
Karmo, E. A. and Vickery, V. R. (1960). The fruit pollination in Nova Scotia. *Glean. Bee Cult.* **88**, 167–170, 187.

Kashkovskii, V. G. (1958). Economic results of pollinating buckwheat crops by honeybees and by hand. *Byull. nauch.-tekh. Inf. Kemerovo* **2**, 59–61.
Kaziev, I. P. and Seidova, S. S. (1965). The nectar yield of flowers of some Curcurbitaceae under Azerbaidjan conditions. *XX International Beekeeping Jubilee Congress* (II: 31).
Kaziev, T. I. (1956). The working of cotton by bees in western Azerbaidzhan. *Uchen. Zap. kirovabad. ped. Inst.* (4), 153–156.
Kaziev, T. I. (1958). Supplementary pollination of cotton by bees. *Pchelovodstvo, Mosk.* **35**, 28–34.
Kaziev, T. I. (1959). The influence of fertilizers and watering on the nectar secretion of cotton. *Pchelovostvo, Mosk.* **36**, 25–28.
Kaziev, T. I. (1960). The influence of bee pollination on cotton quality. *Pchelovodstvo, Mosk.* **37**, 33–35.
Kaziev, T. I. (1961). Influence of bee pollination on the progeny of cotton varieties. *Uchen. Zap. kirovabad ped. Inst.* (8), 100–112.
Kearney, T. H. (1923a). Self-fertilisation and cross-fertilisation in Pima Cotton. *Bull. U.S. Dep. Agric.* **1134**.
Kearney, T. H. (1923b). Segregation and correlation of characters in an Upland-Egyptian cotton hybrid. *Bull. U.S. Dep. Agric.* **1164**.
Kelty, R. H. (1948). Help the bees to help you. *Am. Fruit Grow.* **68**, 16–31.
Kennerley, A. B. (1961). You can't fool the bees (cucurbit pollination). *Am. Veg. Grow.* **9**, 12.
Kerner, A. (1895). "The Natural History of Plants." Vol. II, transl. by F. Oliver. Blackie, London.
Kerr, E. A. and Kribs, L. (1955). Electrical vibrators as an aid in greenhouse tomato production. *Agric. Inst. Rev.* **10**, 34.
Khan, K. S. A. R. (1929). Pollination and fruit formation in litchi (*Nephelium litchi*, Camb). *Agric. J. India* **24**, 183–187.
Kiechle, H. (1961). Die soziale Regulation der Wassersammeltätigkeit im Bienenstaat und deren physiologische Grundlage. *Z. vergl. Physiol.* **45**, 154–192.
Kikuchi, T. (1963). Studies on the coaction among insects visiting flowers. III. Dominance relationship among flower-visiting flies, bees and butterflies. *Sci. Rep. Tôhoku Univ.*, Ser. 4, **29**, 1–8.
Killinger, G. B. and Haynie, J. D. (1951). Honeybees in Florida's pasture development. *Spec. Ser. Fla Dep. Agric.* No. 66, 112–115.
Killion, C. E. (1939). The cucumber and the bee. *Am. Bee J.* **79**, 399–400.
King, G. E. and Burrel, A. B. (1933). An improved device to facilitate pollen distribution by bees. *Proc. Am. Soc. hort. Sci.* **29**, 156–159.
Kinsman, G. B. (1957). The low bush blueberry in Nova Scotia. Publication 1036. Dept. Agriculture and Marketing, Halifax, Nova Scotia.
Kirk, E. E. (1925). Self-pollination of sweet clover. *Scient. Agric.* **6**, 109–112.
Kirk, E. E. and Stevenson, T. M. (1931). Factors which influence spontaneous self-fertilization in sweet clover (*Melilotus*). *Can. J. Res.* **5**, 313–326, 660–664.
Kirk, L. E. and White, W. J. (1933). Autogamous alfalfa. *Scient. Agric.* **13**, 591–593.
Klämbt, H. D. (1958). Untersuchungen über die Befruchtungsverhältnisse bei Schwarzen und Roten Johannisbeeren. *Gartenbauwissenschaft* **23**, 9–28.
Kleber, Elisabeth (1935). Hat das Zeitgedächtnis der Bienen biologische Bedeutung? *Z. vergl. Physiol.* **22**, 221–262.
Kloavev, A. M. and Ulanichev, E. M. (1961). Attempt to estimate the increase in harvest of sunflowers as a result of pollination by bees. *Pchelovodstvo, Mosk.* **38**, 7–11.

Kloet, G. S. (1943). Some improved rearing devices. Hymenopterist's Handbook. *Amat. Ent.* **7,** 94–101.
Knaap, W. P. van der (1955). Observations on the pollination of cacao flowers. *Int. hort. Congr.* **14,** 1287–1293.
Knee, W. J. and Moeller, F. E. (1967). Comparative study of pollen sources of honeybees and bumblebees. *J. apic. Res.* **6,** 133–138.
Knight, R. and Rogers, II. H. (1955). Incompatibility in *Theobroma Cacao*. *Heredity* **9,** 69–77.
Knoll, F. (1926). Insekten und Blumen. *Abh. Zool.-bot. Ges. Wein* **12,** 1–646.
Knuth, P. Handbook of Flower Pollination (trans. by J. R. Ainsworth-Davis). Clarendon Press, Oxford.
Knysh, A. N. (1958). Pollination by bees of varieties of cucumber. *Sadȳ Ogorodȳ* (6), 13–16.
Kobel, F. (1942). Obstbau und Bienenzucht. *Beih. schweiz. Bienenztg* **1,** 111–154.
Koot, Y. van (1960). De invloed van bijen op het onstaan van zaadkoppen bij komkommers. *Meded. Dir. Tuinb.* **23,** 735–749.
Kopelkievskii, G. V. (1953). Timely locations of bees for pollination of buckwheat and the honey crop. *Pchelovostvo, Mosk.* **30,** 28–31.
Kopelkievskii, G. V. (1955). Pollination of buckwheat by bees. *Pchelovodstvo, Mosk.* **32,** 41–48.
Kopelkievskii, G. V. (1959). Annual sweet clover (*Melilotus*). *Pchelovodstvo, Mosk.* **36,** 32–35.
Kopelkievskii, G. V. (1960). Bees and the buckwheat seed crop. *Pchelovodstvo, Mosk.* **37,** 36–39.
Kopelkievskii, G. V. (1964a). Nectar productivity and buckwheat yields under various conditions of cultivation. *Trud. nauch.-issled. Inst. Pchelov.* 150–175.
Koperzinskii, V. V. (1949). The cause of sterility in Lucerne, *Medicago sativa*, and means of checking it. *Sov. Agron.* (3), 68–76.
Korolev, A. M. (1936). The biology of the feijoa blossoming. *Sov. Subtrop.* **11,** 68–70.
Kostylev, A. D. and Vinogradov, S. I. (1934). Role of the honey bee in raising seed production in lucerne. Rostov on Don.
Kottur, G. L. (1930). Use of self seed in maintaining the purity of improved cottons. *Agr. J. India* **25,** 39.
Koutenský, J. (1958). The results of the pollinating work of Bees. *Včelařství* **11,** 72–73.
Koutenský, J. (1959). Opylovací účinek včely medonosné (*Apis mellifera* L.) na zýšsení hektarových výnosů u řepky olejné a hořčice bílé. *Sborn. čsl. Akad. zemed. Ved.* **5,** 571–582.
Kovalev, A. A. (1958). Apparatus for the artificial pollination of vines. *Sadȳ Ogorodȳ* **5,** 67–69.
Kozin, R. B. (1954). Effect of bees on the seed crop of flax. *Pchelovodstvo, Mosk.* (6), 41–43.
Kozin, R. B. (1967). Influence of bee (*Apis mellifera*) pollination of lupine *Lupinus* L. on seed yield and quality. *XXI Int. Beekeep. Congr. Summ.* Paper 158.
Kraai, A. (1954). Het gebruik van bijen bij het veredelingswerk. *Zaadbelangen* **8,** 121–122, 132–133, 144–146.
Kraai, A. (1958). Bijen en hommels bij het veredelingswerk. *Meded. Dir. Tuinb.* **21,** 291–297.
Kraai, A. (1962). How long do honeybees carry germinable pollen on them? *Euphytica* **11,** 53–56.

Kratky, E. (1931). Morphologie und Physiologie der Drüsen im Kopf und Thorax der Honigbiene. *Z. wiss. Zool.* **139**, 120–200.
Kremer, J. C. (1945). Influence of Honey Bee Habits on Radish Seed Yield. *Q. Bull. Mich. agric. Exp. Stn.* **27**, 413–420.
Kremer, J. C. (1948). Traps for the collection and distribution of pollen in orchards. *Bull. Mich. agric. Exp. Stn.* **31**, 12–21.
Kremer, J. C. (1949). Germination tests of the viability of apple pollen gathered in pellets. *Proc. Am. Soc. hort. Sci.* **53**, 153–157.
Kremer, J. C. (1950). The dandelion and its influence on bee behaviour during the fruit blossoming period. *Proc. Am. Soc. hort. Sci.* **55**, 140–146.
Kremer, J. C. and Hootman, H. D. (1930). Greater efficiency in pollination. *Glean. Bee Cult.* **58**, 219–221.
Krezdorn, A. H. (1959). Factors affecting the unfruitfulness of tangelos. *Rep. Fla. agric. Exp. Stn.* 1958–1959, 228–229.
Kristofferson, K. B. (1921). Spontaneous crossing in the garden bean, *Phaseolus vulgaris*. *Hereditas* **2**, 395–400.
Kroll, U. (1961). The influence of fertilization on the production of pyrethrins in the pyrethrum flower. *Pyrethrum Post* **6**, 19–21.
Kronenberg, H. G., Braak, J. P. and Zeilinga, A. E. (1959). Poor fruit setting in strawberries. II. Malformed fruits in Jucunda. *Euphytica* **8**, 245–251.
Kropáčová, Sylvie (1963). Vztahy mezi klimatickými faktory, nektarem vojtěšky a náletem včel (*Apis mellifera* L.) na vojtěšku. *Sborn. vys. Skoly Zeměd. Brně* **4**, 603–611.
Kropáčová, Sylvie (1964). Studie o činnosti včely medonosné (*Apis mellifera* L.) jako opylovace vojtěšky seté (*Medicago sativa* L.). *Zool. Listy, Fol. zool.* **13**, 143–154.
Kropáčová, Sylvie (1965). Možnosti zlepšeni práce včely medonosné (*Apis mellifera* L.) při opylování vojtěšky seté (*Medicago sativa* L.). *Sborn. vys. Skoly Zeměd. Brně* (1), 111–122.
Kropáčová, Sylvie (1969). The relationship of the honeybee to Sainfoin (*Onobrychis sativa*). *Proc. 22nd Int. Beekeep. Cong. Munich*, 141.
Kropáčová, Sylvie and Laitová, Libuše, (1965). Rozbor Obsahu Cukra V Nektaru Vojtěšky. *Sborn. vys. Skoly Zeměd. Brné* **3**, 425–431.
Krug, C. A. and Costa, A. S. (1947). Criação de variedades melhoradas de café. *Fazenda* **42**, 35, 46–47.
Kugler, H. (1932). Blütenökologische Untersuchungen mit Hummeln. IV. *Planta* **16**, 534–553.
Kugler, H. (1943). Hummeln als Blütenbesuchen. *Ergebn. Biol.* **19**, 143–323.
Kuliev, A. M. (1958). The use of bees to increase cotton yield. *XVII Int. beekeep. Congr.*
Kulinčevic, J. (1959). Intenzitet posete pcela kokocu (*Melilotus albus*) u uslovima Metohije. *Zborn. nauč. Rad. Inst. poljopriv. Istraž.* **2**, 389–394.
Kundu, B. C., Basak, K. C. and Sarcar, P. B. (1959). "Jute in India." Leonard Hill, London.
Kurennoi, N. M. (1957). An experiment in increasing effectiveness of bee pollination of sunflowers. *Pchelovodstvo, Mosk.* **34**, 42–48.
Kurennoi, N. M. (1965). Particularities of the activity of bees on the blossoms of reciprocally pollinating apple varieties. *XX Int. Congr. Beekeep.* (II/I).
Kurennoi, N. M. (1967). Apple yield as a function of the multiplicity of bee visits to the flowers. *XXI Int. Beekeep. Congr. prelim. sci. Meet.* Papers 39, 162; 175–181.

Kushman, L. J. and Beattie, J. H. (1946). Natural hybridisation in peanuts. *J. Am. Soc. Agron.* **38,** 755–756.

Kushnir, L. G. (1958). The biological effectiveness of sunflower pollination by various methods. *Dokl. TSKhA* **36,** 81–88.

Kushnir, L. G. (1960). Economic evaluation of sunflower pollination by bees and by hand. *Pchelovodstvo, Mosk.* **37,** 22–25.

Laere, O. van (1957). L'importance des abeilles pour la fructification des arbres fruitiers. *Belg. apic.* **21,** 231–235.

Laere, O. van and Martens, N. (1962). L'importance de l'abeille domestique dans la production de semences de trèfle commun. *Revue Agric. Brux.* **15,** 1383–1395.

Lagassé, F. S. (1928). Proper pollination of fruit blossoms. *Ext. Bull. Univ. Del. Ext. Serv. Agric.* **15.**

Lambeth, V. N. (1951). Some factors influencing pod set and yield of the lima bean. *Res. Bull. Mo. agric. Exp. Stn.* No. 466.

Lambeth, V. N. (1952). Some factors influencing pod set and yield of the lima bean. *Bull. Mo. agric. Exp. Stn.* No. 466.

Langham, D. G. (1944). Natural and controlled pollination in Sesame. *J. Hered.* **35,** 255–256.

Langridge, D. F. (1956). Bees and Pollination. *J. Agric., Victoria* **54,** 462–463.

Lansdell, J. and Macself, A. J. (1948). Grapes, Peaches, Nectarines and Melons. Collingridge, London.

Larsen, P. and Tung, S. M. (1950). Growth-promoting and growth-retarding substances in pollen from diploid and triploid apple varieties. *Bot. Gaz.* **111,** 436–447.

Lathrop, F. H. (1954). Honey bees and blueberry pollination. *Glean. Bee Cult.* **82,** 331.

Latif, A., Qayyum, A. and Manzoor-ul-Haq (1956). Role of *Apis indica* F. in the pollination of Egyptian clover (*Trifolium alexandrinum* Linn.) *Pakist. J. scient. Res.* **8,** 48–50.

Latif, A., Qayyum, A. and Abbas, M. (1960). The role of *Apis indica* in the pollination of "toria" and "sarson" (*Brassica campestris* var. *toria* and *dichotoma*). *Bee Wld* **41,** 283–286.

Latimer, L. P. (1936). Can bees retain pollen of early apple varieties for effective pollination of later blooming sorts. *Proc. Am. Soc. hort. Sci.* **34,** 16–18.

Lavie, P. (1967). Influence de l'utilization du piège à pollen sur le rendement en miel des colonies d'abeilles. *Annls Abeille* **10,** 83–95.

Lecomte, J. (1950). Sur le determinisme de la formation de la grappe chez les abeilles. *Z. vergl. Physiol.* **32,** 499–506.

Lecomte, J. (1955). Observations sur le comportement d'abeilles vivant en serres chaudes durant l'hiver. *Apiculteur* (Sect. sci.) **99,** 39–42.

Lecomte, J. (1959). Luzerne et Apiculture. *Annls Abeille* **2,** 211–221.

Lecomte, J. (1960). Observations sur le comportement des abeilles butineuses. *Annls Abeille* **3,** 317–327.

Lecomte, J. (1961). Observations sur la pollinisation de l'avocatier aux Antilles françaises. *Fruits* **16,** 411–414.

Lecomte, J. (1962a). Techniques d'étude des populations d'insectes pollinisateurs. *Annls Abeille* **5,** 201–213.

Lecomte, J. (1962b). Observations sur la pollinisation du tournesol (*Helianthus annuus* L.). *Annls Abeille* **5,** 69–73.

Lecomte, J. and Tirgari, S. (1965). On some pollinators of the fodder legumes. *Annls Abeille* **8,** 83–93.

Ledeboer, M. and Rietsema, I. (1940). Unfruitfulness in black currants. *J. Pomol.* **18,** 177–180.
Lederhouse, R. C., Caron, D. M. and Morse, R. A. (1968). Onion pollination in New York. *N. Y. Fd Life Sci.* **1,** 8–9.
Lee, W. R. (1958). Pollination studies on low-bush blueberries. *J. econ. Ent.* **51,** 544–545.
Lee, W. R. (1961). The Nonrandom Distribution of Foraging Honey Bees between Apiaries. *J. econ. Ent.* **54,** 928–933.
Lee, W. R. (1965). Relation of distance to foraging intensity of honeybees on natural food sources. *Ann. ent. Soc. Am.* **58,** 94–100.
Lehmann, U. (1952). Die Abhängigkeit der Rotklee-Samenerträge von der Witterung. *Saatgut-Wirt.* **4,** 284–285.
Lepage, M. and Boch, R. (1968). Pollen Lipids attractive to honeybees. *Lipids* **3,** 530–534.
Leppik, E. E. (1953). The ability of insects to distinguish number. *Am. Nat.* **87,** 229–236.
Leshchev, V. (1952). Modern agriculture increases nectar production in buckwheat. *Pchelovodstvo, Mosk.* **29,** 23–26.
Lesik, F. L. (1953). Microfertilization increases the nectar secretion of honey plants. *Pchelovodstvo, Mosk.* **30,** 42–45.
Lesins, K. (1950). Investigations into seed setting of lucerne at Ultuna, Sweden, 1945–1949. *LantbrHögsk. Ann.* **17,** 441–483.
Lesins, K., Akerberg, E. and Böjtös, Z. (1954). Tripping in alfalfa flowers. *Acta Agric. scand.* **4,** 239–256.
Lesley, J. W. (1924). Cross-pollination of tomatoes. *J. Hered.* **15,** 233–235.
Lesley, J. W. and Bringhurst, R. S. (1951). Environmental conditions affecting pollination of avocados. *Yb. Calif. Avocado Soc.* 169–173.
Leuck, D. B. and Hammons, R. O. (1965a). Pollen-collecting activities of bees among peanut flowers. *J. econ. Ent.* **58,** 1028–1030.
Leuck, D. B. and Hammons, R. O. (1965b). Further evaluation of the role of bees in natural cross-pollination of the peanut, *Arachis hypogaea* L. *Agron. J.* **57,** 94.
Leuck, D. B., Forbes, I., Burns, R. E. and Edwardson, J. R. (1968). Insect visitors to flowers of blue lupine, *Lupinus angustifolius*. *J. econ. Ent.* **61,** 573.
Levchenko, I. A. (1959). The distance bees fly for nectar. *Pchelovodestvo, Mosk.* **36,** 37–38.
Levin, M. D. (1955). A technique for estimating the percentage of honey bees visiting alfalfa. *J. econ. Ent.* **48,** 484–485.
Levin, M. D. (1957). Artificial nesting burrows for *Osmia lignaria* Say. *J. econ. Ent.* **50,** 506–507.
Levin, M. D. (1959). Distribution Patterns of Young and Experienced Honey Bees foraging on Alfalfa. *J. econ. Ent.* **52,** 969–971.
Levin, M. D. (1960). A comparison of two methods of mass-marking foraging honey bees. *J. econ. Ent.* **53,** 696–698.
Levin, M. D. (1961a). Distribution of Foragers from Honey Bee Colonies placed in the middle of a large field of Alfalfa. *J. econ. Ent.* **54,** 431–434.
Levin, M. D. (1961b). Interactions among foraging honey bees from different apiaries in the same field. *Insectes soc.* **8,** 195–201.
Levin, M. D. (1966). Orientation of honeybees in alfalfa with respect to landmarks. *J. apic. Res.* **5,** 121–125.
Levin, M. D. and Bohart, G. E. (1955). Selection of pollens by honey bees. *Am. Bee J.* **95,** 392–393, 402.

Levin, M. D. and Bohart, G. E. (1957). The effect of prior location on alfalfa foraging by honey bees. *J. econ. Ent.* **50**, 629–632.

Levin, M. D. and Butler, G. D., Jr. (1966). Bees associated with safflower in south central Arizona. *J. econ. Ent.* **59**, 654–657.

Levin, M. D. and Glowska-Konopacka, S. (1963). Responses of foraging honeybees in alfalfa to increasing competition from other colonies. *J. apic. Res.* **2**, 33–42.

Levin, M. D., Butler, G. D., Jr. and Rubis, D. D. (1967). Pollination of safflower by insects other than honey bees. *J. econ. Ent.* **60**, 1481–1482.

Levin, M. D., Kuehl, R. O. and Carr, R. V. (1968). Comparison of three sampling methods of estimating honey bee visitation to flowers of cucumbers. *J. econ. Ent.* **61**, 1487–1489.

Lewis, C. I. and Vincent, C. C. (1909). Pollination of the apple. *Bull. Ore. agric. Coll.* No. 104.

Lewis, T. and Smith, B. D. (1969). The insect faunas of pear and apple orchards and the effect of windbreaks on their distribution. *Ann. appl. Biol.* **64**, 11–20.

Lindauer, M. (1948). Über die Einwirkung von Duft- und Geschmacksstoffen sowie anderer Faktoren auf die Tänze der Bienen. *Z. vergl. Physiol.* **31**, 348–412.

Lindauer, M. (1952). Ein Beitrag zur Frage der Arbeitsteilung im Bienenstaat. *Z. vergl. Physiol.* **34**, 299–345.

Lindauer, M. (1954). Temperaturregulierung und Wasserhaushalt im Bienenstaat. *Z. vergl. Physiol.* **36**, 391–432.

Lindauer, M. (1955). The water economy and temperature regulation of the honeybee colony. *Bee Wld* **36**, 62–72, 81–92, 105–111.

Lindauer, M. (1961). Communication among social bees. Harvard University Press, Cambridge, Mass.

Linsley, E. G. (1960). Observations on some matinal bees at flowers of *Cucurbita*, *Ipomoea* and *Datura* in desert areas of New Mexico and South-eastern Arizona. *J.N.Y. ent. Soc.* **68**, 13–20.

Loew, E. (1885). Beobachtungen über den Blumenbesuch von Insekten an Freilandpflanzen des Botanisches Gartens zu Berlin. *Jb. Bot. Gart. Berl.* **3**, 69–118.

Lindhard, E. (1911). Om Rødkløverens Bestøvning og de Humlebiarter, som herved er virksomme. *Tidsskr. Landbr. PlAvl.* **18**, 719–737.

Lindhard, E. (1921). Der Rotklee, *Trifolium pratense* L. bei natürlicher und künstlicher Zuchtwahl. *Z. PflZücht* **8**, 95–120.

Linsley, E. G. (1946). Insect pollinators of alfalfa in California. *J. econ. Ent.* **39**, 18–28.

Linsley, E. G. (1961). The role of flower specificity in the evolution of solitary bees. *XI Int. Congr. Ent., Wien*, 1960, 593–596.

Linsley, E. G. and McSwain, J. W. (1942). The parasites, predators, and inquiline associates of *Anthophora linsleyi*. *Am. Midl. Nat.* **27**, 402–417.

Linsley, E. G. and McSwain, J. W. (1947). Factors influencing the effectiveness of insect pollinators of Alfalfa in California. *J. econ. Ent.* **40**, 349–357.

Loden, H. D. and Richmond, T. R. (1951). Hybrid vigor in cotton—cytogenetic aspects and practical applications. *Econ. Bot.* **5**, 387–408.

Loewel, E. L. (1943). Ueber die Bedeutung der Bienenfrage im Obstbaugebiet an der Niederelbe. *Dt. Imkerführer* **16**, 160.

Löken, A. (1958). Pollination studies in apple orchards of Western Norway. *X Int. Congr. Ent.* 1956, **4**, 961–965.

Lötter, J. de V. (1960). Recent Developments in the Pollination Techniques of Deciduous Fruit Trees. *Decid. Fruit Grow.* **10**, 182–190; 212–224; 304–311.

Louveaux, J. (1952). Recherches sur la pollinisation chez la navette (*Brassica rapa* var. *oleifera*). *Apiculteur* (Sect. sci.), **96**, 15–18.

Louveaux, J. (1954). Études sur la récolte du pollen par les abeilles. *Apiculteur* (Sect. sci.), **98**, 43–50.

Louveaux, J. (1959). Recherches sur la récolte du pollen par les abeilles. *Annls Abeille* **2**, 13–111.

Lovell, H. B. (1955). Bird's-foot trefoil—*Lotus corniculatus*. *Glean. Bee Cult.* **83**, 719–720, 759.

Luce, W. A. and Morris, O. M. (1928). Pollination of deciduous fruits. *Bull. Wash. St. Coll.* No. 223.

Lukoschus, F. (1957). Quantitative Untersuchungen über den Pollentransport im Haarkleid der Honigbiene. *Z. Bienenforsch.* **4**, 3–21.

Lundie, A. E. (1925). The flight activities of the honey-bee. *Bull. U.S. Dep. Agric.* **1328**.

Luttso, V. P. (1956). The pollination of sunflowers. *In* "The pollination of agricultural crops" (in Russian). (I. V. Krischumas and A. F. Gubin, eds.) pp. 45–52. State Publishing House for Agricultural Literature, Moscow.

Luttso, V. P. (1957). The pollination of flax by honeybees. *Dokl. TSKhA* **30**, 327–331.

McCollum, J. P. (1934). Vegetable and reproductive responses associated with fruit development in the cucumber. *Mem. Cornell Univ. agric. Exp. Stn.* 163.

McCulloch, J. W. (1914). The relation of the honeybee to other insects in cross-pollination of the apple blossom. *A. Mtg Kans. St. hort. Soc.* **46**, 85–88.

MacDaniels, L. H. (1930a). Practical aspects of the pollination problems. *Proc. N.Y. St. hort. Soc.* **75**, 195–202.

MacDaniels, L. H. (1930b). The possibilities of hand pollination in the orchard on a commercial scale. *Proc. Am. Soc. hort. Sci.* **27**, 370–373.

MacDaniels, L. H. (1931). Further experience with the pollination problem. *Proc. N.Y. St. hort. Soc.* **76**, 32–37.

MacDaniels, L. H. and Heinicke, A. J. (1929). Pollination and other factors affecting the set of fruit with special reference to the apple. *Bull. Cornell agric. Exp. Stn.* No. 497.

McDonald, J. H. (1930). "Coffee growing: with special reference to East Africa." East Africa Ltd., London.

McDonald, J. L. and Levin, M. D. (1965). An improved method for marking bees. *J. apic. Res.* **4**, 95–97.

Macfie, J. W. S. (1944). Ceratopogonidae collected in Trinidad from cacao flowers. *Bull. ent. Res.* **35**, 297–300.

McGregor, S. E. (1950). Activity of honey bees on cantaloupes. *Rep. Iowa St. Apiarist*, 1950, 140–142.

McGregor, S. E. (1952). Collection and utilization of propolis and pollen by caged honey bee colonies. *Am. Bee J.* **92**, 20–21.

McGregor, S. E. (1959). Cotton-flower visitation and pollen distribution by honey bees. *Science, N.Y.* **129**, 97–98.

McGregor, S. E. and Todd, F. E. (1952). Cantaloupe production with honeybees. *J. econ. Ent.* **45**, 43–47.

McGregor, S. E. and Todd, F. E. (1956). Honeybees and cotton production. *Glean. Bee Cult.* **84**, 649–652, 701.

McGregor, S. E., Rhyne, C., Worley, S. and Todd, F. E. (1955). The role of honey bees in cotton pollination. *Agron. J.* **47**, 23–25.

McGregor, S. E., Levin, M. D. and Foster, R. E. (1965). Honey bee visitors and fruit set of cantaloups. *J. econ. Ent.* **58**, 968–970.

McIlroy, R. J. (1963). An introduction to tropical cash crops. Ibadan University Press, Nigeria.

Mackensen, O. and Nye, W. P. (1966). Selecting and breeding honeybees for collecting alfalfa pollen. *J. apic. Res.* **5**, 79–86.

Mackensen, O. and Nye, W. P. (1969). Selective breeding of honeybees for alfalfa pollen collection; sixth generation and outcrosses. *J. apic. Res.* **8**, 9–12.

Mackerras, M. J. (1933). Observations on the life-histories, nutritional requirements and fecundity of blowflies. *Bull. ent. Res.* **24**, 353–362.

Mackie, W. W. and Smith, F. L. (1935). Evidence of field hybridization in beans. *J. Am. Soc. Agron.* **27**, 903–909.

McMahon, H. A. (1954). Pollination of alfalfa by honeybees. *Can. Bee J.* **62**, 4–6.

MacVicar, R. M., Braun, D. R. and Jamieson, C. A. (1952). Studies in red clover seed production. *Scient. Agric.* **32**, 67–80.

Maeta, Y. and Kitamura, T. (1965). Studies on the apple pollination by *Osmia*. II. Characteristics and underlying problems in utilizing *Osmia*. *Kontyû* **33**, 17–34.

Magruder, R. and Wester, R. E. (1940). Natural crossing in lima beans in Maryland. *Proc. Am. Soc. hort. Sci.* **37**, 731–736.

Mahadevan, V. and Chandy, K. C. (1959). Preliminary studies on the increase in cotton yield due to honey bee pollination. *Madras agric. J.* **46**, 23–26.

Mahta, D. N. and Dave, B. E. (1931). Studies in *Cajanus indicus*. *Mem. Dep. Agric. India Bot.* **19**, 1–25.

Maksymiuk, I. (1958). Nectar secretion in winter rape. *Pszcel. Zesz. Nauk.* **2**, 49–54.

Mallik, P. C. (1957). Morphology and biology of the Mango flower. *Indian J. hort.* **14**, 1–23.

Mann, L. K. (1943). Fruit shape of watermelon as affected by placement of pollen on stigma. *Bot. Gaz.* **105**, 257–262.

Mann, L. K. (1953). Honey bee activity in relation to pollination and fruit set in the cantaloupe (*Cucumis melo*). *Am. J. Bot.* **40**, 545–553.

Mann, L. K. and Robinson, Jeanette (1950). Fertilisation, seed development and fruit growth as related to fruit set in the cantaloupe (*Cucumis Melo* L.). *Am. J. Bot.* **37**, 685–697.

Manning, A. (1956). Bees and flowers. *New Biol.* **21**, 56–73.

Markov, I. and Romanchuk, I. (1959). Pollination of cucumbers by honeybees. *Sel. Khoz. Sibiri* No. 2, 53–54.

Marshall, J. (1934). Fertility in cacao. *Rep. Cacao Res.* **3**, 34.

Martin, F. W. and Gregory, L. E. (1962). Mode of pollination and factors affecting fruit set *Piper nigrum* L. in Puerto Rico. *Crop Sci.* **2**, 295–299.

Martin, J. N. (1938). Why the high price of red clover seed? *Am. Bee J.* **78**, 102–104.

Martin, J. N. (1941). The work of bees not fully appreciated. *Rep. Iowa St. Apiarist*, 1940, 68–71.

Marucci, P. E. (1967a). Pollination of the cultivated highbush blueberry in New Jersey. International Soc. for Hort. Sci. Symposium I. Blueberry Culture in Europe, 155–162.

Marucci, P. E. (1967b). Cranberry pollination. *Am. Bee J.* **107**, 212–213.

Marucci, P. E. and Filmer, R. S. (1957). Cranberry blossom blast is not caused by disease. *New Jers. Agric.* **39**, 8–9.

Marucci, P. E. and Filmer, R. S. (1964). Preliminary Cross Pollination Tests on Cranberries. *Proc. Am. Cranberry Grow. Ass.* 1961–64, 48–51.

Massee, A. M. (1936). Notes on some interesting mites and insects observed on fruit trees in 1936. *Rep. E. Malling Res. Stn.* 1936, 222–228.

Materikina, E. I. (1956). Supplementary pollination of buckwheat. *Zemledelie, Mosk.* **4,** 127.
Mather, K. (1947). Species crosses in *Antirrhinum*. I. Genetic isolation of the species *majus, glutinosum* and *orontium*. *Heredity* **1,** 175–186.
Maurizio, Anna (1949). Pollenanalytische Untersuchungen an Honig und Pollenhöschen. *Beih. Schweiz. Bztg.* **2,** 320–455.
Maurizio, Anna (1950). Untersuchungen uber den Einfluss der Pollenahrung und Brutpflege auf die Lebensdauer und den physiologischen Zustand von Bienen. *Schweiz. Bienenztg.* **73,** 58–64.
Maurizio, Anna (1951). Untersuchungen über den Einfluss der Pollenernährung und Brutpflege auf die Lebensdauer und den physiologischen Zustand der Beinen *XIV Int. Beekeep. Congr.*
Maurizio, Anna (1953). Weitere Untersuchungen an Pollenhöschen. Beitrag zur Erfassung der Pollentrachtverhältnisse in verschiedenen Gegenden der Schweiz. *Beih. Schweiz. Bienenztg.* **2,** 485–556.
Maurizio, Anna (1954). Untersuchungen über die Nektarsekretion einiger polyploider Kulturpflanzen. *Arch. Julius Klaus-Stift. Vererb-Forsch.* **29,** 340–346.
Maurizio, Anna and Pinter, L. (1961). Beobachtungen über die Nektarabsonderung und den Insektenbesuch bei einigen schweizerischen Mattenklee-Hofsorten (*Trifolium pratense* L.). *Arb. Futterbau.* (2), 41–46.
Meade, R. M. (1918). Beekeeping may increase the cotton crop. *J. Hered.* **9,** 282–285.
Meader, E. M. and Darrow, G. M. (1944). Pollination of the rabbiteye blueberry (*Vaccinium ashei*) and related species. *Proc. Am. Soc. hort. Sci.* **45,** 267–274.
Meader, E. M. and Darrow, G. M. (1947). Highbush blueberry pollination experiments. *Proc. Am. Soc. hort. Sci.* **49,** 196–204.
Medler, J. T. (1962). Effectiveness of domiciles for bumblebees. *Proc. I int. Symp. Poll.* 1960, 126–133.
Medler, J. T. (1967). Biology of *Osmia* in trap nests in Wisconsin (Hymenoptera: Megachilidae). *Ann. ent. Soc. Am.* **60,** 338–344.
Medler, J. T. and Koerber, T. W. (1958). Biology of *Megachile relativa* Cresson (Hymenoptera, Megachilidae) in trap-nests in Wisconsin. *Ann. ent. Soc. Am.* **51,** 337–344.
Meier, F. C. and Artschwager, E. (1938). Airplane collection of sugar beet pollen. *Science, N.Y.* **88,** 507–508.
Mel'nichenko, A. N. (1962). Biological basis for increasing the yield of buckwheat by different sowing dates and degrees of saturation of bee pollination. *Uchen. Zap. Gor'kov. Univ.* **55,** 5–43.
Menke, H. F. (1950). Apple pollination in Washington State. *Rep. Iowa St. Apiarist,* 1950, 71–79.
Menke, H. F. (1951). Insect pollination of apples in Washington State. *XIV Int. Beekeep. Congr.*
Menke, H. F. (1952a). A six million dollars native bee in Washington State. *Am. Bee J.* **92,** 334–335.
Menke, H. F. (1952b). Alkali bee helps set seed records. *Crops soils* **4.**
Menke, H. F. (1952c). Behaviour and population of some Insect Pollinators of apples in Eastern Washington. *Rep. Iowa St. Apiarist,* 1952, 66–93.
Menke, H. F. (1954). Insect pollination in relation to alfalfa seed production in Washington. *Bull. Wash. agric. Exp. Stn.* No. 555.
Menon, K. K. (1949). The survey of polu and root diseases of pepper. *Indian J. agric. Sci.* **19,** 89–136.

Menon, K. V. P. and Pandalai, K. M. (1958). "The Coconut Palm—A Monograph." Indian Central Coconut Committee, Ernakulam, S. India.

Merrill, J. H. (1924). The relation of stores to brood rearing. *Am. Bee J.* **64**, 508–509.

Merrill, T. A. (1936). Pollination of highbush blueberry. *Tech. Bull. Mich. (St. Coll.) agric. Exp. Stn.* 151.

Merrill, T. A. and Johnston, S. (1940). Further observations on pollination of the highbush blueberry. *Proc. Am. Soc. hort. Sci.* **37**, 617–619.

Meyer, W. (1954). Die "Kittharzbienen" und ihre Tätigkeiten. *Z. Bienenforsch.* **2**, 185–200; (1956). *Bee Wld* **37**, 25–36.

Meyerhoff, G. (1954). Untersuchungen über die Wirkung des Bienenbefluges auf den Raps. *Arch. Geflügelz. Kleintierk.* **3**, 259–306.

Meyerhoff, G. (1958). Zum Sammelverhalten der Bienen im Raps. *Leipzig. Bienenztg.* **72**, 164–165.

Michelbacher, A. E., Hurd, P. D. and Linsley, E. G. (1968). The feasibility of introducing squash bees (*Peponapis* and *Xenoglossa*) into the old world. *Bee Wld.* **49**, 159–167.

Mikitenko, A. S. (1959). Bees increase the seed crop of sugar beet. *Pchelovodstvo, Mosk.* **36**, 28–29.

Miller, J. D. and Amos, J. M. (1965). Use of honeybees to pollinate trefoil in the greenhouse. *Am. Bee J.* **105**, 50–51.

Miller, M. D., Jones, L. G., Osterli, V. P. and Reed, A. D. (1951). Seed production of Ladino clover. *Circ. Calif. agric. Ext. Serv.* **182**.

Milum, V. G. (1940). Bees and Soyabeans. *Am. Bee J.* **80**, 22.

Minderhoud, A. (1931). Untersuchungen über das Betragen der Honigbiene als Blütenbestäuberin. *Gartenbauwissenschaft* **4**, 342–362.

Minderhoud, A. (1948). Over het leiden van bijen naar bepaalds drachtplanten. *Meded. Dir. Tuinb.* **11**, 381–392.

Minderhoud, A. (1950). Het gebruik van bijen en hommels voor bestuiving in afgesloten minuten. *Meded. Dir. Tuinb.* **17**, 32–39.

Minderhoud, A. (1954). The direct pollen consumption of the honey-bee. *15th Int. Bee Congress* (Denmark). XI (E) 2 pp.

Minessy, F. A. (1959). Effect of different pollinizers on yield and seediness in clementine tangerine. *J. agric. Res., Alexandria* **7**, 279–287.

Ministry of Agriculture, Fisheries and Food (1958a). Apples and pears. H.M.S.O. Bull. 133.

Ministry of Agriculture, Fisheries and Food (1958b). The importance of bees in orchards. *Adv. Leafl. Minist. Agric. Fish, Lond.* No. 328.

Ministry of Agriculture, Fisheries and Food (1961). Plums and Cherries. H.M.S.O. Bull. 119.

Min'kov, S. G. (1953). Honeybees and Cotton. *Pchelovodstvo, Mosk.* **30**, 41–44.

Min'kov, S. G. (1956). Nectar productivity of cotton and the role of bees in cross pollination. *Trud. kazakh. opȳt Sta. Pchelov.* (1), 119–150.

Min'kov, S. G. (1957). Nectar productivity of cotton in South Kazakhstan region. *Pchelovodstvo, Mosk.* **34**, 35–40.

Mitchener, A. V. (1950). Honeybees and sunflowers. *Ann. Con. Man. Agron.* p. 27.

Mittler, T. E. (1962). Preliminary studies on apple pollination. *Proc. 1st Int. Symp. Pollination, Copenhagen,* 1960, 173–178.

Młyniec, W. (1962). The mechanism of pollination and generative reproduction in *Vicia villosa* Roth. *Genet. pol.* **3**, 285–299.

Młyniec, W. and Wójtowski, F. (1962). Zastosowanie trzmieli (Bombinae) w badaniach biologii kwitnienia ozimej wyki omszonej (*Vicia villosa* Roth.). *Ekol. pol.* (Ser. B) **8**, 59–65.

Móczár, L. (1961). The distribution of wild bees in the lucerne fields of Hungary (Hymenoptera, Apoidea). *Annls hist.-nat. Mus. hung.* **53**, 451–461.

Moeller, F. E. (1958). Relation between egg-laying capacity of queen bee and populations and honey production of their colonies. *Am. Bee J.* **98**, 401–402.

Moeller, F. E. (1961). The relationship between colony populations and honey production as affected by honey bee stock lines. *Prod. Res. Rep. U.S. Dept. Agric.* **55**, 20.

Mohammed, A. (1935). Pollination studies in tara (*Brassica napus* L. var. dichotsma Prain) and sarson (*Brassica campestris* L. var. sarson, Prain). *Indian J. agric. Sci.* **5**, 125–154.

Mommers, J. (1948). Over het aandeel van de honingbijen in de bestuiving van het fruit. *Meded. Dir. Tuinb.* **11**, 252–259.

Mommers, J. (1951). Honeybees as pollinators of fruit trees. *Bee Wld* **32**, 41–44.

Mommers, J. (1952). De betekenis van de honingbij voor de bestuiving. *Meded. Dir. Tuinb.* **15**, 586–593.

Mommers, J. (1961). De bestuiving van aardbeien onder glas. *Meded. Dir. Tuinb.* **24**, 353–355.

Mommers, J. (1966). The concentration and composition of nectar in relation to honeybee visits to fruit trees. *Bee Wld* **47** (Suppl.), 91–94.

Montgomery, B. E. (1951). The status of bumble bees in relation to the pollination of red clover in New Zealand. *Proc. 6th Ann. Mtg. N.C. States Branch Am. Assoc. econ. Ent.*, 51–55.

Montgomery, B. E. (1958). Preliminary studies of the composition of some Indiana nectars. *Proc. Indiana Acad. Sci.* **68**, 159–163.

Morley, F. H. W. (1963). The mode of pollination in strawberry clover (*Trifolium fragiferum*). *Aust. J. exp. Agric. Anim. Husb.* **3**, 5–8.

Morris, L. E. (1929), Field observations and experiments on the pollination of *Hevea brasiliensis*. *J. Rubb. Res. Inst. Malaya.* **1**, 41–49.

Morrow, E. B. (1943). Some effects of cross-pollination versus self-pollination in the cultivated blueberry. *Proc. Am. Soc. hort. Sci.* **42**, 469–472.

Morse, R. A. (1958). The pollination of bird's-foot trefoil. *Proc. 10th Int. Congr. Ent.* **4**, 951–953.

Morse, W. J. and Cartter, J. L. (1937). Improvement in Soyabeans. *Yb. U.S. Dep. Agric.* 1154–1159.

Moskovljevíc, V. Z. (1936). Reported in *Bee Wld* **18**, 35.

Mound, L. A. (1962). Extra-floral nectaries of cotton and their secretions. *Emp. Cott. Grow. Rev.* **39**, 254–261.

Mukherjee, S. K. (1953). The Mango—its botany, cultivation, uses and future improvement, especially as observed in India. *Econ. Bot.* **1**, 130–162.

Müller, H. (1882). Versuche über die Farbenliebhaberie der Honigbiene. *Kosmos* **12**, 273–299.

Murneek, A. E. (1929). The use of bees in apple sterility studies. *Proc. Am. Soc. hort. Sci.* **26**, 39–42.

Murneek, A. E. (1930). Fruit pollination. *Bull. Mo. agric. Exp. Stn.* **283**.

Murneek, A. E. (1937). Pollination and fruit setting. *Bull. Miss. agric. Exp. Stn.* No. 379.

Murthy, N. S. R. and Murthy, B. S. (1962). Natural cross-pollination in chilli. *Andhra agric. J.* **9**, 161–165.

Musgrave, A. J. (1950). A note on the dusting of crops with fluorescein to mark visiting bees. *Can. Ent.* **82**, 195–196.

Mustard, Margaret J., Su-Ling, L. and Nelson, R. O. (1953). Observations of floral biology and fruit-setting in lychee varieties. *Proc. Fla St. hort. Soc.* **66**, 212–220.

Mustard, Margaret J., Lynch, S. J. and Nelson, R. O. (1956). Pollination and floral studies of the Minneola tangelo. *Proc. Fla St. hort. Soc.* **69**, 277–281.

Muzik, T. J. (1948). What is the pollinating agent for *Hevea brasiliensis*? *Science, N.Y.* **108**, 540.

Naghski, J. (1951). No honey from tartary buckwheat. *Am. Bee J.* **91**, 513.

Naik, K. C. and Rao, M. M. (1943). Studies on blossom biology and pollination in mangoes (*Mangifera indica* L.). *Indian J. hort.* **1**, 107–119.

Nandpuri, K. S. and Brar, J. S. (1966). Studies on floral biology in muskmelon (*Cucumis melo* L.). *J. Res., Ludhiana* **3**, 395–399.

Narayanan, E. S., Sharma, P. L. and Phadke, K. G. (1960). Studies on requirements of various crops for insect pollinators. I. Insect pollinators of saunf (Foeniculum vulgare) with particular reference to the honeybees at Pusa, (Bihar). *Indian Bee J.* **22**, 7–11.

Narayanan, E. S., Sharma, P. L. and Phadke, K. G. (1961). Studies on requirements of various crops for insect pollinators. Insect pollinators of berseem—Egyptian clover (*Trifolium alexandrinum* L.) with particular reference to honeybees and their role in seed setting. *Indian Bee J.* **23**, 23–30.

National Agricultural Advisory Service (1962). Honeybee management in pollination cage experiments. *Bee Craft* **44**, 61–62, 73.

National Agricultural Advisory Service (1964). Pollination of field beans. N.A.A.S., Derby. Leaflet no. 21.

Neiswander, R. B. (1954). Honey bees as pollinators of greenhouse tomatoes. *Glean. Bee. Cult.* **82**, 610–613.

Nekrasov, V. U. (1949). The drifting of bees. *Pchelovodstvo, Mosk.* **26**, 177–184.

Nelson, F. C. (1927). Adaptability of young bees under adverse conditions. *Am. Bee J.* **67**, 242–243.

Nemirovich-Danchenko, E. N. (1964). Concerning the nectar yield and floral biology of cucumbers. *Izv. tomsk. Otd. vses. bot. Obshch.* **5**, 127–132.

Neumann, U. (1955). Die Bedeutung der Befruchtungsverhältnisse und Pflegemassnahmen für den vorzeitigen Früchtefall bei Schwarzen Johannisbeeren. *Arch. Gartenb.* **3**, 339–354.

Nevkryta, A. N. (1937). On the fauna and ecology of insect pollinators of the family Cucurbitaceae. *Acad. Sci. R.S.S. Ukraine Inst. Zool. Biol.* **14**, 231–258.

Nevkryta, A. N. (1953). Insects pollinating cucurbit crops. (In Russian.) *Izdatel'stvo Akademii nauk ukrainskoi S.S.R. Kiev.*

Nevkryta, A. N. (1957). Distribution of apiaries for pollinating cherries. *Pchelovodstvo, Mosk.* **34**, 34–38.

Nicholls, H. A. and Holland, J. H. (1940). A textbook of tropical agriculture. Macmillan, London.

Nielsen, H. M. (1958). Studies on fertility in lucerne and the relation between self-fertility and crop yield. *K. Vet.-og Landbohøjsk Arsskr.* 1958, 48–63.

Nielsen, H. M. (1960). Floral modifications in lucerne. *Proc. Int. Symp. Pollination* **1**, 60–63.

Nilsson, E. (1927). Försök med sjahv—och korspollinerung hos *Raphanus sativus*. *Bot. Notiser* **2**, 128–136.

Nishida, T. (1958). Pollination of the passion fruit in Hawaii. *J. econ. Ent.* **51**, 146–149.

Nishida, T. (1963). Ecology of the pollinators of passion fruit. *Tech. Bull. Hawaii agric. Exp. Stn*, **55**.

Nitsch, J. P. (1952). Plant hormones in the development of fruits. *Q. Rev. Biol.* **28**, 33–57.

Nitsch, J. P., Kurtz, E. B., Liverman, J. L. and Went, F. W. (1952). The development of sex expression in Cucurbit flowers. *Am. J. Bot.* **39**, 32–43.

Nixon, H. L. and Ribbands, C. R. (1952). Food transmission within the honeybee community. *Proc. R. Soc.* (B) **140**, 43–50.

Nogueira-Neto, P., Carvalho, A. and Antunes, H. (1959). Efeito da exclusão dos insectos pollinizadores na producão do café Bourbon. *Bragantia* **18**, 441–468.

Nolan, W. J. (1925). The brood-rearing cycle of the honeybee. *Bull. U.S. Dept. Agric.* No. 1349, 1–56.

Noro, K. and Yago, M. (1934). Studies on sterility of the Japanese pear Chojuro with special reference to hand pollination, dehiscence of anthers and insect visitors. *Bull. Shizuokaken agric. Exp. Stn.* **29**, 13.

Norris, K. R. (1965). The bionomics of blow flies. *Ann. Rev. Ent.* **10**, 47–68.

Northwood, P. J. (1966). Some observations on flowering and fruit-setting in Cashew, *Anacardium occidentale* L. *Trop. Agric. Trin.* **43**, 35–42.

Nye, W. P. (1962). Management of honeybee colonies for pollination in cages. *Bee Wld* **43**, 37–40.

Nye, W. P. and Bohart, G. E. (1964). Nesting holes for the alfalfa leaf-cutting bee. *Circ. Utah agric. Exp. Stn.* **145**.

Nye, W. P. and Mackensen, O. (1965). Preliminary report on solution and breeding of honeybees for alfalfa pollen collection. *J. apic. Res.* **4**, 43–48.

Nye, W. P. and Mackensen, O. (1968). Selective breeding of honeybees for alfalfa pollen: fifth generation and backcrosses. *J. apic. Res.* **7**, 21–27.

Nye, W. P. and Pedersen, M. W. (1962). Nectar sugar concentration as a measure of pollination of alfalfa (*Medicago sativa* L.). *J. apic. Res.* **1**, 24–27.

Oberle, G. D. and Moore, R. C. (1957). Hand pollination in Virginia apple orchards. *Res. Rep. Va. agric. Exp. Stn.* 1953–1957, 189–190.

Odland, M. L. and Porter, A. W. (1941). A study of natural crossing in peppers (*Capsicum frutescens*). *Proc. Am. Soc. hort. Sci.* **38**, 585–588.

Oertel, E. (1934). White clover and honeybees in Louisiana. *Glean. Bee Cult.* **62**, 462–464.

Oertel, E. (1939). Honey and pollen plants of the United States. *Circ. U.S. Dep. Agric.* No. 554.

Oertel, E. (1954). The value of honey bees to white clover seed growers in Louisiana. *Am. Bee J.* **94**, 460–462.

Oertel, E. (1960). Honey bees in production of white clover seed in the southern States. *Publ. U.S. Dep. Agric.* ARS–33–60.

Oertel, E. (1961). Honeybees in production of white clover seed in the southern States. *Am. Bee J.* **101**, 96–99.

Oettingen-Spielberg, T. (1949). Ueber das Wesen der Suchbiene. *Z. vergl. Physiol.* **31**, 454–489.

Oganjan, V. N. (1938). Methods of pollinating soybean. *Selek. Semenoved* (1), 31–35. (Institute of Northern Grain Husbandry and Grained Legumes).

Olmo, H. P. (1943). Pollination of the Almeria grape. *Proc. Am. Soc. hort. Sci.* **42**, 401–406.

Olsson, G. (1960). Self-incompatibility and outcrossing in rape and white mustard. *Hereditas* **46**, 241–252.

Olsson, G. and Persson, Brita (1958). Inkorsningsgrad och självsterilitet hos raps. *Sver. Utsädeför. Tidskr.* **68,** 74–78.
Oppenheimer, H. R. (1948). Experiments with unfruitful "Clementine" Mandarins in Palestine. *Bull. agric. Res. Stn. Rehovot,* No. 48.
Orösi-Pál, Z. (1956). Az épitö alkalom hatása a viaszmirigy müködéséne. *Méhészet* **4,** 105.
Oschmann, H. (1957). Bedeutung des Honigbienenbefluges für den Samenertrag bei Ackerbohnen. *Dt. Landwirt* **8,** 302–303.
Ostaščenko-Kudrjavceva, A. (1941). The importance of nectar production in the pollination of lucerne. *Sov. Bot.* (3), 53–59.
Ostendorf, F. W. (1938). Spontane kruisbestuiving en selectie op zaadlobkleur bij cacao. *Bergcultures* **12,** 552–558.
Otto, F. T. (1928). Verfliege Beobachtungen der Schleswig-Holsteinischen Imkerschule. *Leipzig. Bienenztg* **10,** 203–206; 225–228; 245–249.
Overley, F. L. and Bullock, R. M. (1947). Pollen diluents and application of pollen to fruit trees. *Proc. Am. Soc. hort. Sci.* **49,** 163–169.
Overley, F. L. and O'Neill, W. J. (1946). Experiments with the use of bees for pollination of fruit trees. *Proc. Wash. St. hort. Ass.* **42,** 203–214.
Overley, F. L. and Overholser, E. L. (1938). Commercial hand pollination of apples in Washington. *Proc. Am. Soc. hort. Sci.* **35,** 39–42.
Overseas Food Corporation (1950). Bees and Sunflowers. Rep. Overseas Food Corporation 1949/50, 93–94 and Appendix 5 (part 2), 105–109.
Paddock, F. B. (1951). Apiarist Report for 1950. *Rep. Iowa St. Apiarist* 5–19.
Pain, Janine (1955). Influence des reines mortes sur le développement ovarien de jeunes ouvrières d'abeilles (*Apis mellifica*). *Insectes Soc.* **2,** 35–43.
Pain, Janine (1968). Les phénomènes supérieurs du comportement Régulations sociales complexes. *In* "Traité de Biologie de L'Abeille." Vol. 2, Chap. 4, pp. 201–240. (R. Chauvin, Ed.). Masson et Cie, Paris.
Palmer, S. (1959). A nectar source *par excellence*. *Glean. Bee Cult.* **87,** 460–461.
Palmer-Jones, T. (1959). Effect on honeybees of thiodan applied to broad beans in a cage. *N.Z. Jl agric. Res.* **2,** 229–233.
Palmer-Jones, T. and Clinch, P. G. (1968). Observations on the pollination of apple trees (*Malus sylvestris* Mill.). III. Varieties Granny Smith, Kidd's Orange Red, and Golden Delicious. *N.Z. Jl agric. Res.* **11,** 149–154.
Palmer-Jones, T. and Forster, I. W. (1965). Observations on the pollination of lucerne (*Medicago sativa* Linn.). *N.Z. Jl agric. Res.* **8,** 340–349.
Palmer-Jones, T. Forster, I. W. and Jeffery, G. L. (1962). Observations on the role of the honey bee and bumble bee as pollinators of white clover (*Trifolium repens* Linn.) in the Timaru district and Mackenzie country. *N.Z. Jl agric. Res.* **5,** 318–325.
Palmer-Jones, T., Forster, I. W. and Clinch, P. G. (1966). Observations on the pollination of Montgomery red clover (*Trifolium pratense* L.). *N.Z. Jl agric. Res.* **9,** 738–747.
Pammel, L. H. and King, C. M. (1911). Pollination of red clover. *Proc. Iowa Acad. Sci.* **18.**
Pammel, L. H. and King, C. M. (1930). "Cucurbitaceae, Gourd Family." *Bull. geol. Surv. Iowa* **7,** 650–662.
Pandey, R. S. and Yadava, R. P. S. (1970). Pollination of litchi (*Litchi chinensis* Sonn.) bloom by insects with special reference to honey bees. *J. apic. Res.* **9,** 103–105.
Pankiw, P. (1967). Floral mutants of alfalfa—honeybee preference and seed production. *XXI Int. Beekeep. Congr.* prelim. sci. Meet. Summ. Paper 33, pp. 67.

Pankiw, P. and Bolton, J. L. (1965). Note on a floral mutant of alfalfa. *Can. J. Pl. Sci.* **45,** 228.
Pankiw, P. and Elliot, C. R. (1959). Alsike clover pollination by honey bees in Peace River region. *Can. J. Pl. Sci.* **39,** 505–511.
Pankiw, P. and Goplen, B. P. (1967). The confinement period required to rid honeybees of foreign sweetclover pollen. *Can. J. Pl. Sci.* **47,** 653–656.
Pankiw, P., Bolton, J. L., McMahon, H. A. and Foster, J. R. (1956). Alfalfa pollination by honeybees on the Regina plains of Saskatchewan. *Can. J. agric. Sci.* **36,** 114–119.
Pankratova, E. P. (1958). Data on the biology of flowering and pollination of carrots. *Dokl. TSKha* **36,** 118–123.
Park, O. W. (1922). Time and labour factors involved in gathering pollen and nectar. *Am. Bee J.* **62,** 254–255.
Park, O. W. (1929). The influence of humidity upon sugar concentration in the nectar of various plants. *J. econ. Ent.* **22,** 534–544.
Park, O. W. (1932). Studies on the changes in nectar concentration produced by the honeybee, *Apis mellifera*. Part 1. Changes which occur between the flower and the hive. *Res. Bull. Iowa agric. Exp. Stn.* 151.
Parker, R. L. (1926). The collection and utilization of pollen by the honeybee. *Mem. Cornell agric. Exp. Stn.* No. 98.
Parks, H. B. (1921). The cotton plant as a source of nectar. *Am. Bee J.* **61,** 391.
Parks, H. B. (1925). Critical temperatures in beekeeping. *Beekeep. Item* **9,** 125–127.
Pashchenko, T. E. (1940). Biology of tomato flowering. *Dokl. vses. Akad. sel'.-khoz. Nauk* **12,** 15–19.
Pausheva, Z. P. (1961). Change in intercellular reactions in the buckwheat stigma under the effect of pollination. *Zh. obshch. Biol.* **22,** 220–225.
Pearson, O. H. (1932a). Incompatibility in Broccoli and the production of seed under cages. *Proc. Am. Soc. hort. Sci.* **29,** 468–471.
Pearson, O. H. (1932b). Breeding plants of the cabbage group. *Bull. Calif. agric. Exp. Stn.* 532.
Peck, O. and Bolton, J. L. (1946). Alfalfa seed production in northern Saskatchewan as affected by bees, with a report on means of increasing the populations of native bees. *Scient. Agric.* **26,** 338–418.
Pedersen, A. (1935). Rödklöverens bestövning og angreb af snudebiller (Apion) paa rödklöver. *Nord. Jordbrforsk.* (4–7), 498–507.
Pedersen, A. (1945). Rødkløverens blomstring og bestovning. *Årsskr. Kgl. Vet.- Landbohøjsk.* 59–141.
Pedersen, A. and Sørensen, N. A. (1934). Undersøgelser over rødkløverens bestøvning og angreb af snudebiller på rødkløver. *Tidsskr. Frøavl* **12,** 30–36, 60–63.
Pedersen, A. and Sørensen, N. A. (1935). Undersøgelser over rødkløverens bestøvning og angreb af snudebiller på rødklover. *Tidsskr. Frøavl* **12,** 288–300.
Pedersen, A. and Sørensen, N. A. (1936). Undersøgelser over rødkløverens bestøvning og angreb af snudebiller på rødkløver. *Tidsskr. Frøavl* **12,** 549–554.
Pedersen, A. and Stapel, C. (1944). Undersøgelser over Lucernens bestøvning i 1943. *Tidsskr. Frøavl* No. 386.
Pedersen, A. and Stapel, C. (1945). Undersøgelser over Lucernens bestøvning i 1944. *Tidsskr. Frøavl* No. 394.
Pedersen, M. W. (1953). Seed production in alfalfa as related to nectar production and honey bee visitation. *Bot. Gaz.* **115,** 129–138.
Pedersen, M. W. (1956). Alfalfa seed production in relation to stand density. *Rep. 15th Alfalfa Improvement Conf.*, 33–37.
Pedersen, M. W. (1961). Lucerne pollination. *Bee Wld* **42,** 145–149.

Pedersen, M. W. and Bohart, G. E. (1950). Using bumblebees in cages as pollinators for small seed plots. *Agron. J.* **42**, 523.
Pedersen, M. W. and Bohart, G. E. (1953). Factors responsible for the attractiveness of various clones of alfalfa to pollen-collecting bumble bees. *Agron. J.* **45**, 548–551.
Pedersen, M. W. and McAllister, D. R. (1956). What's new in Alfalfa. *Fm. Home Sci.*
Pedersen, M. W. and Todd, F. E. (1949). Selection and Tripping in Alfalfa Clones by Nectar-Collecting Honey Bees. *Agron. J.* **41**, 247–249.
Pedersen, M. W., Todd, F. E. and Lieberman, F. V. (1950). A portable field cage. *U.S. Bur. Ent.* ET–289.
Pedersen, M. W., Bohart, G. E., Levin, M. D., Nye, W. P., Taylor, S. A. and Haddock, J. L. (1959). Cultural practices for alfalfa seed production. *Bull. Utah agric. Exp. Stn.* 408.
Pederson, M. W., Hurst, R. L., Levin, M. D. and Stoker, G. L. (1969). Computer analysis of the genetic contamination of Alfalfa seed. *Crop Sci.* **9**, 1–4.
Peebles, R. H. (1956). First attempt to produce Hybrid Cotton seed. *Am. Bee J.* **96**, 51–52, 75.
Peens, J. F. (1958). Natuurlike kruisbestuiwing by tabak (*Nicotiana tabacum*). *S. Afr. J. agric. Sci.* **1**, 237–244.
Peer, D. (1955). The foraging range of the honey bee. Part I. Ph.D. Thesis. Univ. of Wisconsin.
Pelerents, C. (1957). L'arachide à Yangambi. *Bull. Inst. Étud. agron. Congo. belge* **6**, 243–255.
Pellett, F. C. (1923). "American honey plants". American Bee Journal, Hamilton, Ill. 2nd ed.
Pellet, F. C. (1948). Bird's-foot trefoil—the coming legume. *Am. Bee J.* **88**, 588; **89**, 595.
Penfold, A. R. and Willis, J. L. (1961). "The Eucalypts." Leonard Hill, London.
Pengelly, D. H. (1953). Alfalfa pollination in S. Ontario. *Rep. ent. Soc. Ont.* **84**, 101–118.
Percival, Mary S. (1946). Observations on the flowering and nectar secretion of *Rubus fructicosus* (Agg.). *New Phytol.* **45**, 111–123.
Percival, Mary S. (1947). Pollen collection by *Apis mellifera*. *New Phytol.* **46**, 142–173.
Percival, Mary S. (1950). Pollen presentation and pollen collection. *New Phytol.* **49**, 40–63.
Percival, Mary S. (1955). The presentation of pollen in certain angiosperms and its collection by *Apis mellifera*. *New Phytol.* **54**, 353–368.
Percival, Mary S. (1961). Types of nectar in Angiosperms. *New Phytol.* **60**, 235–281.
Percival, Mary S. (1965). "Floral biology." Pergamon, Oxford.
Perrin, M. E. B. and Duggan, J. B. (1965). Growing highbush blueberries in England. *Exp. Hort.* **13**, 81–88.
Persson, Brita (1951). Orienterande frekvemsräkiningar av bin i Raps 1949. *Medd. Växtskyddsanst. Stockh.* (59), 32.
Persson, Brita (1956). Undersökningar rörande intektspollingerade Kulturväxter III Korsbefruktningens betydelse och omfattning hos raps. *Medd. Växtskyddsanst. Stockh.* **70**, 36.
Petersen, H. L. (1954). Pollination and seed setting in lucerne. Åskr. k. Vet.-Landbohøjsk. 138–160.
Peterson, A. G., Furgala, B. and Holdaway, F. G. (1960). Pollination of red clover in Minnesota. *J. econ. Ent.* **53**, 546–550.
Peterson, P. A. (1955). Dual cycle of avocado flowers. *Calif. Agric.* **9**, 6–7, 13.

Peterson, P. A. (1956). Flowering types in the avocado with relation to fruit production. *Yb. Calif. Avocado Soc.* 174–179.
Petkov, V. (1958). An investigation on the nectar yield of sainfoin (*Onobrychis sativa*), phacelia (*Phacelia tanacetifolia*), borage (*Borago officinalis*) and buckwheat (*Fagopyrum esculentum*). *Nauchni Trud. Minist. Zemed. Gorite* **1**, 211–246.
Petkov, V. (1963a). Nectar and honey productivity of winter rape. *Izv. Inst. Ovoshcharstvo* **5**, 181–188.
Petkov, V. (1963b). Nectar production in cultivated raspberry. *Sel. Nauk.* (2), 201–207.
Petkov, V. (1965). Contribution of honeybees to the pollination of strawberries. *Gradinar. lozar. Nauk.* **2**, 421–431.
Petkov, V. and Panov, V. (1967). Study on the efficiency of apple pollination by bees. *XXI Int. Beekeep. Congr.* Summ. Paper 149, pp. 97.
Petkov, V. and Simidchiev, T. (1965). The role played by bees in the pollination of lucerne. *XX Int. Beekeep. Congr.* (II/3).
Peto, H. B. (1950). Pollination of cucumbers, watermelons and cantaloupes. *Rep. Iowa St. Apiarist.* 72–87.
Pew, W. D., Marlatt, R. B. and Hopkins, L. (1956). Growing cantaloupes in Arizona. *Bull. Ariz. agric. Exp. Stn.* **275**.
Pharis, R. L. and Urnrau, J. (1953). Seed setting of alfalfa flowers tripped by bees and mechanical means. *Can. J. agric. Sci.* **33**, 74–83.
Phillips, E. F. (1933). Insects collected on apple blossoms in Western New York. *J. agric. Res.* **46**, 851-862.
Phillips, E. F. and Demuth, G. S. (1922). Beekeeping in the Buckwheat Region. *Fmrs'. Bull. U.S. Dept. Agric.* **1216**.
Philp, G. L. and Vansell, G. H. (1932). Pollination of deciduous fruits by bees. *Circ. Calif. agric. Exp. Stn. Ext. Serv.* **62**.
Philp, G. L. and Vansell, G. H. (1944). Pollination of deciduous fruits by bees. *Circ. Calif. agric. Ext. Serv.* No. 62 rev.
Phipps, C. R. (1930). Blueberry and huckleberry insects. *Maine Agric. Expt. Stn. Bull.* **356**, 107–232.
Pieters, A. J. and Hollowell, E. A. (1937). Clover improvement. *U.S. Dept. Agric. Yb.* 1190–1214.
Pinthus, M. J. (1959). Seed set of self-fertilised sunflower heads. *Agron. J.* **51**, 626.
Piper, C. V. and Morse, W. J. (1910). The Soy Bean; history, varieties and field studies. *Bull. U.S. Dep. Agric. Bur. Plant Ind.* **197**.
Piper, C. V. and Morse, W. J. (1923). "The Soybean." McGraw-Hill Co., London.
Plass, F. (1952). Versuche zur Feststellung des Einflusses der Mineraldüngung auf die Nektarabsonderung der Obstgewächse. *Bienenzucht* **5**, 270–271.
Plath, O. E. (1934). "Bumblebees and Their Ways." MacMillan, New York.
Plowright, R. C. and Jay, S. C. (1966). Rearing bumblebee colonies in captivity. *J. apic. Res.* **5**, 155–165.
Poehlman, J. M. (1959). Breeding Field Crops. Henry Holt, New York.
Pohjakallio, O. (1938). Kimalainen puna-apilan polyyttäjänä. *Luonnon ystävä* **42**, 61–67.
Ponomareva, A. A. (1959). Pollinators of alfalfa in the western Kopet-Dogh Mt. Range, Turkmenia. *Trudȳ Inst. Zool. Parazit., Akad. Nauk Turk. SSR* **4**, 34–46.
Poole, C. F. (1937). Improving the root vegetables. *Yb. Agric. U.S. Dep. Agric.* 300–325.
Pope, O. A., Simpson, D. M. and Duncan, E. N. (1944). Effect of corn barriers on natural crossing in cotton. *J. agric. Res.* **68**, 347.
Pope, W. T. (1935). The edible passion fruit in Hawaii. *Bull. Hawaii agric. Exp. Stn.* **74**.

Popenoe, P. B. (1913). Date Growing. (Altadena, California, West India Gardens).
Popov, V. V. (1952). Apidae pollinators of Chenopodiaceae. *Zool. Zh.* **31,** 494–503.
Popov, V. V. (1956). Bees, their relations to Melittophilous Plants and the problem of Alfalfa Pollination. *Proc. 10th Int. Congr. Ent.* **4,** 983–990.
Posnette, A. F. (1938). Incompatibility and pollination in cacao. *Rep. Cacao Res.* **7,** 19–20.
Posnette, A. F. (1942a). Natural pollination of cocoa *Theobroma leiocarpa*, on the Gold Coast. I. *Trop. Agric. Trin.* **19,** 12–16.
Posnette, A. F. (1942b). Natural pollination of cocoa *Theobroma leiocarpa* Bern. on the Gold Coast. II. *Trop. Agric. Trin.* **19,** 188–191.
Posnette, A. F. (1944). Pollination of cocao in Trinidad. *Trop. Agric., Trin.* **21,** 115–118.
Posnette, A. F. (1950). The pollination of cacao in the Gold Coast. *J. hort. Sci.* **25,** 155–163.
Potter, G. F. (1959). The domestic tung industry. I. Production and improvement of the tung tree. *Econ. Bot.* **13,** 328–342.
Potter, J. M. S. (1963). The National Fruit Trials. *Rep. East Malling Res. Stn.* 1962, 40–45.
Pound, F. J. (1932). Studies in fruitfulness in cacao. II. Evidence for partial sterility. *Rep. Cacao Res.* **1,** 26–28.
Pouvreau, A. (1965). Sur une méthode d'élevage des Bourdons (*Bombus* Latr.) à partir de reines capturées dans la nature. *Annls Abeille* **8,** 147–159.
Preston, A. P. (1949). An observation on apple blossom morphology in relation to visits from honeybees (*Apis mellifera*). *Rep. E. Malling Res. Stn.* 1948, 64–67.
Priestley, G. (1954). Use of honey bees as pollinators in unheated glasshouses. *N.Z. Jl Sci. Tech. A.* **36,** 232–236.
Pritsch, G. (1961). Samenertragssteigerung bei Rotklee durch Duftlenkung der Bienen. *Wiss. tech. Fortschr. Landw.* **2,** 226–230.
Pritsch, G. (1965). Untersuchungen über die Steigerung der Ölfruchterträge durch Honigbienen. *Věd. Práce výzkum. Ústav. včelař. ČSAZV* **4,** 157–163.
Pritsch, G. (1966). Untersuchungen über die Bedeutung der Honigbienen für die Sicherung und Steigerung der Samenerträge des Rotklees. *Mitt. zentr. soz. Arbeitsgemeinsch.* (5), 11–26.
Pryce-Jones, J. (1948). The honeybee and pollination. *Welsh Bee J.* **3,** 11–12.
Pryor, L. D. (1951). Controlled pollination of *Eucalyptus*. *Proc. Linn. Soc. N.S.W.* **76,** 135–139.
Purseglove, J. W. (1968). "Tropical Crops. Dicotyledons I and II." Longmans, London.
Putt, E. D. (1940). Observations on morphological characters and flowering processes in the sunflower (*Helianthus annuus* L.). *Scient. Agric.* **21,** 167–179.
Pynaert, L. (1953). Le faux-cotonnier. *Zooleo.* **18,** 453–456.
Radaeva, E. N. (1954). Bee pollination increases the yield of sunflower seeds (*Helianthus annuus*). *Pchelovodstvo, Mosk.* **31,** 33–38.
Radchenko, T. G. (1964). The influence of pollination on the crop and the quality of seed of winter rape. *Bdzhil'nitstvo* (1), 68–74.
Radchenko, T. G. (1966). Role of honeybees as pollinators in increasing the seed crop from cabbage and radish. *Bdzhil'nitstvo* (2), 72–75.
Radoev, L. (1954). Study on the role of honeybee and wild insects in sunflower pollination. *Spis. nauch.-izsled. Inst. Min. Zemed.* (4), 3–16.
Radoev, L. and Bozhinov, M. (1961). Opitna Stantzia po Pchelarstro, Sofiya, Bulgaria. *Izv. kompl. sel. Inst., Chirpan* **1,** 87–108.

Rahman, K. A. (1940). Insect pollinators of toria (*Brassica napus* Linn. var. *dichotoma* Prain) and sarson (*Brassica campestris* Linn. var. sarson Prain) at Lyallpur. *Indian. J. agric. Sci.* **10**, 422–447.

Rahman, K. A. (1945). Progress of beekeeping in the Punjab. *Bee Wld* **26**, 42–44, 50–52.

Rakhmankulov, F. (1955). Controlling the flight activity of honeybees for the pollination of red clover. *Pchelovodstvo, Mosk.* **32**, 44–47.

Rane, F. W. (1898). Fertilization of the muskmelon. *Proc. Soc. Promot. agric. Sci.* **19**, 150–151.

Rao, B. S. (1961). Pollination of hevea in Malaya. *J. Rubb. Res. Inst. Malaya* **17**, 14–18.

Rao, V. N. M. and Hassan, M. V. (1957). Variations in the seed characters of cashew (*Anacardium occidentale* L.). *Indian J. agric. Sci.* **26**, 211–216.

Raphael, T. D. and Cunningham, D. G. (1960). Bees and pollination. *Tasm. J. Agric.* **31**, 287–294.

Rashad, S. E. and Parker, R. L. (1958). Pollen as a limiting factor in brood rearing and honey production during three drought years, 1954, 1955, and 1956. *Trans. Kans. Acad. Sci.* **61**, 237–248.

Reece, P. C. and Register, R. O. (1961). Influence of pollinators on fruit set on "Robinson" and "Osceola" tangerine hybrids. *Proc. Fla St. hort. Soc.* **74**, 104–106.

Reese, C. A. (1951). Package bees for honey production and pollination. *Bull. Ohio St. agric. Ext. Serv.* No. 159.

Reimer, F. C. and Detjen, L. R. (1910). Self-sterility of the Scuppernong and other muscadine grapes. *Bull. N. Carol. agric. Exp. Stn.* **209**.

Reinhardt, J. F. (1952). Some responses of honey bees to alfalfa flowers. *Am. Nat.* **86**, 257–275.

Renard, E. J. (1930). The origin and nature of Rogues in Canning Peas. *Bull. Wis. agric. Exp. Stn.* **101**.

Rhein, W. von (1952). Über die Duftlenkung der Bienen beim Raps in Jahre 1952 und ihre Ergebrisse. *Hess. Biene* **88**, 192–194.

Rhein, W. von (1954). Ueber die Bedeutung der Honigbiene für die Saatzuchtwirtschaft. *Saatgut-Wirt.* **6**, 30–32.

Ribbands, C. R. (1949). The foraging method of individual honeybees. *J. Anim. Ecol.* **18**, 47–66.

Ribbands, C. R. (1950). Changes in the behaviour of honey-bees following their recovery from anaesthesia. *J. exp. Biol.* **27**, 302–310.

Ribbands, C. R. (1951). The flight range of the honey bee. *J. anim. Ecol.* **20**, 220–226.

Ribbands, C. R. (1952). Division of labour in the honeybee community. *Proc. R. Soc.* (B), **140**, 32–43.

Ribbands, C. R. (1953). "The behaviour and social life of honeybees." Bee Research Association, London.

Ribbands, C. R. (1955). The scent perception of the honeybee. *Proc. R. Soc.* (B), **143** 367–379.

Ribbands, C. R. and Speirs, Nancy (1955). Communication between honeybees. II. The recruitment of trained bees, and their response to improvement of crop. *Proc. R. ent. Soc. Lond.* (A), **30**, 26–32.

Richards, O. W. (1953). "The social insects." Macdonald, London.

Richardson, R. W. and Alvarez, E. (1957). Pollination relationships among vegetable crops in Mexico. I. Natural cross-pollination in cultivated tomatoes. *Proc. Am. Soc. hort. Sci.* **69**, 366–371.

Richmond, R. G. (1932). Red clover pollination by honey bees in Colorado. *Bull. Colo. agric. Exp. Stn.* No. 391.

Rick, C. M. (1947). The effect of planting design upon the amount of seed produced by male-sterile tomato plants as a result of natural cross-pollination. *Proc. Am. Soc. hort. Sci.* **50**, 273–284.

Rick, C. M. (1949). Rates of natural cross-pollination of tomatoes in various localities in California as measured by the fruits and seeds set on male-sterile plants. *Proc. Am. Soc. hort. Sci.* **54**, 237–252.

Rick, C. M. (1958). The role of natural hybridisation in the derivation of the derivation of the cultivated tomatoes of Western South America. *Econ. Bot.* **12**, 346–367.

Riedel, I. B. M. and Wort, D. A. (1960). The pollination requirement of the field bean (*Vicia faba*). *Ann. appl. Biol.* **48**, 121–124.

Rives, M. (1957). Études sur la sélection du colza d'hiver. *Annls Inst. natn. Rech. agron., Paris*, Sér. B (*Annls Amél. Pl.*), **7**, 61–107.

Roach, F. A. (1965). The effect of distance from pollinating varieties on the cropping of Cox's Orange Pippin apples. Unnumbered report (Woodstock Agric. Res. Centre, Sittingbourne).

Roach, F. A. (1967). Highbush blueberry production in England. International Society for Hort. Sci. Symposium I. Blueberry Culture in Europe, pp. 22–33.

Roach, F. A. and Perrin, M. E. B. (1959). Highbush blueberries. *Agriculture, Lond.* **66**, 336–339.

Robbins, W. W. (1931). "The Botany of Crop Plants." McGraw-Hill, New York.

Roberts, D. (1956). Sugar sprays aid fertilisation of plums by bees. *N.Z. Jl Agric.* **93**, 206–207, 209, 211.

Roberts, R. H. (1945). Blossom structure and setting of Delicious and other apple varieties. *Proc. Am. Soc. hort. Sci.* **46**, 87–90.

Roberts, R. H. (1947). Delicious apple needs near-by source of suitable pollen. *Bull. Wis. agric. Exp. Stn.* No. 472, 21–23.

Roberts, R. H. and Struckmeyer, B. Ester (1942). Growth and fruiting of the cranberry. *Proc. Am. Soc. hort. Sci.* **40**, 373–379.

Robinson, B. B. (1937). Natural cross-pollination Studies in Fiber-Flax. *J. Am. Soc. Agron.* **29**, 644–649.

Robinson, D. H. (1937). "Leguminous Forage Plants." Edward Arnold, London.

Robinson, F. A. (1958). Factors affecting the unfruitfulness of Tangelo. *Rep. Fla agric. Exp. Stn.* 1957/58, 102.

Robinson, F. A. and Krezdorn, A. H. (1962). Pollination of the Orlando tangelo. *Am. Bee J.* **102**, 132–133.

Robinson, F. A. and Nation, J. L. (1968). Substances that attract caged honeybee colonies to consume pollen supplements and substitutes. *J. apic. Res.* **7**, 83–88.

Roemer, T. (1916). Über die Befruchtungsverhältnisse verschiedener Formens des Gartenbohls (*Brassica oleracea* L.). *Z. Pflanzenzucht.* **4**, 125–141.

Rosa, J. T. (1924). Fruiting habit and pollination of cantaloupe. *Proc. Am. Soc. hort. Sci.* **21**, 51–57.

Rosa, J. T. (1925). Pollination and fruiting habit of the watermelon. *Proc. Am. Soc. hort. Sci.* **22**, 331–333.

Rosa, J. T. (1926). Direct effects of pollen on fruit and seeds of melons. *Proc. Am. Soc. hort. Sci.* **23**, 243–248.

Rosa, J. T. (1927). Results of inbreeding melons—preliminary report. *Proc. Am. Soc. hort. Sci.* **24**, 79–84.

Rösch, G. A. (1925). Untersuchungen über die Arbeitsteilung im Bienenstaat, I. Die Tätigkeiten im normalen Bienenstaate und ihre Beziehungen zum Alter der Arbeitsbienen. *Z. vergl. Physiol.* **2,** 571–631.

Rösch, G. A. (1927). Uber die Bautätigkeit im Bienenvolk und das Alter der Baubienen. Weiterer Beitrag zur Frage nach der Arbeitsteilung im Bienenstaat. *Z. vergl. Physiol.* **6,** 265–298.

Rösch, G. A. (1930). Untersuchungen über die Arbeitsteilung im Bienenstaat. 2. Die Tätigkeiten der Arbeitsbienen unter experimentell veränderten Bedingungen. *Z. vergl. Physiol.* **12,** 1–71.

Rösch, G. A. (1931). Neue Beobachtungen und Versuche über die Arbeitsteilung im Bienenstaat. *Forsch. Fortschr. dt. Wiss.* **7,** 86.

Ross-Craig, Stella (1949). "Drawings of British Plants, Part III." Bell, London.

Ross-Craig, Stella (1957). "Drawings of British Plants." Part X. Bell, London.

Ross-Craig, Stella (1959). "Drawings of British Plants." Part XIII. Bell, London.

Ross-Craig, Stella (1963). "Drawings of British Plants." Part XIX. Bell, London.

Rothenbuhler, W. C., Thompson, V. C. and McDermott, J. J. (1968). Control of the environment of honeybee observation colonies by the use of hive-shelters and flight cages. *J. apic. Res.* **7,** 151–155.

Rowlands, D. G. (1958). The nature of the breeding system in the field bean (*V. faba* L.) and its relationship to breeding for yield. *Heredity* **12,** 113–126.

Rozov, S. A. (1933). Role of the bees in the cross-pollination of the sunflower. *Dokl. Akad. Nauk SSSR* **6,** 303–305.

Rozov, S. A. (1952). Sainfoin—one of the best honey plants in crop rotation. *Pchelovodstvo, Mosk.* **29,** 46–52.

Rozov, S. A. (1957). More bees for flowering orchards. *Pchelovodstvo, Mosk.* **34,** 29–34.

Rozov, S. A. and Skrebtsova, N. D. (1958). Honeybees and selective plant pollination. *XVII Int. Beekeep. Congr.* **2,** 494–501.

Rubis, D. D., Levin, M. D. and McGregor, S. E. (1966). Effects of honey bee activity and cages on attributes of thin-hull and normal safflower lines. *Crop Sci.* **6,** 11–14.

Rudnev, V. Z. (1941a). The effect of pollination by bees on yield of sunflower seeds. *Soc. Zern. Hoz.* (2), 134–140.

Rudnev, V. Z. (1941b). The technique of utilization of bees for the pollination of sunflower. *Soc. Zern. Hoz.* (3), 77–84.

Ruppolt, W. von (1961). Extraflorale Nektarien und ihre Demonstration in der Schule. *Z. Praxis Naturw.*, 147–150.

Ryle, Margaret (1954). The influence of nitrogen, phosphate and potash on the secretion of nectar. *J. agric. Sci.* **44,** 400–419.

Rymashevskii, V. K. (1956). Pollinating activity of bees on the flowers of fruit trees and bushes. *Pchelovodstvo, Mosk.* **33,** 51–52.

Rymashevskii, V. K. (1957). Nectar productivity and nectar sugar concentration of fruit trees and bushes. *Pchelovodstvo, Mosk.* **34,** 39–41.

Rymashevskii, V. K. and Rymashevskaya, R. S. (1960). The importance of nectar production in various bush and tree fruits. *Agrobiologiya* (1), 143.

Sadamori, S., Yoshida, Y., Murakami, H. and Ishizuka, S. (1958). Studies on commercial hand pollination methods for apple flowers. I. Examination of pollen diluents, of degree of pollen dilution and of pollinating methods. *Bull. Tohoku nat. agric. Exp. Stn.* **14,** 74–81.

Sakagami, S. F. (1953). Untersuchungen über die Arbeitsteilung in einem Zwergvolk der Honingbiene. Beiträge zur Biologie des Bienenvolkes, *Apis mellifera* L. I. *Jap. J. Zool.* **11,** 117–185.

Sakharov, M. K. (1958). Pollination by bees on vegetable seed plots. *Sad Ogorod* **96,** 21–23.
Sampson, D. R. (1957). The genetics of self- and cross-incompatibility in *Brassica oleracea*. *Genetics* **42,** 252–263.
Sampson, H. C. (1936). Cultivated crop plants of the British Empire and Anglo-Egyptian Sudan. (Tropical and Sub-Tropical.) *Addit. Ser. Kew* **12.**
Sanders, T. W. and Lansdell, J. (1924). "Grapes, peaches and melons and how to grow them." Collingridge, London.
Sănduleac, E. V. (1959). Data on the entomophilous pollination and the selection of *Cucurbitaceae*. *Lucr. sti. Stat. cent. Seri. Apic.* **1,** 129–132.
Sănduleac, E. V. (1962). La sélection convergente, méthode practique pour l'augmentation de l'efficacité de la pollinisation par les insectes. *Annls Abeille* **5,** 135–143.
Sănduleac, E. and Baculinschi, H. (1959). Date preliminare privind valoarea meliferă a zmeurului sizburătoarei în muntii Buzăului. *Lucr. sti. Stat. cent. Seri. Apic.* **1,** 126–128.
Sartorius, O. (1926). Zur Entwicklung und Physiologie der Rebblüte. *Angew. Bot.* **8,** 29–62, 65–89.
Šaškina, L. M. (1950). Cross-pollination in strawberries. *Agrobiologija* (5), 45–47.
Savitsky, H. (1950). A method of determining self-fertility and self-sterility in sugar beets, based upon the stage of ovule development shortly after flowering. *Proc. Am. Soc. Sug. Beet Technol.* **6,** 198–201.
Sax, K. (1922). The sterility relationship in Maine apple varieties. *Bull. Me agric. Exp. Stn* No. 307, 61–76.
Sazykin, Y. V. (1952). Raspberry—an important honey plant. *Pchelovodstvo, Mosk.* **29,** 43–45.
Sazykin, Y. V. (1953). Nectar secretion in gooseberries and black currants. *Pchelovodstvo, Mosk.* **30,** 43–46.
Sazykin, Y. V. (1955). Orchard trees and bushes as honey plants. *Pchelovodstvo, Mosk.* **32,** 30–34.
Schaefer, C. W. and Farrar, C. L. (1946). The use of pollen traps and pollen supplements in developing honeybee colonies. *Circ. U.S. Bur. Ent.* E–531.
Schander, H. (1955). Über die Veränderlichkeit der Fruchtgestalt bei der Birnesorte "Conference". *Mitt. ObstVersAnst. Jork.* **10,** 271–277.
Schander, H. (1956a). Die schwarze Johannisbeere. *Hess. Obstb.* **11,** 24–25.
Schander, H. (1956b). Experimentelle Untersuchungen über den Einfluss der Honigbienen auf den Ertrag der Kultursorten von *Ribes nigrum*. *Gartenbauwissenschaft* **3,** 284–291.
Schaub, I. O. and Baver, L. D. (1942). Blueberries earlier and larger when cross-pollinated. *Rep. N. Carol. agric. Exp. Stn.* **65,** 53.
Schelhorn, M. von (1942). Blütenbiologische studien an der zottelwicke. *Pflanzenbau* **18,** 311–320.
Schelhorn, M. von (1946). Blütenbiologie und Samenansatz bei *Vicia Villosa*. *Züchter* **17/18,** 22–24.
Schlecht, F. (1921). Untersuchungen über die Befructungsverhältnisse bei Rotklee. *Arch. Bienenk.* **7,** 37–39.
Schneck, H. W. (1928). Pollination of greenhouse tomatoes. *Bull. Cornell agric. Exp. Stn.* **470.**
Schroeder, C. A. (1947). Pollination requirements of the feijoa. *Proc. Am. Soc. hort. Sci.* **49,** 161–162.
Schroeder, C. A. (1951). Heterostyly and sterility in *Carissa grandiflora*. *Proc. Am. Soc. hort. Sci.* **57,** 419–422.

Schuster, C. E. (1925). Pollination and growing of the cherry. *Bull. Ore. agric. Exp. Stn.* No. 212.
Schwan, B. (1953). Iakttagelser rörande rödklöverpollinerande insekter aren 1942–1946. *Medd. Sverig. FröodlFörb.* (2), 34–61.
Schwan, B. (1962). Swedish investigations on pollinating insects. Proc. 1st Int. Symp. on Pollination, Copenhagen, 1960. *Medd. Sverig. FröodleFörh.* (7), 75–77.
Schwan, B. and Martinovs, A. (1954). Studier över binas (*Apis mellifica*) pollendrag i Ultuna. *Medd. Husdjursförsök* (57).
Schwanwitsch, B. N. (1956). The work of the honeybee on vetch and *Lotus corniculatis*. *Vest. leningr. Univ.*, 55–61.
Scriven, W. A., Cooper, B. A. and Allen, H. (1961). Pollination of field beans. *Outl. Agric.* **3**, 69–75.
Scullen, H. A. (1952). Ladino clover seed production in the Pacific Northwest. *Am. Bee J.* **92**, 287–288.
Scullen, H. A. (1956). Bees ... for legume seed production. *Circ. Inf. Ore. agric. Exp. Stn.* No. 554.
Seaney, R. R. (1962). Evaluation of methods for self- and cross-pollinating birdsfoot trefoil, *Lotus corniculatus* L. *Crop Sci.* **2**, 81.
Seaton, H. L. and Kremer, J. C. (1939). The influence of climatological factors on anthesis and anther dehiscence in the cultivated cucurbits. A preliminary report. *Proc. Am. Soc. hort. Sci.* **36**, 627–631.
Šedivý, J., Taimr, L., Veselý, V., Haragsim, O., Dočkal, J. and Balcar, J. (1966). Evaluation of activity of honeybee colonies moved to a lucerne field. *Sb. ent. Odd. nár. Mus. Praze* **63**, 1–9.
Sein, F. (1959). Ayudan las abejas al cafetalero? *Hacienda* **55**, 36–50.
Ševčenko, A. Ja. (1939). Preliminary results of breeding and vernalization of clovers. *Mater. Sovešč. Korm. Trav.* Jan. 1938, 20–28.
Shanks, C. H. (1969). Pollination of Raspberries by honeybees. *J. apic. Res.* **8**, 19–21.
Sharma, P. L. (1958). Sugar concentration of nectar of some Punjab honey plants. *Indian Bee J.* **20**, 86-91.
Sharma, P. L. (1961). The honeybee population among insects visiting temperate-zone fruit flowers and their role in setting fruit. *Bee Wld* **42**, 6–8.
Sharma, P. L. and Sharma, A. C. (1950). Influence of numbers in a colony on the honey-gathering capacity of bees. *Indian Bee J.* **12**, 106–107.
Sharples, G. C., Todd, F. E., McGregor, S. E. and Nilne, R. L. (1965). The importance of insects in the pollination and fertilization of the Cardinal grape. *Proc. Am. Soc. hort. Sci.* **86**, 321–325.
Shaw, F. R. (1953). The sugar concentration of the nectar of some New England honey plants. *Glean. Bee Cult.* **81**, 88–89.
Shaw, F. R. and Bailey, J. S. (1937a). The honeybee as a pollinator of cultivated blueberries. *Am. Bee J.* **77**, 30.
Shaw, F. R. and Bailey, J. S. (1937b). "Insect pollination of cultivated blueberries." *Am. Fruit Grow.* **57**, 8, 23.
Shaw, F. R. and Turner, M. (1942). Preference of bees for certain varieties of apples. *Am. Bee J.* **82**, 521.
Shaw, F. R., Bailey, J. S. and Bourne, A. I. (1939). The comparative value of honeybees in the pollination of cultivated blueberries. *J. econ. Ent.* **32**, 872–874.
Shaw, F. R., Savos, M. and Shaw, W. M. (1954). Some observations on the collecting habits of bees. *Am. Bee J.* **94**, 422.

Shaw, F. R., Shaw, W. M. and Weidhaas, J. (1956). Observations on sugar concentrations of cranberry nectar. *Glean. Bee Cult.* **84**, 150–151.

Shaw, H. B. (1914). Thrips as pollinators of beet flowers. *Bull. U.S. Dep. Agric.* **104**.

Shaw, H. B. (1916). Self, close and cross-fertilization of beets. *Mem. N. Y. bot. Gdn.* **6**, 149–152.

Shchibrya, A. A. and Mart'yanova, A. I. (1960). (Special features of pollination of *Lotus corniculatus*). *Agrobiologiya* (5), 694–697.

Shearer, D. A. and Boch, R. (1966). Citral in the Nassanoff pheromone of the honey bee. *J. Insect Physiol.* **12**, 1513–1521.

Shemetkov, M. F. (1957a). Feeding pollen substitutes to bees in greenhouses. *Pchelovodstvo, Mosk.* **34**, 17–20.

Shemetkov, M. F. (1957b). The use of bees for pollinating cucumbers in hothouses and forcing-beds. *Byull. nauch.-tekh. Inf. Inst. Pchelovodstva* (2), 21–24.

Shemetkov, M. F. (1960). Pollinating activity of bees in greenhouses. *Pchelovodstvo, Mosk.* **37**, 28–31.

Shimanuki, H., Lehnert, T. and Stricker, M. (1967). Differential collection of cranberry pollen by honey bees. *J. econ. Ent.* **60**, 1031–1033.

Shishikin, E. A. (1946). Honeybees in the service of cotton pollination. *Pchelovodstvo, Mosk.* **23**, 31–32.

Shishikin, E. A. (1952). Effect of pollination by honeybees on increasing the productivity of cotton. Chapter of book by Krishchunas, pp. 95–103. Gosundarstvennoe Izdatel'stvo Sel'skokhoz, Moskva.

Shoemaker, J. S. (1931). "Premier" as a pollinator for "Sample". *Rep. Ohio agric. Exp. Stn.* **49**, 112.

Shoemaker, J. S. (1955). "Small Fruit Culture." (3rd edition). McGraw-Hill Book Co., London.

Shoemaker, J. S. and Teskey, B. J. E. (1959). Tree fruit production. Wiley, New York.

Shuel, R. W. (1952). Some factors affecting nectar secretion in red clover. *Plant Physiol.* **27**, 95–110.

Shuel, R. W. (1955a). Nectar Secretion. *Am. Bee J.* **95**, 229–234.

Shuel, R. W. (1955b). Nectar secretion in relation to nitrogen supply, nutritional status, and growth of the plant. *Can. J. agric. Sci.* **35**, 124–138.

Shuel, R. W. (1956). Studies of nectar secretion in excised flowers. I. The influence of cultural conditions on quantity and composition of nectar *Can. J. Bot.* **34**, 142–153.

Shuel, R. W. (1957). Some aspects of the relation between nectar secretion and nitrogen, phosphorus, and potassium nutrition. *Can. J. Pl. Sci.* **37**, 220–236.

Shuel, R. W. (1961). Influence of reproductive organs on secretion of sugars in flowers of *Streptosolen jamesonii*, Miers. *Plant Physiol.* **36**, 265–271.

Sidhu, A. S. and Singh, S. (1961). Studies on agents of cross pollination of cotton. *Indian Cott. Grow. Rev.* **15**, 341–353.

Sidhu, A. S. and Singh, S. (1962). Role of honeybees in cotton production. *Indian Cott. Grow. Rev.* **16**, 18–23.

Sigfrids, A. G. (1947). Humlar och bin—i lantmannens tjänst. *Tidskr. Lantm.* **29**, 121.

Silow, R. A. (1931). Self-fertility of Lotus spp. *Bull. Welsh. Pl. Breed. Stn.* Ser. H. **12**, 234–240.

Silversides, W. H. and Olsen, P. J. (1941). Seed setting in alfalfa 2. *Scient. Agric.* **22**, 129–134.

Simao, S. and Maranhao, Z. C. (1959). Os insetos como agentes polinizadores da mangueira. Anais Esc. sup. Agric. "Luiz Queiroz" **16**, 299–304.

Simmonds, N. W. (1957). "Bananas." Longmans, London.
Simonov, I. N, (1949). Using mixed pollen in raising black currants. *Agrobiologija* (5), 133–134.
Simpson, D. M. (1954). Natural Cross-Pollination in Cotton. *Tech. Bull. U.S. Dep. Agric.* **1094**.
Simpson, D. M. and Duncan, E. N. (1956). Cotton pollen dispersal by insects. *Agron. J.* **48**, 305–308.
Simpson, J. (1961). Nest climate regulation in honey bee colonies. *Science, N.Y.* **133**, 1327–1333.
Simpson, J. (1964). Dilution by honeybees of solid and liquid food containing sugar. *J. apic. Res.* **3**, 37–40.
Singh, L. B. (1960). "The Mango." Leonard Hill, London.
Singh, S. (1950). Behaviour studies of honeybees in gathering nectar and pollen. *Mem. Cornell agric. Exp. Stn.* No. 288.
Singh, S. (1953). Activities of Insect Pollinators and Orchard Planning. *Indian J. Hort.* **10**.
Singh, S. (1954). Insect Pollinators and the Breeding of Fruit Varieties. *Indian J. Hort.* **11**.
Singh, S. (1962). Studies on the morphology and viability of the pollen grains of mango (*Mangifera indica*). *Hort. Adv. Saharanpur* **5**, 121–144.
Singh, S. and Boynton, D. (1949). Viability of apple pollen in pollen pellets of honeybees. *Proc. Am. Soc. hort. Sci.* **53**, 148–152.
Sirks, M. J. (1923). Die Verschiebung genotypischer Verhältniszahlen innerhalb Polulationen laut Mathematischer Berechnung und Experimenteller Prüfung. *Meded. LandbHoogesch. Wageningen* **26**.
Sivori, E. M. (1941). Biología floral de girasol. *Rev. Argent. Agron.* **8**, 150–154.
Skirde, W. (1961). Blüten- und Nekarunterscuhungen an tetraploiden Frühkleeformen (*Trifolium pratense praecox*). *Grünland* **10**, 75–78.
Skirde, W. (1963). Morphologische und Sekretorische Untersuchungen an diund tetraploidem Klee. *Ann. agric. Fenn.* **2**, 73–90.
Skovgaard, O. S. (1936). Rødkloverens bestøvning, humlebier og humleboer. *K. danske Vidensk. Selsk. Skr. Nat. Mat.* Afd.9. rk.6. VI, 1–140.
Skovgaard, O. S. (1952). Humlebiers og honningbiers arbejdshastighed ved bestøvningen af rødkløver. *Tidsskr. PLAvl.* **55**, 449–475.
Skovgaard, O. S. (1956). Den kaukasiske honningbi som rødkløverbestøver. *Tidsskr. PLAvl.* **59**, 877–887.
Skrebtsov, M. F. (1964). The problem of abundant pollination of cotton by honeybees. *Trud. nauch.-issled Inst. Pchelov.* 246–264.
Skrebtsova, N. D. (1957a). Amounts of pollen on the body of bees. *Pchelovodstvo, Mosk.* **34**, 39–42.
Skrebtsova, N. D. (1957b). Pollination of buckwheat flowers by bees. *Pchelovodstvo, Mosk.* **34**, 48–50.
Skrebtsova, N. D. (1957c). The role of bees in pollinating strawberries. *Pchelovodstvo, Mosk.* **34**, 34–36.
Skrebtsova, N. D. (1959). Bees increase the crop of black currants. *Pchelovodstvo, Mosk.* **36**, 26–27.
Skrebtsova, N. D. (1964). The use of the pollinating activity of honeybees for producing vegetable seed by heterosis. *Trud. nauch. issled. Inst. Pchelov.* 223–245.
Sladen, F. W. L. (1912). "The humble bee. Its life history and how to domesticate it." Macmillan, London.

Smaragdova, N. P. (1933). Utilization of the responses of the bee for the pollination of plants. *In* "Red clover pollination and methods of clover seed production". pp. 219–227. (A. F. Gubin and G. I. Romashov, eds.) In Russian. Zhizn i Znanie, Moscow.

Smaragdova, N. P. (1957). Coloration of the corolla and the sugar content of nectar. *Pchelovodstvo, Mosk.* **34,** 42–43.

Smirnov, V. M. (1954). Cross-pollination of flax by bees. *Pchelovodstvo, Mosk.* (9), 53–55.

Smit, A. G. (1950). Pollination of cacao in Costa Rica. Unpublished thesis, 1950. I.I.C.A. Turrialba, Costa Rica.

Smith, B. D. and Williams, R. R. (1967). Methods of pollen transfer at Long Ashton. *Rep. agric. hort. Res. Stn. Univ. Bristol* 1966, 120–125.

Smith, F. G. (1958). Beekeeping operations in Tanganyika, 1949–1957. *Bee Wld* **39,** 29–36.

Smith, M. V. (1952). Honeybees for pollination. *Circ. Ont. Dep. Agric.* No. 133.

Smith, M. V. (1958). The use of fluorescent markers as an aid in studying the foraging behaviour of honeybees. *X Int. Congr. Ent.* 1956, **4,** 1063.

Smith, M. V. (1963). The O.A.C. pollen trap. *Can. Bee J.* **74,** 4–5, 8.

Smith, M. V. and Townsend, G. F. (1951). Techniques useful in pollination studies. *14th Int. beekeep. Congr.* Paper 12.

Smith, M. V. and Townsend, G. F. (1952). A method of measuring pollination populations on field crops. *Can. Ent.* **134,** 314.

Smith, Ora (1935). Pollination and Life-history of the Tomato (*Lycopersicum esculentum* Mill). *Mem. Cornell agric. Exp. Stn.* **184.**

Smith, Ora (1932). Relation of temperature to Anthesis and Blossom drop of the Tomato together with a histological study of the Pistils. *J. agric. Res.* **44,** 183–190.

Smith, Ora and Cochran, H. L. (1935). Effect of temperature on pollen germination and tube growth in the Tomato. *Mem. Cornell agric. Exp. Stn.* **175.**

Smith, R. F., McSwain, J. W., Linsley, E. G. and Platt, F. R. (1948). The effect of DDT dusting on honeybees. *J. econ. Ent.* **41,** 960–971.

Sneep, J. (1952). Selection and breeding of some Brassica plants. *Proc. 13th Int. hort. Congr.,* 422–426.

Snodgrass, R. E. (1956). Anatomy of the honey bee. Comstock, Ithaca, New York.

Snyder, J. C. (1942). Commercial hand-pollination methods for apples in the northwest. *Proc. Am. Soc. hort. Sci.* **41,** 183–186.

Snyder, J. C. (1946). Pollination of tree fruits and nuts. *Ext. Bull. Wash. St. Coll. Ext. Serv.* No. 342.

Snyder, J. C. (1947). Pollination results. *Bett. Fruit* **41,** 10–12.

Soetardi, R. G. (1950). De betekenis van insecten bij de bestuiving van *Theobroma cacao* L. *Arch. Koffiecult.* **17,** 1–31.

Solov'ev, G. M. (1951). Sources of bee forage and possibilities of improving them. *Pchelovodstvo, Mosk.* **28,** 42–45.

Solov'ev, G. M. (1960). Use of some characteristics of the interrelation between bees and entomophilous plants. *Agrobiologiya* (6), 939–942.

Soost, R. K. (1956). Unfruitfulness in the Clementine mandarin. *Proc. Am. Soc. hort. Sci.* **67,** 171–175.

Soper, M. H. R. (1952). A study of the principal factors affecting the establishment and development of the field bean (*Vicia faba*). *J. agric. Sci.* **42,** 335–346.

Sorokin, V. (1958). Pollination by bees in greenhouses. *Sad Ogorod* **96,** 24–25.

Sosunkov, V. I. (1953). Cross-pollination in vines. *Sad Ogorod* **5**, 26–27.
Sovoleva, E. M. (1952). Bees and lucerne production. *Pchelovodstvo, Mosk.* **29**, 39–41.
Spencer, J. L. and Kennard, W. C. (1955). Studies on mango (*Mangifera indica* L.) fruit set in Puerto Rica. *Trop. Agric. Trin.* **32**, 323–330.
Spencer-Booth, Yvette (1960). Feeding pollen, pollen substitutes and pollen supplements to honeybees. *Bee Wld* **41**, 253–263.
Spencer-Booth, Yvette (1965). The collection of pollen by bumblebees and its transport in the corbiculae and the proboscidial fossa. *J. apic. Res.* **4**, 185–190.
Sprenger, A. M. (1916). Het nut der bijen voor de fruitteelt. *Maandschr. Bijent.* **19**, 121–124.
Srinivasalu, N. and Chandrasekaran, N. R. (1958). A note on natural crossing in groundnut, *Arachis hypogaea* L. *Sci. Cult.* **23**, 650.
Stadhouders, P. J. (1949). De overdracht van het stuifmeel bij de kruisbestuiving van vruchtbomen. *Meded. Dir. Tuinb.* **12**, 821–830.
Stahel, G. (1928). Beiträge zur Kenntnis der Blüten-biologie von Kakao (*Theobroma cacao* L.). *Verh. K. Akad. Wet.* (Tweede Sectie) **25**.
Stählin, A. and Bommer, D. (1958). Ueber die Wege zu einer besseren Befruchtung des Rotklees. *Angew. Bot.* **32**, 165–185.
Stanford, E. E. (1937). "Economic plants." New York.
Stapel, C. (1933). Undersøgelser over humblebier (*Bombus* Latr.) deres udbredelse, traekplanter og betydning for bestøvingen af rødkløver (*Trifolium pratense* L.). *Tidsskr. PlAvl.* **39**, 193–294.
Stapel, C. (1934). Om rødkløverens bestøvning i Czechoslovakiet. *Tidsskr. PlAvl.* **40**, 148–159.
Stapel, C. (1935). Rødkløverens bestøvning, saerlig med henblik paa honningbiernes betydning. *Nord. JorbrForskn.* **17**, 508–517.
Stapel, C. (1943). Uber die Befruchtung der Luzerne durch Insekten in Dänemark. *Ent. Meddr* **23**, 224–239.
Stapel, C. (1952). Undersøgelser over lucernens bestøvning. *Tidsskr. Frøavl* **20**, 385–391.
Stapel, C. and Eriksen, K. M. (1936). Pollenavalytiske Undersøgelser over Honningbiernes betydning. *Nord. Jordbr. Forskn.* 489–497.
Starling, T. M., Wilsie, C. P. and Gilbert, N. W. (1950). Corolla tube length studies in red clover. *Agron. J.* **42**, 1–8.
Stephen, W. P. (1955). Alfalfa pollination in Manitoba. *J. econ. Ent.* **48**, 543–548.
Stephen, W. P. (1958). Pear pollination studies in Oregon. *Tech. Bull. Ore. agric. Exp. Stn.* No. 43.
Stephen, W. P. (1959). Maintaining alkali bees for alfalfa seed production. *Stn Bull. Ore. agric. Exp. Stn.* **568**.
Stephen, W. P. (1960). Artificial bee beds for the propagation of the alkali bee, *Nomia melanderi. J. econ. Ent.* **53**, 1025–1030.
Stephen, W. P. (1961). Artificial nesting sites for the propagation of the leaf-cutter bee, *Megachile* (*Eutricharaea*) *rotundata*, for alfalfa pollination. *J. econ. Ent.* **54**, 989–993.
Stephen, W. P. (1962). Propagation of the leaf-cutter bee for alfalfa seed production. *Stn. Bull. Ore. agric. Exp. Stn.* **586**.
Stephen, W. P. (1965). Temperature effects on the development and multiple generation in the alkali bee (*Nomia melanderi*). *Ent. exp. appl.* **8**, 228–240.
Stephen, W. P. and Osgood, C. E. (1965). Influence of tunnel size and nesting medium on sex ratios in a leaf-cutter bee, *Megachile rotundata. J. econ. Ent.* **58**, 965–968.

Stephens, S. G. and Finkner, M. D. (1953). Natural crossing in Cotton. *Econ. Bot.* **7**, 257–269.

Stereva, R. (1962). On certain questions of the biology of lucerne flowering and fertilization. *Izv. Dobrudž selskestop. nauč.—izsled. Inst. Tolbuhin, Bulgaria.* **3**, 177–189.

Steshenko, F. N. (1958). The role of honeybees in cross-pollinating grape vines. *Pchelovodstvo, Mosk.* **35**, 37–40.

Steuckardt, R. (1961). Der Anteil der Honigbienen an der Lucernebefruchtung. *Leipzig. Bienenztg.* **75**, 244–246.

Steuckardt, R. (1962). Untersuchungen über die Wirksamkeit von Honigbienen *Apis mellifica* bei der Luzerne-bestäubung. *Z. PflZücht.* **47**, 15–50.

Steuckardt, R. (1963). Der Einsatz von Honigbienen bei diallelen Kreuzungen insektenblütiger Fremdbefruchter. *Z. PflZücht.* **49**, 161–172.

Steuckardt, R. (1965). Pollensammelnde Honigbienen (*Apis mellifica*) als wirksame Bestäuber bei der Züchtung und im Samenbau von Luzerne, Rotklee und Ackerbohnen. *Züchter* **35**, 66–72.

Stevenson, F. J. and Clark, C. F. (1937). Breeding and genetics in potato improvement. *Yb. Agric. U.S. Dep. Agric.* 405–444.

Stevenson, T. M. and Bolton, J. L. (1947). An evaluation of the self-tripping character in breeding for improved seed-yield in alfalfa. *Emp. J. exp. Agric.* **15**, 82–88.

Stewart, D. (1946). Insects as a Minor Factor in Cross Pollination of Sugar Beets. *Proc. Am. Soc. Sug. Beet Technol.* **4**, 256–258.

Stokes, W. E. and Hull, F. H. (1930). Peanut breeding. *J. Am. Soc. Agron.* **22**, 1004–1019.

Storey, W. B. (1958). Modifications of sex expression in Papaya. *Hort. Adv., Sahranpur* **2**, 49–60.

Stout, A. B. (1921). Types of Flowers and Intersexes in grapes with reference to Fruit Development. *Tech. Bull. N.Y. St. agric. Exp. Stn.* **82**.

Stout, A. B. (1933). The pollination of avocados. *Bull. Fla. agric. Exp. Stn.* **257**.

Stroman, G. N. and Mahoney, C. H. (1925). Heritable Chlorophyll deficiencies in seedling cotton. *Bull. Tex. agric. Exp. Stn.* **333**.

Sturtevant, A. P. and Farrar, C. L. (1935). Further observations on the flight range of the honeybee in relation to honey production. *J. econ. Ent.* **28**, 585–589.

Sun, V. G. (1937). Effects of self-pollination in rape. *J. Am. Soc. Agron.* **29**, 555–567

Svendsen, O. (1964). Om nogle bifamiliers pollentraek i rødkløverens blomstringstid. *Nord. Bitidskr.* **15/16**, 22–27.

Swart, F. W. J. (1960). De bestuiving van vlinderbloemige voedergewassen. *Wet. Pratijk*, **17**, Abstract: *Herb. Abstr.* No. 1639, 30.

Synge, A. D. (1947). Pollen collection by honeybees (*Apis mellifera*). *J. Anim. Ecol.* **16**, 122–138.

Taber, S. (1954). The frequency of multiple mating of queen honeybees. *J. econ. Ent.* **47**, 995–998.

Taber, S. (1960). Estimation of total honey bee populations using a known population of marked bees. *J. econ. Ent.* **53**, 993–995.

Taber, S. (1963a). Why bees collect pollen. *XIX Int. Beekeep. Congr., Prague* 114.

Taber, S. (1963b). The effect of a disturbance on the social behaviour of the honeybee colony. *Am. Bee J.* **103**, 286–288.

Tamargo, M. A. and Jones, M. D. (1954). Natural cross-fertilization in Kenaf. *Agron. J.* **46**, 456–459.

Tammes, P. M. L. (1937). Over den bloei en de bestiuving van den klapper. *Landbouw, Buitenz.* **13**, 74–89.
Taranov, G. F. (1952). Rules governing the flight activities of bees. *Zool. Zh.* **31**, 61–71.
Taschdjian, E. (1932). Beobachtung über Variabilität, Dominanz und Vizinismus bei *Coffea arabica. Z. Zücht.* **17**, 341–354.
Taylor, E. A. (1955). Cantaloupe production increased with honey bees. *J. econ. Ent.* **48**, 327.
Teaotia, S. S. and Luckwill, L. C. (1956). Fruit drop in black currants: Factors affecting "running-off". *Rep. agric. hort. Res. Stn. Univ. Bristol* 1955, **64**, 64–74.
Tedin, O. (1931). Korsningsfaran vid fröodling av rovor. *Landtmannen.—Svenskt-Land* **14**, 454–455.
Tedoradze, S. G. (1959). The role of bees in the production of new varieties of beans in the conditions of Georgia. *Pchelovodstvo, Mosk.* **36**, 40–42.
Thies, S. A. (1953). Agents concerned with natural crossing of cotton in Oklahoma. *Agron. J.* **45**, 481–484.
Thomas, W. (1951). Bees for pollinating red clover. *Glean Bee Cult.* **79**, 137–141.
Thompson, D. J. (1962). Natural Cross-Pollination in Carrots. *Proc. Am. Soc. hort. Sci.* **81**, 332–334.
Thompson, F. (1940). The importance of bees in agriculture. *Bee Craft* **22**, 6–7.
Thompson, J. B. (1925). Production of sweet-potato seedlings at the Virgin Islands Experiment Station. *Bull. Virg. Is (U.S.) agric. Exp. Stn.* **5**.
Thompson, R. C. (1933). Natural cross-pollination in lettuce. *Proc. Am. Soc. hort. Sci.* **30**, 545–547.
Thompson, R. C., Whitaker, T. W., Bohn, G. W. and, Horn, C. W. van, (1958). Natural cross-pollination in lettuce. *Proc. Am. Soc. hort. Sci.* **72**, 403–409.
Thomson, G. M. (1922). "The naturalisation of animals and plants in New Zealand." Cambridge University Press, London.
Thomson, J. R. (1938). Cross- and self-fertility in sainfoin. *Ann. appl. Biol.* **25**, 695–704.
Tidbury, G. E. (1949). "The clove tree." Crosby Lockwood, London.
Tiedjens, V. A. (1928). Sex ratios in cucumber flowers as affected by different conditions of soil and light. *J. Agric. Res.* **36**, 721–746.
Tobgy, H. A. and Said, I. (1956). Some factors influencing seed setting in berseem or Egyptian clover (*Trifolium alexandrinum* L.). *1st Meeting Working Party on Grazing and Fodder Research Near East, Cairo Univ.*
Todd, F. E. (1951). The community approach to pollination problems. *Res. Iowa St. Apiarist* 1950, 104–108.
Todd, F. E. and Bishop, R. K. (1941). The role of pollen in the economy of the hive. *Circ. U.S. Bur. Entomol.* No. E–536, 1–9.
Todd, F. E. and Bretherick, O. (1942). The composition of pollens. *J. econ. Ent.* **35**, 312–317.
Todd, F. E. and McGregor, S. E. (1960). The use of honey bees in the production of crops. *A. Rev. Ent.* **5**, 265–278.
Todd, F. E. and Reed, C. B. (1970). Brood measurement as a valid index to the value of honey bees as pollinators. *J. econ. Ent.* **63**, 148–149.
Todd, F. E. and Vansell, G. H. (1942). Pollen grains in nectar and honey. *J. econ. Ent.* **35**, 728–731.
Todd, F. E. and Vansell, G. H. (1952). The role of pollinating insects in legume-seed production. *Proc. 6th Int. Grassld. Congr.* 835–840.
Townsend, G. F. and Adie, A. (1952). Moving bees. *Ont. Dept. Agric. Circ.* No. 130.

Townsend, G. F. and Shuel, R. W. (1962). Some recent advances in apicultural research. *A. Rev. Ent.* **7,** 481–500.
Townsend, G. F., Riddell, R. T. and Smith, M. V. (1958). The use of pollen inserts for tree fruit pollination. *Can. J. Pl. Sci.* **38,** 39–44.
Traub, H. P., Pomeroy, C. S., Robinson, T. R. and Aldrich, W. W. (1941). Avocado production in the United States. *U.S.D.A. Circ.* **620.**
Traynor, J. (1966). Increasing the pollinating efficiency of honeybees. *Bee Wld* **47,** 101–110.
Treherne, R. C. (1923). The relation of insects to vegetable seed production. *Rep. Queb. Soc. Prot. Pl.* **15,** 47–59.
Trofimec, N. K. (1940). The biology and fertilisation in Allium. *Vest. Sots Rasteniev. Rasten.* **5,** 76–86.
Trought, T. (1930). Notes on certain facts on vicinism and artificial pollination in Egypt. *Emp. Cott. Grow. Rev.* **7,** 13–18.
Trushkin, A. V. (1960). Use of bees for cotton pollination. *Agrobiologiya* (5), 787–788.
Tsygankov, S. K. (1953). Pollination by bees increases the number and improves the quality of fruit. *Pchelovodstvo, Mosk.* **30,** 36–38.
Tufts, W. P. (1919). Almond pollination. *Univ. Calif. Bull.* No. 306, 337–366.
Tukey, H. B. (1924). An experience with pollenizers for cherries. *Proc. Am. Soc. hort. Sci.* **21,** 69–73.
Tukey, H. B. (1936). A relation between seed attachment and carpel symmetry and development in *Prunus*. *Science, N.Y.* **84,** 513–515.
Tyler, F. J. (1908). The Nectaries of Cotton. *Bull. Bur. Pl. Ind. U.S. Dep. Agric.* **131,** 45–54.
Tysdal, H. M. (1940). Is tripping necessary for seed setting on alfalfa? *J. Am. Soc. Agron.* **32,** 570–585.
Tysdal, H. M. (1946). Influence of tripping, soil moisture, plant spacing on alfalfa seed production. *J. Am. Soc. Agron.* **38,** 515–535.
Ufer, M. (1930). Untersuchungen über die Befruchtungsverhältnisse einiger Melilotusarten (Steinklee). *Züchter* **2,** 341–354.
Ufer, M. (1933). Untersuchungen über die den Samensatz der Luzerne beeinflussenden klematischen Faktoren. *Züchter* **5,** 217–221.
Umaerus, M. and Åkerberg, E. (1959). Pollination and seed setting in red clover and lucerne under Scandinavian conditions. *Herb. Abstr.* **29,** 157–164.
Umaerus, M. and Åkerberg, E. (1967). Undersökningar åren 1960–62 över förekomst av pollinerande insekter inom Svalöfsomradet. *Medd. Sverig. FröodlFörb.* (2), 7–25.
Valle, O. (1938). Cuna-ja alsikeapilan siementuotannon kehittämisestä. *Siemenjulkaisu* 180–190.
Valle, O. (1946). Puna-apilan siemenen satotoiveista. Koetoim. Käyt. (9).
Valle, O. (1947). Mehiläishoidon merkityksestä eri apilalajien siemenviljelyksessä. *Maatalrus* (5), 67–71.
Valle, O. (1948). De olika humlearternas samt binas betydelse för rödklöverns pollination. *Ann. Ent. Fennici,* **14,** Liite-suppl., 225–31.
Valle, O. (1955). Untersuchungen zur Sicherung der Bestäubung von Rotklee. *Suom. Maatal. Seur. Julk.* **83,** 205–220.
Valle, O. (1959). Pollination and seed setting in tetraploid red clover in Finland. I. *Suom. maatal. Seur. Julk.* **95,** 1–35.
Valle, O., Salminen, M. and Huokuna, E. (1960). Pollination and seed setting in tetraploid red clover in Finland. II. *Suom. maatal. Seur. Julk.* **97,** 1–64.

Valle, O., Sarisalo, M. and Paatela, J. (1962). Pollination studies on red clover in Finland. Proc. 1st Intern. Symp. on Pollination, Copenhagen, 1960. *Medd. Sverig. FröodlFörb.* No. 7, 98–105.
Valleau, W. D. (1918). Sterility in the strawberry. *J. agr. Res.* **12**, 613–669.
Vansell, G. H. (1934). Relations between the nectar concentrations in fruit blossoms and the visits of honeybees. *J. econ. Ent.* **27**, 943–945.
Vansell, G. H (1942). Factors affecting the usefulness of honeybees in pollination. *Circ. U.S. Dep. Agric.* No. 650.
Vansell, G. H. (1944a). Cotton nectar in relation to bee activity and honey production. *J. econ. Ent.* **37**, 528–530.
Vansell, G. H. (1944). Some western nectars and their corresponding honeys. *J. econ. Ent.* **37**, 530–536.
Vansell, G. H. (1951). Honeybees activity on ladino florets. *J. econ. Ent.* **44**, 103.
Vansell, G. H. (1952). Variations in nectar and pollen sources affect bee activity. *Am. Bee J.* **92**, 325–326.
Vansell, G. H. (1955). Alfalfa pollen on nectar-collecting honey bees. *J. econ. Ent.* **48**, 477.
Vansell, G. H. and Griggs, W. H. (1952). Honey bees as agents of pollination. *Yb. Agric. U.S. Dep. Agric.* 88–107.
Vansell, G. H. and Reinhardt, J. F. (1948). Do honeybees help pollinate baby lima beans? *Glean. Bee Cult.* **76**, 678–679.
Vansell, G. H. and Todd, F. E. (1946). Alfalfa tripping by insects. *J. Am. Soc. Agron.* **38**, 470–488.
Vansell, G. H. and Todd, F. E. (1947). Honeybees and other bees pollinate the alfalfa seed crop in Utah. *Glean. Bee Cult.* **75**, 136–138.
Vansell, G. H., Watkins, W. G. and Bishop, R. K. (1942). Orange nectar and pollen in relation to bee activity. *J. econ. Ent.* **35**, 321–323.
Väre, A. (1960). Puna-apilan siemensaden riippuvaisuudesta kimalaisbiotoopeista. *Lounais-Hämeen Luonto* **8**, 34–41.
Veerman, A. and Van Zon, J. C. J. (1965). Insect Pollination of Pansies (*Viola* spp.). *Ent. exp. appl.* **8**, 123–134.
Velichkov, V. (1961). Sweet clover (*Melilotus*). *Pchelarstvo*, **38**, 12–15.
Velthuis, H. H. W. and Es, J. van (1964). Some functional aspects of the mandibular glands of the queen honeybee. *J. apic. Res.* **3**, 11–16.
Veprikov, P. N. (1936). "Die Bestäubung der Landwirtschaftlichen Kulturpflanzen." Sel'skogis, Moskow.
Verdieva, M. G. and Ismailova, M. K. (1960). The influence of bee pollination on the increase of the crop from feed squash. *Pchelovodstvo, Mosk.* **37**, 40–41.
Verkerk, K. (1957). The pollination of Tomatoes. *Neth. J. agric. Sci.* **5**, 37–54.
Vestad, R. (1962). Pollination by honey- and bumble-bees in diploid and tetraploid red clover. *Medd. Sverig. FröodlFörb.* **7**, 106–113.
Vieira, C. (1960). Sôbre a hibridação natural em *Phaseolus vulgaris* L. *Rev. Ceres.* **11**, 103–107.
Vinson, R. (1935). Growing healthy strawberries. *In* "Cherries and Soft Fruits." pp. 25–51. Royal Horticultural Society, London.
Voelcker, O. J. (1937). Self-incompatibility in cacao. *Rep. Cacao Res.* **6**, 2.
Voelcker, O. J. (1938). Self-incompatibility in cacao II. *Rep. Cacao Res.* **7**, 2.
Voelcker, O. J. (1940). The degree of cross-pollination in cacao in Nigeria. *Trop. Agric. Trin.* **17**, 184–186.
Vriend, S. (1953). Zaadkommers als gevolg van bestuiving door bijen. *Meded. Dir. Tuinb.* **16**, 811–816.

Wafa, A. K. (1956). Contribution to the study of factors affecting the amount of pollen grains gathered by honeybees, *Apis mellifera* L. *Bull. Fac. Agric. Ain Shams Univ.* No. 99.

Wafa, A. K. and Ibrahim, S. H. (1957a). Temperature as a factor affecting pollen-gathering activity by the honeybee in Egypt. *Bull. Fac. Agric. Ain Shams Univ.* No. 163.

Wafa, A. K. and Ibrahim, S. H. (1957b). The honeybee as an important insect for pollination. *Bull. Fac. Agric. Ain Shams Univ.* No. 162.

Wafa, A. K. and Ibrahim, S. H. (1958). Temperature as a factor affecting nectar-gathering activity in Egypt. *Bull. Fac. Agric. Ain Shams Univ.* No. 164.

Wafa, A. K. and Ibrahim, S. H. (1959a). Pollinators of the chief sources of nectar and pollen grain plants, in Egypt. *Bull. Soc. ent. Égypte* **43**, 133–154.

Wafa, A. K. and Ibrahim, S. H. (1959b). Effect of the honeybee as a pollinating agent on the yield of beans and clover in Egypt. *Cairo*, 28.

Wafa, A. K. and Ibrahim, S. H. (1960a). The effect of the honeybee as a pollinating agent on the yield of broad bean. *Bull. Fac. Agric. Ain Shams Univ.* **205**.

Wafa, A. K. and Ibrahim, S. H. (1960b). The effect of the honeybee as a pollinating agent on the yield of clover and cotton. *Bull. Fac. Agric. Ain Shams Univ.* **206**.

Wafa, A. K. and Ibrahim, S. H. (1960c). Effect of the honeybee as a pollinating agent on the yield of orange. *Elfelaha* (Jan./Feb.).

Waite, M. B. (1895). The pollination of pear flowers. *Bull. Div. Veg. Physiol. Path. U.S. Dep. Agric.* No. 5.

Waite, M. B. (1898). Pollination of pomaceous fruits. *Yb. U.S. Dep. Agric.* 167–180.

Wallace, A. T., Hanson, W. D. and Phares, D. (1954). Natural cross pollination in blue and yellow lupines. *Agron. J.* **46**, 59–60.

Walstrom, R. J. (1958). Effects of flight distances from honey bee colonies on red clover seed yields. *J. econ. Ent.* **51**, 64–67.

Wanic, D. and Mostowska, I. (1964). Cukrowce w nektarze i miodzie. *Zesz. nauk. wyzsz. Szk. roln. Olsztyn.* **17**, 543–551.

Ward, J. F. (1961). "Pimento." Govt. Printer, Kingston.

Ware, J. O. (1927). The inheritance of Red Plant Colour in Cotton. *Bull. Ark. agric. Exp. Stn.* **220**.

Warmke, H. E. (1951). Studies on pollination on *Hevea brasiliensis* in Puerto Rico. *Science, N.Y.* **113**, 646–648.

Warmke, H. E. (1952). Studies on natural pollination of *Hevea brasiliensis* in Brazil. *Science, N.Y.* **116**, 474–475.

Warren, L. O. (1961). Pollinating cucumbers with honeybees. *Arkans. Fm Res.* **10**, 7.

Watts, F. H. and Marshall, P. R. (1961). Pollination of field beans. Yield response due to bees. *Rep. Fld. Exps. Obsn. Stud. Hort. E. Midl. Reg.* 76–79.

Watts, L. E. (1958). Natural cross-pollination in lettuce, *Lactuca sativa* L. *Nature, Lond.* **181**, 1084.

Watts, L. E. (1963). Investigations into the breeding system of cauliflower *Brassica oleracea* var. *Botrytis* (L.). I. Studies of self-incompatibility. *Euphytica* **12**, 323–340.

Weatherley, P. E. (1946). Uganda: Serere Area Progress Report. *Prog. Rep. Exp. Stns. Emp. Cott. Grow. Corp.* 1944/45, 72–77.

Weaver, N. (1954). Pollination of hairy vetch by honeybees. *Prog. Rep. Texas agric. Exp. Stn.* No. 1649.

Weaver, N. (1956a). The foraging behaviour of honeybees on hairy vetch. I. Foraging methods and learning to forage. *Insectes Soc.* **3**, 537–549.

Weaver, N. (1956b). The pollination of hairy vetch by honey bees. *J. econ. Ent.* **49**, 666–671.

Weaver, N. (1957a). The foraging behaviour of honeybees on hairy vetch. II. The foraging area and foraging speed. *Insectes Soc.* **4**, 43–57.

Weaver, N. (1957b). Pollination of white clover. *Tex. Agric. Exp. Stn. Prog. Rep.* 1926.

Weaver, N. (1965a). Foraging behaviour of honeybees on white clover. *Insectes soc.* **12**, 231–240.

Weaver, N. (1965b). The foraging behaviour of honeybees on hairy vetch. III. Differences in the vetch. *Insectes soc.* **12**, 321–325.

Weaver, N. and Ford, R. N. (1953). Pollination of crimson clover by honeybees. *Tex. Agric. Exp. Stn. Prog. Rep.* 1557.

Weaver, N. and Weihing, R. M. (1960). Pollination of several clovers by honeybees. *Agron. J.* **52**, 183–185.

Weaver, N., Alex, A. H. and Thomas, F. L. (1953). Pollination of Hubam clover by honeybees. *Prog. Rep. Texas agric. Exp. Stn.* No. 1559.

Weaver, N., Weaver, E. C. and Law, J. H. (1964). The attractiveness of citral to foraging honeybees. *Prog. Rept.* 2324, Texas A and M University.

Webber, H. J. and Batchelor, L. D. (1948). "The Citrus Industry." 2 vols. University of California Press, Los Angeles.

Webster, C. C. and Wilson, P. M. (1966). "Agriculture in the Tropics." Longmans, London.

Webster, J. L. (1944). Pea and bean seed production. *Brit. Columbia Dept. Agric. Seed Prod.* Ser. 13.

Webster, R. L. (1946). The role of insects in fruit tree pollination with special reference to the honeybee. *Proc. Wash. St. hort. Ass.* **42**, 199–202.

Webster, R. L. (1947). Beekeeping in Washington. *Ext. Bull. St. Coll. Wash.* No. 289.

Webster, R. L., Telford, H. S. and Menke, H. F. (1949). Bees and pollination problems. *Stn. Circ. St. Coll. Wash.* No. 75.

Weinstein, A. I. (1926). Cytological studies on *Phaseolus vulgaris. Am. J. Bot.* **13**, 245–263.

Weiss, K. (1957). Die Abhängigkeit der Kirschenernte vom Bienenbesatz im "Alten Land". *Dtsch. Bienenw.* **8**, 124–126.

Welch, J. E. and Grimball, E. L. (1951). Natural crossing in lima beans in South Carolina. *Proc. Am. Soc. hort. Sci.* **58**, 254–256.

Wellensiek, S. J. (1932). Bloembiologische waarnemigen aan cacao, *Theobroma cacao* L. *Archief. Kofficult. Indonesië* **6**, 87–101.

Wellington, R. (1930). Pollination of pears and small fruits. 75th Ann. Mtg. New York State Hort. Soc., pp. 216–220.

Wellington, R. (1956). Artificial pollination of Eumelan grape. *Fruit var. hort. Dig.* **11**, 21–22.

Wellington, R., Hatton, R. G. and Amos, J. M. (1921). The "running off" of black currants. *J. Pomol.* **2**, 160–198.

Wellman, F. L. (1961). "Coffee." Leonard Hill, London.

Wenholz, H. (1933). Plant breeding in New South Wales. (Sixth year of progress 1931–32). *Sci. Bull. Dep. Agric. N.S.W.* **41**.

Wenner, A. M., Wells, P. H. and Johnson, D. L. (1969). Honey bee recruitment to food sources; olfaction or language. *Science, N.Y.* **164**, 84–86.

Wertheim, I. S. J. (1968). Niewe inzichten in de bestuiving van appel en peer. *Meded. Dir. Tuinb.* **31**, 438–447.

Westgate, J. M. and Coe, H. S. (1915). Red clover seed production: pollination studies. *Bull. U.S. Dep. agric.* **289**.

Wexelsen, H. (1940). Selection and inbreeding in red clover and timothy. *Jt. Publs. Commonw. agric. Bur.* **3**, 93–114.

Wexelsen, H. and Vestad, R. (1954). Observations on pollination and seed setting in diploid and tetraploid red clover. *European Grassland Conf.*, Paris, 1954, 64–68.

Whitaker, T. W. and Bohn, G. W. (1952). Natural cross-pollination in muskmelon. *Proc. Am. Soc. hort. Sci.* **60**, 391–396.

Whitaker, T. W. and Davis, G. N. (1962). "Cucurbits." Leonard Hill, London.

Whitaker, T. W. and Jagger, I. C. (1937). Breeding and improvements of cucurbits. *Yb. Agric. U.S. Dep. Agric.* 207–232.

Whitaker, T. W. and Pryor, D. E. (1946). Effect of plant-growth regulators on the set of fruit from hand-pollinated flowers in *Cucumis melo* L. *Proc. Am. Soc. hort. Sci.* **48**, 417–422.

White, Elizabeth and Clarke, J. H. (1939). Some results of self-pollination of the highbush blueberry at Whitesbog, N.J. *Proc. Am. Soc. hort. Sci.* **36**, 305–309.

White, T. H. (1918). The pollination of greenhouse tomatoes. *Bull. Md. agric. Exp. Stn.* **222**, 93–101.

Whitehead, R. A. (1963). The processing of coconut pollen. *Euphytica* **12**, 167–177.

Whitehead, R. A. (1965). The flowering of *Cocos nucifera* Linn. in Jamaica. *Trop. Agric. Trin.* **42**, 19–29.

Wiering, D. (1964). The use of insects for pollinating Brassica crops in small isolation cages. *Euphytica* **13**, 24–28.

Wigglesworth, V. B. (1964). "The Life of Insects." Weidenfeld and Nicolson, London.

Williams, R. D. (1925). Studies concerning the pollination, fertilization and breeding of red clover. Welsh Plant Breeding Station, Aberystwyth.

Williams, R. D. (1930). Some of the factors influencing yield and quality of red clover seeds. *Bull. Welsh Pl. Breed. Stn.* Ser H, **11**, 60–91.

Williams, R. D. and Evans, G. (1935). The efficiency of spatial isolation in maintaining the purity of red clover. *Welsh J. Agric.* **11**, 164–171.

Williams, R. R. (1966). Pollination studies in fruit trees. III. The effective pollination period for some apple and pear varieties. *Rep. agric. hort. Res. Stn. Univ. Bristol* 1965, 136–138.

Williams, R. R. (1969). Pollination Studies in fruit trees. VIII. Hand Pollination. *Rep. agric. hort. Res. Stn. Univ. Bristol* 1968, 142–145.

Williams, R. R. and Child, R. D. (1963). Some preliminary observations on the development of self- and cross-pollinated flowers of black currants. *Rep. agric. hort. Res. Stn. Univ. Bristol* 1962, 59–64.

Williams, R. R. and Smith, B. D. (1967). Pollination studies in fruit trees. VII. Observations on factors influencing the effective distance of pollinator trees in 1966. *Rep. agric. hort. Res. Stn. Univ. Bristol* 1966, 126–134.

Williams, W. (1951). Genetics of Incompatibility in Alsike Clover, *Trifolium hybridum*. *Heredity* **5**, 51–73.

Wilsie, C. P. (1949). Producing alfalfa and red clover seed in Iowa. *Agron. J.* **41**, 545–550.

Wilsie, C. P. and Gilbert, N. W. (1940). Preliminary results on seed setting in red clover strains. *J. Am. Soc. Agron.* **32**, 231–234.

Wilson, D. (1964). Cross pollination can be hard to achieve. *Grower* (Jan.), 159–160.

Wilson, G. F. (1926). Insect visitors to fruit blossoms. *J. R. hort. Soc.* **51**, 225–251.
Wilson, G. F. (1929). Pollination of hardy fruits: insect visitors to fruit blossoms. *Ann. appl. Biol.* **16**, 602–629.
Winkler, A. J. (1962). "General Viticulture." University of California Press, Berkley and Los Angeles.
Wójtowski, F. (1964). Z doświadczeń nad tworzeniem przenośnych kolonii porobnic (*Anthophora parietina* F.). *Roczn. wyż. Szk. roln. Poznan.* **19**, 177–184.
Wójtowski, F. (1965). Zastosowanie blonkówek pszczolowatych z rodzaju *Bombus* Latr. oraz *Anthophora* Latr. (Hymenoptera, Apoidea) do zapylania plantacji nasiennych roślin motylkowych. *Roczn. wyż. Szk. roln. Poznan.* **24**, 223–274.
Wolfenbarger, D. O. (1957). Insects in relation to fruit set of mangoes. *Proc. 17th ann. Mtg. Fla. Mango Forum* 11–13.
Wolfenbarger, D. O. (1962). Honey bees increase squash yields. *Res. Rep. Fla. agric. exp. Stn.* **7**, 15, 19.
Wood, G. W. (1961). The influence of honeybee pollination on fruit set of the lowbush blueberry. *Can. J. Pl. Sci.* **41**, 332–335.
Wood, G. W. (1965a). Note on the activity of native pollinators in relation to the bloom period of lowbush blueberry. *J. econ. Ent.* **58**, 777.
Wood, G. W. and Wood, F. A. (1963). Nectar production and its relation to fruit set in the lowbush blueberry. *Can. J. Bot.* **41**, 1675–1679.
Wood, G. W., Craig, D. L. and Hall, I. V. (1967). Highbush blueberry pollination in Nova Scotia. International Society for Horticultural Sci. Symposium I. Blueberry Culture in Europe. 163–168.
Woodhouse, E. J. and Taylor, C. S. (1913). The varieties of soybeans found in Bengal, Bihar and Orissa and their commercial possibilities. *Mem. Dep. Agric. India. bot. ser.* **5**, 103–175.
Woodrow, A. W. (1934). The effect of colony size on the flight rates of honeybees during the period of fruit blossom. *J. econ. Ent.* **27**, 624–629.
Woodrow, A. W. (1952). Effect of time of pollination by honeybees on red clover seed yields. *J. econ. Ent.* **45**, 517–519.
Woodworth, C. M. (1922). The extent of natural cross pollination in soy beans. *J. Am. Soc. Agron.* **14**, 278–283.
Wykes, G. R. (1952a). An investigation of the sugars present in the nectar of flowers of various species. *New Phytol.* **51**, 210–215.
Wykes, G. R. (1952b). The preferences of honeybees for solutions of various sugars which occur in nectar. *J. exp. Biol.* **29**, 511–518.
Wykes, G. R. (1953). The sugar content of nectars. *Biochem. J.* **53**, 294–296.
Yakovlev, A. S. (1959). Pollination of orchards by bees increases the productivity of the work of horticulturists. *Pchelovodstvo, Mosk.* **36**, 22–25.
Yamane, G. M. and Nakasone, K. Y. (1961). Pollination and fruit set studies of acerola *Malpighia glabra* L. in Hawaii. *Proc. Am. Soc. hort. Sci.* **78**, 141–148.
Yu, C. P. and Hsieh, L. C. (1937). A discussion of the methods of studying the percentage of natural crossing in cotton. *J. agric. Ass. China* **160**, 1–16.
Zadražil, K. (1955). Pestování semenné vojtesky. *Za Soc. Zemed.* **5**, 355–363.
Zakharov, G. A. (1958). Bees in the pollination of black currants and gooseberries. *Pchelovodstvo, Mosk.* **35**, 29–33.
Zakharov, G. A. (1960a). Role of supplementary pollination with pollen of a different species in increasing the yield of black currant. *Agrobiologiya* (3), 461–462.
Zakharov, G. A. (1960b). About the visitation of black currants by bees. *Pchelovodstvo, Mosk.* **37**, 39–40.

Zander, E. (1936). "Bienenkunde im Obstbau." Eugen Ulmer, Stuttgart.
Zander, E. (1951). Raps und Biene. *Z. Bienenforsch.* **1,** 135–140.
Zavrashvili, R. M. (1967). Influence of bees on the yield of citrus trees on the commercial plantations of Georgia. *XXI Int. Beekeep. Congr.* Summ. Paper 156, 159–163.
Zobel, M. P. and Davis, G. N. (1949). Effect of the number of fruit per plant on the yield and quality of cucumber seed. *Proc. Am. Soc. hort. Sci.* **53,** 355–358.

Author Index

Numbers in italics refer to the page on which the reference is listed.

A

Aalders, L. E., 340, 341, 342, *463*
Abbas, M., 143, *474*
Abdur Rahman, K., 145, 148, 150, 166, 167, 168, 179, 181, 267, 268, 269, 271, 331, 332, 333, 356, 431, 432, 433, 437, *467*
Adie, A., 70, *499*
Adlerz, W. C. 312, 313, *444*
Afzal, M., 48, 155, 162, 163, 165, *444*
Agati, G., 427, 428, *444*
Akamine, E. K., 293, 294, 295, *444*
Åkerberg, E., 82, 199, 201, 205, 206, 208, 211, 215, 223, 225, 226, 227, 228, 229, 230, 231, *444*, *475*, *500*
Akhter, A. R., 143, 148, 149, *444*
Aldrich, W. W., 368, *499*
Alex, A. H., 181, 182, 255, 275, 302, 303, 309, 325, 373, *444*, *503*
Allan, P., 435, *445*
Allard, H. A., 155, 156, 157, 162, *444*
Allen, H., 122, 124, 125, 244, 248, *492*
Allen, M. Delia, 61, 63, 65, 84, *444*, *445*, *469*
Allman, S. L., 201, 202, *455*
Alpatov, V. V., 226, *445*
Alvarez, E., 352, *489*
Amaral, E., 319, *445*
Ammen, H. J., 370, *445*
Amos, J. M., 124, 236, 258, 278, 279, 281, 282, 285, *445*, *480*, *503*
Anandan, M., 438, *445*
Anderson, E. J., 83, 221, 227, 268, *445*
Anderson, S. R., 272, *446*
Angelo, E., 370, *445*
Ankinovič, G. B., 303, *445*
Antles L. C., 83, *445*
Antsiferova, T. A., 210, *445*

Antunes, H., 319, *483*
Archimowitsch, A., 360, 361, 362, *445*
Arizona University, 305, 310, *445*
Armstrong, J. M., 199, 222, 226, *445*
Armstrong, W. D., 192, *445*
Arrand, J. C., 409, *453*
Artschwager, E., 359, 360, *445*
Atkinson, W. T., 121, 122, 124, *445*
Atwood, C. E., 387, 388, 402, *445*
Atwood, S. S., 232, *445*
Auchter, E. C., 191, 384, 386, 391, *445*
Avetisyan, G., 157, 160, 161, *446*

B

Baculinschi, H., 328, 424, *446*, *492*
Bader, K. L., 272, *446*
Bailey, J. S., 338, 342, 343, *446*, *493*
Bailey, M. E., 233, *446*
Bailey, N. T. J., 243, *461*
Bailey, S. F., 127, *470*
Bain, H. F., 346, *456*
Baker, H. G., 37, 432, *446*
Balcar, J., 41, 76, 205, *464*, *493*
Baldini, E., 282, *446*
Balfour-Browne, F., 102, *446*
Balls, W. L., 46, 162, 163, *446*
Banks, C. J., 302, *446*
Barbier, E., 185, *446*
Barrons, K. C., 258, *446*
Barskii, Ya. S., 192, *446*
Basak, K. C., 432, 433, *473*
Batchelor, L. D., 185, 186, *503*
Bateman, A. J., 37, 46, 49, 149, *446*
Batjer, L. P., 410, *446*
Battaglini, M. B., 299, *446*
Baver, L. D., 339, *492*
Bawolski, S., 230, *446*
Beach, S. A., 190, 191, *446*

Beard, D. F., 223, 229, 231, *446*
Beattie, J. H., 267, *473*
Beckham, C. M., 236, *446*
Beckham, W., 342, 343, *447*
Beckwith, C. S., 337, *447*
Beghtel, F. E., 317, *447*
Behlen, F., 390, *447*
Beling, Ingeborg, 18, *447*
Belozerova, E. I., 138, 139, 140, 142, *447*
Benoit, P., 218, 219, *447*
Bergman, H. F., 347, *447*
Berkel, N. van, 302, *447*
Bethell, R. S., 410, *463*
Betts, A. D., 34, 35, 36, 400, *447*
Beutler, Ruth, 16, 18, 391, *447*
Bewley, W. F., 190, 298, 306, 421, *447*
Bhambure, C. S., 298, 299, 304, 312, 313, 333, *447*
Bieberdorf, G. A., 211, 252, *447*
Bigger, M., 197, *447*
Billes, D. J., 172, 173, 174, 175, 176, *447*
Bingefors, S., 220, 227, 228, 231, *447*, *460*
Bird, J. N., 231, *447*
Bishop, R. K., 19, 31, 63, 65, 184, *499*, *501*
Bitkolov, R. Sh., 328, *448*
Blackwall, F. L. C., 259, 261, *448*
Blagoveschchenskaya, N. N., 207, *448*
Blake, G. H., 236, 237, *448*
Blasberg, C. H., 412, *448*
Blinov, N. M., 83, *448*
Boch, R., 21, 40, 41, 74, 76, 331, 332, *448*, *467*, *475*, *493*
Bogoyavlenskii, S. G., 19, 84, 212, 276, *448*
Bohart, G. E., 4, 19, 21, 75, 103, 108, 111, 112, 114, 117, 120, 126, 202, 205, 206, 207, 209, 210, 212, 213, 214, 215, 218, 225, 228, 271, 315, 316, 354, 387, 389, *448*, *449*, *465*, *475*, *483*, *485*, *486*
Bohn, G. W., 309, 310, 311, 335, *449*, *499*, *503*
Böjtös, Z., 201, *475*
Bolhuis, G. G., 267, *449*
Bolton, J. L., 102, 122, 199, 200, 201, 202, 209, 210, *449*, *484*, *485*, *498*
Bommer, D., 226, 227, *497*
Bond, D. A., 249, 250, *449*
Bonnier, G., 42, *449*
Booth, N. O., 190, *449*

Borthwick, H. A., 129, 314, *449*
Bourne, A. I., 338, 342, 343, *493*
Boynton, D., 411, *495*
Bozhinov, M., 161, *488*
Braak, J. P., 83, 419, *473*
Bradford, F. C., 381, 403, 410, *459*
Bradner, N. R., 200, *447*
Braines, L., 82, *447*
Brande, J., 218, 219, *447*
Brandenburg, W., 245, 251, *449*
Brar, J. S., 306, *482*
Braun, D. R., 83, 215, 224, 225, *478*
Braun, E., 41, 76, 222, 227, 230, *449*
Bretherick, O., 19, 389, *499*
Brewer, J. W., 339, 340, 343, *449*
Brian, A. D., 25, 26, 33, 35, 225, 388, *449*
Bringhurst, R. S., 368, *449*, *475*
Brittain, W. H., 34, 38, 77, 384, 386, 387, 388, 391, 392, 395, 396, 398, 399, 402, 404, 406, 407, 408, 410, 412, *450*
Brown, A. G., 387, 391, 395, 402, 404, 407, *450*
Brown, H. B., 46, 153, 162, 163, 165, *450*
Brown, R. T., 370, *450*
Buffet, M., 164, *450*
Bulatovic, S., 386, *450*
Bullock, R. M., 410, 411, 412, *450*
Bunting, E. S., 137, *450*
Burchill, R. T., 384, *450*
Burk, E. F., 350, *450*
Burkart, A., 207, *450*
Burminstrov, A. N., 328, *450*
Burns, R. E., 273, *475*
Burrel, A. B., 413, *450*, *471*
Butcher, F. G., 434, *450*
Butenant, A., 59, *470*
Butler, C. G., 7, 15, 18, 29, 39, 40, 41, 42, 43, 46, 58, 61, 74, 89, 95, 119, 121, 122, 145, 207, 220, 222, 273, 284, 331, 332, 364, 388, 391, 400, 407, 408, 424, *450*, *451*, *459*
Butler, G. D., 156, 207, 331, *451*, *475*, *476*
Buzzard, 42, *451*

C

Calam, D. H., 74, *451*
Calder, R. A., 199, 202, *463*

AUTHOR INDEX 509

Cale, G. H., 41, 63, 386, 388, *451, 467*
Callow, R. J., 56, *451*
Canadian Department of Agriculture 142, 302, *451*
Carlson, J. W., 199, *451*
Carmin, J., 266, *451*
Caron, D. M., 427, 428, *474*
Carr, R. V., 8, *476*
Cartter, J. L., 270, *481*
Carvalho, A., 319, 320, *451, 483*
Cejtlin, M. G., 190, *451*
Chambers, V. H., 24, *452*
Chandler, S. E., 333, *452*
Chandrasekaran, N. R., 267, *496*
Chandy, K. C., 159, 441, 442, *478*
Chapman, G. P., 290, 441, 442, *452*
Chauvin, R., 61, *452*
Child, R. D., 283, *504*
Chittenden, F. J., 384, *452*
Christensen, S. A., 351, *452*
Christopher, E. P., 192, *452*
Cîrnu, I., 326, 331, *452*
Ciaassen, C. E., 332, *452*
Clark, C. F., 357, 358, *498*
Clark, K. K., 386, 409, *452*
Clark, O. I., 289, *452*
Clarke, J. H., 338, *504*
Clements, F. E., 15, 34, *452*
Clinch, P. G., 219, 221, 222, 223, 387, *484*
Cobley, L. S., 149, 152, 193, 277, 297, 350, 369, 431, *452*
Cochran, H. L., 350, 354, *452, 496*
Coe, H. S., 216, 221, 229, 230, 274, 275, *452, 503*
Coggshall, W. L., 72, 80, 397, 408, 409, *452*
Coit, J. E., 186, *452*
Colby, A. S., 285, *452*
Collins, J. L., 441, *452*
Condit, I. J., 374, 378, *452*
Congdon, N. B., 404, *452*
Constable, E. F., 121, 122, 124, *445*
Cook, O. F., 166, *452*
Cooke, D. A., 275, *462*
Cooper, B. A., 122, 124, 125, 244, 248, *492*
Copaitici, M., 275, *452*
Cope, F. W., 170, 171, 173, *453*
Copeland, E. B., 442, *453*
Corley, W. L., 311, *468*

Corner, J., 409, *453*
Correll, D. S., 440, *453*
Costa, A. S., 320, *473*
Cottrell-Dormer, W., 351, *453*
Courtois, G., 9, *453*
Couston, R., 423, *453*
Coville, F. V., 336, 337, 338, *453*
Cox, I. E., 293, 295, *453*
Craig, D. L., 342, *505*
Crane, M. B., 46, 47, 149, 395, 407, 422, *453*
Cross, C. E., 345, 347, *453*
Cross, E. A., 112, *448*
Cruchet, P., 257, *453*
Crum, C. P., 221, 239, *453*
Cumakov, V., 83, 328, *453*
Cunningham, D. G., 66, 409, *489*
Currence, T. M., 353, 354, *453*
Cuypers, J., 124, 125, 126, *453*

D

Dade, H. A., 17, 52, *453*
D'Aevilar, I., 126, *453*
Damisch, W., 226, 230, *453*
Dann, B., 274, *454*
Darchen, P., 61, *452*
Darrow, G. M., 339, 342, 417, 422, *454, 479*
Darwin, C., 15, 34, 42, 49, 145, 146, 149, 215, 232, 236, 245, 249, 256, 258, 259, 262, 271, 274, 356, 364, 366, 391, 428, 430, *454*
Dave, B. E., 268, *478*
Davey, V. M., 135, *454*
Davis, G. N., 298, 304, 309, *449, 503, 505*
Davis, J. H., 271, *454*
Davydova, N. S., 363, 364, *454*
Degman, E., 412, 414, 416, *469*
Della Giustina, W., 126, *453*
Demianowicz, Zofia, 139, 273, 364, *454*
Demuth, G. S., 364, *487*
Dennis, B. A., 223, 228, 229, *454*
Dessart, P., 178, *454*
Detjen, L. R., 191, *489*
Dickson, G. H., 389, 398, 403, 410, *454*
Dillman, A. C., 179, 180, 181, *454*
Dirks, C. O., 397, *454*
Dobson, R. C., 339, 340, 343, *449*
Dočkal, J., 41, 76, 205, *464, 493*
Doehlert, C. A., 346, 347, *456*

Dorr, J., 339, 342, 343, *454*
Doull, K. M., 203, 208, 214, *454*
Downey, R. K., 142
Drake, C. J., 213, *454*
Drayner, Jean M., 243, 245, *454*
Duggan, J. B., 336, 340, *486*
Dunavan, D., 235, 236, *454*
Duncan, E. N., 162, 164, 165, *487*, *494*
Dunham, W. E., 70, 221, 222, 223, 224, 229, 231, 239, 240, 388, 397, *446*, *455*
Dunne, T. C., 190, *455*
Dunning, J. W., 218, *455*
Durham, G. B., 299, *455*
Durrant, A. J., 18, 122, 402, *459*
Dwyer, R. E., 201, 202, *455*
Dyce, E. J., 387, 396, *455*

E

Eaton, F. M., 166, *455*
Eaton, G. W., 286, 339, 346, 383, 417, *455*
Eckert, J. E., 31, 67, 68, 69, 331, 332, 429, *455*
Eden, T., 431, *455*
Edgecombe, S. W., 303, 304, *455*
Edwards, A. J., 249, *455*
Edwardson, J. R., 273 *475*
Einset O. 190, 191, *455*
Elagin, I., 365, *456*
El Banby, M.A., 24, 184, *464*
Elders, A. T., 274, *456*
Elliot, C. R., 240, *484*
Emsweller, S. L., 127, 128, 129, 314, 426, *449*, *469*, *470*
Enderud, J., 374, *452*
Entwistle, Helen, 170, 176, 177, *456*
Eriksen, K. M., 66, 226, 231, *497*
Ermakova, I. A., 144, 145, *456*
Es, J. van, 61, *501*
Eskilsson, L., 220, 227, 228, 231, *447*, *460*
Evans, G., 47, *504*
Evans, R. J., 199, *451*
Eves, J., 110, *469*
Ewert, R., 137, 138, 385, 390, *456*

F

Fabre, J. H., 102, *456*
Faegri, K., 37, *456*

Fahn, A., 184, *456*
Farrar, C. L., 19, 65, 68, 69, 84, 121, 346, 398, *456*, *492*, *498*
Faulkner, G. J., 129, 130, *456*
Fechner, E., 138, *456*
Ferwerda, F. P., 318, 320, 321, *456*
Fieger, E. A., 233, *446*
Filmer, R. S., 84, 87, 339, 342, 343, 345, 346, 347, 395, 398, 407, 408, 409, *456*, *457*, *478*
Finkner, M. D., 155, 164, 165, *457*, *497*
Fischer, R. L., 206, 239, 240, *457*
Fisher, E. G., 370, *450*
Fletcher, S. W., 350, 384, *457*
Folsam, J. W., 218, *457*
Fomina, K. Ya., 81, *457*
Fontanilla-Barroga, S., 178, *457*
Foot Guimarães, R., 289, *457*
Forbes, I., 273, *475*
Ford, R. N., 121, 236, 237, *502*
Forshey, C. G., 395, *457*
Forster, I. W., 121, 124, 199, 200, 203, 207, 219, 221, 222, 223, 229, 233, 234, *457*, *484*
Foster, J. R., 200, 202, *485*
Foster, R. E., 307, 311, *457*, *477*
Frakes, R. V., 200, *449*
Françon, J., 41, 76, *457*
Franklin, H. J., 347, *457*
Franklin, W. W., 206, 209, 211, *457*
Free, J. B., 7, 9, 10, 18, 23, 24, 25, 26, 29, 30, 31, 32, 34, 35, 38, 39, 40, 42, 45, 46, 55, 57, 58, 59, 61, 62, 63, 64, 65, 66, 70, 71, 72, 73, 74, 77, 78, 79, 82, 83, 85, 86, 87, 89, 95, 117, 119, 121, 122, 124, 125, 126, 131, 138, 139, 140, 142, 148, 149, 207, 217, 220, 222, 223, 245, 246, 247, 248, 249, 250, 258, 259, 261, 262, 263, 264, 283, 284, 285, 325, 326, 328, 329, 330, 384, 385, 387, 388, 389, 392, 394, 396, 397, 398, 399, 400, 401, 402, 403, 404, 405, 406, 408, 409, 410, 411, 412, 419, 420, 421, 425, *451*, *457*, *458*, *459*, *460*
Free, Nancy W., 9, 72, *460*
Fresnaye, J., 71, *457*
Frey-Wyssling, A., 16, *460*
Frick, K. E., 114, *460*
Fridén, F., 220, 228, *441*, *460*
Frisch, K. von, 15, 17, 18, 26, 28, 29, 30, 39, 74, 81, 82, *460*

Frison, T. H., 96, 97, *460*
Fronk, W. D., 114, 299, 313, *460*
Frost, S. W., 102, *460*
Fryxell, P., 163, *460*
Fujita, K., 185, *460*
Fujita, M., 139, *460*
Furgala, B., 17, 36, 204, 223, 325, 326, 327, 330, *461*, *486*
Fye, R. E., 96, *461*
Fyfe, J. L., 243, 250, *449*, *461*

G

Gaag, H. C. van der, 131, *461*
Gagnard, J. M., 395, 397, *461*
Garber, R. J., 364, *461*
Gardner, E. J., 333, 334, *461*
Gardner, V. R., 381, 403, 410, *461*
Gary, N. E., 61, *461*
Gates, B. N., 385, *461*
Gibson, D. A., 41, 76, 222, 224, 227, 230, *449*
Gilbert, N. W., 225, 226, *497*, *504*
Gill, N. T., 136, 323, *461*
Gillard, A., 218, 219, *447*
Giltay, E., 42, *461*
Girardeau, J. H., 236, 266, *446*, *461*
Girolami, G., 293, 294, 295, *444*
Gladwin, F. E., 190, 191, 192, *461*
Glasgow, S. K., 290, *452*
Glendinning, D. R., 172, *461*
Glowska, Z., 407, *461*
Glowska-Konopacka, S., 66, 77, 79, 80, *475*
Glukhov, M. M., 327, *461*
Glushkov, N. M., 82, 154, 158, 161, 282, 286, *461*, *462*
Glynne-Jones, G. D., 138, 146, 147, *469*
Gochnauer, T. A., 17, *461*
Goetze, G., 37, 221, 222, *462*
Goff, C. G., 312, 313, *462*
Gooderham, C. B., 65, 239, *462*
Goplen, B. P., 122, 123, 124, 275, *462*, *484*
Gordienko, V. A., 270, *462*
Gould, H. P., 381, *462*
Grant, V., 37, *462*
Green, H. B., 233, 234, *462*
Greenway, P. J., 435, *462*
Gregg, O. I., 350, *457*

Gregory, L. E., 437, *478*
Griffin, H. H., 305, *462*
Griggs, W. H., 382, 383, 390, 397, 403, 409, 410, 411, 414, 416, *462*, *463*, *501*
Grimball, E. L., 258, *503*
Groom, P., 322, 324, 430, *463*
Groot, A. P. de, 61, *463*
Grout, R. A., 154, 157, 158, 204, 397, 398, 407, *463*
Grubb, N. H., 122, *463*
Gubin, A. F., 82, 159, 179, 181, 222, 223, 226, 229, 303, *463*
Guppy, J., 41, 76, 222, 224, 227, 230, *449*

H

Haarer, A. E., 320, *463*
Haas, H., 98, 99, 223, 228, 229, *454*, *463*, *466*
Haddock, J. L., 212, *486*
Hadfield, J. W., 199, 202, *463*
Hadfield, W. V., 219, 222, 229, *457*
Hagberg, A., 273, *463*
Hahlin, M., 229, 231, *444*
Haigh, J. C., 121, 122, 136, 148, *451*, *463*
Hale, C. R., 190, *463*
Hall, C. J. J. van, 172, 320, *500*
Hall, I. V., 340, 341, 342, *463*, *505*
Hambleton, J. I., 23, 407, 408, *463*
Hamilton, R., 70, *463*
Hammer, O., 69, 139, 140, 144, 230, 231, *463*, *464*
Hammons, R. O., 266, 267, *464*, *475*
Hansen, P., 218, *464*
Hanson, W. D., 273, *502*
Hansson, A., 83, *464*
Haragsim, O., 41, 76, 205, *464*, *493*
Haragsimová-Neprašová, L., 140, 145, 207, 276, 424, *464*
Hardy, M. B., 422, *464*
Hare, Q. A., 207, *464*
Harland, S. C., 172, 174, *464*
Harle, A., 138, *464*
Harris, B. J., 432, *446*
Harris, W. B., 382, *464*
Harrison, K. A., 384, *466*
Hartman, F. O., 383, *464*
Hartwig, E. E., 275, 276, *464*
Haskell, G., 136, 277, 315, 317, *464*
Hasler, A., 140, *464*

Hassan, M. V., 197, *489*
Hassanein, M. H., 24, 181, 182, 184, 185, 237, 239, *464*
Hasselrot, T. B., 98, 99, *464*
Hatton, R. G., 278, 279, 281, 282, 285, *503*
Hawkins, R. P., 125, 218, 219, 223, 225, 227, 249, 250, *449*, *464*, *465*
Haws, B. A., 275, *465*
Hawthorn, L. R., 258, 304, 315, 317, 354, 362, 366, 429, *465*
Haydak, M. H., 56, 85, 124, *465*
Haynie, J. D., 221, 236, 275, *471*
Hector, J. M., 257, 364, *465*
Heide, F. F. R., 266, *465*
Heinicke, A. J., 384, 386, 404, 410, *477*
Heinrichs, D. H., 127, *465*
Hendrickson, A. H., 7, 386, *465*
Henry, A. W., 179, 181, *465*
Henslow, G., 262, *465*
Heran, H., 61, *465*
Hilkenbäumer, F., 283, *465*
Hills, K. L., 224, *466*
Hill-Cottingham, D. G., 383, *465*
Hirashima, Y., 116, *466*
Hitchings, J. M., 125, 126, *466*
Hobbs, G. A., 96, 100, 108, 110, 111, 201, 203, 205, 206, 207, 208, 209, 230, *466*
Hockey, J. F., 384, *466*
Hodges, Dorothy, 19, 20, *466*
Holdaway, F. G., 17, 36, 204, 223, 275, *461*, *465*, *486*
Holland, J. H., 196, 434, 439, 440, 443, *482*
Holm, S. N., 33, 95, 98, 99, 127, 209, 225, 229, 230, *466*
Holowell, E. A., 216, *487*
Hooker, H. D., 381, 403, 410, *459*
Hooper, C. H., 284, 286, 384, 386, 387, 399, 421, 423, *467*
Hootman, H. D., 41, 386, 388, 410, *467*, *473*
Hopkins, C. Y., 21, *467*
Hopkins, L., 310, *487*
Horber, E., 98, 99, *467*
Horn, C. W. van, 186, 188, 335, *499*, *500*
Horticultural Education Association 381, 382, 397, 417, *467*
Howard, A., 145, 148, 150, 166, 167, 168, 179, 181, 267, 268, 269, 271, 331, 332, 333, 356, 431, 432, 433, 437, *467*

Howard, Gabrielle L.C., 145, 148, 150, 166, 167, 168, 179, 181, 267, 268, 269, 271, 331, 332, 333, 356, 431, 432, 433, 437, *467*
Howlett, F. S., 383, 384, 385, 386, 388, 398, *464*, *467*
Hsieh, L. C., 162, *505*
Hua, H., 243, *467*
Hudson, R., 87, *467*
Hughes, Hilary M., 121, 190, 282, 283, 286, 287, 417, 419, *467*, *468*
Hughes, J. H., 201, *468*
Hull, F. H., 267, *498*
Hume, H. H., 183, *468*
Huokuna, E., 228, 230, 231, *500*
Hurd, P. D., 37, 300, *446*, *468*, *480*
Hurst, R. L., 48, *486*
Hurt, E. F., 326, *468*
Hutchinson, J., 437, *468*
Hutson, R., 70, 77, 346, 347, 381, 386, 387, 388, 396, 398, 406, *468*
Hutt, R., 341, *469*
Huyskes, J. A., 429, *468*
Hyams, E., 417, 421, *468*

I

Ibrahim, M. M., 184, 185, *464*
Ibrahim, S. H., 23, 121, 156, 157, 160, 184, 185, 238, 239, 240, 245, 246, 247, 250, 288, 387, *468*, *501*, *502*
Ishizuka, S., 412, *491*
Ismailova, M. K., 298, *501*
Istomina-Tsvetkova, K. P., 74, 82, 449, *468*
Ivanoff, S. S., 310, *468*
Iwakiri, B. T., 383, 403, 410, 411, 414, 416, *462*, *463*
Izmaĭlova, A. V., 201, *468*

J

Jabloński, B., 139, 150, 273, *454*, *468*
Jagger, I. C., 298, *503*
James, E., 311, *468*
Jamieson, C. A., 83, 215, 222, 225, 226, 229, *445*, *468*, *478*
Jany, E., 262, 263, *468*
Jay, S. C., 9, 55, 71, 72, 98, *460*, *468*, *469*, *487*
Jaycox, E. R., 64, 415, *469*

Jeffery, G. L., 121, 124, 233, 234, *484*
Jeffree, E. P., 39, 41, 42, 43, 63, 65, 84, 397, *445*, *451*, *469*
Jenkins, J. M., 304, 353, 354, *453*, *469*
Jenkinson, J. G., 138, 146, 147, *469*
Jevans, A. W., 21, *467*
Johannson, T. S. K., 165, *469*
Johansen, B. R., 200, *469*
Johansen, C. A., 110, 227, 231, 347, 390, 395, 411, 412, 414, 415, 416, *469*
Johnson, A. P., 270, *469*
Johnson, D. L., 29, *503*
Johnson, W. C., 236, *469*
Johnston, N. C., 56, *451*
Johnston, S., 338, 423, *469*, *480*
Jones, D. F., 350, 352, *469*
Jones, G. A., 172, *469*
Jones, H. A., 127, 128, 129, 149, 257, 300, 312, 315, 335, 358, 426, 427, 428, 429, *469*, *470*
Jones, L. G., 236, *480*
Jones, L. T., 190, *463*
Jones, M. D., 167, 168, *470*, *498*
Jorgensen, C. O., 217, *470*
Julén, G., 225, 228, *470*

K

Kadam, B. S., 332, *470*
Kakizaki, Y., 146, 149, *470*
Kalashnikov, V. M., 333, *470*
Kalmus, H., 39, 41, 42, 43, *451*
Karlson, P., 59, *470*
Karmo, E. A., 72, 73, 74, 75, 86, 341, 342, 343, 388, 390, 397, 398, 402, 406, 408, 411, 414, 415, *470*
Kasai, T., 149, *470*
Kashkovskii, V. G., 365, *470*
Kaziev, I. P., 303, 307, *470*
Kaziev, T. I., 154, 157, 158, 159, 166, *471*
Kearney, T. H., 152, 156, 158, 162, 165, *471*
Kelty, R. H., 398, *471*
Kennard, W. C., 194, 195, *496*
Kennerley, A. B., 309, *471*
Kerner, A., 37, *471*
Kerr, E. A., 351, *471*
Kerr, W. E., 289, *457*
Khan, H., 48, 155, 162, 163, 165, *446*
Khan, K. S. A. R., 433, 434, *471*

Kiechle, H., 62, *471*
Kikuchi, T., 45, *471*
Killinger, G. B., 221, 236, 275, *471*
Killion, C. E., 123, *471*
King, C. M., 215, 217, 300, *484*
King, G. E., 413, *450*, *471*
Kinsman, G. B., 344, *471*
Kirk, E. E., 274, *471*
Kirk, L. E., 201, *471*
Kitamura, T., 116, *478*
Klümbt, H. D., 283, *465*, *471*
Kleber, Elizabeth, 18, *471*
Kloavev, A. M., 331, *471*
Kloet, G. S., 102, *471*
Knaap, W. P. van der, 171, 176, 177, *471*
Knapp, H. B., 191, 384, *445*
Knee, W. J., 25, *472*
Knight, R., 170, *472*
Knoll, F., 15, 374, *472*
Knowlton, G. F., 114, *448*
Knuth, P., 145, 149, 241, 247, 249, 256, 259, 262, 268, 271, 276, 334, 357, 363, 417, 430, *472*
Knysh, A. N., 304, *472*
Kobel, F., 384, *472*
Koerber, T. W., 103, *479*
Konstantinovic, B., 386, *450*
Koot, Y. van, 302, *472*
Kopelkievskii, G. V., 84, 275, 364, 365, *472*
Koperzinskii, V. V., 202, *472*
Korolev, A. M., 289, *472*
Kostylev, A. D., 208, *472*
Kottur, G. L., 162, *472*
Koutenský, J., 138, 139, 142, *472*
Kovalev, A. A., 190, *472*
Kovarskaya, A. L., 19, *448*
Kozin, R. B., 181, 273, *472*
Kraai, A., 120, 122, 123, 125, *472*
Kratky, E., 56, *472*
Kremer, J. C., 149, 298, 301, 305, 312, 383, 408, 410, 411, 414, *472*, *472*, *439*
Krezdorn, A. H., 187, 188, *473*, *490*
Kribs, L., 351, *471*
Kristofferson, K. B., 258, 259, *473*
Kroll, U., 333, *473*
Krombein, K. V., 266, *464*
Kronenberg, H. G., 83, 419, *473*
Kropáčová, Sylvie, 203, 204, 206, 211, 212, 276, 277, *473*
Krug, C. A., 320, *451*, *473*

L

Kuehl, R. O., 8, *476*
Kugler, H., 15, *473*
Kuliev, A. M., 158, *473*
Kulinčevic, J., 275, *473*
Kundu, B. C., 432, 433, *473*
Kurennoi, N. M., 327, 330, 331, 397, 401, *473*
Kurtz, E. B., 298, *482*
Kushman, L. J., 267, *473*
Kushnir, L. G., 325, 328, *473*

L

Laere, O. van, 221, 226, 381, *474*
Lagassé, F. S., 191, *474*
Laitová, Libuše, 212, *473*
Lambeth, V. N., 258, *474*
Langham, D. G., 437, *474*
Langridge, D. F., 233, *474*
Lansdell, J., 306, *474*, *491*
Lapins, K. O., 409, *453*
Larsen, P., 389, *474*
Lathrop, F. H., 344, *474*
Latif, A., 143, 237, *474*
Latimer, L. P., 143, *474*
Lavie, P., 85, *484*
Law, J. H., 74, *503*
Lawrence, W. J. C., 422, *453*
Lecomte, J., 9, 11, 58, 68, 125, 126, 214, 234, 327, 328, 368, 369, *453*, *474*
Ledeboer, M., 282, *474*
Lederhouse, R. C., 427, 428, *474*
Lee, W. R., 9, 76, 340, 342, *474*, *475*
Lehmann, U., 226, *475*
Lehnert, T., 73, 348, *494*
Lepage, M., 21, *475*
Leppik, E. E., 25, *475*
Leshchev, V., 364, *475*
Lesik, F. L., 328, *475*
Lesinš, K., 200, 201, 202, 205, 206, 208, 211, *444*, *475*
Lesley, J. W., 352, 368, *475*
Leuck, D. B., 266, 267, 273, *461*, *464*, *475*
Levchenko, I. A., 75, *475*
Levin, M. D., 8, 9, 19, 21, 36, 45, 48, 66, 75, 76, 77, 79, 80, 117, 121, 204, 205, 212, 307, 311, 315, 331, 332, *449*, *451*, *457*, *465*, *475*, *476*, *477*, *486*, *491*
Lewis, C. I., 384, 391, *476*
Lewis, T., 81, 387, 400, *476*
Lieberman, F. V., 121, *486*
Lilly, C. E., 201, 203, 205, 206, *466*
Lindauer, M., 26, 29, 30, 40, 56, 57, 58, 61, 62, 70, 85, *476*
Lindhard, E., 119, *476*
Linsley, E. G., 8, 9, 24, 207, 208, 209, 299, 300, *466*, *476*, *480*, *496*
Liverman, J. L., 298, *482*
Ljubimov, A. A., 303, *445*
Loden, H. D., 166, *476*
Loew, E., 24, *476*
Loewel, E. L., 386, *476*
Löken, A., 395, 396, 407, *476*
Long, F. L., 15, 34, *452*
Loper, G. M., 84, *474*
Lötter, J. de V., 387, 410, 411, 412, 414, 415, *476*
Louveaux, J., 19, 21, 31, 32, 139, 389, *476*
Lovell, H. B., 273, *477*
Luce, W. A., 388, *477*
Luckwill, L. C., 279, 280, 281, *498*
Lukoschus, F., 24, 35, 49, *477*
Lundie, A. E., 23, *477*
Luttso, V. P., 181, 325, 328, *477*
Lynch, S. J., 187, 188, *481*

M

McAllister, D. R., *485*
McCollum, J. P., 301, *477*
McCulloch, J. W., 396, *477*
MacDaniels, L. H., 42, 384, 386, 399, 403, 404, 406, 410, *477*
McDermott, J. J., 124, 125, *491*
McDonald, J. H., 318, 319, 320, *477*
McDonald, J. L., 9, *477*
Macfie, J. W. S., 175, *477*
McGregor, S. E., 70, 71, 79, 82, 85, 121, 124, 154, 156, 158, 159, 150, 161, 191, 192, 305, 307, 308, 309, 331, 332, *451*, *477*, *491*, *493*, *499*
McIlroy, R. J., 321, *477*
Mackensen, O., 11, 87, 88, *477*, *483*
Mackerras, M. J., 129, *478*
Mackie, W. W., 257, 258, *478*
McMahon, H. A., 200, 202, 206, 207, 211, *478*, *485*
Macself, A. J., 306, *474*
McSwain, J. W., 9, 24, 207, 209, *476*, *496*
MacVicar, R. M., 41, 76, 83, 215, 222, 224, 225, 227, 230, *449*, *478*

Maeta, Y., 116, *478*
Magruder, R., 258, *478*
Mahadevan, V., 159, 441, 442, *478*
Mahoney, C. H., 162, *498*
Mahta, D. N., 268, *478*
Maksymiuk I., 140, *478*
Mallik, P. C., 194, *478*
Mann, L. K., 128, 129, 305, 306, 307, 308, 309, 312, 428, *449*, *469*, *478*
Manning, A., 15, 35, *478*
Manzoor ul Haq, 237, *474*
Maranhao, Z. C., 195, *494*
Markov, I., 303, *478*
Marlatt, R. B., 310, *487*
Marshall, J., 170, *478*
Marshall, P. R., 244, *502*
Martens, N., 221, 226, *474*
Martin, E. C., 339, 342, 343, *454*
Martin, F. W., 437, *478*
Martin, G. C., 410, *446*
Martin, J. N., 221, 225, *478*
Martin, J. R., 215, 274, 275, *452*
Martinovs, A., 31, *492*
Mart'yanova, A. I., 272, 273, *493*
Marucci, P. E., 74, 339, 342, 343, 344, 345, 346, 347, 348, *457*, *478*
Massee, A. M., 388, *478*
Massey, J. H., 311, *468*
Materikina, E. I., 365, *478*
Mather, K., 15, 36, 37, 46, 47, 149, *453*, *478*
Maurizio, Anna, 16, 19, 21, 31, 32, 34, 140, 215, 218, 227, 228, 389, *460*, *464*, *478*, *479*
Meade, R. M., 158, 166, *479*
Meader, E. M., 339, 342, *479*
Medler, J. T., 96, 103, 117, *461*, *479*
Meier, F. C., 360, *479*
Mel'nichenko, A. N., 365, *479*
Menke, H. F., 72, 80, 112, 114, 115, 206, 207, 209, 210, 386, 387, 388, 389, 390, 395, 396, 397, 398, 400, 402, 403, 407, 414, *479*, *503*
Menon, K. K., 438, *479*
Menon, K. V. P., 442, *479*
Merrill, J. H., 85, *479*
Merrill, T. A., 336, 337, 338, 343, *479*, *480*
Meyer, W., 62, *480*
Meyerhoff, G., 139, 140, *480*
Michelbacher, A. E., 300, *480*

Mikitenko, A. S., 362, *480*
Miller, J. D., 124, *480*
Miller, M. D., 236, *480*
Milum, V. G., 270, *480*
Minderhoud, A., 36, 42, 47, 83, 120, 123, 399, *480*
Minessy, F. A., 186, *480*
Ministry of Agriculture, Fisheries & Food, 382, 397, *480*
Minkov, S. G., 153, 157, 159, 166, *480*
Mitchener, A. V., 328, *480*
Mittler, T. E., 401, *480*
Mlyniec, W., 252, 253, *480*
Móczár, L., 207, 208, 211, *480*
Moeller, F. E., 25, 65, *472*, *480*, *481*
Mohammed, A., 142, 144, *481*
Mommers, J., 76, 385, 388, 391, 398, 399, 401, 419, *481*
Montgomery B. E., 207, 218, 233, 273, 275, 307, 328, *481*
Moore, R. C., 411, *483*
Moradeshagi, M. J., 202, 210, *449*
Morley, F. H. W., 240, *481*
Morris, L. E., 371, 372, *481*
Morris, O. M., 388, *477*
Morrow, E. B., 338, 339, 340, *481*
Morse, R. A., 271, 273, 427, 428, *474*, *481*
Morse, W. J., 270, *481*, *487*
Moskovljevíc, V. Z., 57, *481*
Mostowska, I., 140, *502*
Mound, L. A., 153, 154, *481*
Mukherjee, S. K., 193, 194, 195, *481*
Müller, H., 42, 357, *481*
Murakami, H., 412, *491*
Murneek, A. E., 121, 381, 383, 388, 398, *481*
Murthy, B. S., 355, *481*
Murthy, N. S. R., 355, *481*
Musgrave, A. J., 9, *481*
Mustard, Margaret J., 187, 188, 443, 434, *481*
Muzik, T. J., 372, *482*

N

Naghski, J., 365, *482*
Naik, K. C., 194, *482*
Nakasone, H. Y., 433, *505*
Nandpuri, K. S., 306, *482*
Narayanan, E. S., 237, 239, 317, *482*

Nation, J. L., 21, *490*
National Agricultural Advisory Service 122, 124, 251, *482*
Needham, P. H., 249, *460*
Neiswander, R. B., 124, 351, *482*
Nekrasov, V. U., 71, *482*
Nelson, F. C., 56, *482*
Nelson, J. W., 339, 340, 343, *449*
Nelson, R. O., 187, 188, 433, 434, *481*
Nemirovich-Danchenko, E. N., 303, *482*
Nettles, W. C., 236, *469*
Neumann, U., 282, 283, *482*
Nevkryta, A. N., 41, 83, 299, 399, *482*
Newton, Dorothy E., 34, 38, 402, *450*
Nicholls, H. A., 196, 434, 439, 440, 443, *482*
Nielsen, H. M., 199, 211, *482*
Nilne, R. L., 191, 192, *493*
Nilsson, E., 149, *482*
Nishida, T., 118, 293, 294, 295, 296, *482*
Nitsch, J. P., 298, 417, *482*
Nixon, H. L., 59, *483*
Nogueira-Neto, P., 319, *483*
Nolan, W. J., 63, 65, *483*
Noro, K., 387, *483*
Norris, K. R., 129, *483*
Northwood, P. J., 196, 197, *483*
Nummi, W. O., 96, 100, 230, *466*
Nuttall, P. M., 23, 73, 138, 139, 140, 142, 409, *459*
Nye, W. P., 11, 87, 88, 108, 122, 124, 205, 212, 213, 315, 316, *449, 465, 477, 483, 486*

O

Oberle, G. D., 411, *483*
Odland, M. L., 355, *483*
Oertel, E., 232, 233, 234, 390, *446, 483*
Oettingen-Spielberg, T., 40, 62, *483*
Oganjan, V. N., 270, *483*
Olmo, H. P., 190, 191, 192, *483*
Olsen, P. J., 202, *494*
Olsson, G., 137, 138, 139, 148, *483*
O'Neill, W. J., 81, 395, 407, 413, 414, *484*
Oppenheimer, H. R., 186, *483*
Orösi-Pál, Z., 58, *483*
Oschmann, H., 244, *484*
Osgood, C. E., 108, *497*
Osraščenko-Kudrjavcava, A., 212, *484*
Ostendorf, F. W., 170, *484*
Osterli, V. P., 236, *480*
Otto, F. T., 71, *484*
Overholser, E. L., 411, *484*
Overley, F. L., 81, 395, 407, 410, 411, 412, 413, 414, *450, 484*
Overseas Food Corporation 325, 326, 329, 330, *484*
Owen, F. W., 415, *469*

P

Paatela, J., 226, 228, *500*
Paddock, F. B., 86, *484*
Pain, Janine, 61, *452, 484*
Painter, L. I, 114, *460*
Palmer, S., 139, *484*
Palmer-Jones, T., 121, 122, 124, 199, 200, 203, 207, 219, 221, 222, 223, 233, 234, 387, *484*
Pammel, L. H., 215, 217, 300, *484*
Pandalai, K. M., 442, *479*
Pandey, R. S., 433, 434, *484*
Pankiw, P., 41, 76, 122, 123, 124, 199, 200, 202, 212, 222, 224, 227, 230, 240, *449, 462, 484, 485*
Pankratova, E. P., 315, 316, *485*
Panov, V., 397, *487*
Park, O. W., 17, 18, 19, 24, 390, *485*
Parker, R. L., 19, 21, 24, 85, 390, 392, 395, *485, 489*
Parks, H. B., 23, 154, 157, *485*
Pashchenko, T. E., 350, *485*
Patankar, V. K., 332, *470*
Pausheva, Z. P., 364, *485*
Pearson, O. H., 121, 123, 144, 145, 146, *485*
Peck, O., 102, 209, 210, *485*
Pedersen, A., 201, 202, 219, 226, 229, 230, *485*
Pedersen, M. W., 18, 19, 48, 83, 84, 120, 121, 126, 199, 200, 205, 212, 213, *446, 451, 483, 485, 486*
Peebles, R. H., 166, *486*
Peens, J. F., 356, *486*
Peer, D., 9, 67, *486*
Pelerents, C., 267, *486*
Pellet, F. C., 144, 273, 364, 390, *486*
Penfold, A. R., 288, *486*
Pengelly, D. H., 202, 208, 210, *486*

Percival, Mary S., 15, 16, 18, 22, 23, 34, 35, 36, 38, 39, 62, 249, 285, 389, 390, 391, 408, 421, 424, 425, *486*
Perrin, M. E. B., 336, 340, *486*, *490*
Persson, Brita, 137, 138, 139, 141, *483*, *486*
Petersen, H. L., 199, 200, 202, 206, 211, *486*
Peterson, A. G., 223, *486*
Peterson, P. A., 368, 369, *486*
Petkov, V., 140, 203, 206, 212, 276, 397, 419, 420, 421, 425, *486*, *487*
Peto, H. B., 130, 303, 307, 427, 428, *487*
Pew, W. D., 310, *487*
Phadke, K. G., 237, 239, 317, *482*
Phares, D., 273, *502*
Pharis, R. L., 200, 201, 202, *487*
Phillips, E. F., 127, 364, *487*
Philp, G. L., 77, 398, 403, 407, *487*
Phipps, C. R., 337, 342, *487*
Pieters, A. J., 216, *487*
Pijl, L. van der, 37, *456*
Pinter, L., 215, 218, 227, *479*
Pinthus, M. J., 325, *487*
Piper, C. V., 270, *487*
Pisani, P., 282, *446*
Plass, F., 390, *487*
Plath, O. E., 89, *481*
Platt, F. R., 9, *496*
Plowright, R. C., 98, *487*
Poehlman, J. M., 179, 180, 216, 356, *487*
Pohjakallio, O., 231, *487*
Pollard, L., 258, 304, 315, 317, 354, 362, 366, 429, *465*
Pomeroy, C. S., 368, *499*
Ponomareva, A. A., 210, *487*
Poole, C. F., 360, *487*
Pope, O. A., 164, 165, *487*
Pope, W. T., 293, 295, *487*
Popenoe, P. B., 443, *487*
Popov, V. V., 210, 362, *487*
Porter, A. W., 355. *483*
Posnette, A. F., 170, 171, 172, 173, 174, 175, 176, *488*
Potter, G. F., 371, *488*
Potter, H., 114, *460*
Potter, J. M. S., 282, *488*
Pound, F. J., 170, 176, *488*
Pouvreau, A., 100, *488*

Preece, D. A., 66, *459*
Preston, A. P., 394, 407, *488*
Priestley, G., 120, 125, *488*
Pritsch, G., 83, 139, 223, 229, *488*
Pryce-Jones, J., 233, *488*
Pryor, D. E., 306, *504*
Pryor, L. D., 288, *488*
Purdie, J. D., 214, *454*
Purseglove, J. W., 168, 257, 270, 277, 289, 290, 292, 296, 301, 321, 349, 355, 367, 369, 373, 431, 432, 433, 436, *488*
Putt, E. D., 325, 326, 329, *488*
Pynaert, L., 432, *488*

Q

Qayyum, A., 143, 237, *474*
Quisenberry, K. S., 364, *461*

R

Racey, P. A., 29, 61, 65, 124, 125, 126, 131, 249, 261, 397, *459* *460*
Radaeva, E. N., 325, 330, 331, *488*
Radchenko, T. G., 139, 140, 146, 149, *488*
Radoev, L., 161, 328, 330, *488*
Rahman, K. A., 144, *488*
Rakhmankulov, F., 85, *489*
Rane, F. W., 308, *489*
Rao, B. S., 372, 373, *489*
Rao, M. M., 194, *482*
Rao, V. N. M., 197, *489*
Raphael, T. D., 66, 409, *489*
Rashad, S. E., 85, *489*
Reece, P. C., 186, *489*
Reed, A. D., 236, *480*
Reed, C. B., 63, 66, *499*
Reese, C. A., 223, 229, 231, 398, 446, *489*
Register, R. O., 186, *489*
Reimer, F. C., 191, *489*
Reinhardt, J. F., 204, 258, 411, *462*, *489*, *501*
Renard, E. J., 277, *489*
Rhein, W. von, 83, *489*
Rhyne, C., 156, 160, *477*
Ribbands, C. R., 10, 15, 17, 23, 30, 38, 40, 42, 44, 59, 62, 64, 69, 400, *483*, *489*
Richards, O. W., 218, *489*
Richardson, R. W., 352, *489*

Richmond, R. G., 221, *489*
Richmond, T. R., 166, *476*
Rick, C. M., 352, 353, 354, *489*, *490*
Riddell, R. T., 399, 400, 403, 413, 415, *499*
Riedel, I. B. M., 244, *490*
Rietsema, I., 282, *474*
Rives, M., 137, *490*
Roach, F. A., 336, 403, *490*
Robbins, W. W., 135, 151, 168, 189, 198, 199, 265, 277, 298, 336, 337, 350, 355, 357, 359, 366, 375, 380, 381, 417, 418, 422, 323, 429, 436, *490*
Roberts, D., 83, 399, 407, 408, *490*
Roberts, R. H., 345, 381, 392, 395, 398, 403, 404, *490*
Robinson, B. B., 180, 241, 242, *490*
Robinson, D. H., 241, 242, *490*
Robinson, F. A., 21, 187, 188, *490*
Robinson, Jeanette, 306, *478*
Robinson, T. R., 368, *499*
Roemer, T., 146, *490*
Rogers, H. H., 170, *472*
Romanchuk, I., 303, *478*
Rosa, J. T., 129, 149, 257, 300, 305, 306, 307, 310, 312, 315, 335, 358, 427, 428, 429, *470*, *490*
Rösch, G. A., 29, 55, 57, 58, 61, 62, *460*, *490*, *491*
Ross-Craig, Stella, 137, 286, 314, 317, 345, *491*
Rothenbuhler, W. C., 124, 125, *491*
Rowlands, D. G., 243, *491*
Rozov, S. A., 276, 328, 331, 363, 389, *491*
Rubis, D. D., 121, 331, 332, *476*, *491*
Rudnev, V, Z., 325, 327, 329, 330, *491*
Ruppolt, W., von, 34, *491*
Rust, R. W., 202, 210, *449*
Ruszkowska, B., 364, *454*
Ryle, Margaret, 83, 84, 227, 391, *491*
Rymashevskaya, R. S., 285, *491*
Rymashevskii, V. K., 285, 287, 391, 396, 401, 424, *491*

S

Sadamori, S., 412, *491*
Said, I., 237, *499*
Sakagami, S. F., 56, *491*
Sakharov, M. K., 146, *491*
Salminen, M., 228, 230, 231, *500*
Sampson, D. R., 136, *491*
Sampson, H. C., 277, 355, *491*
Sanders, T. W., 306, *491*
Sănduleac, E. V., 299, 326, 328, 424, 452, *492*
Sarcar, P. B., 432, 433, *473*
Sarisalo, M., 226, 228, *500*
Sartorius, O., 191, *492*
Šaškina, L. M., 420, *492*
Savitsky, H., 360, *492*
Savos, M., 233, 240, 337, 343, 420, 424, 425, *493*
Sax, K., 41, 385, 396, *492*
Sazykin, Y. V., 284, 287, 391, 424, *492*
Schaefer, C. W., 19, *492*
Schander, H., 281, 282, 382, *492*
Schaub, I. O., 339, *492*
Schelhorn, M. von, 251, 253, *492*
Schlecht, F., 220, *492*
Schneck, H. W., 354, *492*
Schöntag, A., 391, *447*
Schroeder, C. A., 289, 435, *492*
Schuster, C. E., 398, 399, 403, 409, 410, *492*
Schwan, B., 31, 220, 222, 226, 227, 229, 231, *492*
Schwanwitsch, B. N., 255, 271, 272, *492*
Scriven, W. A., 122, 124, 125, 244, 248, *492*
Scullen, H. A., 235, 236, 239, 240, *493*
Seaney, R. R., 271, *493*
Seaton, H. L., 298, 301, 305, 312, 383, *493*
Šedivý, J., 41, 76, 205, *464*, *493*
Seidova, S. S., 303, 307, *470*
Sein, F., 319, 320, *493*
Selim, H. A., 288, *468*
Ševčenko, A. Ja., 221, *493*
Shanks, C. H., 423, *493*
Sharma, A. C., 65, *493*
Sharma, P. L., 65, 144, 145, 237, 239, 317, 346, 347, 386, 391, *482*, *493*
Sharples, G. C., 191, 192, *493*
Shaw, F. R., 18, 207, 215, 233, 240, 275, 298, 337, 338, 342, 343, 347, 348, 364, 406, 420, 424, 425, *493*
Shaw, H. B., 360, 361, *493*
Shaw, W. M., 233, 240, 337, 343, 347, 348, 420, 424, 425, *493*

Shchibrya, A. A., 272, 273, *493*
Shearer, D. A., 74, *448*, *493*
Shemetkov, M. F., 124, 125, 126, 302, 304, *493*, *494*
Shimanuki H., 73, 348, *494*
Shishikin, E. A., 159, 161, *494*
Shoemaker, J. S., 190, 347, 382, 417, 420, 422, 424, *494*
Shuel, R. W., 16, 58, 83, 84, 227, *494*, *499*
Sidhu, A. S., 155, 156, 159, 161, 164, *494*
Sigfrids, A. G., 231, *494*
Silow, R. A., 271, *494*
Silversides, W, H., 202, *494*
Simao, S., 195, *494*
Simidchiev, T., 203, 206, 212, *487*
Simmonds, N. W., 441, *494*
Simonov, I. N., 282, *494*
Simpson, D. M., 162, 164, 165, *487*, *494*
Simpson, J., 7, 18, 56, 61, 207, 325, 326, 400, *451*, *459*, *494*
Singh, L. B., 193, 194, *494*
Singh, S., 44, 45, 127, 144, 155, 156, 159, 161, 164, 195, 347, 391, 399, 400, 404, 407, 411, *494*, *495*
Sirks, M. J., 243, *495*
Sivori, E. M., 330, *495*
Skirde, W., 227, 228, *495*
Skovgaard, O. S., 218, 220, 222, 226, 229, *495*
Skrebtsov, M. F., 74, 154, 158, 160, 161, *461*, *462*, *468*, *495*
Skrebtsova, N. D., 24, 35, 285, 304, 363, 364, 417, 420, *491*, *495*
Sladen F. W. L., 89, 90, 91, 95, *495*
Slater, J. A., 299, 313, *460*
Smaragdova, N. P., 37, 81, 82, *463*, *495*
Smirnov, V. M., 181, *495*
Smit, A. G., 177, *495*
Smith, B. D., 81, 384, 385, 387, 400, *476*, *495*
Smith, F. G., 333, *496*
Smith, F. L., 257, 258, *478*
Smith, M. V., 8, 9, 10, 72, 75, 287, 387, 399, 400, 403, 410, 413, 415, 417, *454*, *455*, *459*, *496*, *499*
Smith, Ora, 350, *496*
Smith, R. F., 9, *496*
Sneep, J., 120, *496*

Snodgrass, R. E., 16, 19, *496*
Snyder, J. C., 72, 382, 383, 397, 409, 410, 411, 412, *450*, *496*
Soetardi, R. G., 172, 173, 177, *496*
Solov'ev, G. M., 84, 202, 364, 365, *49*
Soost, R. K., 186, *496*
Soper, M. H. R., 242, 248, *496*
Sørensen, N. A., 219, 229, *485*
Sorokin, V., 125, *496*
Sosunkov, V. I., 190, *496*
Sovoleva, E. M., 202, *496*
Speirs, Nancy, 30, 40, *489*
Spencer, J. L., 194, 195, *496*
Spencer-Booth, Yvette, 35, 36, 45, 55 61, 70, 77, 78, 79, 85, 86, 121, 148, 149, 394, 395, 396, 399, 401, 403, 406, 409, 412, *459*, *496*
Sprenger, A. M., 384, 385, *496*
Srinivasalu, N., 267, *496*
Stadhouders, P. J., 402, *497*
Stahel, G., 172, 174, *497*
Stählin, A., 226, 227, *497*
Stanford, E. E., 158, 440, *497*
Stapel, C., 26, 66, 82, 85, 201, 202, 206, 208, 215, 219, 222, 226, 228, 229, 230, 231, *444*, *485*, *497*
Starling, T. M., 225, 226, *497*
Stephen, W. P., 83, 103, 106, 108, 110, 112, 114, 115, 116, 200, 201, 206, 208, 209, 384, 385, 392, 396, 403, 408, *449*, *497*
Stephens, S. G., 164, 165, *497*
Stereva, R., 201, *497*
Steshenko, F. N., 192, *497*
Steuckardt, R., 122, 203, 206, 249, *497*, *498*
Stevenson, F. J., 357, 358, *498*
Stevenson, J. H., 249, *460*
Stevenson, T. M., 261, 274, *471*, *498*
Stewart, D., 360, *498*
Stoker, G. L., 48, 199, *451*, *486*
Stokes, W. E., 267, *498*
Storey, W. B., 435, *498*
Stout, A. B., 190, 368, 369, *498*
Stricker, M., 73, 348, *494*
Stroman, G. N., 162, *498*
Struckmeyer, B. Ester, 345, *490*
Sturtevant, A. P., 68, 69, *498*
Su-Ling, L., 433, 434, *481*
Sun, V. G., 139, *498*
Svendsen, O., 69, *498*

Swart, F. W. J., 219, 230, *498*
Synge, A. D., 19, 22, 23, 31, 222, 328, 389, *498*

T

Taber, S., 9, 21, 85, *498*
Taimr, L., 41, 76, 205, *464*, *493*
Tamargo, M. A., 167, 168, *470*, *498*
Tammes, P. M. L., 442, *498*
Tannenbaum, L., 342, 343, *447*
Taranov, G. F., 66, *498*
Taschdjian, E., 320, *498*
Taylor, C. S., 270, *505*
Taylor, E. A., 309, *498*
Taylor, S. A., 212, *486*
Teaotia, S. S., 279, 280, 281, *498*
Tedin, O., 149, *498*
Tedoradze, S. G., 259, 262, *498*
Telford, H. S., 72, 80, 387, 388, 389, 397, 398, 407, 414, *503*
Teskey, B. J. E., 382, *494*
Thies, S. A., 155, 156, 164, *499*
Thomas, F. L., 255, 275, *444*, *503*
Thomas, W., 66, 223, 231, *499*
Thompson, D. J., 315, *499*
Thompson, F., 125, *499*
Thompson, J. B., 436, *499*
Thomspon, R. C., 335, *499*
Thompson, V. C., 124, 125, *491*
Thomson, G. M., 101, *499*
Thomson, J. R., 276 *499*
Tidbury, G. E., 289, *499*
Tiedjens, V. A., 301, *499*
Tirgari, S., 234, *474*
Tobgy, H. A., 237, *499*
Todd, F. E., 7, 8, 18, 19, 31, 63, 65, 70, 71, 79, 82, 85, 121, 156, 158, 159, 160, 186, 188, 191, 192, 202, 203, 204, 205, 206, 207, 210, 213, 214, 305, 307, 308, 309, 354, 389, *449*, *451*, *477*, *486*, *493*, *499*, *500*, *501*
Toole, E. H., 315, *465*
Townsend, G. F., 8, 9, 58, 70, 399, 400, 403, 413, 415, *496*, *499*
Toynbee-Clarke, G., 250, *449*
Traub, H. P., 368, *499*
Traynor, J., 80, 87, 412, *499*
Treherne, R. C., 316, 317, 361, 427, *499*
Trofimec, N. K., 428, *499*
Trought, T., 163, *499*

Trushkin, A. V., 159, *500*
Tsygankov, S. K., 386, 387, *500*
Tu, C. 179, 181, *465*
Tucker, K. W., 36, 204, *461*
Tufts, W. P., 7, 389, 397, 403, 409, 410, *500*
Tukey, H. B., 403, *500*
Tung, S. M., 389, *474*
Turner, M., 406, *493*
Tuttle, D. M., 7, 207, *451*
Tyler, F. J., 153, *500*
Tysdal, H. M., 199, 201, 209, *500*

U

Ufer, M., 201, 274, 276, *500*
Ulanichev, E. M., 331, *471*
Umaerus, M., 199, 206, 223, 226, 228, 229, 230, 231, *444*, *500*
Urnrau, J., 200, 201, 202, *487*

V

Valle, O., 98, 226, 228, 230, 231, 240, *500*
Valleau, W. D., 419, *500*
Vansell, G. H., 8, 15, 18, 41, 63, 75, 77, 84, 153, 154, 184, 185, 202, 203, 204, 206, 207, 210, 214, 235, 258, 387, 388, 390, 391, 392, 396, 398, 403, 407, 408, 409, 410, 411, 414, *462*, *463*, *464*, *487*, *499*, *500*, *501*
Väre, A., 218, *501*
Vear, K. C., 136, 323, *461*
Veerman, A., 42, *501*
Velichkov, V., 275, *501*
Velthuis, H. H. W., 61, *501*
Veprikov, P. N., 64, 85, *501*
Verdieva, M. G., 159, 298, *463*, *501*
Verkerk, K., 350, 351, *501*
Veselý, V., 41, 76, 205, *464*, *493*
Vestad, R., 227, *501*, *503*
Vickery, V. R., 73, 74, 75, 86, 390, 397, 398, 408, 411, 414, 415, *470*
Vieira, C., 258, *501*
Vincent, C. C., 384, 391, *476*
Vinogradov, S. I., 208, *472*
Vinson, R., 420, *501*
Virostek, J. F., 96, 100, 230, *466*
Voelcker, O. J., 170, 172, *501*

Voogd, S., 61, *463*
Vriend, S., 302, *447, 501*

W

Wafa, A. K., 19, 23, 121, 156, 157, 160, 184, 185, 238, 239, 240, 245, 246, 247, 250, 387, *501, 502*
Waite, M. B., 385, *502*
Wallace, A. T., 273, *502*
Wallace, M. M., 135, *462*
Waller, G. D., 84, *474*
Walstrom, R. J., 224, *502*
Wanic, D., 140, *502*
Ward, J. F., 290, *502*
Ware, J. O., 153, 162, 165, *450, 502*
Warmke, H. E., 372, *502*
Warne, B., 255, *444*
Warren, L. O., 303, *502*
Watkins, W. G., 184, *501*
Watts, F. H., 244, *502*
Watts, L. E., 131, 147, 335, *502*
Weatherley, P. E., 164, *502*
Weaver, E. C., 74, *503*
Weaver, H., 114, *460*
Weaver, N., 45, 46, 74, 75, 121, 122, 232, 233, 234, 236, 237, 240, 252, 253, 254, 255, 256, 275, *502, 503*
Webber, H. J., 185, 186, *503*
Webster, C. C., 81, *503*
Webster, J. L., 258, *503*
Webster, R. L., 72, 80, 387, 388, 389, 391, 396, 397, 398, 407, 414, *503*
Weidhaas, J., 347, 348, *493*
Weihing, R. M., 240, *503*
Weinstein, A. I., 257, *503*
Weiss, K., 385, *503*
Welch, J. E., 258, *503*
Wellensiek, S. J., 172, *503*
Wellington, R., 190, 278, 279, 281, 282, 285, 423, *503*
Wellman, F. L., 320, *503*
Wells, P. H., 29, *503*
Wenholz, H., 350, *503*
Wenner, A. M., 29, *503*
Went, F. W., 298, *482*
Werner, F. G., 156, 331, *451*
Wertheim, I. S. J., 385, *503*
Wester, R. E., 258, *478*
Westgate, J. M., 216, 221, 229, 230, *503*
Wexelsen, H., 226, 227, *503*
18*

Whitaker, T. W., 298, 306, 310, 311, 335, *499, 503, 504*
White, Elizabeth, 338, *504*
White, T. H., 350, *504*
White, W. J., 199, 201, *445, 471*
Whitehead, R. A., 442, *504*
Wiering, D., 130, *504*
Wigglesworth, V. B., 376, *504*
Williams, Ingrid H., 29, 64, 117, *459*
Williams, M. W., 410, *446*
Williams, R. D., 47, 119, 121, 215, 216, 217, 218, 219, 221, 229, 230, *504*
Williams, R. R., 283, 383, 384, 385, 403, 404, 406, *465, 495, 504*
Williams, W., 239, *504*
Willis, J. L., 288, *486*
Wilsie, C. P., 66, 225, 226, *497, 504*
Wilson, D., 279, *504*
Wilson, G. F., 121, 387, 388, 392, 396, 397, 402, 407, *504*
Wilson, P. M., 81, *503*
Winkler, A. J., 192, *504*
Wójtowski, F., 118, 208, 252, *480, 504*
Wolfenbarger, D. O., 195, 196, 299, *504, 505*
Wood, F. A., 342, 344, *505*
Wood, G. W., 341, 342, 343, 344, *505*
Wood, M., 83, 221, 227, *445*
Woodhead, C. E., 404, *452*
Woodhouse, E. J., 270, *505*
Woodrow, A. W., 36, 65, 217, 398, *505*
Woodworth, C. M., 270, *505*
Worley, S., 156, 160, *477*
Wort, D. A., 244, *490*
Wykes, G. R., 16, 17, 18, *505*

Y

Yadava, R. P. S., 433, 434, *484*
Yago, M., 387, *483*
Yakovlev, A. S., 69, *505*
Yamane, G. M., 433, *505*
Yoshida, Y., 412, *491*
Yu, C. P., 162, *505*

Z

Zadražil, K., 202, *505*
Zakharov, G. A., 281, 282, 283, 284, 285, *505*

Zander, E., 139, 248, 396, *505*
Zavrashvili, R. M., 185, *505*
Zeilinga, A. E., 83, 419, *473*

Zimmermann, M., 16, *460*
Zobel, M. P., 304, *505*
Zon, J. C. J. van, 42, *501*

Animal Index

Numbers in italics refer to pages with most detailed entries. Numbers followed by an asterisk refer to pages where an illustration occurs.

A

Acalymma vittata, 299
Acari, 184
Agapostemon, 331
Agapostemon texanus, 335
Agrobombus, 225
Andrena, 34, 35, 208, 239, 273, 316, 342, 362
Andrena armata, 25
Andrena convexiuscula, 208
Andrena erincia, 184
Andrena haemorrhoa, 409
Andrena flavipes, 140
Andrena ilerda, 144
Andrena labialis, 208
Andrena ovatula, 208
Andrena tibialis, 140
Andrena varians, 25
Andrena wilkella, 218
Andrenidae, 24, 139, 144, 387
Ants, Formicidae, 173, 174, 175, 197, 249, 277, 303, 355, 362, 433, 442
Anthidium, 239, 271
Anthidium punctatum, 234
Anthomyidae, 388
Anthophoridae, 24, 362
Anthophora, 209
Anthophora confusa, 155
Anthophora linsleyi, 24
Anthophora parietina, 118
Anthophora ursina, 273
Aphids, Aphididae, 173, 174, 254
Aphis fabae, 361, 362
Apis cerana, 3, 50, 143, 144, 155, 157, 159, 160, 237, 239, 269, 270, 271, 299, 304, 312, 313, 317, 333, 347, 433
Apis dorsata, 3, 50, 155, 157, 239, 268, 270, 313, 317, 433

Apis florea, 3, 50, 144, 155, 157, 159, 160, 239, 269, 271, 304, 312, 313, 317, 433
Atrichopogan, 372
Augochloropsis ignata, 352

B

Bats, fruit, Pteropodidae, 432
Bee, alkali, *Nomia melanderi*, 25, *111–116*, 112*, 113*, 115*, 209, 210, 213, 216
Bee, bumble, *Bombus*, 2, 3, 15, 25, 26, 34, 35, 39, 45, 46, 47, *89–101*, *119–120*, 155, 184, 208, 209, 212, 218–221, 228–232, 240, 247, 258, 259, 261–264, 266, 271, 273, 274, 277, 284, 295, 301, 313, 325, 335, 336, 342, 343, 347, 358, 387–389, 421, 425
Bee, carpenter, *Xylocopa*, 156, 239, 247, 255, 295, 301
Bee, halictid, *Halictus*, 35, 144, 155, 208, 239, 290, 303, 316, 330, 331, 335, 342, 362
Bee, leaf cutter, *Megachile rotundata*, 25, *103–111*, 104*, 105*, 107*, 109*, 116, *126–127*, 209, 211
Bee, solitary, 102–118, 126–127
Bee, stingless, *Melipona*, 304, 312, 313, 433, 440
Bibio, 2
Bibionidae, 387
Blastophaga psenes, fig wasp, 374–379, 375*, 376*
Blowflies, 3, 45, *127–131*, 261, 314, 315, 421, 427
Bombidae, 266
Bombus, 2, 3, 15, 25, 26, 34, 35, 39, 45, 46, 47, *89–101*, *119–120*, 155, 184, 208,

Bombus (contd.)
  209, 212, 218–221, 228–232, 240, 247, 258, 259, 261–264, 266, 271, 273, 274, 277, 284, 295, 301, 313, 325, 335, 336, 342, 343, 347, 358, 387–389, 421, 425
Bombus affinis, 94, 218
Bombus agrorum, 25, 26, 92*, 97*, 120, 218, 247, 251–253, 262, 273
Bombus americanorum, 26, 155, 156, 208
Bombus appositus, 120
Bombus auricomus, 155
Bombus borealis, 96, 208, 230
Bombus distinguendus, 231
Bombus fervidus, 26, 96, 120, 208, 230
Bombus fraternus, 155
Bombus griseocollis, 26
Bombus hortorum, 25, 91*, 97*, 120, 218, 247, 250, 251, 260*, 261
Bombus humilis, 120
Bombus huntii, 96, 230
Bombus impatiens, 347
Bombus lapidarius, 94, 120, 218, 252, 284
Bombus lucorum, 25, 26, 33, 39, 90*, 93*, 94, 208, 219*, 247, 250, 252, 262, 263, 284
Bombus mormonorum, 120
Bombus morrisoni, 93*, 120
Bombus nevadensis, 26, 96, 230
Bombus occidentalis, 230
Bombus pratorum, 25, 261, 263
Bombus ruderatus, 101, 218, 251
Bombus rufocinctus, 230
Bombus subterraneus, 101, 251, 252
Bombus sylvarum, 25, 120
Bombus ternarius, 98*
Bombus terrestris, 33, 94, 101, 208, 218, 219, 221, 247, 250, 251, 252, 262, 263, 284
Bombus terricola, 94, 208, 218
Bombus vagans, 208
Bombycidae, 139
Bombyliidae, 387
Bombylius, 2, 342
Brachygastra augustii, 319
Butterflies, Lepidoptera, 45, 239, 335

C

Calliphora, 2, 342
Calliphora vomitoria, 129
Calliphorinae, 144, 387
Callitroga macellaria, 433
Campsomeris, 156
Camposmeris trifasciata, 167
Cataglyphis bicolor, 247
Cecidomyidae, 177
Ceratina, 290
Ceratina bispinosa, 266
Ceratinidae, 266
Ceratopogonidae, 175–178
Cerceris, 316
Cerocoma, 362
Chauliognathus marginatus, 434
Chironomidae, 177, 195, 295, 315, 372, 373, 387
Chloralictus, 316
Coccinella septempunctata, 362
Coccinellidae, 307, 316, 334, 362
Coleoptera, 184, 195, 239, 275, 295, 299, 333, 372, 423
Collembola, 438
Colletes, 316, 342
Conocephalus saltator, 296
Crematogaster, 174
Culicidae, 315
Culicoides, 373

D

Dasyhelea, 372
Diabrotica, 312
Diabrotica soror, 334
Dilophus, 2
Diptera, 2, 127, 184, 195, 239, 275, 295, 299, 316, 317, 333, 366, 369, 372, 373, 387, 388, 389, 423, 427, 428
Dolichoderus bituberculatus, 177
Drosophila, 177
Drosophila hydei, 334
Drosophila melanogaster, 334

E

Eidolon helvum, 432
Elis plumipes, 156
Elis thoracica, 155, 157
Epicharis, 295
Epomophorus gambianus, 432
Eristalis, 2, 127, 144, 184, 342
Eristalis arvorum, 295, 296
Eucera clypeata, 208

Eucera longicornis, 208
Eulalia, 316
Exomalopsis, 167, 290

F

Fannia canicularis, 131
Fig wasp, *Blastophaga psenes*, 374–379, 375*, 376*
Forcipomyia, 177, 372
Forcipomyia ashantii, 177
Forcipomyia ingrami, 177
Forcipomyia quasi-ingrami, 175
Formicidae, 173, 174, 175, 197, 249, 277, 303, 355, 362, 433, 442
Frankliniella fusca, 267, 361
Frankliniella occidentalis, 257, 259
Frankliniella parvula, 175
Frankliniella tritici, 361

G

Gnats, Culicidae, 315

H

Halictus, 35, 144, 155, 208, 239, 267, 290, 303, 315, 316, 330, 331, 335, 342, 352, 362
Halictus confusus arapahonum, 315
Halictus ligatus, 332
Halictus tripartitus, 332
Hawkmoths, Sphingidae, 432
Heliothrips fasciatus, 361
Hemiptera, 195, 299, 316
Hippodamia convergens, 334
Hortobombus, 225
Hoverflies, Syrphidae, 335
Hummingbirds, Trochilidae, 168, 295, 356, 440, 441
Hylemya, 334
Hymenoptera, 184, 195, 239, 271, 273, 295, 299, 373, 423, 428, 436

L

Ladybird beetles, Coccinellidae, 307, 316, 334, 362
Lapidariobombus, 225, 226
Lasioglossum, 273, 331
Lasioglossum pectoraloides, 332

Lasiohelea litoraurea, 177
Lasiohelea nana, 175
Lasiophthicus pyrastri, 192
Lepidoptera, 45, 184, 195, 239, 275, 295, 299, 335, 372, 432, 435
Leptura, 362
Lindenius, 316
Lucilia, 2
Lucilia sericata, 128
Lycaenidae, 266
Lygus hesperus, 334

M

Megachile, 34, 102, 103, 208, 217, 240, 271, 331
Megachile anthracine, 269
Megachile brevis, 208, 209
Megachile frigida, 102, 208
Megachile inermis, 102
Megachile lanata, 268
Megachile latimanus, 208
Megachile nevalis, 102
Megachile perihirta, 208
Megachile quinquelineata, 208
Megachile rotundata, 25, *103–111*, 104*, 105*, 107*, 109*, 116, *126–127*, 209–211
Megachile willughbiella, 208
Megachilidae, 24, 144, 362, 387
Melipona, 304, 312, 313, 433, 440
Melipona quadrifasciata anthidioides, 319
Melissodes, 155, 156, 218, 266, 331
Melissodes communis, 303
Melissodes foxi, 436
Melissodes rufodentata, 436
Melitta leporina, 208, 211, 234
Melittidae, 24
Midges, Chironomidae, 195, 295, 315, 372, 373, 387
Mites, *Acari*, 184
Moths, Lepidoptera, 295, 372, 432, 435
Musca, 184, 195, 433
Musca autumnalis, 131
Musca domestica, 131
Muscidae, 144, 270, 387

N

Nannotrigona testaceicornis, 319
Nanonycteris veldkampii, 432

ANIMAL INDEX

Neuroptera, 184
Nomada, 342
Nomadidae, 144
Nomadopsis, 240
Nomia australiaca, 208
Nomia cognata, 270
Nomia melanderi, 25, *111–116*, 112*, 113*, 115*, 209, 210, 213, 316
Nysson, 316

O

Orthoptera, 295
Osmia, 102, 116, 217, 239, 271
Osmia cornifrons, 116
Osmia excavata, 116
Osmia lignaria, 117
Osmia rufa, 116*, 117*
Osmia seclusa, 116

P

Panurgidae, 24
Peponapis, 299, 300
Peponapis pruinosa, 300
Philanthus, 316
Phormia regina, 128
Phormia terranovae, 129, 130
Pieris rapae, 184
Platycheirus, 2
Plebeia, 319
Polistes, 295
Polistes exclamans, 332
Psithyrus, 100, 120
Psychonosma, 195
Pteropodidae, 432

R

Rhingia, 2

S

Sarcophaga, 2
Scatopsinae, 373
Sceliphron, 316
Scolia, 155, 156
Scolia avreipennis, 157
Sepsis, 144
Sphaerophoria scripta, 362
Sphaerularia bombi, 99

Sphecidae, 316
Sphingidae, 432
Stilobezzia, 372
Stratiomyidae, 316
Stratiomys, 316
Syritta, 316
Syrphidae, 45, 144, 192, 316, 317, 335, 347, 387
Syrphus, 2, 184, 195

T

Terrestribombus, 226
Tetragona jaty, 319
Tetralonia, 218
Tetralonia dubitata, 273
Thrips, Thysanoptera, 173, 175–178, 181, 184, 195, 257, 259, 267, 270, 295, 307, 355, 361, 373, 427, 435
Thrips tabaci, 361
Thysanoptera, 173, 175–178, 181, 184, 195, 257, 267, 270, 295, 307, 355, 361, 373, 427, 435
Trichometallea pollinosa, 144
Trigona, 295
Trigona hyalinata, 319
Trigona ruficrus, 319
Trochilidae, 168, 295, 356, 440, 441
Tubifera, 316
Tubifera pertinax, 302

V

Vespa, 433

X

Xenoglossa, 299, 300
Xenoglossa strenua, 299, 300
Xylocopa, 156, 239, 247, 255, 295, 301
Xylocopa aestuans, 184, 247, 251
Xylocopa amethystina, 269
Xylocopa cubaecola, 167
Xylocopa sonorina, 118
Xylocopa varipuncta, 295, 296
Xylocopa virginica, 273, 343
Xylocopidae, 266

Z

Zonabris, 362

Plant Index

Numbers in italics refer to pages with most detailed entries. Numbers followed by an asterisk refer to pages where an illustration occurs.

A

Acacia mearnsii, 277
Acer pseudoplatanus, sycamore, 25
Aleurites, 370–371
Aleurites fordii, tung, 370–371
Aleurites montana, tong, 370–371
Alfalfa, *Medicago sativa*, 3, 17, 19, 25, 26, 32, 36, 41, 45, 48, 66, 72, 75, 76, 79, 82, 83, 87, 88, 102, 104*, 109*, 120, 122, 126, 127, *198–214*, 199*, 203*, 230
Algaroba, *Prosopis juliflora*, 277
Allium ampeloprasum, leek, 426
Allium cepa, common onion, shallot, 69, 127, 128*, 129, *426–428*, 427*
Allium chinese, rakkyo, 426
Allium fistulosum, Welsh onion, 46, *426*, 428
Allium odorum, 46
Allium sativum, garlic, 426
Allium schoenoprasum, chives, 426
Allspice, *Pimenta dioica*, 290, 291
Almond, *Prunus amygdalus*, 381–416
Anacardiaceae, 193–197
Anacardium occidentale, cashew, 196*–197
Ananas sativus, pineapple, 440–441
Anethum graveolens, dill, 317
Angelica, angelica, 131
Anise, *Pimpinella anisum*, 317
Antirrhinum, 37
Antirrhinum glutinosum, 36, 37
Antirrhinum majus, snapdragon, 36, 37
Apium graveolens, celery, 123, 130, *317*
Apocynaceae, 435–436
Apple, *Pyrus malus*, 32, 34, 38, 44, 45, 69, 72, 73, 76, 77, 79, 82, 83, 116, 121, *381*–*416*, 382*, 393*, 394*
Apple, rose, *Eugenia jambosi*, 289

Apricot, *Prunus armeniaca*, 381–416
Arachis hypogaea, groundnut, peanut, monkeynut, 265*–267
Arctostaphylos manzanita, manzanita, 18, 407, 408
Armoracia rusticana, horse-radish, 135
Arracacha, *Arracacia xanthorrhiza*, 317
Arracacia xanthorrhiza, arracacha, 317
Artichoke, globe, *Cynara scolymus*, 322
Artichoke, Jerusalem, *Helianthus tuberosus*, 322
Asparagus officinalis, asparagus, 123, 426, *428–429**
Aster, 44
Aubrieta, 31
Avaram, *Cassia auriculata*, 277
Avocado, *Persea americana*, 367*–369

B

Bamboo, *Bambusae*, 116
Bambuseae, bamboo, 116
Banana, *Musa*, 81, *441*
Bean, adzuki, *Phaseolus angularis*, 257
Bean, broad, *Vicia faba*, 242, *245–247*
Bean, butter, *Phaseolus lunatus*, 257–259
Bean, carab, *Ceratonia siliqua*, 277
Bean, cluster, *Cyamopsis tetragonoloba*, 277
Bean, common, *Phaseolus vulgaris*, 219, *257–258*
Bean, field, *Vicia faba*, 1, 21, 22, 31, 33, 39, 72, 73, 82, 83, 86, 95, 121, *242*–*251*, 246*, 247*, 248*
Bean, French, *Phaseolus vulgaris*, 219, *257–258*
Bean, goa, *Psophocarpus tetragonolobus*, 277
Bean, haricot, *Phaseolus vulgaris*, 219, *257–258*

Bean, honey locust, *Prosopis juliflora*, 277
Bean, horse, *Canavalia ensiformis*, 277
Bean, hyacinth, *Lablab niger*, 277
Bean, jack, *Canavalia ensiformis*, 277
Bean, kidney, *Phaseolus vulgaris*, 219, *257–258*
Bean, lima, *Phaseolus lunatus*, 257–259
Bean, locust, *Ceratonia siliqua*, 277
Bean, mat, *Phaseolus aconitifolius*, 257
Bean, mesquit, *Prosopis juliflora*, 277
Bean, moth, *Phaseolus aconitifolius*, 257
Bean, potato, *Pachyrrhizus tuberosus*, 277
Bean, rice, *Phaseolus calcaratus*, 257
Bean, runner, *Phaseolus multiflorus*, 125, 131, 257, *259–264*, 260*, 263*
Bean, scarlet runner, *Phaseolus multiflorus*, 125, 131, 257, *259–264*, 260*, 263*
Bean, sieva, *Phaseolus lunatus*, 257–259
Bean, sword, *Canavalia gladiata*, 277
Bean, tepary, *Phaseolus acutifolius* var. *latifolius*, 257
Bean, tonka, *Dipteryx odorata*, 277
Bean, velvet, *Mucuna deeringiana*, 277
Bean, white pea, *Phaseolus vulgaris*, 219, *257–258*
Bean, yam, *Pachyrrhizus erosus*, 277
Beet, sugar,- garden-, spinach-, *Beta vulgaris*, 48, *359*–362*
Begonia, begonia, 122
Berberis, 31
Beta, 359
Beta vulgaris, sugar-, garden-, spinach-, beet and mangold, 48, *359*–362*
Betula, 31
Blackberry, *Rubus fruticosus*, 22, 23, *422–425*
Blueberry, *Vaccinium*, 74, *336–344*, 337*
Blueberry, dryland, *Vaccinium pallidium*, 336
Blueberry, high bush, *Vaccinium atrococeum*, 336–340, 342–344
Blueberry, high bush, *Vaccinium australe*, 336–340, 342–344
Blueberry, highbush, *Vaccinium corymbosum*, 336–340, 342–344
Blueberry, low bush, *Vaccinium angustifolium*, 75, *336–337*, *340–344*
Blueberry, low bush, *Vaccinium boreale*, 336–337, 340–344
Blueberry, low bush, *Vaccinium brittonii*, 336–337, 340–344
Blueberry, low bush, *Vaccinium lamarkii*, 76, *336–337*, *340–344*
Blueberry, low bush, *Vaccinium myrtilloides*, 336–337, 340–344
Blueberry, low bush, *Vaccinium vacillans*, 336–337, 340–344
Blueberry, mountain, *Vaccinium membranaceum*, 336
Blueberry, rabbit eye, *Vaccinium ashei*, 336–337, 342–344
Blueberry, western evergreen, *Vaccinium ovatum*, 336
Bombacaceae, 431–432
Borage, *Borago officinalis*, 104*
Borago officinalis, borage, 104*
Brassica, 48, *135–149*, 250, 408
Brassica alba, white mustard, 18, 19, 21, 31, 47, 76, 121, 135, 145, *148*, 230, 408
Brassica campestris, field mustard, 116, 135, *137*–142
Brassica campestris var. *oleifera*, turnip rape, 137
Brassica campestris var. *sarson*, sarson, Indian colza, 137, *142–144*
Brassica campestris var. *toria*, toria, Indian rape, 142–144
Brassica carinata, Abyssinian cabbage, 135
Brassica chinensis, chinese cabbage, pak-choi, 135
Brassica eruca, rocket cress, 149
Brassica juncea, trowse mustard, 42, 135, 136, 137, 145*, *148–149*
Brassica napobrassica, rutabaga, swede, 135
Brassica napus, rape, 23, 31, 69, 82, 83, 120, 130, 135, 136, *137*–142*, 141*, 230
Brassica napus var. *oleifera*, swede rape, 137
Brassica nigra, black mustard, 121, 135, *148*, 207
Brassica oleracea, wild cabbage and its domestic forms, 22, 69, 73, 120, 123, 130, 135, 136, *145–148*
Brassica pekinensis, 149

PLANT INDEX 529

Brassica rapa, turnip, 18, 46, 47, 82, 123, 135, 136, *149*
Brassica sinapis, wild mustard, 31
Bromeliaceae, 440–441
Broom, *Sarothamnus scoparius*, 38
Brussels sprouts, *Brassica oleracea*, 130
Buckwheat, *Fagopyrum esculentum*, 42, 44, 82, 230, *363*–365*
Buckwheat, tartary, *Fagopyrum tataricum*, 365
Buxus, 31

C

Cabbage, Abyssinian, *Brassica carinata*, 135
Cabbage, Chinese, *Brassica chinensis*, 135
Cabbage, wild, and its domestic forms, *Brassica oleracea*, 22, 69, 73, 120–123, 130, 135, 136, *145–148*
Cactaceae, cacti, 24
Cajanus indicus, pigeon pea, 267–268
Calluna vulgaris, ling, heather, 69
Camellia sinensis, tea, 81, *431**
Canavalia ensiformis, jack bean, horse bean, 277
Canavalia gladiata, sword bean, 277
Canavalia plagiosperma, 277
Cannabis sativus, hemp, 33
Capsicoum annuum, sweet pepper, chillies, 349, *354, 355**
Capsicom frutescens, bird chilli, 349, *354, 355*
Caraway, *Carum carvi*, 317
Carica papaya, pawpaw, *434*–435*
Caricaceae, 434–435
Carissa grandiflora, Natal plum, African carissa, *435*–436*
Carrot, *Daucus carota*, 123, 129, 230, *314*–316*
Carthamus tinctorius, safflower, 121, 322, *331–332*
Carum carvi, caraway, 317
Cashew, *Anacardium occidentale*, *196*–197*
Cassia angustifolia, Indian senna, Tinnevelly senna, 277
Cassia auriculata, avaram, 277
Cassia senna, Alexandrian senna, 277

Castor, *Ricinus communis*, 33, *373*
Cauliflower, *Brassica oleraceae*, 147
Ceiba pentandra, kapok, silk-cotton tree, 431–432
Celery, *Apium graveolens*, 123, 130, *317*
Centaurea cyanus, cornflower, 69, 122
Ceratonia siliqua, carab bean, locust bean, 277
Cheiranthus cheiri, wallflower, 122
Chenopodiaceae, 359–362
Cherry, Barbados, *Malpighia glabra*, 433
Cherry, sour, *Prunus cerasus*, 73, 380*, *381–416*
Cherry, sweet, *Prunus avium*, 18, 24, 41, 72, 86, 283, 285, *381–416*
Cherry, West Indian, *Malpighia glabra*, 433
Chicory, *Cichorium intybus*, 120, 123, 322
Chilli, bird, *Capiscum frutescens*, 349, *354, 355*
Chillies, *Capiscum annuum*, 349, *354, 355**
Chives, *Allium schoenoprasum*, 426
Chrysanthemum cinerariaefolium, 322, 333
Chukpea, *Cicer arietinum*, 268,
Cicer arietinum, chukpea, gram, 268
Cichorium endivia, endive, 120, 123, 322
Cichorium intybus, chicory, 120, 123, 322
Cinchona, quinine, 321
Cinchona calisaya, quinine, 321
Cinchona ledgeriana, quinine, 321
Cinchona officinalis, quinine, 321
Cinchona succirubra, quinine, 321
Cinnamomum zeylanicum, cinnamon, 369
Citron, *Citrus medica*, 183, 184
Citrus, 81, *183*–188*, 369
Citrus aurantifolia, lime, 183
Citrus aurantium, sour orange, Seville orange, 183, 184
Citrus grandis, pummelo, shaddock, 183
Citrus limon, lemon, 183, 184
Citrus medica, citron, 183, 184
Citrus paradisi, grapefruit, 183, 184, 186

Citrus reticulata, mandarin, tangerine, 183–187
Citrus sinensis, sweet orange, 18, *183–185*, 187
Clove, *Eugenia caryophyllus*, 289
Clover, alsike, *Trifolium hybridum*, 17, 44, 45, 82, 230, *239–240*
Clover, ball, *Trifolium nigrescens*, 241
Clover, bitter, *Melilotus indica*, 276
Clover, crimson, *Trifolium incarnatum*, 236–237
Clover, Dutch, *Trifolium repens*, 7, 18, 19, 21, 25, 31, 69, 73, 82, 120, 230, *232–235*, 237
Clover, Egyptian, *Trifolium alexandrinum*, 237–239
Clover, hare's foot, *Trifolium arvense*, 241
Clover, Kura, *Trifolium ambigium*, 241
Clover, ladino, *Trifolium repens latum*, 235–236
Clover, lappa, *Trifolium lappacaeum*, 241
Clover, large hop, *Trifolium procumbens*, 241
Clover, Persian, *Trifolium resupinatum*, 241
Clover, red, *Trifolium pratense*, 11*, 17–19, 21, 25, 31, 36, 37, 39, 41, 42, 47, 69, 72, 76, 81–83, 86, 94, 101, 119, 120, 127, 144, 207, *215–232*, 216*, 217*, 219*, 220*
Clover, rose, *Trifolium hirtum*, 241
Clover, small hop, *Trifolium dubium*, 241
Clover, small yellow annual sweet, *Melilotus indica*, 276
Clover, strawberry, *Trifolium fragiferum* 240–241
Clover, subterranean, *Trifolium subterranean*, 241
Clover, sweet, *Melilotus alba*, 17, 82, 122, 123, 255, 268, *274*–276*
Clover, white, *Trifolium repens*, 7, 18, 19, 21, 25, 31, 69, 73, 82, 120, 230, *232–235*, 237
Clover, zig zag, *Trifolium medium*, 241
Clubmoss, *Lycopodium*, 414
Cochlearia armoracia, horse radish, 149
Cocoa, *Theobroma cacao*, 2, 81, *169*–178*

Coconut palm, *Cocos nucifera*, 299, *441–442**
Cocos nucifera, coconut palm, 299, *441–442**
Coffea, coffee, 81, *318–321*
Coffea arabica, 318*–321
Coffea canephora, 318–321
Coffea dewevrei, 319
Coffea excelsa, 318–321
Coffea liberica, 318–321
Coffee, *Coffea*, 81, *318–321*
Colocynthis citrullus, watermelon, 311–313
Colza, Indian, *Brassica campestris* var *sarson*, 137, *142–144*
Compositae, 322–335
Convolvulaceae, 24, *436*
Corchorus capsularis, white jute, round podded jute, 432
Corchorus olitorius, tossa jute, long podded jute, 432, 433
Coriander, *Coriandrum sativum*, 317
Coriandrum sativum, coriander, 317
Cornflower, *Centaurea cyanus*, 69, 122
Coronilla varia, crown vetch, 268
Cotoneaster horizontalis, 42
Cotton, Asiatic, *Gossypium herbaceum*, 151
Cotton, Egyptian, *Gossypium barbadense*, 151–166
Cotton, long staple, *Gossypium barbadense*, 151–166
Cotton, sea island, *Gossypium barbadense*, 151–166
Cotton, short staple, *Gossypium hirsutum*, 48, 67, *151*–166*
Cotton, tree, *Gossypium arboreum*, 151
Cotton, upland, *Gossypium hirsutum*, 151*–166
Cowpea, *Vigna unguiculata*, 277
Cranberry, *Vaccinium* spp., 344–348, 345*
Cranberry, *Vaccinium oxycoccus*, 344–345*
Cranberry, American, *Vaccinium macrocarpum*, 73, *344–348*
Cranberry, large, *Vaccinium macrocarpum*, 73, *344–348*
Crataegus monogyna, hawthorn, 38, 408
Crotalaria juncea, sann hemp, sunn hemp, 268–269*

Cruciferae, 69, *135*–150*
Cucumber, *Cucumis sativus*, 8, 82, 123, 125, 126, 127, *301*–304*
Cucumis melo, melon, 1, 7, 83, 126, *304–311*, 305*
Cucumis sativus, cucumber, 2, 82, 123, 125, 126, 127, *301*–304*
Cucurbita, 297–301
Cucurbita maxima, pumpkin, 297–301
Cucurbita mixta, pumpkin, 297–301
Cucurbita moschata, squash, pumpkin, 297–301
Cucurbita pepo, marrow, 21, 22, *297*–301*
Cucurbitaceae, 24, *297–313*
Cuminum cyminum, cumin, 317
Currant, black, *Ribes nigrum*, 21, 22, 24, 73, 121, *278*–285*, 286, 287
Currant, red, *Ribes rubrum*, 286*, 287
Cyamopsis tetragonoloba, cluster bean, 277
Cynara scolymus, globe artichoke, 322

D

Dandelion, *Taraxacum officinale*, 23, 42, 44, 73, 408, 409
Date palm, *Phoenix dactylifera*, 443
Daucus carota, carrot, 123, 129, 230, *314*–316*
Derris elliptica, derris, tuba root, 277
Dewberry, *Rubus caesius*, 422–425
Dill, *Anethum graveolens*, 317
Dipteryx odorata, tonka bean, 277
Dolichos uniflorus, horse gram, 277
Doryenium, 42
Duan, *Eruca sativa*, 150

E

Echinops sphaerocephalus, ornamental thistle, 42
Echium vulgare, 15
Egg-plant, *Solanum melongena*, 349, *358*
Elder, *Sambucus*, 102
Endive, *Cichorium endivia*, 120, 123, 322
Epilobium angustifolium, willow herb, 31, 42
Erica, 25
Eruca sativa, rocket cress, taramira, duan, 150

Eschscholtzia, 44
Eucalyptus, 19, 84, *288–289*
Eucalyptus alba, 289
Eucalyptus bicostata, 288
Eucalyptus blakelyi, 288
Eucalyptus citriodora, 288
Eucalyptus dives, 288
Eucalyptus fruticetorum, 288
Eucalyptus globulus, 288
Eucalyptus maidenii, 288
Eucalyptus radiata, 288
Eucalyptus smithii, 288
Eucalyptus staigeriana, 288
Eugenia, 288, 289
Eugenia caryophyllus, clove, 289
Eugenia jambosi, rose apple, 289
Euphorbiaceae, 370–373

F

Fagopyrum, 363–365
Fagopyrum emarginatum, 24, 365
Fagopyrum esculentum, buckwheat, 42, 44, 82, 230, *363*–365*
Fagopyrum tataricum, tartary buckwheat, 365
Fagopyrum tetratataricum, tetraploid variety of tartary buckwheat, 365
Feijoa, 288, 289
Feijoa sellowiana, feijoa, 289
Fennel, *Foeniculum vulgare*, 317
Ficus carica, wild fig, capri fig, 374*–379, 375*
Fig, capri, wild, *Ficus carica*, 374*–379, 375*
Flax, *Linum usitatissimum*, 179*–182, 180*
Foeniculum vulgare, fennel, saunf, 317
Fragaria, 417–421
Fragaria × *ananassa*, strawberry, 1, 22, 24, 42, 73, 83, 123, 126, 131, *417–421*, 418*
Freesia refracta, freesia, 126
Furze, *Ulex europaeus*, 408

G

Garlic, *Allium sativum*, 426
Giant granadilla, *Passiflora quadrangularis*, 292, 296
Gilia capitata, 37

Glycine max, soyabean, 269–271, *270**
Goldenrod, *Solidago virgaurea*, 44, 230
Gooseberry, *Ribes grossularia*, 24, 283, *285–287*, 286*
Gorse, *Ulex*, 31
Gossypium, 33, 82, 84, 94, *151–166*
Gossypium arboreum, tree cotton, 151
Gossypium barbadense, long staple cotton, Egyptian cotton, sea island cotton, 151–166
Gossypium herbaceum, Asiatic cotton, 151
Gossypium hirsutum, short staple cotton, upland cotton, 48, 67, *151*–166*
Gourd, bottle, *Lagenaria siceraria*, 299, *313*
Gourd, ridge, *Luffa acutangula*, 313
Gram, black, *Phaseolus mungo*, 257
Gram, golden, green, *Phaseolus aureus*, 257
Gram, horse, *Dolichos uniflorus*, 277
Grape, *Vitis vinifera*, 189
Grapefruit, *Citrus paradisi*, 183, 184, 186
Grossulariaceae, 278–287
Groundnut, *Arachis hypogaea*, 265*–267
Groundnut, bambara, *Voandzeia subterranea*, 277
Guava, *Psidium guajava*, 290
Guayale, *Parthenium argentatum*, 322, *333–334*
Guizotia abyssinica, niger, 322, *332–333*

H

Hakea saligna, 408
Hawthorn, *Crataegus monogyna*, 38, 408
Heather, ling, *Calluna vulgaris*, 69
Hedera helix, ivy, 31
Helianthemum, 42
Helianthus annuus, sunflower, 23, 33, 38, 42, 46, 73, 81, 103, 207, *322*–331*, 323*, 324*
Helianthus tuberosus, Jerusalem artichoke, 322
Hemp, *Cannabis sativus*, 33
Hemp, sann, sunn, *Crotalaria juncea*, 268–269*

Heracleum lanatum, cow parsnip, 102
Hevea brasiliensis, para rubber tree, 371*–373
Hibiscus, 151, *166–168*
Hibiscus cannabinus, kenaf, 166–168, 167*
Hibiscus esculentus, okra, 168
Hibiscus sabdariffa, roselle, 168

I

Indigo, Java, *Indiofera arrecta*, 271
Indigo, Sumatrana, *Indiofera sumatrana*, 271
Indiofera arrecta, Java indigo, 271
Indiofera sumatrana, Sumatrana indigo, 271
Ipomoea batatas, sweet potato, 436*
Ivy, *Hedera helix*, 31

J

Juncus, reed, 116
Jute, long podded, *Corchorus olitorius*, 432, 433
Jute, round podded, *Corchorus capsularis*, 432
Jute, tossa, *Corchorus olitorius*, 432, 433
Jute, white, *Corchorus capsularis*, 432

K

Kale, *Brassica oleracea*, 130
Kapok, *Ceiba pentandra*, 431–432
Kenaf, *Hibiscus cannabinus*, 166–168, 167*
Kowhai, *Sophora microphylla*, 408

L

Lablab niger, hyacinth bean, 277
Lactuca sativa, lettuce, 322, *334–335*
Lagenaria siceraria, bottle gourd, 299, *313*
Lathyrus sativus, grass pea, chickling pea, 277
Lauraceae, 367–369
Leek, *Allium ampeloprasum*, 426

Lemon, *Citrus limon*, 183, 184
Lens esculenta, lentil, 277
Lentil, *Lens esculenta*, 277
Leptospermum scoparium, manuka, 408
Lettuce, garden, *Lactuca sativa*, 322, *334–335*
Liliaceae, 426–429
Lime, *Citrus aurantifolia*, 183
Limnanthes douglasii, 38
Linaceae, 179–182
Linaria vulgaris, toadflax, 15
Linseed, *Linum usitatissimum*, 179*–182, 180*
Linum usitatissimum, flax, linseed, 179*–182, 180*
Litchi chinensis, lychee, 433–434
Loofah, angled, *Luffa acutangula*, 313
Loofah, smooth, *Luffa aegyptiaca*, 313
Lotus, trefoil, 230, *271–273*
Lotus corniculatus, birdsfoot trefoil, 25, 44, 72, 119, 127, 268, *271–273*, 272*
Lucerne, *Medicago sativa*, 3, 17, 19, 25, 26, 32, 36, 41, 45, 48, 66, 72, 75, 76, 79, 82, 83, 87, 88, 102, 104*, 109*, 120, 122, 126, 127, *198–214*, 199*, 203*, 230
Luffa acutangula, angled loofah, ridge gourd, 313
Luffa aegyptiaca, smooth loofah, 313
Lupin, blue, *Lupinus angustifolius*, 273–274
Lupin, yellow, *Lupinus luteus*, 273–274
Lupinus, 273–274
Lupinus angustifolius, blue lupin, 273–274
Lupinus luteus, yellow lupin, 273–274
Lychee, *Litchi chinensis*, 433–434
Lycopersicon esculentun, tomato, 126, *349*–354*, 353*
Lycopersicon pimpinellifolium, wild tomato, "currant" tomato, 352
Lycopodium, clubmoss, 414

M

Mace, *Myristica fragrans*, 438–439*
Mallow, Jew's, *Corchorus olitorius*, 432, 433
Malpighia glabra, Barbados cherry, West Indian cherry, 433

Malpighiaceae, 433
Malvaceae, 24, *151–168*
Mandarin, *Citrus reticulata*, 183–187
Mangifera indica, mango, 193*–196
Mango, *Mangifera indica*, 193*–196
Mangold, *Beta vulgaris*, 48, *359*–362*
Manuka, *Leptospermum scoparium*, 408
Manzanita, *Arctostaphylos manzanita*, 18, 407, 408
Marrow, *Cucurbita pepo*, 21, 22, *297*–301*
Medicago, 198–214
Medicago sativa, lucerne, alfalfa, 3, 17, 19, 25, 26, 32, 36, 41, 45, 48, 66, 72, 75, 76, 79, 82, 83, 87, 88, 102, 104*, 109*, 120, 122, 126, 127, *198–214*, 199*, 203*, 230
Melilot, common, *Melilotus officinalis*, 122, *274–276*
Melilot, white, *Melilotus alba*, 17, 82, 122, 123, 255, 268, *274*–276*
Melilot, yellow, *Melilotus officinalis*, 122, *274–276*
Melilotus, 42, 207, 230, *274–276*
Melilotus alba, sweet clover, white melilot, 17, 82, 122, 123, 255, 268, *274*–276*
Melilotus dentata, 276
Melilotus indica, small yellow annual sweet clover, bitter clover, 276
Melilotus officinalis, yellow melilot, common melilot, 122, *274–276*
Melilotus sauveolens, 276
Melilotus taurica, 276
Melon, *Cucumis melo*, 1, 7, 83, 126, *304–311*, 305*
Mesquit bean, *Prosopis juliflora*, 277
Mignonette, *Reseda* spp., 47
Momordica, 313
Monkeynut, *Arachis hypogaea*, 265*–267
Moraceae, 374–379
Mucuna deeringiana, velvet bean, 277
Mung, *Phaseolus aureus*, 257
Musa, banana, 81, *441*
Musaceae, 441
Mustard, *Brassica* spp., 18, 47, 207
Mustard, black, *Brassica nigra*, 121, 135, *148*, 207
Mustard, field, *Brassica campestris*, 116, 135, *137–142*

534 PLANT INDEX

Mustard, Indian, *Brassica juncea*, 42, 135, 137, 145*, *148, 149*
Mustard, wild, *Brassica sinapis*, 31
Myristica fragrans, nutmeg, mace, 438–439*
Myristicaceae, 438–439
Myrtaceae, 288–291

N

Nasturtium, *Tropaeolum majus*, 44
Nectarine, *Prunus persica*, 23, 72, 73, 121, 126, *381–416*, 382*
Nicotiana rustica, yellow flowered tobacco, nicotine tobacco, 356–357
Nicotiana tabacum, tobacco, 349, *355–356**
Niger seed, *Guizotia abyssinica*, 322, *332–333*
Nutmeg, *Myristica fragrans*, 438–439*

O

Oak, *Quercus*, 31
Okra, *Hibiscus esculentus*, 168
Onagraceae, 24
Onion, common, *Allium cepa*, 69, 127, 128*, 129, *426–428*, 427*
Onion, Welsh, *Allium fistulosum*, 46, *426*, 428
Onobrychis, 276–277
Onobrychis sativa, sainfoin, 81
Onobrychis viciifolia, sainfoin, 276*–277
Orange, *Citrus* spp., 18, *183–188*
Orange, Seville, *Citrus aurantium*, 183, 184
Orange, sour, *Citrus aurantium*, 183, 184
Orange, sweet, *Citrus sinensis*, 18, *183–185, 187*
Orchidaceae, 439–440
Oyster plant, *Tragopogon porrifolius*, 123, *322*

P

Pachyrrhizus erosus, yam bean, 277
Pachyrrhizus tuberosus, yam bean, potato bean, 277

Pak-choi, *Brassica chinensis*, 135
Palmae, 441–443
Papaver rhoeas, shirley poppy, 31, 38, 44
Papaver somniferum, medicinal poppy, opium poppy, oil poppy, 72, *430**–*431*
Papaveraceae, 42, *430–431*
Papilionaceae, 198*–277
Para rubber tree, *Hevea brasiliensis*, 371*, 373
Parsley, *Petroselinum crispum*, 317
Parsnip, *Pastinaca sativa*, 130, *317**
Parsnip, cow, *Heracleum lanatum*, 102
Parthenium argentatum, guayale, 322, *333**–*334*
Passiflora edulis, passion fruit, 118, *292**–*296*
Passiflora quadrangularis, giant granadilla, 292, *296*
Passifloraceae, 292–296
Passion fruit, *Passiflora edulis*, 118, *292**–*296*
Pastinaca sativa, parsnip, 130, *317**
Pawpaw, *Carica papaya*, 434*–435
Pea, chickling, *Lathyrus sativus*, 277
Pea, field, *Pisum sativum*, 277
Pea, garden, *Pisum sativum*, 277
Pea, grass, *Lathyrus sativus*, 277
Pea, pigeon, *Cajanus indicus*, 267–268
Peach, *Prunus persica*, 23, 72, 73, 121, 126, *381–416*, 382*
Peanut, *Arachis hypogaea*, 265*–267
Pear, *Pyrus communis*, 23, 41, 72–74, 82, 83, *381–416*, 382*
Pedaliaceae, 437
Pepper, *Piper nigrum*, 437*–438
Pepper, sweet, *Capiscum annuum*, 349, *354, 355**
Persea americana, avocado, 367*–369
Petroselinum crispum, parsley, 317
Petunia hybridae, 83
Phaseolus, 257–264
Phaseolus aconitifolius, mat bean, moth bean, 257
Phaseolus acutifolius var. *latifolius*, tepary bean, 257
Phaseolus angularis, adzuki bean, 257
Phaseolus aureus, green gram, golden gram, mung, 257
Phaseolus calcaratus, rice bean, 257

Phaseolus lunatus, butter bean, sieva bean, lima bean, 257–259
Phaseolus multiflorus, runner bean, scarlet runner bean, 125, 131, 257, *259–264*, 260*, 263*
Phaseolus mungo, black gram, urd, 257
Phaseolus vulgaris, common bean, french bean, kidney bean, haricot bean, white pea bean, 219, *257*, *258*
Phoenix dactylifera, date palm, 443
Pimenta, 288, *290*, *291*
Pimenta dioica, allspice, pimento, 290, 291
Pimento, *Pimenta dioica*, 290, 291
Pimpinella anisum, anise, 317
Pineapple, *Ananas sativus*, 440–441
Piper nigrum, pepper, 437*–438
Piperaceae, 437–438
Pisum, 277
Pisum sativum, field pea, garden pea, 277
Plum, *Prunus americana*, 381–416
Plum, *Prunus domestica*, 381–416
Plum, *Prunis insititia*, 381–416
Plum, *Prunus salicina*, 381–416
Plum, natal, *Carissa grandiflora*, 435*–436
Polygonaceae, 363–366
Poppy, medicinal, *Papaver somniferum*, 72, *430*–431*
Poppy, oil, *Papaver somniferum*, 72, *430*–431*
Poppy, opium, *Papaver somniferum*, 72, *430*–431*
Poppy, shirley, *Papaver rhoeas*, 31, 38, 44
Potato, *Solanum tuberosum*, 349, *357–358*
Potato sweet, *Ipomoea batatas*, 436
Prosopis juliflora, mesquit bean, algaroba, honey locust, 277
Prunus, 380–416
Prunus americana, 381–416
Prunus amygdalus, almond, 381–416
Prunus armeniaca, apricot, 381–416
Prunus avium, sweet cherry, 18, 24, 41, 72, 86, 283, 285, *381–416*
Prunus cerasus, sour cherry, 73, 380*, *381–416*
Prunus domestica, plum, 18, 77, 78, 126, *381–416*
Prunus insititia, plum, 381–416
Prunus persica, peach, nectarine, 23, 72, 73, 121, 126, *381–416*, 382*
Prunus salicina, plum, 381–416
Psidium, 288, 290
Psidium guajava, guava, 290
Psophocarpus tetragonolobus, goa bean, 277
Pummelo, *Citrus grandis*, 183
Pumpkin, *Cucurbita maxima*, 297–301
Pumpkin, *Cucurbita mixta*, 297–301
Pumpkin, *Cucurbita moschata*, 297–301
Pyrus, 380–416
Pyrus communis, pear, 23, 41, 72, 73, 74, 82, 83, *381–416*, 382*
Pyrus malus, apple, 32, 34, 38, 44, 45, 69, 72, 73, 76, 77, 79, 82, 83, 116, 121, *381*–416*, 382*, 393*, 394*

Q

Quercus, oak, 31
Quinine, *Cinchona*, 321
Quinine, *Cinchona calisaya*, 321
Quinine, *Cinchona ledgeriana*, 321
Quinine, *Cinchona officinalis*, 321
Quinine, *Cinchona succirubra*, 321

R

Radish, *Raphanus sativus*, 46, 47, 120, 122, 123, 135, *149*
Radish, horse, *Armoracia rusticana*, 135
Radish, horse, *Cochlearia armoracia*, 149
Rakkyo, *Allium chinese*, 426
Ranunculus, 31
Ranunculus acris, 22
Rape, *Brassica napus*, 23, 31, 69, 82, 83, 120, 130, 135, 136, *137*–142*, 141*, 230
Rape, Indian, *Brassica campestris* var. *toria*, 142–144
Rape, swede, *Brassica napus* var. *oleifera*, 137
Rape, turnip, *Brassica campestris* var. *oleifera*, 137
Raphanus sativus, radish, 46, 47, 120, 122, 123, 135, *149*
Raspberry, *Rubus idaeus*, 23, 24, *422–425*, 424*

Raspberry, *Rubus occidentalis*, 422–425
Raspberry, American red, *Rubus strigosus*, 422*–425
Reeds, *Juncus*, 116
Reseda spp., mignonette, 42
Rheum rhaponticum, rhubarb, 130, 365–366*
Rhubarb, *Rheum rhaponticum*, 130, 365–366*
Ribes, 278–287
Ribes grossularia, gooseberry, 24, 283, 285–287, 286*
Ribes nigrum, black currant, 21, 22, 24, 73, 121, 278*–285, 286, 287
Ribes rubrum, red currant, 286*, 287
Ricinus communis, castor, 33, *373*
Rocket cress, *Brassica eruca*, 149
Rocket cress, *Eruca sativa*, 150
Rosa, 31
Rosaceae, 380–425
Roselle, *Hibiscus sabdariffa*, 168
Rubiaceae, 318–321
Rubus, 422–425
Rubus caesius, dewberry, 422–425
Rubus fruticosus, blackberry, 22, 23, *422–425*
Rubus idaeus, raspberry, 23, 24, *422–425*, 424*
Rubus occidentalis, raspberry, 422–425
Rubus strigosus, American red raspberry, 422*–425
Rutabaga, *Brassica napobrassica*, 135
Rutaceae, 183–188

S

Safflower, *Carthamus tinctorius*, 121, 322, *331–332*
Sage, annual, *Salvia carduacea*, 24
Sainfoin, *Onobrychis sativa*, 81
Sainfoin, *Onobrychis viciifolia*, 276*–277
Salix, 31
Salsify, *Tragopogon porrifolius*, 123, 322
Salsify, black, *Scorzonera hispanica*, 322
Salvia, 42
Salvia carduacea, thistle sage, annual sage, 24
Sambucus, elder, 102

Sapindaceae, 433–434
Sarothamnus scoparius, broom, 38
Sarson, *Brassica campestris* var. *sarson*, 137, *142–144*
Saunf, *Foeniculum vulgare*, 317
Scorzonera, 130
Scorzonera hispanica, black salsify, 322
Sechium, 313
Senna, Alexandrian, *Cassia senna*, 277
Senna, Indian, *Cassia angustifolia*, 277
Senna, tinnevelly, *Cassia angustifolia*, 277
Sesame, *Sesamum indicum*, 437
Sesamum indicum, sesame, 437
Shaddock, *Citrus grandis*, 183
Shallot, *Allium cepa*, 69, 127, 128*, 129, *426–428*, 427*
Silk-cotton tree, *Ceiba pentandra*, 431–432
Snapdragon, *Antirrhinum majus*, 36, 37
Solanaceae, 349–358
Solanum melongena, egg-plant, 349, *358*
Solanum tuberosum, potato, 349, *357–358*
Solidago virgaurea, goldenrod, 44, 230
Sophora microphylla, kowhai, 408
Soyabean, *Glycine max*, 269–271, 270*
Spinach, *Spinacia oleracea*, 362
Spinach, New Zealand, *Tetragonia expansa*, 362
Spinacia, 359
Spinacia oleracea, spinach, 362
Squash, *Cucurbita moschata*, 297–301
Stellaria, 408
Sterculiaceae, 169–178
Strawberry, *Fragaria* × *ananassa*, 1, 22, 24, 42, 73, 83, 123, 126, 131, 417–*421*, 418*
Sunflower, *Helianthus annuus*, 23, 33, 38, 42, 46, 73, 81, 103, 207, *322*–331*, 323*, 324*
Swede, *Brassica napobrassica*, 135
Sycamore, *Acer pseudoplatanus*, 25

T

Tangelo, 186–188
Tangerine, *Citrus reticulata*, 183–187
Taramira, *Eruca sativa*, 150
Taraxacum officinale, dandelion, 23, 42, 44, 73, 408, 409

Tare, *Vicia sativa*, 219, 242, *256*
Tea, *Camellia sinensis*, 81, *431**
Tetragonia, 359
Tetragonia expansa, New Zealand spinach, 362
Theaceae, 431*
Theobroma cacao, cocoa, 2, 81, *169*–178*
Thistle, ornamental, *Echinops sphaerocephalus*, 42
Tilia, 69
Tiliaceae, 432–433
Toadflax, *Linaria vulgaris*, 15
Tobacco, *Nicotiana tabacum*, 349, *355–356**
Tobacco, nicotine, *Nicotiana rustica*, 356–357
Tobacco, yellow flowered, *Nicotiana rustica*, 356–357
Tomato, *Lycopersicon esculentum*, 126, *349*–354*, 353*
Tomato, currant, wild, *Lycopersicon pimpinellifolium*, 352
Toria, *Brassica campestris* var. *toria*, 142–144
Tragopogon porrifolius, salsify, oyster plant, 123, *322*
Trefoil, *Lotus*, 230, *271–273*
Trefoil, birdsfoot, *Lotus corniculatus*, 25, 44, 72, 119, 127, 268, *271–273*, *272**
Trefoil, lesser yellow, *Trifolium dubium*, 241
Trichosanthes, 313
Trifolium, 48, *215–241*, 250
Trifolium alexandrinum, Egyptian clover, 237–239
Trifolium ambiguum, Kura clover, 241
Trifolium arvense, hare's foot clover, 241
Trifolium dubium, small hop clover, 241
Trifolium fragiferum, strawberry clover, 240–241
Trifolium hirtum, rose clover, 241
Trifolium hybridum, alsike clover, 17, 44, 82, 230, *239*–240
Trifolium incarnatum, crimson clover, 236–237
Trifolium isthmocarpum, 241
Trifolium lappacaeum, lappa clover, 241
Trifoilium medium, zig zag clover, 241

Trifolium michelianum, 241
Trifolium nigrescens, ball clover, 241
Trifolium pratense, red clover, 11*, 17, 18, 19, 21, 25, 31, 36, 37, 39, 41, 42, 47, 69, 72, 76, 81, 82, 83, 86, 94, 101, 119, 120, 127, 144, 207, *215–232*, 216*, 217*, 219*, 220*
Trifolium procumbens, large hop clover, 241
Trifolium repens, Dutch clover, white clover, 7, 18, 19, 21, 25, 31, 69, 73, 82, 120, 123, *232–235*, 237
Trifolium repens latum, ladino clover, 235–236
Trifolium resupinatum, Persian clover, 241
Trifolium subterranean, subterranean clover, 241
Trifolium xerocephalum, 241
Tropaeolum majus, nasturtium, 44
Trowse, *Brassica juncea*, 42, 135–137, 145*, *148–149*
Tuba root, *Derris elliptica*, 277
Tung, *Aleurites fordii*, 370–371
Tung, *Aleurites montana*, 370–371
Turnip, *Brassica rapa*, 18, 46, 47, 82, 123, 135, 136, *149*

U

Ulex, gorse, 31
Ulex europaeus, gorse, furze, whin, 408
Umbelliferae, 314–317
Urd, *Phaseolus mungo*, 257

V

Vacciniaceae, 336–348
Vaccinium, 336–348, 337*
Vaccinium angustifolium, low bush blueberry, 75, *336–337*, *340–344*
Vaccinium ashei, rabbit eye blueberry, 336–337, 342–344
Vaccinium atrococeum, highbush blueberry, 336–340, 342–344
Vaccinium australe, highbush blueberry, 336–340, 342–344
Vaccinium boreale, low bush blueberry, 336–337, 340–344
Vaccinium brittonii, low bush blueberry, 336–337, 340–344

Vaccinium canadense, lowbush blueberry, 336–337, 340–344
Vaccinium corymbosum, highbush blueberry, 336–340, 342–344
Vaccinium lamarikii, low bush blueberry, 76, *336–337, 340–344*
Vaccinium macrocarpum, "large" or "American" cranberry, 73, *344–348*
Vaccinium membranaceum, mountain blueberry, 336
Vaccinium myrtilloides, low bush blueberry, 336–337, 340–344
Vaccinium myrtillus, 15
Vaccinium ovatum, western evergreen blueberry, 336
Vaccinium oxycoccus, cranberry, 344–345*
Vaccinium pallidium, dryland blueberry, 336
Vaccinium pennsylvanicium, lowbush blueberry, 336–337, 340–344
Vaccinium vacillans, lowbush blueberry, 336–337, 340–344
Vanilla, 439*–440*
Vanilla planifolia, 439–440
Vanilla pompona, 439–440
Vanilla tahitensis, 439–440
Vetch, common, *Vicia sativa*, 219, 242, *256*

Vetch, crown, *Coronilla varia*, 268
Vetch, hairy, *Vicia villosa*, 45, 46, 75, 118, 122, 242, *251–256*
Vicia, 25, 230, *242–256*
Vicia faba, broad bean, field bean, 1, 21, 22, 31, 33, 39, 72, 73, 82, 83, 86, 94, 121, *242*–251*, 246*, 247*, 248*
Vicia sativa, common vetch, tare, 219, 242, *256*
Vicia villosa, hairy vetch, 45, 46, 75, 118, 122, 242, *251–256*
Vigna unguiculata, cowpea, 277
Viola, 42
Vitaceae, 189–192
Vitis, grape, 189*–192
Vitis munsoniana, grape, 189
Vitis rotundifolia, grape, 189
Vitis vinifera, grape, 189
Voandzeia subterranea, bambara ground, nut, 277

W

Wallflower, *Cheiranthus cheiri*, 122
Watermelon, *Colocynthis citrullus*, 311–313
Whin, *Ulex europaeus*, 408
Willow-herb, *Epilobium angustifolium*, 31, 42

General Index

A

Adaptability of worker bees
  and age, 92, 94
  to duties, 56–57
  to forage, 38
Anaesthetics
  marking bees, 10
Androecium
  attractive organ, 15
Antennae, 58
Anther
  dehiscence, 21–23 (see also under crop concerned)
Anthesis, 21–23, (see also under crop concerned)
Artificial aids to pollination, 201, 202, 266, 345, 351, 352, 377, 410–416, 443
Attractants, 21, 29, 30, 40, 74
Attractiveness of crops
  attempts to increase, 84, 212, 213
  breeding to increase, 4, 83, 84, 201, 211, 212, 226, 227, 234
Auxins, 351

B

Beehive, 50
Behaviour on flowers
  and pollination, 33, 84, 138, 145, 157, 181, 184, 195, 203–212, 218–228, 230–231, 233, 248, 253, 266–269, 271, 273, 296, 328, 331, 332, 343, 345, 361, 364, 369, 372, 373, 375–377, 389, 392–395, 420, 421, 424, 425
Blowflies
  in enclosed spaces, 127–131
Breeding
  crops for attractiveness, 4, 83, 84, 201, 211, 212, 226, 227, 234

honeybees for pollination, 87, 88, 211, 232
Brood
  of bumblebees, 89–92
  of honeybees, 51–58
  of solitary bees, 103, 105–106, 110, 112–114, 117
Brood food glands, 21, 56–58
Brood rearing
  age of nurse bees, 55–57
  and pollen collection, 85
  and proportion of bees to brood, 65, 66, 91, 92
  effect of colony size on, 65, 91, 92
  effect of feeding syrup on, 86
  effect of food stores on, 86
  seasonal fluctuations in, 63, 65
Bumblebees, 3, 89–101
  artificial nest sites of, 95–97, 100, 101
  colony founding, 89–91, 95–100
  colony growth, 91–92
  colony maturity, 93, 94
  colony size, 94
  foraging preferences of, 25, 26
  foraging constancy of, 34, 35, 39
  foraging areas of, 46
  hibernating queens, 89, 99–101
  pollinating in enclosed spaces, 119–120
  importing beneficial species of, 101
  laboratory domiciles, 96–100
  value as pollinators, 94, 155, 156, 218–221, 225, 226, 228–232, 244, 247, 250, 259, 285, 388, 389

C

Cages, 4, 5, 6, 49, 121, 245
Caste
  of bumblebees, 89, 94
  of honeybees, 50
Cell
  capping, 55

Cell (*contd.*)
  cleaning, 55, 61
  preparing for eggs, 61
  preparing for pollen, 64, 90
Colour
  and foraging, 15
  of beehives, 71
  of flowers, 15
  vision, 15
Comb building, 54–58, 89–94
Communication
  between bumblebees, 26
  food transfer, 26, 27, 58–61, 86
  honeybee dances, 26–30, 38, 41, 62, 81, 82, 285
  pheromones, 29, 30, 40, 58, 59, 61, 64, 66, 74, 87
Competition
  between crops, 4, 18, 66, 142, 144, 149, 207, 230, 250, 302, 343, 399–402, 404, 405
  between foragers, 45, 210, 211, 343, 344
Constancy
  at new sites, 40, 72
  during consecutive trips, 37–40, 404, 405
  during single trip, 34–36, 404, 405
  to area of crop, 41–46
  to collecting nectar and pollen, 38, 39, 62, 63, 204, 248, 249, 254, 264, 328, 329, 395
  to flower species, 34–36, 37, 126, 311
Contamination of seed crops, 45–49, 119, 120, 122, 126, 128, 135, 136, 162–166, 172, 180, 200, 258, 277, 298, 304, 311, 332, 335, 353–355, 360, 362
Corbicula, 19, 20
Cordovan strain, 9
Corolla
  depth of, 33, 94, 224–226, 228, 230, 231, 247
Cross-pollination, 1, 41–43, 46–49 (see also under crop concerned)
Cultural practices, 2, 3, 4, 364, 383, 390, 391

D

Dance language, 26–30, 38, 41, 62, 81, 82, 285

Dehiscence of anthers, 21–23 (see also under crop concerned)
Deposition of pollen, 63–64, 90
Discrimination between species, 36, 37
Development stages
  of blowflies, 128, 129
  of bumblebees, 89, 90
  of honeybees, 50–55
  of solitary bees, 103, 105–106, 110, 112–114
"Directing" bees to crops, 81–83
Direction communication, 26–29
Distance communication, 26–29
Distribution of colonies, 3, 75–81, 126
Division of labour
  adaptability and, 56–64
  age and occupation and, 92, 94
  among bumblebees, 92, 94
  among honeybees, 50–64
  in abnormal colonies, 56, 57
  body size and, 92, 94
Domiciles, 95–101
Drifting, 71
Drones, 50

E

Egg-laying
  by blowflies, 128, 129
  by bumblebees, 89–92
  by honeybees, 52, 55, 65
  by solitary bees, 103, 105, 112, 113
Enclosures, pollination in, 119–131
  by blowflies, 127–131
  by bumblebees, 119, 120
  by honeybees, 121–126
  by solitary bees, 126, 127
Extra-floral nectary, 33, 34, 153, 154, 167, 248, 249, 251, 258, 296, 330, 437

F

Fat body, 94
Feeding
  food transfer, 26, 27, 58–61, 86
Feeding bumblebee colonies, 120
Feeding honeybee colonies, 160
  in enclosures, 123, 124
  pollen, 84, 85
  pollen substitutes, 84, 85
  sugar syrup, 86, 87

GENERAL INDEX

Flowers
  attraction of, 7, 15, 25
  colour of, 15
  constancy to, 34–37, 126
  density of, 7, 44
  preference for kinds of, 24–26, 30. 31, 165
  rate of visiting, 140, 144, 149, 155, 156, 167, 177, 181, 184, 229, 234, 235, 237, 248, 254, 266, 272, 273, 275, 284, 311, 330, 343, 388, 395–397, 425
  shape, 15, 36, 37
  size, 36, 46
  structure of (see under crop concerned)
Flying activity
  effect of brood rearing on, 63, 64
  effect of colony size on, 65, 66
  effect of foraging range on, 41, 69
  effect of light intensity on, 74
  effect of weather conditions on, 41, 67, 73, 74, 76, 80, 398, 399
Food
  larval, 56–58, 89, 103, 105, 112, 113
  storage of, 58, 63, 72, 84, 85, 90
  transfer of, 26, 27, 58, 59, 62, 86
Foraging
  areas, 41–49, 75–81, 126, 164, 284, 329, 330, 402, 420
  constancy, 34–46, 62, 63, 72, 126
  distance, 41–46, 67–71, 327
  method, 39, 40
  of neighbouring colonies, 31–33
  in new site, 40, 72, 75, 76
  preferences, 24–26, 30, 41, 43
  range of colonies, 56–71, 75, 76
Fruit
  shape of, 302, 312, 381, 382, 418, 419
  size of, 185, 282, 302, 339, 341, 346, 350, 351, 418, 423

G

Glands
  hypopharangeal, 21, 56–58
  labial, 18
  mandibular, 56
  Nasanov, 29, 30, 40, 74
  wax, 56–58
Glasshouse
  pollination in, 49, 119–131

H

Hand pollination
  commercial, 410–412, 440
  for assessing pollination need of, 6, 7, 261, 278, 293, 410
Herbicides
  3, 95, 409
Hibernation
  of bumblebees, 89, 99–101
Hives, 50
  colour of, 71
  location of, 71, 80
Honeybee colonies
  concentration needed, 7, 8, 66, 67, 161, 162, 188, 213, 214, 229, 231, 234–237, 251, 255, 256, 275, 276, 299, 309, 310, 313, 330, 331, 344, 365, 369, 397–399
  distribution of, 3, 75–81
  foraging range of, 67–69, 70, 71, 75, 76
  management of, for pollination, 3, 65–88
  organization of, 50–64
  siting of, 80
  size of, 50, 65, 66
Honeybees
  pollinating value of, 3 (see also under crop concerned)
Honey production, 70 (see also under crop concerned)
  effect of foraging range on, 67–69
  effect of size of colony on, 65
Honeystomach, 16, 17, 26
Honeystores, 58
Housebee, 55–61
Hybrid seed,
  formation of, 1, 2, 36, 37, 49, 136, 146, 160, 162, 166, 250, 304, 315, 352, 354, 428
Hybrid vigour, 1, 166
Hypopharyngeal glands, 21, 56, 57, 58

I

Importing beneficial species, 4, 101, 103, 210, 300
Insecticides, 3–5, 73, 95, 108, 127, 173, 174, 195
Isolating mechanisms, 36, 37

Isolation
  distances, 45–49, 119, 120, 122, 126, 128, 135, 162–166, 172, 180, 200, 258, 277, 298, 304, 311, 332, 335, 353–355, 360, 362
  from bees, 2, 302

L

Labial glands, 18
Landmarks, 45
Larvae
  bumblebee, 89–91
  feeding, 55, 56
  honeybee, 52–55
  solitary bee, 105, 110, 112, 113, 117
  stimulating foraging, 63
Larva/worker ratio, 65, 91, 92
Longevity,
  of worker honeybees, 55

M

Male
  bumblebees, 94, 120
  honeybees, 50, 61
  solitary bees, 103, 106, 112
Mandibular glands, 56
Marking bees, 9, 10
Mating
  of bumblebees, 94, 99
  of honeybees, 61
  of solitary bees, 103, 112
Monotropic, 24
Mouth parts, 16, 17, 33
Moving
  colonies to crops, 70–75, 100, 205, 211
  solitary bee nests, 108

N

Nasanov gland, 29, 30, 40, 74
Nectar
  collection of, 16–18, 39, 40
  composition of, 16, 17, 18, 140, 233
  concentration of, 17, 18, 43, 62, 140, 144, 145, 153, 154, 184, 212, 215, 234, 235, 237, 249, 255, 258, 273, 275, 277, 284, 285, 287, 298, 307, 319, 328, 331, 337, 347, 364, 391, 424
  loads of, 17, 18, 24, 69, 184, 233, 253, 254
  secretion of, 16, 18, 19, 140, 145, 154, 184, 202, 212, 234, 235, 255, 276, 284, 285, 287, 308, 390, 431, 432
  unloading of, 56
  yield of, 42, 43, 145, 149, 153, 249, 255, 275, 277, 287, 288, 292, 303, 304, 309, 328, 348, 364, 391, 424
Nectary, 16 (see also under crop concerned)
  extrafloral, 33, 34, 153, 154, 167, 248, 249, 251, 258, 296, 330, 437
Nest,
  box, 96–99
  material of, 89
  site of, 89
  of bumblebee,
  of solitary bee, 102–116
Nurse bees, 55–57

O

Odour (see scent)
Oligotropic, 24
Orientation
  to flowers, 45
  to hive, 71, 125
  to nest, 100, 108

P

Parthenocarpy, 184–187
Petals,
  attractiveness of, 15–16
Pheromones, 29, 40, 58, 59, 61, 64, 66, 87
Plant, bearing capacity, 148, 197, 243, 345, 361, 419
Plant form
  recognition of, 15
Pollen
  amount colonies collect, 23, 31, 32, 142, 249, 262, 288, 347, 348, 389, 425
  amount produced, 23, 24, 222, 234, 284, 305, 389
  amount on bees, 24, 48, 49, 122, 329, 369, 405
  attractant, 21
  composition of, 19, 21, 389

Pollen (contd.)
  food value of, 21
  fossil, 36, 204
  identification of, 10
  presentation of, 21–23, 38, 73, 249, 298, 301, 305, 306, 312, 318, 324, 334, 367, 368, 383, 426
  quantity required, 19
  scent of, 21
  stores of, 51, 63, 84
  substitutes for, 21, 89, 124, 158
  viability of, 185, 199, 288, 290, 300, 306, 325, 411, 412 (see also under crop concerned)
Pollen collection, 19–26, 63, 64
  age of bees, 62
  by neighbouring colonies, 31–33
  brood rearing and, 63, 84, 85, 94
  choice of, 19
  deliberate and incidental, 23, 62, 140, 328, 329, 425
  effect of anaethetics on, 10
  factors causing, 63, 64
  increasing, 84–87, 211
  number of species collected, 32
  packing, 19–21
  queen's presence and, 64
Pollen dispensers, 86, 188, 413–416
Pollen loads, 19, 69
  constancy, 34–36
  mixed, 34, 35, 39
  segregated, 35, 39
  size of, 24, 184, 233, 253
Pollen traps, 10, 11, 31, 85
Pollen tubes
  growth rate of, 146, 170, 171, 200, 253, 258, 283, 319, 342, 350, 356, 360, 383, 384
Pollinating efficiency per visit, 223, 271–273, 307, 312, 329, 350, 351, 394, 395
Pollinating insects
  abundance of, 2, 3, 66, 67, 94, 95, 244, 347, 388
  measurement of population of, 7, 8
  types of, 2 (see also under crop concerned)
  value of honeybees as, 5–7, 66, 67, 121 (see also under crop concerned)
Pollination
  contracts, 8
  determining need for, 4–6, 19
  hand, 6, 7, 261, 278, 293, 410–412, 440
  mechanical, 201, 202, 266, 351, 352, 377, 410–416, 443
  requirements of crops, 1, 66, 67 (see also under crop concerned)
Pollinia, 440
Pollinizer varieties
  arrangement of, 171, 186–188, 191, 192, 284, 291, 342, 368, 369, 377, 378, 406–410, 420
Polytropic, 24, 34
Population needed,
  of blowflies, 130
  of honeybees, 7, 8, 66, 67, 112, 113, 161
  of solitary bees, 111, 116
  variations, 2, 37, 347
Propolis, 62
Pupa, 52–54, 90, 91, 106, 112–114

Q

Queen bumblebee
  egg laying, 89
  foraging, 90, 91, 120
  founding colony, 89–91, 95–100
  hibernation, 89, 99–101
  incubating, 91, 92
Queen honeybee
  behaviour of workers toward, 59–61
  egg laying, 60, 61
  food of, 60, 61
  foraging and, 64, 66, 123
  pheromones of, 59, 61, 64, 66, 87

R

Rain
  foraging and, 18, 23, 259
  pollination, by, 438
Recruitment to crops
  threshold for, 39, 41
Relative humidity, 70
  nectar secretion and, 18
  pollen presentation and, 23
Rhythm
  of nectar secretion, 18
  of pollen presentation, 21, 22, 23, 402 (see also under crop concerned)
Rotating colonies, 74–75

S

Scent
  as repellant, 15
  communication of, 26–30
  of flowers, 15, 37
  of Nasonov gland, 29, 30, 40, 74
  of pollen, 21
  perception, 15
Scout bees, 31, 40, 41
Scrabbling for pollen, 23
Size
  of bees, 92, 94
  of colonies, 65, 66, 94, 122, 123
  of flowers, 36, 46
  of nectar loads, 24, 184, 233
  of pollen loads, 24, 184, 233, 253
Social life, 50
Solitary bees, 3, 25, 104, 105
  artificial nests and management, 102–116, 295, 319
  in enclosed spaces, 126–127
  search for suitable species of, 116–118
Stigma
  receptivity of, 67, 142, 146, 152, 170, 174, 194–196, 274, 288, 293, 306, 318, 325, 336, 344, 345, 350, 359, 365, 367, 368, 375, 383, 384, 418, 423, 426, 436, 437
Sugars, 16, 17

T

Techniques for studying foraging, 9–11, 164
Temperature
  effect of, on
    flight, 23, 74, 111, 145, 285, 287, 387, 389, 410, 421
    nectar secretion, 18, 154, 390
    pollen presentation, 23

Temperature regulation
  brood nest, in, 61, 73
  high temperatures, at, 50
  low temperatures, at, 50
Time of release of colonies, 72–74, 402
Tongue length and accessibility of nectar, 33, 94, 218–228, 230, 231, 247, 248, 251, 253, 255, 261–264, 343, 423

V

Varieties, discrimination between, 311, 404, 405
Vision
  acuity of, 15
  colour and, 15

W

Water collection, 29, 62
Wax production, 56, 57, 58
Weather and flight, 23, 41, 44, 67, 73, 76, 80, 398, 399
Wild insect pollinators
  abundance of, 2, 3, 94, 95, 347, 388
  types of (see under crop concerned)
Wind breaks, 4, 45, 80, 81
Wind pollination, 138, 139, 192, 202, 286, 290, 320, 352, 357, 360, 362, 426, 427, 438
  tests for, 155, 172, 173, 181, 182, 190, 191, 195, 197, 237, 282, 293, 320, 333, 334, 337, 338, 345, 346, 370, 371, 372, 384, 385
Worker bees
  adaptability to duties, 56
  body size of, 92, 94
  coordination of activities, 57–61
  division of labour, 55
  duties of foragers, 62, 63
  duties of house-bees, 55, 56, 58
  longevity of, 55
  number per colony, 50, 51, 94
  starting to forage, 61, 62